Food microstructures

Related titles:

Understanding and controlling the microstructure of complex foods
(ISBN 978-1-84569-151-6)

Designing functional foods: measuring and controlling food structure breakdown and absorption
(ISBN 978-1-84569-432-6)

Texture in food Volume 1: Semi-solid foods
(ISBN 978-1-85573-673-3)

Details of these books and a complete list of titles from Woodhead Publishing can be obtained by:

- visiting our web site at www.woodheadpublishing.com
- contacting Customer Services (e-mail: sales@woodheadpublishing.com; fax: +44 (0) 1223 832819; tel.: +44 (0) 1223 499140 ext. 130; address: Woodhead Publishing Limited, 80, High Street, Sawston, Cambridge CB22 3HJ, UK)
- in North America, contacting our US office (e-mail: usmarketing@ woodheadpublishing.com; tel.: (215) 928 9112; address: Woodhead Publishing, 1518 Walnut Street, Suite 1100, Philadelphia, PA 19102-3406, USA)

If you would like e-versions of our content, please visit our online platform: www.woodheadpublishingonline.com. Please recommend it to your librarian so that everyone in your institution can benefit from the wealth of content on the site.

We are always happy to receive suggestions for new books from potential editors. To enquire about contributing to our Food Science, Technology and Nutrition series, please send your name, contact address and details of the topic/s you are interested in to nell.holden@woodheadpublishing.com. We look forward to hearing from you.

The team responsible for publishing this book:

Commissioning Editor: Sarah Hughes
Publications Coordinator: Emily Cole
Project Editor: Elizabeth Moss
Editorial and Production Manager: Mary Campbell
Production Editor: Mandy Kingsmill
Project Manager: Annette Wiseman, RefineCatch Ltd
Copyeditor: Jo Egré
Proofreader: Eileen Power
Cover Designer: Terry Callanan

Woodhead Publishing Series in Food Science, Technology and Nutrition:
Number 254

Food microstructures

Microscopy, measurement and modelling

**Edited by
V. J. Morris and K. Groves**

WOODHEAD
PUBLISHING

Oxford   Cambridge   Philadelphia   New Delhi

Published by Woodhead Publishing Limited,
80 High Street, Sawston, Cambridge CB22 3HJ, UK
www.woodheadpublishing.com
www.woodheadpublishingonline.com

Woodhead Publishing, 1518 Walnut Street, Suite 1100, Philadelphia,
PA 19102-3406, USA

Woodhead Publishing India Private Limited, 303, Vardaan House, 7/28 Ansari Road, Daryaganj,
New Delhi – 110002, India
www.woodheadpublishingindia.com

First published 2013, Woodhead Publishing Limited

British Library Cataloguing in Publication Data
A catalogue record for this book is available from the British Library.

Library of Congress Control Number: 2013944631

ISBN 978-0-85709-525-1 (print)
ISBN 978-0-85709-889-4 (online)
ISSN 2042-8049 Woodhead Publishing Series in Food Science, Technology and Nutrition (print)
ISSN 2042-8057 Woodhead Publishing Series in Food Science, Technology and Nutrition (online)

Typeset by RefineCatch Ltd, Bungay, Suffolk
Printed by TJ International Ltd, Padstow, Cornwall, UK

Contents

Contributor contact details

(* = main contact)

Editors

Victor J. Morris*
Institute of Food Research
Norwich Research Park
Colney
Norwich, NR4 7UA, UK

E-mail: vic.morris@ifr.ac.uk

Kathy Groves
Leatherhead Food Research
Randalls Road
Leatherhead, KT22 7RY, UK

E-mail: KGroves@LeatherheadFood.
  com

Chapter 1

Debbie J. Stokes
FEI Company
Achtseweg Noord 5
5651 GG Eindhoven, The Netherlands

E-mail: Debbie.Stokes@fei.com

Chapter 2

Victor J. Morris
Institute of Food Research
Norwich Research Park
Colney
Norwich, NR4 7UA, UK

E-mail: vic.morris@ifr.ac.uk

Chapter 3

Paul A. Gunning
Surface Analysis Department
Smith & Nephew Research Centre
York Science Park
Heslington
York, YO10 5DF, UK

E-mail: Paul.Gunning@Smith-
  Nephew.com

Chapter 4

Mark A. E. Auty
Food Chemistry and Technology
  Department
Teagasc Food Research Centre
Moorepark
Fermoy
Co. Cork, Ireland

E-mail: Mark.auty@teagasc.ie

Chapter 5

Alessandro Torricelli
Dipartimento di Fisica
Politecnico di Milano
Piazza Leonardo da Vinci 32
I-20133 Milan, Italy

E-mail: alessandro.torricelli@polimi.it

Chapter 6

Nikolaus Wellner
Institute of Food Research
Norwich Research Park
Colney
Norwich, NR4 7UA, UK

E-mail: klaus.wellner@ifr.ac.uk

Chapter 7

Malcolm J. W. Povey* and
    Nicholas Watson
School of Food Science and Nutrition
University of Leeds
Woodhouse Lane
Leeds, LS2 9JT, UK

E-mail: m.j.w.povey@leeds.ac.uk

Nicholas G. Parker
School of Mathematics and Statistics
Newcastle University
Newcastle upon Tyne, NE1 7RU, UK

Chapter 8

Peter S. Belton
School of Chemistry
University of East Anglia
Norwich, NR4 7TJ, UK

E-mail: p.belton@uea.ac.uk

Chapter 9

Mostafa Barigou* and Maëlle Douaire
School of Chemical Engineering
University of Birmingham
Edgbaston
Birmingham, B15 2TT, UK

E-mail: M.Barigou@bham.ac.uk

Chapter 10

M. A. Rao
Food Process Engineering
Department of Food Science
Cornell University
Geneva
NY 14456-1447, USA

E-mail: mar2@cornell.edu

Chapter 11

T. B. Mills* and Ian T. Norton
School of Chemical Engineering
University of Birmingham
Edgbaston
Birmingham, B15 2TT, UK

E-mail: millstb@bham.ac.uk;
    I.T.Norton@bham.ac.uk

Chapter 12

Simon J. Cox
Institute of Mathematics and Physics
Aberystwyth University
Aberystwyth, SY23 3BZ, UK

E-mail: foams@aber.ac.uk

Chapter 13

Gary C. Barker
Institute of Food Research
Norwich Research Park
Colney
Norwich, NR4 7UA, UK

E-mail: gary.barker@ifr.ac.uk

Chapter 14

Stephen R. Euston
School of Life Sciences
Heriot-Watt University
Edinburgh, EH14 4AS, UK

E-mail: S.R.Euston@hw.ac.uk

Appendix

Kathy Groves*
Leatherhead Food Research
Randalls Road
Leatherhead, KT22 7RY, UK

E-mail: KGroves@LeatherheadFood.
    com

Mary L. Parker
Institute of Food Research
Norwich Research Park
Colney
Norwich, NR4 7UA, UK

E-mail: mary.parker@ifr.ac.uk

Woodhead Publishing Series in Food Science, Technology and Nutrition

Dedication to Brian Hills

11 June 1949–29 October 2012

I first met Brian when he joined the Institute of Food Research (IFR) to help develop the use of NMR in food science. Brian was an undergraduate and postgraduate at Oxford. He also did postdoctoral research at MIT and Cambridge prior to joining IFR in 1987. At IFR he became a key research leader whose work was recognised internationally. From the time he joined IFR, it was clear that Brian was very knowledgeable, highly motivated and extremely innovative. He had a 'hands-on' approach to science, the ability to identify key problems and to devise novel solutions at both the theoretical and practical level. His models for molecular transport in complex media underpin research on food processing, the physical, chemical and microbial stability of foods, the modelling of flavour encapsulation and release, and the structure and structural changes in cellular materials such as starch, emulsions or plant tissue. In terms of NMR applications Brian developed novel pulse sequences to interrogate materials, new theoretical models to interpret the data, and experimented with new types of spectrometers, such as field cycling NMR. Recently he invested considerable effort into the use of his knowledge of water in foods, to develop novel imaging methods directed towards industrial challenges requiring high throughput, low-field, low cost methods, that can be used as sensors in real industrial environments. Just prior to his death he had been working on new advances in the acquisition and interpretation of MRI images, which substantially reduced the acquisition times and enhanced the quality of the images: research of value in the food area but also with wider clinical applications.

Brian published alone and with numerous co-authors over 100 peer-reviewed articles on NMR and its applications, plus standard textbooks such as *Magnetic Resonance Imaging in Food Science* and *Advances in Magnetic Resonance in Food Science*, together with numerous book chapters on NMR methods and applications. Brian was deeply religious and saw physics as a way of glimpsing what he regarded as the wonders of God's creation. His interests in physics were widespread and knowledgeable and he published in other areas outside NMR and food science. This broader aspect of his interests is perhaps illustrated by his book *Origins: cosmology, evolution and creation*.

It was a pleasure to know and work with Brian. His 'hands-on' approach made him well respected and liked by co-workers. He did not suffer fools gladly but, in all the time I knew him, I never heard him say a bad word about anyone. Sadly he left us before some of the ideas which he spawned and nurtured were able to blossom, as I hope they will do in the future. Brian was to have written a chapter on NMR and MRI applications in food science. Because of his illness this proved impossible. I wish to thank Peter Belton for taking on this task at the last minute. I hope this book conveys some of the interest and enthusiasm Brian had for understanding food structure and might inspire others to continue in this area of research.

V. J. Morris, IFR, 2013

Preface

The knowledge of food structure has advanced considerably over the last 30 years. Aside from the academic interest in understanding the complex structure of foods, these advances are enabling the design of new foods to improve their safety and quality, and to enhance the nutritional and health benefits of natural and processed foods. This improvement in our understanding has resulted from the continued development of new methods for visualising and modelling food microstructure.

This book is not intended to provide a detailed description of all the wide and varied types of food structures. Rather the intention is to introduce the methodologies available to probe food microstructure and to indicate the type of information that can be obtained through their use. By choice the focus is on microscopy and modelling techniques that yield direct information on structure, ranging over different hierarchical levels from the molecular to the macroscopic. The level of coverage in different chapters varies, depending on the maturity of the techniques under discussion. In all cases, the hope is that sufficient information is provided to indicate how the technique works, the type of information obtainable, and the advantages and disadvantages of this method.

The literature on food microstructure is vast and continues to expand rapidly. The coverage in the chapters is not meant to be exhaustive but rather to emphasise particular points or to provide a route to the literature in particular areas. The choice is not meant to indicate priority and omissions are simply a result of the restrictions on space rather than any reflection on the quality of the publications.

What do we wish to achieve in editing this book? The aim is to introduce the methods available for visualising and modelling food structure. In general terms, the intention is to convey the sort of information that has been obtained, to indicate the progress being made at the present time, and to speculate on what can be achieved in the future. Who do we feel should benefit from reading this book? We hope the book will be of interest to researchers and participants in industry, research institutes and universities, and to those with major interests not just in food structure but also in the application and design of natural microstructures for pharmaceutical, nutritional or health benefits. It is hoped that this book will be useful for researchers interested in developing microscopic and modelling techniques, and that it might foster greater collaboration between these two schools, particularly in the food area.

We wish to thank the contributors for their participation and for their patience in awaiting completion of this volume. Both of us wish to acknowledge the financial support from our respective organisations and the guidance and patience of the editors at Woodhead Publishing during the construction of this volume.

V. J. Morris and K. Groves

Introduction

This book is about the methods available to allow food scientists to monitor, visualize or simulate food structures, and consequently understand the changes that occur to such food structures during the cooking, processing or digestion of natural or processed foods.

Food structures are complex: they range from the intricate self-assembled structures present in plant and animal tissue to the prefabricated structures generated in processed foods. Food structures change with time. Most people will be familiar with the softening of strawberries or tomatoes on ripening, or the loss of the crunchiness of pears and apples on storage, associated with the gradual breakdown of their structures. All processed foods have finite shelf lives, often associated with the deterioration of their internal structure. In the current climate of concerns over health, obesity, environmental issues, waste and food shortages, the shelf-life and quality of processed foods is of key importance. Understanding how the ingredients make up the structure of the foods and how this structure changes during its life or on eating will play an important part in the development and management of the food industry.

Observing how foods flow or deform when subjected to stress can monitor these structural changes. The techniques used to study these properties are called mechanical and rheological methods and these methods are discussed in this book. The instruments range from simple, cheap empirical methods, through comparatively cheap quality control instruments, to sophisticated, expensive machines for detailed characterization of food systems. In addition to defining structure and structural changes, the techniques can be used to extract parameters that reflect textural changes, and can be related to certain perceived sensory attributes of food. However, this latter aspect will not be considered in the present book.

The behaviour of food systems changes when they are confined to small regions between surfaces. The study of such behaviour requires specialized equipment and these techniques are called tribology: this is a new, emerging field of study which is important for studying structural changes in the mouth or during complex processing operations.

Although not new, there is a growing interest and development in methods for using probes to monitor the internal structure of food systems. These micro-rheological methods observe the constrained oscillation or restricted meandering of the probes through the food structure. In their earliest incarnation, these methods were largely restricted to studies on transparent food systems, but nowadays can be used increasingly to investigate complex opaque food systems, and their use will be described and discussed.

Apart from simple studies on dilute molecular solutions, rheological techniques cannot provide direct information on the underlying food structure, at least at the molecular level. The interpretation of the rheological and mechanical properties relies on structural information derived mainly from microscopic and imaging methods and, increasingly, on the use of computer simulations to test models for food structure.

The development of the first (optical) microscope literally opened a window with which to visualize and describe the structure of materials. New microscopic techniques have proved equally significant and nowadays the food scientist has a wide variety of microscopic methods, which are available to probe food structure at different hierarchical levels. Some of these methods are extremely well established but are still benefiting from new instrumental developments, others are still emerging, and some are still in their infancy, at least in their use in food science. The use of new microscopic methods to study foods has revealed additional structural information and new insights and applications in food science and technology.

The major structural components of natural and processed foods are biological molecules: carbohydrates, proteins and lipids (fats). These components can self-assemble, or be induced to assemble into higher-order structures, and these structures themselves can be components of even more complex food structures. Examples of such structures include the self-assembly of proteins to form the fibrous structure of meat, or the self-assembly of polysaccharides to form the granule structure of starch or the cellular network structures of plant cell walls. Association of globular proteins and polysaccharides can be used to create fibrous or particulate gels and the assembly of fats (surfactants) or proteins at interfaces can be used to prepare and stabilize foams and emulsions.

Processing can be used to develop more complex structures in foods such as ice cream, which is a solid milk fat emulsion containing air bubbles, ice, sugar and fat crystals, or the aerated structures of cakes and bread. Thus a range of microscopic and imaging techniques are required to span the different structural regimes that can be present in food materials. The intention of this book is to introduce the range of microscopic and imaging methods currently available to investigate such food structures.

Optical microscopy is the oldest, most established and most versatile method for studying food materials. As early as 1665, Robert Hooke reported images of plant materials in his book 'Micrographia' and introduced the concept of biological cells. The advantages of optical microscopy are the ease of use, relatively low cost, and the wide variety of contrast mechanisms and stains. The development of the phase contrast microscope alone was sufficiently significant to justify the award of the Nobel Prize in Physics to Fritz Zernike in 1953. Extension of the operating range into the infra-red, through the use of Infra-red microscopy (developed first in the mid-1950s) and Raman microscopy (developed in the mid-1970s) has allowed the mapping of different structural components, together with additional information on their physical state within complex food structures. Finally, the arrival of confocal laser scanning microscopy in the late 1980s rejuvenated the use of optical techniques, particularly in the use of the methods to follow processing operations. The resolution achievable with optical and infra-red methods is limited by the wavelength of the radiation used to probe the sample. Thus, although the techniques can identify the presence of different molecular species, it is not possible to image the structures at molecular resolution.

Ernst Ruska constructed the first electron microscope in 1933, although he and Maximillion Knoll obtained the first images earlier, in 1931. Physicists and material scientists were the main users of the technique until 1959, when R. (Bob) W. Horne and Sydney Brenner developed the technique of negative staining, opening up the use of the technique for biologists. The use of transmission electron microscopy (TEM) allowed the molecular structure of foods to be probed for the first time. TEM and the companion technique of scanning electron microscopy (SEM, first marketed in the mid-1960s) require that the samples are imaged under vacuum, and this has led to the development of elegant preparative methods to preserve the 'native' structure of food samples. The development of the environmental scanning electron microscope and similar 'low vacuum' SEMs (the ESEM became available in the late 1980s) is now allowing the imaging of 'wet' food samples, although the resolution achievable still lags behind that obtainable by conventional SEM. This remains a new and developing technique.

Gerd Binnig and Heinrich Rohrer developed the scanning tunnelling microscope (STM) in the early 1980s, a discovery that won them the Nobel Prize in Physics in 1986, together with Ernst Ruska for the development of the electron microscope. This discovery of the STM led to the development of a family of microscopes (probe microscopes), which image by feeling, rather than visualizing a surface. The most versatile member of this family for studying biological samples is the atomic force microscope (AFM), developed by Binning and colleagues in 1986 and first commercialized in the early 1990s. Applications in food science began in that period and have expanded through the development of successive generations of AFMs.

A related technique that can generate images by feeling is that of optical tweezers: a laser beam is used to trap a probe particle and monitor its interaction with the internal structure of complex materials. This can be used to generate a 3D map of the structure and these microscopes are called photonic microscopes. The development of this technique in food science has been restricted by the lack of commercial instruments that are now only starting to become available. The ability of the AFM and optical tweezers to measure the forces between derivatized probes and samples allows them to determine and map the forces involved in the assembly of food structures at the molecular and colloidal level. These techniques of molecular and colloidal force spectroscopy are new and emerging techniques in food science.

A number of non-invasive imaging techniques have been developed principally for clinical use. Examples include acoustic microscopy (introduced in the mid-1970s), optical coherence tomography (developed in the late 1980s to early 1990s) and the related techniques such as space- or time-resolved reflectance spectroscopy, X-ray micro-computed tomography (also introduced in the early 1970s) and magnetic resonance imaging (MRI).

Of these methods, perhaps the most used in food science is MRI. The construction of the first MRI machine is attributed to Raymond Damadian in 1971, with the first MRI image obtained by Paul Christian Lauterbur in 1973. Sir Peter Mansfield developed mathematical procedures and techniques for enhancing the clarity and acquisition of images. The significance of the clinical applications of MRI led to the award of the Nobel Prize in Physiology or Medicine to Lauterbur and Mansfield in 2005. Applications of NMR and MRI for studying food structure began in the early 1970s and have expanded since that time. The intention in this book is to focus on the investigation of food structure, and also the changes in food structure within the body on digestion, rather than the clinical uses for mapping the consequent fat distribution in the body, or changes in brain activity associated with food consumption.

A new and emerging area of research on food structure has been made possible by the dramatic improvement in computing power and speed. This has led to the use of computer simulations, which can be used to run and examine models for food structure. This area is covered by a description of modelling and simulation techniques and their applications to food systems, plus descriptions of the modelling of particular generic food structures, which include granular materials and cellular structures in foods. These methods provide a link between the microscopic elements of food structure and the nature and behavior at the macroscopic level.

What do we wish to achieve in writing this book? One aim is to collect together the large and increasing number of methods available to characterize food structure and the changes that occur on the formation and breakdown of such structures. The coverage of different methods is variable, depending on how new or established these methods are in investigating food structures. However, the hope is that the coverage is sufficient in all cases to introduce the methods, to demonstrate the types of information that can be obtained, and the types of structures that can be studied. In general terms, the intention is to look at what can be done, how it is done and where things may go in the future. We hope the book will provide a good resource base for the literature on techniques for probing food structure and provide a basis for understanding how such techniques have been used to characterize food structure. For the well-established methods, mainly microscopies, which have contributed to our present views on food structure, the intention is to describe how they can now be used routinely (at least in the hands of experts) to characterize foods and food structures. In the case of new and emerging techniques, the hope is that the reader will get a feel for the new insights these methods are providing, and could provide in the future.

During both of our careers, the methods available to study food structure and the new insights and understanding of complex foods they have provided have increased dramatically. We feel this is likely to continue in the future. We hope this book provides a picture of the current state of the art and a springboard to future developments. It is always difficult and perhaps foolish to try to predict the future. The discovery and development of probe microscopes is clearly an example of a technique which could not have been predicted in advance. However, certain aspects of the study of food structure are predictable. It is clear that the development of hybrid instruments combining different forms of microscopy will continue, extending the range and nature of images of food structure.

At present, the resolution achievable with most microscopic techniques is limited by the wavelength of the incident radiation. This limitation can be overcome through the use of near-field methods, where the source is brought to within less than the wavelength from the surface. At present, the signal-to-noise ratios for such methods are generally low, and the acquisition times are long. It is to be expected that these largely technological problems will be reduced in the future, opening up the use of these methods. Computing power and availability is likely to continue to increase. This will probably improve the speed of acquisition, processing and presentation of images. We might expect the use of modelling and simulations to become more routine and widely used in food science. The availability of high power sources of radiation, such as synchrotrons, means that the ability to model kinetics is likely to be complemented by new experimental data.

Nowadays there is an increasing demand for functional foods designed rationally to enhance health and reduce the risks of contracting long-term chronic diseases. Developing such products requires the construction of foods that are acceptable to the consumer in terms of cost, taste, texture and appearance. In order to deliver health benefits it is necessary to tailor the breakdown of the structure during digestion to facilitate release of structural components, and to optimize uptake and transport within the body. The key to success is to understand how to design and construct the correct food matrix.

This book covers the methods available for probing or simulating the assembly and stability of food structure, and for selecting and monitoring the site and mode of breakdown during digestion. Thus we hope that it will be of interest to students and researchers interested in food structure and to food scientists and technologists faced with the continuous and growing demand for the production of safer and healthier functional foods.

V. J. Morris and K. Groves

Part I

Microstructure and microscopy

1

Environmental scanning electron microscopy (ESEM): principles and applications to food microstructures

D. J. Stokes, FEI Company, The Netherlands

DOI: 10.1533/9780857098894.1.3

Abstract: This chapter introduces some basic principles of scanning electron microscopy (SEM) and its extension to environmental scanning electron microscopy (ESEM), describing why ESEM is useful for characterising materials of interest in food research. It first surveys the main techniques of imaging and microanalysis in SEM. The principles of ESEM are then described, explaining how gases can be used to mitigate electrical charging of uncoated insulating materials and contribute to the image formation process, and other ways in which gases and specimen temperature control are useful for expanding the range of available techniques to yield additional information about the structure–property relations of hard, soft and even liquid specimens. Several key application techniques will be covered, from general imaging of dry or moist, uncoated specimens through to *in situ* dynamic experiments.

Key words: scanning electron microscopy (SEM), environmental scanning electron microscopy (ESEM), gases, aqueous and hydrated specimens, *in situ*, dynamic experiments, cryoESEM.

1.1   Introduction

The main themes of this chapter are scanning electron microscopy (SEM) and its extension to environmental scanning electron microscopy (ESEM), with examples demonstrating the ways in which ESEM can be used for characterising materials of interest in food research. These materials range from confectionery and cereal products to fluid-filled vegetable cells and tissues through to emulsions, highlighting the diversity of material types that can be accommodated in the ESEM without the need for extensive specimen preparation traditionally associated with high vacuum electron microscopy.

Section 1.2 surveys the basic components of the SEM, the requirements for placing specimens in high vacuum and the main techniques of imaging and microanalysis. The principles of ESEM are then described in Section 1.3, explaining how gases can be used to mitigate electrical charging of uncoated, insulating materials and contribute to the image formation process. The section also explains other ways in which gases and specimen temperature control are useful for expanding the range of available techniques to yield additional information about the structure–property relationships of hard, soft and even liquid specimens. Several key application techniques are covered in Section 1.4, from general imaging of uncoated specimens, through to *in situ* dynamic experiments such as wetting and drying, mechanical testing and freezing. The chapter concludes with Section 1.5, briefly discussing the outlook for ESEM in the study of food microstructure.

1.2   Scanning electron microscopy (SEM)

SEM has its beginnings in the 1930s (Knoll, 1935; von Ardenne, 1938a,b) and continued its development through the 1940s onwards (Zworykin *et al.*, 1942; McMullen, 1953; Smith and Oatley, 1955; Everhart and Thornley, 1960), becoming commercially available in 1965. With it came the ability to study the microstructural characteristics of bulk materials with large depth of field across length scales ranging from millimetres to nanometres, offering a valuable new addition to the suite of visual characterisation tools such as light microscopy and scanning/transmission electron microscopy (S/TEM).

Typically, an SEM consists of an electron source to generate a beam of primary electrons; a column with electromagnetic lenses for focusing and demagnifying the primary electron beam; coils for scanning the electron beam across the specimen surface; a chamber containing a stage to hold the specimen; vacuum pumps to maintain the system under high vacuum (usually of the order of 10^{-5}–10^{-7} Pa); and one or more detectors for collecting signals generated by electron irradiation of the specimen. Finally, the magnified image is displayed on a monitor, as the beam is scanned pixel-by-pixel across the field-of-view.

The most straightforward specimen types for SEM are metals, primarily since these materials are less prone to the effects of charging and damage under electron irradiation in high vacuum. Methods have evolved to address the issue of imaging electrically insulating materials such as polymers and ceramics, including coating the surface with conductive materials; incorporating heavy metal salts into the specimen to increase bulk conductivity, especially for biological materials; or using low voltages to minimise the accumulation of negative charge within the specimen, making it possible to image uncoated materials (Goldstein, 2003). For materials classes that are not naturally solids or have a tendency to outgas in vacuum, there are methods for conferring rigidity and preventing outgassing so that the specimen is suitable for high vacuum conditions in the SEM. These include critical point drying, freeze drying and the use of cryo-stages for frozen-hydrated specimens (cryoSEM).

Section 1.2.1 gives a brief overview of imaging in the SEM, mainly concentrating on beam–specimen interactions. For further reading on the topic as a whole, the interested reader is referred to Reimer (1985), Newbury *et al.* (1986), Sawyer and Grubb (1987), Goodhew *et al.* (2001) and Goldstein *et al.* (2003).

1.2.1   Beam–specimen interactions

The SEM is primarily used for detecting backscattered electrons (BSEs) and secondary electrons (SEs), as well as X-rays and light, emanating from the surface or near-surface of bulk materials following interaction of the primary electron beam with the specimen. Specimen sizes are usually on the order of a few millimeters thick and can be up to around a centimeter in length or diameter, but the SEM can also be fitted with a detector to receive electrons transmitted through thin samples of up to a few microns in thickness, depending on the density of the specimen. To distinguish the latter technique from the more traditional ultra-high resolution scanning transmission electron microscope (STEM), this approach is often referred to as STEM-in-SEM. The principal signals in the SEM are illustrated in Fig. 1.1.

BSEs are generated via elastic (non energy-absorbing) scattering of primary electrons within the specimen and are defined by convention as having energies from 50 eV all the way up to the primary electron energy of the source, which is usually in the range 1 to 30 keV. BSEs are thus essentially primary electrons that

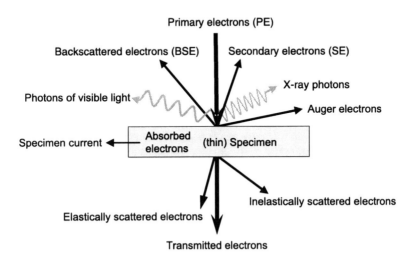

Fig. 1.1   Diagram showing a range of signals in the SEM. For bulk specimens, these include backscattered and secondary electrons, various photons such as X-rays and visible light and Auger electrons. For thin specimens, transmitted electrons provide information from the degree and nature of scattering.

re-emerge from the specimen surface after a series of trajectory-altering interactions with the Coulombic field around atoms in the specimen. Inelastic energy loss mechanisms also come into play, causing electrons to transfer energy to the atoms of the specimen; hence BSEs are emitted from the surface with a range of energies (or are absorbed by the specimen). BSEs can travel from comparatively large depths (10^2–10^3 nm) to reach the surface, and the BSE emission co-efficient increases as a function of atomic number, so the BSE signal generally gives rise to images that reflect compositional information rather than being surface-sensitive.

SEs, of which there are several types, are produced via inelastic interactions with primary electrons. Here, the generated electrons (known as type I, or SE_I) originate from the specimen's atoms as primary electrons excite atomic orbital electrons sufficient to result in ionisation. SEs are defined as having energies up to 50 eV, but are typically emitted at around a few eV. Other types of SE include SE_{II}, generated by BSEs as they exit the specimen surface and interact with atoms as they pass by, and SE_{III}, generated by primary electrons striking the polepiece and chamber walls. The low-energy nature of SEs means that the signal is emitted from very close to the surface ($<$ 50 nm) – SEs are prone to inelastic, energy-absorbing processes and so have a short escape depth – giving rise to images that are characteristically topographic, showing surface relief. SE imaging thus provides complementary information to the BSE signal. SE signals are sensitive to variations in primary electron beam energy, since this affects the depth at which SEs are produced. With increasing beam energy, primary electrons penetrate further into the material, particularly in soft materials such as polymers, resulting in a lower SE yield, compared to lower beam energies where SEs are generated closer to the surface and so a greater proportion can escape. As a general rule, SE imaging of fine surface features is best carried out at lower beam energies.

Inelastic scattering also produces characteristic and continuum (Bremsstralung) X-rays, Auger electrons, electron-hole pairs (excitons), long-wavelength electromagnetic radiation (cathodoluminescence), lattice vibrations (phonons) and collective electron oscillations (plasmons). X-ray photons are often used for microanalysis in the SEM, the most common form of detection being energy dispersive spectroscopy (EDS), and are produced as a result of the relaxation of an excited state in the atom following primary electron irradiation. X-ray energies are thus directly related to the chemical elements in the specimen, thus providing quantitative or qualitative information about chemical composition, either via spectra or mapping of elements present.

In the absence of a conductive coating, electrically insulating materials are especially susceptible to both radiation damage and electrical charging. For example, organic materials can become noticeably damaged during electron excitation and ionisation, and atoms can become displaced, changing the structure and appearance visible at the surface. Other effects include breaking of bonds, cross-linking, mass loss (e.g. through the formation and liberation of volatile components), temperature change and the formation of a carbonaceous deposit (contamination). Care must be taken to minimise these effects by careful choice of electron beam energy and flux, specimen thickness and temperature. More detailed information can be found in Talmon (1987).

Electrical charging is a particular problem: electrons accumulate in the specimen, leading to electric fields that distort or deflect the primary electron beam. This can significantly interfere with the ability to interpret images, or even to obtain an image at all. Again, appropriate choice of beam energy, flux and specimen thickness can be employed to help mitigate the problem (Joy and Joy, 1998; Goldstein *et al.*, 2003), but there are certain inter-dependencies that make it difficult to satisfy the conditions needed to control both radiation damage and charge build-up at the same time, particularly for heterogeneous specimens. Hence, together, these are the main reasons why conductive coatings are applied to electrically insulating materials, especially those of a delicate, organic nature. However, such specimen preparation, including the steps necessary to render the specimen ready for coating, can involve lengthy procedures that may introduce changes to or obscure features of interest.

1.3   Environmental scanning electron microscopy (ESEM)

In order to ease some of the sample preparation and handling requirements for SEM noted above and so allow a wider range of observation conditions, particularly for insulating and non-solid materials, the SEM was adapted for greater flexibility. This involved the introduction of gases into the specimen area. In essence, the use of a gas serves the purposes of mitigating charging effects in insulators, providing an alternative mechanism for electron signal amplification, and enabling hydrated/liquid specimens to be observed directly in their natural state. These attributes make it possible to eliminate many of the specimen preparation steps that are sometimes required in the SEM, enabling the natural surfaces of delicate or otherwise challenging materials to be observed without a conductive coating and allowing for dynamic *in situ* experiments, as well as helping to avoid artefacts that can arise as a result of freezing and/or drying ordinarily required to render specimens solid for high vacuum SEM imaging.

Early demonstrations of such capabilities include Lane (1970), Robinson (1974, 1975), Danilatos and Robinson (1979) and Shah and Beckett (1979). First commercialised around 1980, the technology became more widely available in the 1990s, bringing yet further insights into the structure–property relationships of many materials that were previously unsuitable for study without preparation for conventional SEM.

Given that materials of relevance to food research are predominantly electrically insulating in physical character and are sometimes hydrated or liquid in their natural state, the various derivatives of this type of SEM have a clear place amongst the tools used for the study of food structure and properties. The technology is popularly known by terms such as environmental SEM (ESEM) and variable pressure SEM (VPSEM), amongst others. For a description of these terms, see Stokes (2008). ESEM is one of the more well-known versions of the instrument, which is trademarked by FEI Company, and is the focus of discussion in Section 1.3.1 onwards. Many of the principles also apply to equipment supplied by other manufacturers.

1.3.1   Principles of ESEM

From a hardware perspective, many of the components of an ESEM are the same as for regular high vacuum SEM, and each has similar capabilities when used in high vacuum mode. For example, the electron source is maintained under high vacuum in both cases, as is the main part of the electron column, to protect the source from attack by gaseous species and to minimise scattering of primary electrons. However, the ability to maintain a gaseous chamber environment does call for a few technological differences when operating in ESEM mode. The main differences between high vacuum SEM and ESEM are: (a) the presence of differentially pumped zones, separated by pressure limiting apertures; and (b) SE detectors capable of operating in a gas, since the high electric potential at the traditional Everhart–Thornley SE detector (ETD) would cause arcing in a gaseous environment. The principal components of ESEM are shown in Fig. 1.2.

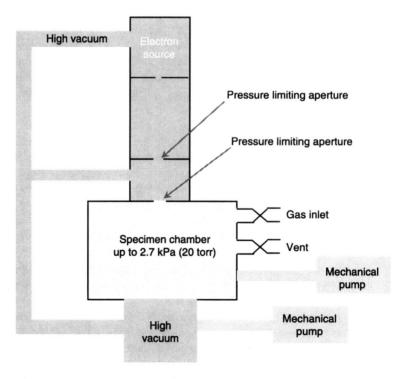

Fig. 1.2 Diagram showing the basic components of the ESEM column. The overall construction is similar to that of a conventional SEM, but with the addition of differentially pumped zones separated by pressure limiting apertures.

As noted previously, the presence of a gas in the vicinity of an ESEM specimen is useful for compensating against negative charge build-up from electron irradiation and also serves as a means of amplifying the SE signals generated from the specimen. In both cases, interactions between gas molecules and SEs are key: ionising collisions between the two species creates positive ions and additional SEs.

The methods of collecting or detecting signals in the gas environment of ESEM generally rely on some form of pre-detection amplification of the signal, utilising the gas itself, in contrast to the traditional SEM scintillator detectors in which collected electrons are converted into a shower of post-acquisition photoelectrons. In the ESEM, electrons emitted from the specimen are accelerated in the electric field between the specimen and detector (a positively biased electrode), colliding with and ionising gas molecules to produce further SEs, which can in turn participate in further ionisations. Provided that the field strength is high enough, this process creates a cascade, increasing the number of signal electrons arriving at the detector. At the same time, positively ionised gas molecules or atoms left behind will be driven in the opposite direction to the signal electrons. In this way, in the simplest case, if negative charge is accumulating in the specimen, a source of positive charge is available to compensate and restore charge neutrality. Figure 1.3 shows how the gas amplification mechanism takes place in the ESEM.

Ideally, the gas pressure should be kept to a minimum in order to reduce scattering of primary electrons, although if the concentration of gas is not sufficient to result in charge balance for a given set of operating parameters, the specimen will still display negative charging artefacts. Some experimentation is therefore

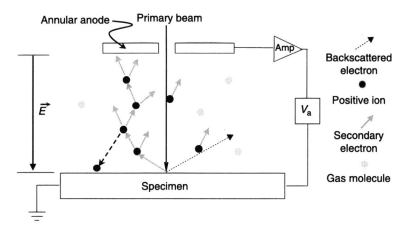

Fig. 1.3 Diagram illustrating the process of gas amplification in the ESEM. Primary electrons impinge on the specimen, creating secondary (SEs) and backscattered electrons (BSEs). SEs are sensitive to the relatively small electric field in the specimen–detector gap and are accelerated towards the detector anode, colliding with gas molecules and producing further SEs to amplify the signal as well as positive ions that are swept towards the specimen and help compensate for negative charge build-up.

needed for different gases and specimens to determine the appropriate gas pressure, sample-detector distance, field strength, beam current, and so on (Thiel *et al.*, 1997). Conversely, if the gas concentration is higher than necessary, the number of positive ions can far exceed the requirements for charge neutrality, leading to a number of scenarios that can similarly affect image interpretation or quality. The specimen itself therefore plays an integral part in the process.

For a grounded, metallic specimen, charges of either sign are easily dealt with and the specimen remains charge neutral. But this is seldom the type of specimen for which ESEM is used. Indeed, even an electrically conductive specimen can acquire a net potential if not properly grounded, as was demonstrated by Craven *et al.* (2002), in which a non-grounded metallic specimen was shown to float up to increasingly higher positive electrical potentials as the gas pressure was increased beyond the charge-neutral point, yielding correspondingly less information about the surface. An excess of positive ions can lead to screening of the detector field, reducing its strength and amplification properties, or to recombination of ions with low energy SEs, reducing the number of signal electrons entering the amplification process in the first place (Toth *et al.*, 2002). Hence, for non-metallic, uncoated specimens, the gas pressure should be kept to an optimal level for charge-free imaging or, if a specific pressure is a necessary component of the experiment, some form of localised grounding should be employed (e.g. placing conductive wire or tape in the vicinity of the specimen surface to act as a charge drain).

Water vapour is commonly used as an imaging gas in ESEM. It has good signal amplification properties and is particularly useful as it can be used to control the thermodynamic stability of moist or liquid specimens, as well as having a role in dynamic hydration and dehydration experiments. This will be outlined in Section 1.4.2. Other gases may be selected either because the specimen is sensitive to water vapour or its constituent hydrogen and oxygen, the thermodynamic conditions are below the freezing point of water and so would cause its precipitation as ice, or because the conditions of a given experiment dictate the use of a specific gas as an active participant in the experiment.

The pressure in the specimen chamber can be up to 2.7 kPa (20 torr): a significant departure from the conditions of a high vacuum SEM. This does result in some scattering of the primary electron beam, compared to high vacuum SEM, although with the pressures used in ESEM, the mean free path of primary electrons is generally long enough to ensure that there are still enough electrons to form a focused probe, while the scattered part of the beam covers a diffuse area, known as the probe 'skirt'. The overall effect is that the beam current in the focused probe is reduced, the resolution stays about the same, but the signal-to-background ratio is decreased. The extent and influence of these effects on image quality is highly dependent on primary electron beam energy and current as well as gas type, but the general principle is illustrated in Fig. 1.4, where electron scattering in high vacuum and in gas is compared.

A further consequence of the probe skirt relates to X-ray microanalysis. Because primary electrons can be scattered outside the central focused probe and

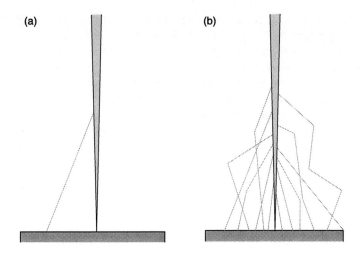

Fig. 1.4 Primary electron scattering in (a) vacuum and (b) a low-pressure gas as may be found in the ESEM. Provided that electrons undergo only a few scattering events (oligo-scattering), a focused probe still reaches the specimen to generate an imaging signal. Some primary electrons are scattered out of the probe, and these form a diffuse probe 'skirt' that adds to the background signal, reducing the current, but not resolution, of the focused probe.

impact the specimen surface in the surrounding area, X-rays contributing to an EDS spectrum or map may actually originate some distance away from the point of interest. In general, X-ray microanalysis in ESEM is therefore taken as a qualitative guide to the elements present, although various numerical methods have been proposed to correct for spurious data (Doehne, 1997; Bilde-Sorensen and Appel, 1997; Gauvin, 1999; Le Berre *et al.*, 2007). Despite this limitation, collection of chemical information from uncoated or hydrated materials at the desired beam energy is still useful.

For further reading on the origins of ESEM technology, refer to Danilatos (1991) and Stokes (2008) and references therein, and for a review of X-ray microanalysis in ESEM and related equipment, see Newbury (2002).

1.4   Key applications of ESEM for the study of food microstructure

ESEM naturally lends itself to the study of food-related materials, since they are most often electrically insulating in character and generally benefit from less invasive specimen preparation and observation in the absence of conductive coating or other treatment. The following sections explore how the extended capabilities of ESEM can be applied to the study of food microstructure in a variety of different ways, depending on specimen characteristics and the features or properties of interest. Section 1.4.1 briefly deals with the most straightforward case of imaging solids in a gaseous environment, where the imaging gas can be water vapour, nitrogen, nitrous oxide, etc.

Water vapour is a commonly used gas in ESEM and, aside from its excellent imaging characteristics, its physico-chemical properties can be exploited to good effect, making it possible to study aqueous solutions and hydrated materials, or to perform *in situ* hydration/dehydration experiments, as will be described in Section 1.4.2.

Section 1.4.3 brings together the use of different gases and studies at sub-zero or cryogenic temperatures for various dynamic studies of materials with, for example, low glass transition temperatures (T_g).

Finally, one of the key properties of food systems is their elastic behavior, and this can be studied *in situ* by employing a straining stage or applying some form of micromanipulation in the ESEM (Section 1.4.4).

An excellent review of ESEM for studying a wide variety of food systems is given by James (2009), which outlines the different approaches covered here along with numerous examples. Another review is that of Donald (2004), which discusses ESEM studies of foods in a materials science context.

1.4.1   Imaging uncoated materials

ESEM can be utilised in so-called 'low vacuum' mode, where the gas participates as a means of charge-reduction at a suitably low pressure, just sufficient to mitigate negative charge build-up. This has the advantage of keeping primary electron scattering to a minimum and helps avoid excessive positive ion production, as described in Section 1.3.1.

Low vacuum mode is most suitable for imaging solid materials, where the vapour pressure of the specimen is not a particular consideration. The requisite gas pressure is therefore primarily a function of the dielectric properties of the specimen: the more insulating the material, the more likely it is to acquire negative charge, hence the higher the gas pressure needed to generate sufficient positive ions to offset the charge. In addition, the amplification properties of the imaging gas come into play, since gas amplification efficiency varies according to the electronic properties of each gas. Therefore some experimentation is needed to find an optimum pressure value for a given experiment, and becomes more complicated when dealing with heterogeneous materials with components of widely differing dielectric character. Aside from water vapour, gases commonly used in ESEM include nitrogen, nitrous oxide and carbon dioxide.

In general, gas pressures in the region of 100 to 300 Pa (0.75–2.25 torr) are sufficient for many electrically insulating specimens. An example, shown in Fig. 1.5, is that of chocolate (James, 2009), in which the gas pressure (nitrous oxide) is just 100 Pa (0.75 torr). The specimen was subjected to thermal cycling in order to assess the effects of temperature on surface roughness, and the imaging conditions are suitable for observing delicate cocoa butter fat crystals that have 'bloomed' from the bulk material.

Fig. 1.5 ESEM micrograph of chocolate to demonstrate the use of 'low vacuum' mode for an uncoated, electrically insulating material. Sample temperature is $-5\,^{\circ}C$ and imaging gas is nitrous oxide, at a pressure of 100 Pa (0.75 torr). During image formation, the concentration of gas is sufficient to produce charge balanced conditions via the gas cascade mechanism. Horizontal field width is approximately 200 μm. Reproduced with permission from James (2009), © Elsevier Ltd.

Further reading on the gas amplification process can be found in Thiel *et al* (1997), while Toth *et al.* (2003) report on image interpretation taking into account imaging conditions and dielectric properties of materials. Information on the amplification properties of different gases can be found in Fletcher *et al.* (1997).

1.4.2   Handling and imaging aqueous and hydrated materials

Many food systems contain moisture, so it is often desirable to maintain some degree of specimen hydration. This is achieved through the combination of specimen cooling, by placing the specimen and holder on a Peltier stage, and appropriate selection of water vapour imaging gas pressure, together with some precautions to avoid unwanted water loss from the specimen along the way.

The phase diagram of water, shown in Fig. 1.6, indicates the pressures and temperatures required for a given set of stable thermodynamic conditions. Points that lie on the curve represent thermodynamic equilibria – water molecules are evaporating and condensing all the time, but the net liquid–vapour ratio remains constant for a given temperature. This is defined as 100% relative humidity (RH). Deviations from the curve result in either evaporation or condensation of water. In the ESEM, these conditions are achieved by setting the microscope to a specific water vapour pressure and cooling the specimen to the relevant temperature.

Many systems containing one or more aqueous phases (e.g. hydrated specimens) consist not of pure water, but of aqueous phases containing dissolved solutes. We should therefore consider what influence these solutes have on the vapour pressure of the specimen. According to Raoult's law, the vapour pressure

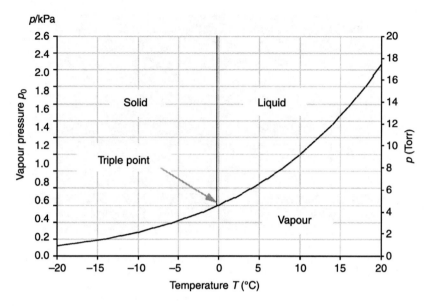

Fig. 1.6 Plot of the phase diagram of water. Points along the curve indicate 100% relative humidity, while changes in temperature or pressure take conditions into the solid, liquid or vapour phases. The conditions shown are attainable in the ESEM. Reproduced with permission from Stokes (2008), © Wiley.

of a solution is proportional to the mole fraction of solute (Tabor, 1991). This implies that the vapour pressure of a solution is *less* than that of the pure solvent, and hence the equilibrium vapour pressure will be less than 100% RH. If the microscope is operated under 100% RH conditions (i.e. using a temperature and pressure of water vapour that lies on the curve in Fig. 1.6), and the specimen has an equilibrium vapour pressure below this, an osmotic driving force arises that can lead to unwanted condensation of the imaging gas onto the specimen. For a more detailed description, refer to Stokes (2008). This effect has been reported for biological specimens in ESEM (Tai and Tang, 2001; Muscariello *et al.*, 2005).

A useful concept, frequently used in food research, is that of water activity a_w, which describes the energy state of the system and is defined as the ratio of the equilibrium vapour pressure p_{eq} of the liquid or substance to the vapour pressure p_0 of pure water at the same temperature:

$$a_w = p_{eq}/p_0 \qquad\qquad [1.1]$$

If we relate the water activity of a solution to its RH (since $RH = a_w \cdot 100$), then the saturated vapour pressure curve for water can thus be modified to reflect the equilibrium vapour pressure of a given aqueous phase using Equation 1.2:

$$p_{eq} = a_w p_0 \qquad\qquad [1.2]$$

where p_0 represents the vapour pressure for $RH = 100\%$ at a specific temperature for pure water and a_w is the water activity of the aqueous, solute-containing phase.

If we take the water activity a_w of a saturated solution of common salt (NaCl) as a guide, with $a_w = 0.75$ (RH = 75%), it becomes clear that the extent to which this factor lowers the vapour pressure of a solution becomes significant. For example, the equilibrium vapour pressure at a temperature $T = 3\,°C$ would be $p_{eq} = 572\,Pa$ (4.3 torr), 25% lower than the pressure for pure water where $p_0 = 758\,Pa$ (5.6 torr).

Another consideration is the behaviour of aqueous phases under non-equilibrium conditions. In particular, what are the kinetic implications – what is the rate of water loss? The phase behaviour of water is a non-linear function of temperature: by analogy with the Maxwell distribution of speeds in gases, the probability that an individual molecule will have a speed much in excess of the average increases with temperature, and so evaporation occurs more readily at higher temperatures. This kinetic behaviour has important implications in the case of controlling water in and around specimens in the ESEM chamber. Typical operating temperatures for hydrated specimens tend to be around 1 to 6 °C, where the rate of moisture loss is comparatively low. It is therefore acceptable to employ pressures somewhat below the equilibrium vapour pressure: real specimens can usually withstand slowly dehydrating conditions for a finite period of time (tens of minutes if the temperature is around 1 °C). Figure 1.7 shows bread mould

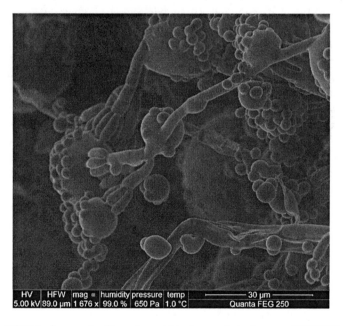

HV	HFW	mag ⫶	humidity	pressure	temp	———————— 30 µm ————————
5.00 kV	89.0 µm	1 676 x	99.0 %	650 Pa	1.0 °C	Quanta FEG 250

Fig. 1.7 ESEM micrograph of bread mould to demonstrate the use of 'ESEM' mode for a hydrated, uncoated, electrically insulating material. Sample temperature is 1 °C and imaging gas is water vapour at a pressure of 650 Pa (4.9 torr), suitable for both image formation and to maintain the specimen's stability via thermodynamic control of its water content. Horizontal field width = 100 µm. Courtesy of Marc Castagna, FEI Company, Hillsboro, USA.

imaged under hydrated conditions, allowing structures to retain their overall morphology without shrinkage or collapse.

A final practical consideration is on achieving the correct starting conditions for imaging a hydrated specimen at thermodynamic equilibrium with the surrounding environment, since the ESEM chamber initially contains air at atmospheric pressure. One way to do this is through 'purge-flood' cycles, in which air is systematically replaced with water vapour. First, the specimen should always be allowed to thermally equilibrate with the Peltier stage before pumping down. Work by Cameron and Donald (1994) graphically showed the importance of using the correct parameters to minimise any water loss during pumpdown. The recommended procedure for a specimen temperature of $3\,°C$ is to pump to a pressure $p = 731.5\,Pa$ (5.5 torr), allow water vapour into the chamber until the pressure rises to $p = 1.3\,kPa$ (9.6 torr) and repeat 8 times, finishing at $p = 731.5\,Pa$ (5.5 torr). Note that the specified values are for conditions of 100% RH.

For a specimen having an equilibrium vapour pressure of 75% RH, the corresponding upper-limit pressure for this temperature would be 975 Pa (7.3 torr) and the lower limit $p = 545\,Pa$ (4.1 torr). However, some specimens may still be vulnerable to the evaporation of water before the purge-flood cycle begins. A simple method to deal with this is to place small drops of water on a non-cooled area near the specimen. When the humidity in the chamber falls, these droplets will sacrificially evaporate, having higher vapour pressure than the cooled specimen, giving a much-needed burst of vapour. Another method involves surrounding the specimen with a medium such as agar gel, again having a higher vapour pressure, which will similarly lose water in preference to the specimen itself (Neděla, 2007). Studies carried out to show that proper control of these environmental conditions is needed for meaningful results include those of Cameron and Donald (1994), Bache and Donald (1998) and Callow et al. (2003).

Materials that contain moisture or fluid-filled compartments therefore need care in the ESEM, but perhaps the most critical type of sample for which precautions are needed is when dealing with aqueous solutions or suspensions, since these are highly prone to evaporation if not handled correctly. Figure 1.8 shows an image of mayonnaise – an oil-in-water emulsion – which is a relatively challenging material to image in its liquid state, partly due to the intrinsically low SE contrast between the phases (Stokes et al., 1998), but which can be successfully handled and observed in the ESEM.

Further examples of utilising water vapour for SE imaging of hydrated materials in the ESEM include studies of vegetable tissue (Thiel and Donald, 1998; Donald et al., 2003; Zheng et al., 2009); fruit tissue (Chen et al., 2006; Sun-Waterhouse et al., 2008); colloidal suspensions (He and Donald, 1996); gels (Cameron et al., 1994); microgel particles (Garcia-Salinas and Donald, 2010); films (Keddie et al., 1996; Donald et al., 2000; Phan The et al., 2008); cereals and cereal products (McDonough et al, 1993, 1996; Bache and Donald, 1998; Stokes and Donald, 2000; Sivam et al., 2011; Guillard et al., 2003; Roman-Gutierrez et al., 2002), as well as cellulose (Astley et al., 2001; Miller and Donald, 2003) and the whey protein β-lactoglobulin (Bromley et al., 2006).

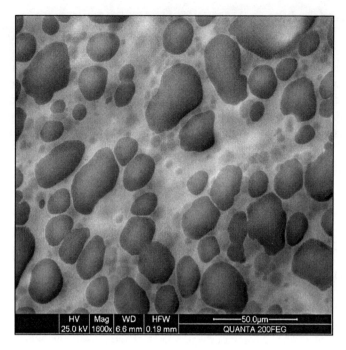

HV	Mag	WD	HFW	⟵50.0µm⟶
25.0 kV	1600x	6.6 mm	0.19 mm	QUANTA 200FEG

Fig. 1.8   ESEM micrograph of an oil-in-water emulsion (mayonnaise) to demonstrate the use of ESEM mode for a fully liquid material. Sample temperature is 2 °C and pressure of water vapour imaging gas is 670 Pa (5 torr). During initial pumpdown it is advisable to keep a small reservoir of water at room temperature near the specimen, to provide a source of preferential evaporation to replace air in the chamber. Horizontal field width = 190 µm. Courtesy of Ellen Baken, FEI Company, Eindhoven, The Netherlands.

1.4.3   Cryogenic and low temperature experiments

In the conventional SEM, specimens can be cooled to sub-zero temperatures using a cryo-transfer system with a plunge-freezing station and SEM cooling stage, typically operating at or near liquid nitrogen temperature (around –195 °C, 78K), known as cryoSEM. Stages employing liquid helium are also available, although are less common (for an ESEM example, see Cartwright et al. (2008)). The rate of freezing is a critical aspect of cryoSEM and, in some cases, a high pressure freezer is employed in preference to plunge-freezing, to prevent the unwanted growth of crystalline ice in bulk frozen-hydrated materials. Specimens can be freeze-fractured or sectioned prior to or during transfer, to reveal inner microstructure, and are often then sputter-coated with metals or carbon to confer electrical conductivity. Further information on low temperature SEM techniques can be found in Echlin (1992).

A similar approach can be used in ESEM and, if the sputter-coating is omitted and a gas used for imaging and charge control, the specimen temperature can be varied to reveal low temperature dynamic behaviour. For example, a material that is amorphous (glassy) only at very low temperatures (i.e. it has a low T_g) can in principle be observed as the temperature is increased and the material passes through the glass transition, where its morphology or the distribution of components may change as the viscosity is lowered or a transformation to an ordered crystalline state occurs. Studies of this kind can be conducted using imaging gases other than water vapour, good examples being nitrogen, nitrous oxide and carbon dioxide, which have higher vapour pressures than water and so remain in the vapour phase at low temperatures. These gases also possess adequate SE-gas cascade amplification properties (Fletcher *et al.*, 1997), making them suitable alternatives to water vapour for ESEM imaging.

The point at which frozen water sublimates as vapour is another example of a dynamic process that can be observed in the ESEM. Referring again to Fig. 1.6, we see that small partial pressures of water vapour are involved in maintaining or changing equilibrium conditions between the solid and vapour phases at sub-zero temperatures, and this can be achieved in practice by varying the pressure of water vapour in the chamber in the same way as described in Section 1.4.2. This has been used to study the dynamic behaviour of ice cream, using the combination of an alternative gas as a background gas for imaging and water vapour to control the thermodynamics (Stokes *et al.*, 2004b). Figure 1.9 illustrates this point, where nitrous oxide serves as the imaging gas at low temperature (–95 °C in this case). Subsequent raising of the temperature requires the introduction of very

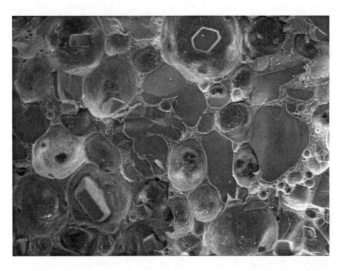

Fig. 1.9   ESEM micrograph of a low-T_g material (ice cream) to demonstrate the use of ESEM under sub-zero thermal conditions. Sample temperature is –95 °C and the imaging gas is nitrous oxide at a pressure of approximately 100 Pa (0.75 torr). In order to control the rate of sublimation for frozen-hydrated specimens, small concentrations of water vapour can be added to the chamber, in accordance with the phase diagram for water. Horizontal field width = 200 µm. Reproduced with permission from Stokes *et al.* (2004), © Blackwell Publishing.

small but increasing concentrations of water vapour if sublimation of ice is to be avoided, allowing the specimen to be observed as it passes through its glass transition. Controlled sublimation can be performed by slowly raising the temperature without the addition of water vapour. This is also a method for carrying out freeze-drying *in situ*, allowing the development of microstructural features such as pores to be followed in real time (Meredith *et al.*, 1996), and can provide valuable insight into the quality and reconstitutive properties of freeze-dried foods.

Alternatively, an aqueous solution can be taken from the liquid to the frozen state, and vice versa, allowing the nucleation and growth of ice to be observed at very small length scales as the solution passes through the triple point at which all three phases (vapour, liquid and solid) are said to co-exist. This has been demonstrated for an aqueous solution containing anti-freeze proteins (Waller *et al.*, 2008), offering a promising method to assess the influence of additives on the freezing point and ice morphology and distribution in food products.

For temperatures down to about $-20\,°C$, a regular Peltier stage can be used. For lower temperatures, it is advisable to use purpose-built cryoSEM apparatus. In all cases, when working with water vapour at low temperature, it is necessary to shield any surfaces that will inevitably become colder than the specimen (i.e. the coolant inlet system and the bulk of the specimen stage) in order to avoid unwanted precipitation of ice (Waller *et al.*, 2008).

1.4.4   Probing mechanical properties *in situ*

Using a strain stage, with the specimen clamped between opposing moveable jaws, mechanical properties can be recorded via load-displacement data and simultaneously imaged, allowing stress-strain characteristics to be correlated with changes in microstructure. For hydrated food materials, the same procedures described in Section 1.4.2 are needed, and so the strain stage should be equipped with a cooling mechanism and the specimen imaged at an appropriate water vapour pressure. Even then, it is not trivial to ensure that test conditions replicate those that would be carried out using more conventional *ex situ* methods, and so it is wise to tailor the ESEM conditions in such a way to mimic *ex situ* results. A certain amount of trial-and-error is involved, and it is useful to perform the intended experiments by first running them with the ESEM chamber vented, so that results can be recorded for calibration purposes. This method has been used to study the mechanical behavior of dry and hydrated bread in the ESEM, revealing the interplay between starch granules and the surrounding gluten matrix and the transition between brittle failure and ductile behaviour as the intrinsic moisture content is varied from sample to sample (Stokes and Donald, 2000). Figure 1.10 shows the fracture surface of breadcrumb following *in situ* brittle failure in the ESEM.

Similar methods have been employed for studying the tensile properties of biopolymer gels and gelatin films (Rizzieri *et al.*, 2003, 2006) and to assess the tenderness of meat following pre-treatment such as marinating, cooking and

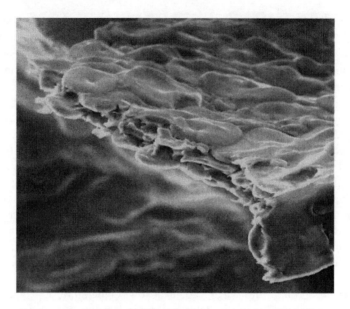

Fig. 1.10   ESEM micrograph of a biopolymer composite (breadcrumb) to demonstrate *in situ* compression testing. Following deformation and brittle failure, intact starch granules can be seen protruding from the fractured gluten matrix. The imaging gas is water vapour at a pressure of 600 Pa (4.5 torr) and the specimen is at room temperature. Horizontal field width is approximately 250 μm. Reproduced with permission from Stokes and Donald (2000), © Kluwer Academic Publishers.

freezing (James and Yang, 2011). Figure 1.11 shows the de-adhesion of onion epidermal cells (Donald *et al.*, 2003) and is a nice demonstration of mechanical testing combined with specimen hydration, with cells retaining their turgor pressure during the experiment.

Sensitive apparatus can be constructed for compression testing and observation of individual polymeric nanoparticles (Liu *et al.*, 2005) and has also been applied to measure the viscoelastic properties of yeast cells (Ren *et al.*, 2008; Ahmad *et al.*, 2010). Meanwhile Dragnevski *et al.* (2008) employed a mechanically driven roller in place of one of the jaws of a tensile testing stage to stretch superelastic specimens, whose extension before fracture exceeds 1000%.

The jaws of the strain stage can also be fitted with attachments to test for specific responses. An example is that of Thiel and Donald (1998), in which a small blade was driven into carrot cells at various stages of ageing or cooking, showing that fresh carrots fracture in an intracellular manner, bursting the cell walls and releasing their contents, while aged or cooked carrots undergo more extensive deformation followed by intercellular failure, leaving cells walls intact. This behavior is related both to the perception of texture and the availability of nutrients released via cell wall rupture.

For a general review of ESEM mechanical testing techniques for hydrated food systems, refer to Donald (2003).

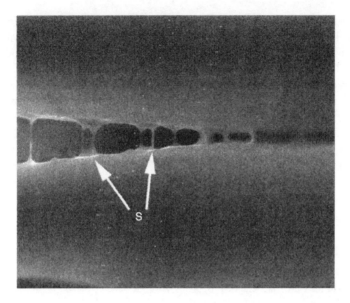

Fig. 1.11   TV-rate ESEM micrograph of hydrated plant cells (onion epidermis) during *in situ* tensile testing under hydrated conditions, showing intercellular debonding prior to failure. Sample temperature is 2 °C and imaging gas is water vapour at a pressure of approximately 670 Pa (5 torr). Horizontal field width = 200 µm. Reproduced with permission from Donald *et al.* (2003), © Oxford University Press.

1.5   Conclusion and future trends

We have seen that ESEM is well suited to the study of features at the microstructural level of interest in foods, with the possibility for making observations under various environmental conditions or in combination with different stimuli, such as thermal and mechanical. Understanding and further refinement of methods for maintaining image quality and specimen integrity will ensure that the technique continues to gain greater use in this field, along with more widespread adoption of methods for complementary *in situ* studies.

Recent advances in detector design allow for additional experimental variations. One of these is improved high pressure detection (e.g. 2.7 kPa, 20 torr) to enable experiments at increased RH without the necessity to cool the specimen (Stokes *et al.*, 2004a). Another is 'wetSTEM' detection to obtain the transmitted electron signal, with the option for specimen cooling and use in water vapour, for high resolution information from liquid films (Bogner *et al.*, 2005), dry or hydrated thin sections and small objects such as nanoparticles, water droplets (Rykaczewski and Scott, 2011; Barkay, 2010) or bacteria (Staniewicz *et al.*, 2012). These relatively new techniques have yet to make a significant impact on food research, but offer much promise.

The advent of focused ion beam (FIB) technology in combination with SEM means that bulk materials can be cross-sectioned using the FIB to mill away material in a localised, site-specific manner. The cross-sectional face can then be imaged with either the electron or ion beam. By continuing this process in a sequential way, a series of 2D images can be collected and used to make a 3D representation of the bulk specimen. This technique has been demonstrated for cells and tissues (Heymann *et al.*, 2006; Hayles *et al.*, 2007). Note that, while FIB milling and ion beam imaging is performed in high vacuum, to avoid excessive scattering of the ion beam, certain systems can be switched to low vacuum/ESEM conditions between sequential FIB slices for electron beam imaging of the freshly exposed surfaces of electrically insulating materials. Either approach offers great potential for increasing our knowledge of the distribution of and relationship between heterogeneous components or pores in complex food systems.

1.6   References

AHMAD, M. R., NAKAJIMA, M., KOJIMA, S., HOMMA, M. and FUKUDA, T. (2010), Nanoindentation methods to measure viscoelastic properties of single cells using sharp, flat, and buckling tips inside ESEM, *IEEE Transactions on NanoBioscience*, **9**, 12–23.

ASTLEY, O. M., CHANLIAUD, E., DONALD, A. M. and GIDLEY, M. J. (2001), Structure of Acetobacter cellulose composites in the hydrated state, *International Journal of Biological Macromolecules*, **29**, 193–202.

BACHE, I. C. and DONALD, A. M. (1998), The structure of the gluten network in dough: a study using Environmental Scanning Electron Microscopy, *Journal of Cereal Science*, **28**, 127–33.

BARKAY, Z. (2010), Wettability study using transmitted electrons in environmental scanning electron microscope, *Applied Physics Letters*, **96**, 183109–3.

BILDE-SORENSEN, J. and APPEL, C. C. (1997), X-ray spectrometry in ESEM and LVSEM: corrections for beam skirt effects, in, Tholen, A. R. (ed.), SCANDEM-97, 12–15.

BOGNER, A., THOLLET, G., BASSET, D., JOUNEAU, P. H. and GAUTHIER, C. (2005), Wet STEM: a new development in environmental SEM for imaging nano-objects included in a liquid phase, *Ultramicroscopy*, **104**, 290–301.

BROMLEY, E., KREBS, M. and DONALD, A. (2006), Mechanisms of structure formation in particulate gels of β-lactoglobulin formed near the isoelectric point, *The European Physical Journal E: Soft Matter and Biological Physics*, **21**, 145–52.

CALLOW, J. A., OSBORNE, M. P., CALLOW, M. E., BAKER, F. and DONALD, A. M. (2003), Use of environmental scanning electron microscopy to image the spore adhesive of the marine alga Enteromorpha in its natural hydrated state, *Colloids and Surfaces B: Biointerfaces*, **27**, 315–21.

CAMERON, R. E. and DONALD, A. M. (1994), Minimising sample evaporation in the Environmental Scanning Electron Microscope, *Journal of Microscopy*, **173**, 227–37.

CAMERON, R. E., DURRANI, C. M. and DONALD, A. M. (1994), Gelation of Amylopectin without long-range order, *Starch–Stärke*, **46**, 285–7.

CARTWRIGHT, J. H. E., ESCRIBANO, B. and SAINZ-DIAZ, C. I. (2008), The mesoscale morphologies of ice films: porous and biomorphic forms of ice under astrophysical conditions, *The Astrophysical Journal*, **687**, 1406.

CHEN, X. D., CHIU, Y. L., LIN, S. X. and JAMES, B. (2006), *In situ* ESEM examination of microstructural changes of an apple tissue sample undergoing low-pressure air-drying followed by wetting, *Drying Technology*, **24**, 965–72.

CRAVEN, J. P., BAKER, F. S., THIEL, B. L. and DONALD, A. M. (2002), Consequences of positive ions upon imaging in low vacuum SEM, *Journal of Microscopy*, **205**, 96–105.

DANILATOS, G. (1991), Review and outline of environmental SEM at present, *Journal of Microscopy*, **102**, 391–402.

DANILATOS, G. D. and ROBINSON, V. N. E. (1979), Principles of scanning electron microscopy at high specimen chamber pressures, *Scanning*, **2**, 72–82.

DOEHNE, E. (1997), A new correction method for high-resolution energy-dispersive X-ray analyses in the Environmental Scanning Electron Microscope, *Scanning*, **19**, 75–8.

DONALD, A. M. (2003), *In situ* deformation of hydrated food samples, in E. Dickinson and T. Van Vliet (eds), *Food Colloids, Biopolymers and Materials*, Cambridge, UK, Royal Society of Chemistry.

DONALD, A. M. (2004), Food for thought, *Nature Materials*, **3**, 579–81.

DONALD, A. M., HE, C., ROYALL, C. P., SFERRAZZA, M., STELMASHENKO, N. A. and THIEL, B. L. (2000), Applications of environmental scanning electron microscopy to colloidal aggregation and film formation, *Colloids and Surfaces A: Physicochemical and Engineering Aspects*, **174**, 37–53.

DONALD, A. M., BAKER, F. S., SMITH, A. C. and WALDRON, K. W. (2003), Fracture of plant tissues and walls as visualized by environmental scanning electron microscopy, *Annals of Botany*, **92**, 73–7.

DRAGNEVSKI, K. I., FAIRHEAD, T. W., BALSOD, R. and DONALD, A. M. (2008), A new tensile stage for in situ electron microscopy examination of the mechanical properties of 'superelastic' specimens, *Review of Scientific Instruments*, **79**, 126107–3.

ECHLIN, P. (1992), *Low Temperature Microscopy and Analysis*, Plenum, New York.

EVERHART, T. E. and THORNLEY, R. F. M. (1960), Wide-band detector for micro-microampere low-energy electron currents, *Journal of Scientific Instruments*, **37**, 246–8.

FLETCHER, A., THIEL, B. and DONALD, A. (1997), Amplification measurements of Potential Imaging Gases in Environmental SEM, *Journal of Physics D: Applied Physics*, **30**, 2249–57.

GARCIA-SALINAS, M. J. and DONALD, A. M. (2010), Use of Environmental Scanning Electron Microscopy to image poly(N-isopropylacrylamide) microgel particles, *Journal of Colloid and Interface Science*, **342**, 629–35.

GAUVIN, R. (1999), Some theoretical considerations on X-ray microanalysis in the environmental or variable pressure scanning electron microscope, *Scanning*, **21**, 388–93.

GOLDSTEIN, J., NEWBURY D., JOY, D., LYMAN, C., ECHLIN, P. *et al.* (2003), *Scanning Electron Microscopy and X-Ray Microanalysis*, 3rd edition, New York, Plenum.

GOODHEW, P. J., HUMPHREYS, F. J. and BEANLAND, R. (2001), *Electron Microscopy and Analysis*, 3rd edition, London and New York, Taylor & Francis.

GUILLARD, V., BROYART, B., BONAZZI, C., GUILBERT, S. and GONTARD, N. (2003), Moisture diffusivity in sponge cake as related to porous structure evaluation and moisture content, *Journal of Food Science*, **68**, 555–62.

HAYLES, M., STOKES, D. J., PHIFER, D. and FINDLAY, K. C. (2007), A technique for improved focused ion beam milling of cryo-prepared life science specimens, *Journal of Microscopy*, **226**, 263–9.

HE, C. and DONALD, A. M. (1996), Morphology of core-shell polymer lattices during drying, *Langmuir*, **12**, 6250–6.

HEYMANN, J. A. W., HAYLES, M., GESTMANN, I., GIANNUZZI, L. A., LICH, B. and SUBRAMANIAM, S. (2006), Site-specific 3D imaging of cells and tissues with a dual beam microscope, *Journal of Structural Biology*, **155**, 63–73.

JAMES, B. (2009), Advances in 'wet' electron microscopy techniques and their application to the study of food structure, *Trends in Food Science and Technology*, **20**, 114–24.

JAMES, B. and YANG, S. W. (2011), Testing meat tenderness using an *in situ* straining stage with variable pressure scanning electron microscopy, *Procedia Food Science*, **1**, 258–66.

JOY, D. C. and JOY, C. S. (1998), A study of the dependence of E2 energies on sample chemistry, *Microscopy and Microanalysis*, **4**, 475–80.

KEDDIE, J. L., MEREDITH, P., JONES, R. A. L. and DONALD, A. M. (1996), Film formation of acrylic latices wit varying concentrations of non-film-forming Latex particles, *Langmuir*, **12**, 3793–801.

KNOLL, M. (1935), Aufladepotentiel und Sekundäremission elektronenbestrahlter Körper, *Z. tech. Phys.*, **16**, 467–75.

LANE, W. C. (1970), The environmental cold stage, in *Proceedings of the Third Scanning Electron Microscopy Symposium*, Chicago, IL, IIT Research Institute, 60616

LE BERRE, J. F., DEMOPOULOS, G. P. and GAUVIN, R. (2007), Skirting: a limitation for the performance of X-ray microanalysis in the variable pressure or environmental scanning electron microscope, *Scanning*, **29**, 114–22.

LIU, T., DONALD, A. M. and ZHANG, Z. (2005), Novel manipulation in environmental scanning electron microscope for measuring mechanical properties of single nanoparticles, *Materials Science and Technology*, **21**, 289–94.

MCDONOUGH, C., GOMEZ, M. H., LEE, J. K., WANISKA, R. D. and ROONEY, L. W. (1993), Environmental Scanning Electron Microscopy evaluation of Tortilla chip microstructure during deep-fat frying, *Journal of Food Science*, **58**, 199–203.

MCDONOUGH, C. M., SEETHARAMAN, K., WANISKA, R. D. and ROONEY, L. W. (1996), Microstructure changes in wheat flour Tortillas during baking, *Journal of Food Science*, **61**, 995–99.

MCMULLEN, D. (1953), An improved scanning electron microscope for opaque specimens, *Proceedings of the Institute of Electrical Engineers*, **100**, 245–59.

MEREDITH, P., DONALD, A. M. and PAYNE, R. S. (1996), Freeze-drying: *in situ* observations using Cryoenvironmental Scanning Electron Microscopy and Differential Scanning Calorimetry, *Journal of Pharmaceutical Sciences*, **85**, 631–7.

MILLER, A. F. and DONALD, A. M. (2003), Imaging of anisotropic cellulose suspensions using Environmental Scanning Electron Microscopy, *Biomacromolecules*, **4**, 510–17.

MUSCARIELLO, L., ROSSO, F., MARINO, G., GIORDANO, A., BARBARISI, *et al.* (2005), A critical review of ESEM applications in the biological field, *Journal of Cellular Physiology*, **205**, 328–34.

NEDELA, V. (2007), Methods for additive hydration allowing observation of fully hydrated state of wet semples in environmental SEM, *Microscopy Rsearch and Techniques*, **70**, 95–100.

NEWBURY, D. E. (2002), X-ray microanalysis in the variable pressure (environmental) scanning electron microscope, *Journal of the National Institute of Standards and Technology*, **107**, 567–603.

NEWBURY, D. E., JOY, D. C., ECHLIN, P., FIORI, C. E. and GOLDSTEIN, J. (1986), *Advanced Scanning Electron Microscopy and X-ray Microanalysis*, New York, Kluwer Academic/Plenum Publishers.

PHAN THE, D., DEBEAUFORT, F., LUU, D. and VOILLEY, A. (2008), Moisture barrier, wetting and mechanical properties of shellac/agar or shellac/cassava starch bilayer bio-membrane for food applications, *Journal of Membrane Science*, **325**, 277–83.

REIMER, L. (1985), *Scanning Electron Microscopy. Physics of Image Formation and Microanalysis*, New York, Springer-Verlag.

REN, Y., DONALD, A. M. and ZHANG, Z. (2008), Investigation of the morphology, viability and mechanical properties of yeast cells in environmental SEM, *Scanning*, **30**, 435–42.

RIZZIERI, R., BAKER, F. S. and DONALD, A. M. (2003), A study of the large strain deformation and failure behaviour of mixed biopolymer gels via *in situ* ESEM, *Polymer*, **44**, 5927–35.

RIZZIERI, R., MAHADEVAN, L., VAZIRI, A. and DONALD, A. (2006), Superficial wrinkles in stretched, drying gelatin films, *Langmuir*, **22**, 3622–6.

ROBINSON, V. N. E. (1974), A wet stage modification to a scanning electron microscope, in J. V. Sanders and D. J. Goodchild (eds), *Proceedings of the 8th International Congress on Electron Microscopy*, Canberra, Australia, Academy of Sciences, 50–1.

ROBINSON, V. N. E. (1975), The elimination of charging artifacts in the scanning electron microscope, *Journal of Physics E: Scientific Instruments*, **8**, 638–40.

ROMAN-GUTIERREZ, A. D., GUILBERT, S. and CUQ, B. (2002), Description of microstructural changes in wheat flour and flour components during hydration by using Environmental Scanning Electron Microscopy, *LWT – Food Science and Technology*, **35**, 730–40.

RYKACZEWSKI, K. and SCOTT, J. H. J. (2011), Methodology for imaging nano-to-microscale water condensation dynamics on complex nanostructures, *ACS Nano*, **5**, 5962–8.

SAWYER, L. C. and GRUBB, D. T. (1987), *Polymer Microscopy*, London, Chapman and Hall.

SHAH, J. S. and BECKETT, A. (1979), A preliminary evaluation of moist environment ambient temperature scanning electron microscopy (MEATSEM), *Micron*, **10**, 13–23.

SIVAM, A., WATERHOUSE, G., ZUJOVIC, Z., PERERA, C. and SUN-WATERHOUSE, D. (2011), Structure and dynamics of wheat starch in breads fortified with polyphenols and pectin: an ESEM and Solid-State CP/MAS ^{13}C NMR spectroscopic study, *Food and Bioprocess Technology*, October, 1–14.

SMITH, K. C. A. and OATLEY, C. W. (1955), The Scanning Electron Microscope and its fields of application, *British Journal of Applied Physics*, **6**, 391–9.

STANIEWICZ, L., *et al.* (2012), The application of STEM and *in-situ* controlled dehydration to bacterial systems using ESEM, *Scanning*, **34**(4), 237–46.

STOKES, D. J. (2008), *Principles and Practice of Variable Pressure/Environmental Scanning Electron Microscopy (VP-ESEM)*, Chichester, UK, John Wiley and Sons.

STOKES, D. J. and DONALD, A. M. (2000), *In situ* mechanical testing of dry and hydrated breadcrumb using Environmental SEM, *Journal of Materials Science*, **35**, 599–607.

STOKES, D. J., THIEL, B. L. and DONALD, A. M. (1998), Direct observations of water/oil emulsion systems in the liquid state by Environmental Scanning Electron Microscopy, *Langmuir*, **14**, 4402–8.

STOKES, D. J., BAKER, F. S. and TOTH, M. (2004a), Raising the pressure: realising room temperature/high humidity applications in ESEM, *Microscopy and Microanalysis*, **10**, 1074–5.

STOKES, D. J., MUGNIER, J. Y. and CLARKE, C. J. (2004b), Static and dynamic experiments in cryo-electron microscopy: comparative observations using high-vacuum, low-voltage and low-vacuum SEM, *Journal of Microscopy – Oxford*, **213**, 198–204.

SUN-WATERHOUSE, D., FARR, J., WIBISONO, R. and SALEH, Z. (2008), Fruit-based functional foods, Part I: Production of food-grade apple fibre ingredients, *International Journal of Food Science and Technology*, **43**, 2113–22.

TABOR, D. (1991), *Gases, Liquids and Solids, and Other States of Matter*, Cambridge, UK, Cambridge University Press.

TAI, S. S. W. and TANG, X. M. (2001), Manipulating biological samples for Environmental Scanning Electron Microscopy observation, *Scanning*, **23**, 267–72.

TALMON, Y. (1987), Electron beam radiation damage to organic and biological cryo-specimens, in R. A. Steinbrecht and K. Zerold (eds), *Cryotechniques in Biological Electron Microscopy*, New York, Springer-Verlag.

THIEL, B. L. and DONALD, A. M. (1998), *In situ* mechanical testing of fully hydrated carrots (*Daucus Carota*) in the Environmental SEM, *Annals of Botany*, **82**, 727–33.

THIEL, B. L., BACHE, I. C., FLETCHER, A. L., MEREDITH, P. and DONALD, A. M. (1997), An improved model for gaseous amplification in the Environmental SEM, *Journal of Microscopy*, **187**, 143–57.

TOTH, M., PHILLIPS, M. R., THIEL, B. L. and DONALD, A. M. (2002), Electron imaging of dielectrics under simultaneous electron-ion irradiation, *Journal of Applied Physics*, **91**, 4479–91.

TOTH, M., THIEL, B. L. and DONALD, A. M. (2003), Interpretation of secondary electron images obtained using low vacuum SEM, *Ultramicroscopy*, **94**, 71–87.

VON ARDENNE, M. (1938a), Das Elektronen-Rastermikroskop. Praktische Ausführung, *Z. tech. Phys.*, **19**, 407–16.

VON ARDENNE, M. (1938b), Das Elektronen-Rastermikroskop. Theoretische Grundlagen, *Z. tech. Phys.*, **109**, 553–72.

WALLER, D., STOKES, D. J. and DONALD, A. M. (2008), Improvements to a cryosystem to observe ice nucleating in a variable pressure scanning electron microscope, *Review of Scientific Instruments*, **79**(10), 103709.

ZHENG, T., WALDRON, K. and DONALD, A. (2009), Investigation of viability of plant tissue in the environmental scanning electron microscopy, *Planta*, **230**, 1105–13.

ZWORYKIN, V. A., HILLIER, J. and SNYDER, R. L. (1942), A scanning electron microscope, *ASTM Bulletin*, **117**, 15–23.

2

Probe microscopy and photonic force microscopy: principles and applications to food microstructures

V. J. Morris, Institute of Food Research, UK

DOI: 10.1533/9780857098894.1.27

Abstract: This chapter discusses the methods and use of the techniques of atomic force microscopy, molecular and colloidal force spectroscopy, and optical tweezers as tools for probing food microstructure. Emphasis is placed on describing the basic methodologies, providing an entry to the background literature, and illustrating the applications of these methods through examples of where they have led to new insights in food science, plus potential future applications or challenges for these tools.

Key words: atomic force microscopy, force spectroscopy, colloidal force spectroscopy, optical tweezers, photonic microscopy, soft matter, gels, foams, emulsions, dispersions, food microstructure.

2.1 Introduction

Probe microscopes are a recent and novel type of microscope, but their impact, particularly in the genesis of nanoscience and nanotechnology, led to the joint award of the Nobel Prize in Physics in 1986 to the discoverers G. Binnig and H. Rohrer, together with the father of electron microscopy, E. Ruska. Probe microscopes are unique in creating images by feeling the surface of a sample with a sharp probe (Binnig *et al.*, 1986). The most used analogy is that of a blind person touching an object. The most common form of probe microscope used for studying biological and food systems is the atomic force microscope (AFM): this instrument generates images by monitoring changes in the force between a sharp probe and the sample surface, as the probe and the sample are scanned relative to each other (Fig. 2.1a).

The earliest types of AFM were designed primarily to image biological molecules, or molecular assemblies, deposited onto flat substrates and often

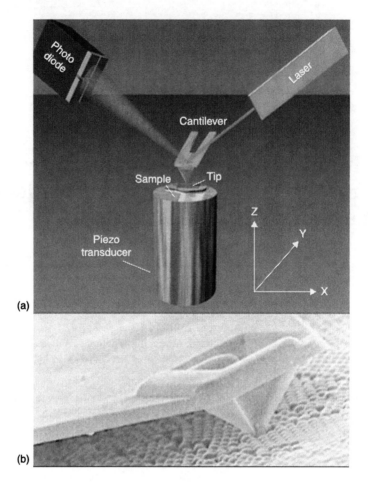

(a)

(b)

Fig. 2.1   The principal features of an AFM: (a) schematic of the microscope; (b) scanning electron micrograph of a typical tip–cantilever assembly, showing the pyramidal tip at the end of a 'V-shaped' cantilever positioned over a substrate surface. The tip is approximately 3 μm in height.

imaged in air. A major challenge to these studies was the need to prevent the probe damaging or displacing the sample. These problems have now been largely overcome, allowing reproducible imaging, which has now been extended to studies of molecules and molecular processes in realistic liquid environments. The development of hybrid AFM/optical microscopes has allowed the use of the AFM to be extended to more complex structures such as cells and tissues. In these 'first generation' instruments, the mechanical coupling of the AFM was poor, restricting resolution to less than that of a stand-alone AFM. In the 'second and subsequent generation' hybrid instruments, these problems have been overcome, allowing molecular resolution on complex biological and food systems. Through the development of sample preparation techniques the use of the AFM as a microscope has matured and, although not yet routine for biological systems, is now increasingly used as a tool to solve biological problems (Morris *et al.*, 2010; Morris, 2012).

The AFM is more than just a microscope and its use as a 'force transducer' is now rapidly emerging as a new area of interest. Measuring the force between the probe and the sample surface, as a function of probe-surface separation, allows a variety of different types of interactions to be investigated: this technique is called 'force spectroscopy'. Selecting different types of probes has made it possible to investigate the mechanical behaviour of individual molecules, to study specific interactions between molecules, and to probe the molecular basis of macroscopic phenomena such as colloidal particle interactions, adhesion and friction (tribology).

This chapter will describe AFM microscopy and force spectroscopy and illustrate how they have evolved to allow solutions to previously intractable problems in food science.

AFM microscopy images are just 3D surface profiles and the use of AFM as a force transducer is restricted to forces in the range 10^2 to 10^4 pN, which may exclude certain molecular or weak colloidal interactions. A complementary technique, known as photonic microscopy or optical tweezers (Neuman and Nagy, 2008), can be used to probe these weaker forces, covering the range 0.1 to 100 pN. These instruments allow particles to be captured and positioned using laser beams. Although originally developed as a tool for manipulating particles (cells), the technique has emerged as a powerful tool for measuring a range of interactions from the molecular to the colloidal level. Optical tweezers have matured as a technique for probing forces but the use of force measurements to generate images is now being developed: monitoring the changes in force caused by probe particles interacting with surfaces or interfaces, or moving inside biological systems, makes it possible to generate 3D images of these structures. Use of optical tweezers in food science has lagged behind that of AFM, due to the lack of availability of commercial instruments. As these instruments become available, it is anticipated that this will lead to new applications in food science (Morris, 2012). Thus this chapter will also look briefly at the potential use of optical tweezers or photonic microscopes and consider potential applications in food science.

2.2   Machines and methods: atomic force microscopes

2.2.1   Microscopy

In the AFM, a sharp probe is pressed against the sample surface (Fig. 2.1a). In the same way that a blind person gains information on the texture of surfaces in addition to their shape and size, the AFM images surface roughness, but also can map material properties such as adhesion, hardness or charge. Images are obtained by scanning the sample relative to a sharp probe: this can be achieved either by scanning the probe in a raster fashion over the surface or more usually by scanning the sample surface beneath the probe. Piezoelectric devices are used to move the sample in three dimensions, in order to generate high-resolution images (Binnig et al., 1986; Morris et al., 2010). The noise level on the electrical signal used to control the expansion and contraction of these piezoelectric devices thus effectively determines the practical resolution of the instrument. The sensing element is a tip-cantilever assembly and the most commonly used tips are microfabricated from silicon nitride (Albrecht and Quate, 1988). They are pyramidal in shape, approximately 3 μm in height, with a typical tip radius of approximately 30 to 50 nm (Fig. 2.1b).

Once a sample has been loaded into the instrument, the probe is lowered towards the sample, and a pre-determined force is applied to the surface. As the sample is scanned beneath the tip, the separation between the tip and the sample surface changes, altering the force on the tip and causing the cantilever to bend and/or twist. Bending and twisting is monitored by reflecting light, produced by a low power laser, from the end of the cantilever onto the surface of a four-quadrant photodiode. This enables the bend or the twist of the cantilever to be converted into a potential difference (error signal). The measured error signal is then used to generate a 3D image of the surface.

AFM imaging usually involves raster scanning the sample beneath the probe: a typical scan will consist of a matrix of 256×256 points and the total image is acquired by scanning the matrix line by line. After moving to the next image point, the system is allowed to dwell for a pre-selected period of time before moving on to the next image point. Conventional images are generated by monitoring changes in the normal force (the up and down motion of the cantilever). If the sample is homogeneous, then the change in force arises just from the change in separation between the tip and the surface, and the resultant error signal can be used to generate a topological map of the surface. For heterogeneous samples, changes in material parameters, such as charge, adhesion or stiffness will contribute to the image contrast, and imaging conditions can be chosen to enhance such contrast, and to map particular material characteristics of the sample (Morris et al., 2010).

2.2.2   Imaging modes

Error signals can be used to create images through the use of a number of pre-set imaging modes (Garcia and Perez, 2002; Holscher and Schirmeisen, 2005; Meyer, 1992; Morris et al., 2010). Ideally, when the probe is moved to the next image point then, during the dwell time, the error signal is fed into a feedback circuit controlling the vertical displacement of the sample by the piezoelectric device. The position of the sample is adjusted to reduce the error signal to zero. The vertical motion of the sample is then amplified to provide an image profile of the sample surface. When the dwell time is long enough, and the sample is homogeneous, then the x, y and z motion of the sample yields a 3D map of surface topography: for heterogeneous samples, additional contrast results from material properties that alter the tip-surface interaction. If the tip is simply pressed against the surface, this is called 'dc contact mode' imaging. An alternative mode of imaging (ac imaging mode) involves exciting the cantilever to oscillate and the most common example of this approach is 'Tapping mode™' (Garcia and Perez, 2002): in this case the cantilever-tip assembly is oscillated close to a resonant frequency, the amplitude of the oscillation is large and the tip only momentarily 'taps' onto the sample surface. In the case of 'ac' methods such as Tapping, it is also possible to monitor the phase shift between the oscillation of the cantilever and the driving excitation: this is determined essentially by energy lost during the interaction with the sample and phase images can be used to enhance the contrast due to different material characteristics of the sample.

Images can also be generated in what is called the 'deflection mode', in which the feedback loop is switched off and the error signal is amplified during scanning. In this case, the image represents changes in force and is not a direct measure of surface topography. Nevertheless, the images give some indication of surface structure, and have the advantage that in this mode the image can be acquired rapidly, and high-speed video rate imaging can be achieved by collecting and processing the analogue signal from the photo-diode (Humphris *et al.*, 2005).

For some samples, it is useful to operate the AFM in an intermediate mode, sometimes called the 'error signal mode', where the feedback circuit is switched on, but the length of the dwell times are insufficient to completely zero the error signal. The residual error signal at each image point is used to generate the final image. This mode can be useful for imaging rough sample surfaces. In general, images are normalised in order to maximise the use of grey levels to display contrast in the images. For rough samples, fine detail can obscured, because most of the grey levels are used to accommodate the relatively large, slowly varying background structure in the image. The feedback circuit is able to correct for this slowly varying background and acts effectively as a filter, removing the background and enhancing the fine detail. The pictorial representation of the structure (not strictly a topographic image) facilitates later processing of true topographical images to remove the background and reveal the fine detail in the image.

2.2.3   Friction, adhesion and sample damage

Friction or localised adhesion between the tip and the sample surface causes the cantilever to twist: for heterogeneous samples, these changes can be used to generate frictional or adhesive maps of the surface. However, if the adhesive forces become too large, then scanning can damage the sample. This can be a serious problem when an AFM is used to image biological structures or single molecules deposited onto flat substrates: imaging in air can distort, destroy or even displace deposited structures. A common origin of such adhesion is 'capillary forces'. When biological samples are allowed to dry in air, then most of the aqueous solvent evaporates, leaving the 'dry' material on the substrate. In practice, a thin aqueous layer remains, coating the sample, substrate and tip. Close 'contact' of the tip with the substrate leads to coalescence of these surface layers, effectively gluing the tip to the substrate. On scanning it becomes difficult to lift the tip over the deposited sample.

Also, during scanning, tips can collect debris from the sample leading to localised adhesion with the substrate and sample: thus tips used to image biological systems, particularly in air, generally have a finite lifetime, after which the image quality begins to decay (Morris *et al.*, 2010). Adhesion of the tip to the substrate or sample will damage the sample, rip through the sample, or displace it across the substrate. Capillary forces can be eliminated if images are acquired under liquids, or through the use of ac imaging modes, which minimise the contact time with the substrate and sample (Garcia and Perez, 2002; Holscher and Schirmeisen, 2005; Meyer, 1992; Morris *et al.*, 2010). Currently, Tapping mode is the best method for imaging samples in air (Garcia and Perez, 2002). Use of stiff cantilevers effectively prevents adhesion, limiting damage to the sample. Tapping mode largely eliminates adhesive forces, but the applied normal force may still lead to some sample damage.

Imaging under liquids requires a liquid cell to contain the sample and substrate. The design of the liquid cell varies for different makes of AFM, but the principles involved are similar. High-resolution images of biopolymers can often be obtained by imaging deposited molecules under a precipitant, which prevents any desorption from the surface, and often induces an ordered conformation that restricts molecular motion and blurring of the image. When the liquid used is a solvent for the molecules, it is necessary to develop methods to pin the molecules onto the substrate surface, preventing desorption or motion of the molecules on the substrate surface. Both contact and ac modes can be used for imaging under various liquids, although it becomes harder to select the optimum imaging conditions for the non-contact modes (Morris *et al.*, 2010).

2.2.4   Image resolution and reconstruction

For routine use of AFM to study food systems, the resolution achieved is equivalent to that obtained with the transmission electron microscope (TEM). Although environmental electron microscopes are becoming more available, TEM studies are generally made in vacuum, requiring specialised sample preparation procedures to preserve the native state of the samples. The mode of operation of the AFM allows images to be obtained under more natural conditions. Although of similar resolution, the nature of the images obtained using the two types of instrument are very different. The AFM delivers a surface profile. In the TEM, passing an electron beam through the object generates images and it is possible to obtain sufficient information to allow a 3D reconstruction of the object. Clearly this is not possible with AFM images. For isolated objects, the surface profile determined by AFM is also modified by an effect called 'probe broadening'. This effect depends on the shape and size of the probe tip used to generate the image, and the size of the object being scanned. During scanning, different parts of the tip interact with the sample, increasing the apparent width of the objects. The effect can be corrected, allowing the determination of true sample widths (Vesenka *et al.*, 1996).

In addition to imaging under natural conditions, the AFM yields 3D images from which heights can be directly measured. Because imaging involves pressing the tip onto the sample, some compression of the sample will occur during scanning, and the measured height may be smaller than the true height. This effect depends on the deformability of the sample and can be reduced by optimising the normal force applied to the sample. As well as generating images, the AFM can be used to measure surface roughness for quality control, or as a scientific tool to understand the effects of processing on surface structure: a useful description of such studies in the case of chocolate manufacture is given by Rousseau (2007). Surface roughness is important because it determines the reflection and scattering of light and thus the appearance of food samples.

2.3   Machines and methods: force spectroscopy

The AFM generates images by feeling the sample surface. At any given image point on the sample surface, the force between the tip and the sample will vary as the tip is brought towards and pulled away from the surface. The measurement of these 'force-distance curves' is called force spectroscopy. The initial slope of the force-distance curve has the dimensions of a modulus and can be used to map the mechanical properties of the surface. The AFM can also be used as a force transducer to investigate the mechanical properties of individual macromolecules, or to probe interactions between biological molecules or surfaces. This can be achieved for single molecules by attaching one end of an individual molecule to the tip and the other end to the surface. In the case of interactions between different molecules, one molecule is attached to the tip and the other to the surface. Attaching particles to the tip and the surface, and measuring the interactions between the particles, can be used to probe colloidal interactions. The experimental procedures for acquiring and interpreting such force data are still being developed, and are not as well advanced as those used for obtaining images. However, the use of force spectroscopy is growing rapidly and may soon rival the use of AFM solely as an imaging tool. Applications of AFM as a microscopic tool in food science have yielded important new insights into the nature of food structure and it is likely that force spectroscopy will have a similar impact in the near future.

2.3.1   Calibration
The generation of a force-distance curve requires absolute measurements of both force and distance. When a hard sample is moved towards and comes in contact with the tip, the resultant plot of error signal against distance will be of the form shown in Fig. 2.2. The point of contact between the tip and the sample is

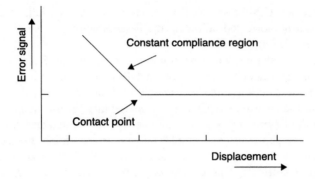

Fig. 2.2 Schematic error signal versus displacement plot: the diagram indicates the constant compliance region and the point of contact between the tip and the substrate surface.

evident from the marked change in slope of the plot. Thus the change in slope defines the zero point and allows the absolute distance of approach or retraction to be determined from the potential difference applied to the piezoelectric device, knowing the calibration constant for this material. The force exerted on the tip-cantilever assembly can be determined by measuring the absolute deflection of the cantilever and the force constant (spring constant) of the cantilever. When the sample is pressed against the tip after the point of contact, the bending of the cantilever is linearly dependent on the change in the distance of approach of the sample towards the tip: this is known as the constant compliance region. From the slope of the plot in the constant compliance region, it is possible to calibrate the error signal in terms of an absolute deflection of the cantilever: the so-called *Inv*OLS or inverse optical lever sensitivity factor (Meyer and Amer, 1988). The calibration has to be performed in the liquid medium in which the force measurements are to be made, because the measured deflection of the optical lever deflection system depends on the refractive index of the liquid.

   This direct approach can be used for robust normal tip-cantilever assemblies, but can lead to damage to modified tips. An alternative approach is to measure and analyse the thermal noise spectrum of the cantilever, as revealed in the photodiode output signal: different vibration modes of the cantilever are excited by Brownian collisions between solvent molecules and the cantilever (Higgens *et al.*, 2006). Measurements of the resonance frequency of the cantilever can be used to determine the force constant of the cantilever: either by exciting the cantilever into resonance or by analysis of thermally driven excitation of the cantilever. The *Inv*OLS factor and the force constant allow the measured error-signal data to be converted into force-distance curves. Nowadays, the calculation of the *Inv*OLS factor and the force constant are usually included in the software modules as semi- or fully automated procedures, allowing the raw output data to be presented directly as force-distance curves.

2.3.2   Molecular force spectroscopy

Molecular force spectroscopy consists of two types of related studies: stretching of individual macromolecules and studies of interactions between different macromolecules.

2.3.3   Stretching single molecules

Stretching of individual molecules usually involves depositing the molecules onto a suitable substrate within a liquid cell and then 'fishing' for adsorbed molecules with the tip. When a molecule is 'hooked', the retract curves reveals the extension of the molecule and the rupture point, at which the molecule breaks or, more usually, at which the molecule detaches from the tip or surface (Fig. 2.3). Proteins and polysaccharides are fairly sticky molecules and will usually bind well to the surface and tip. The majority of the studies on biopolymers are concerned with studies of the unfolding and refolding of multi-domain proteins and model polypeptide structures (Morris *et al.*, 2010; Ng *et al.*, 2007). In this type of study, the main concern is whether this type of forced unfolding/refolding is representative of the processes observed normally in traditional chemical unfolding experiments.

Proteins can also be studied by holding them clamped between the tip and the surface, at a relatively low force, and then recording the thermally-induced

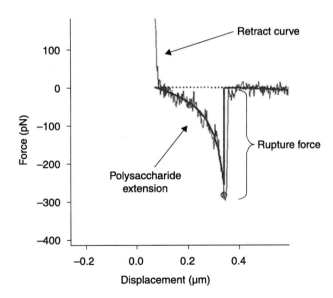

Fig. 2.3   Force spectroscopy data demonstrating the stretching of a polysaccharide chain between the tip and a substrate. The black solid line shows a theoretical fit (worm-like chain model) to the experimental data used to extract information such as the stiffness of the chain.

changes in structure by monitoring the thermal noise spectra of the cantilever. This type of approach may provide a more representative means of monitoring structural changes similar to those seen in solution. Force spectroscopy studies have also been made on a range of polysaccharides (Marszalek *et al.*, 1998, 2001; Morris *et al.*, 2010). The extension at rupture provides a measure of the 'contour length' of the segment of the molecule attached between the tip and the surface. Fishing results in random attachment of the tip to different points along the molecules and hence a range of apparent contour lengths.

The collective measurements can be analysed by normalising the data at a given force value: this normalisation treatment is useful, because it allows discrimination between data for single molecules, and that involving several molecules attached between the tip and surface, or individual molecules bound at several points to the surface. The data can be analysed to give information on the elasticity of the chains, which in turn is sensitive to the primary structure and conformation of the polysaccharide (Fig. 2.3). A concern with this type of analysis is that the values of persistence lengths, a measurement of the stiffness of the polysaccharide chain deduced from force spectroscopy measurements, appear to be substantially lower than those values determined from bulk solution studies, or by an analysis of the shape of the molecules, as determined from AFM or TEM images. The studies on polysaccharides have been reviewed in detail (Marszalek *et al.*, 1998, 2001; Morris, 2012; Morris *et al.*, 2010), but have yet to reveal new insights directly of importance in food science.

2.3.4   Probing intermolecular interactions

Even in the studies of single molecules, it is useful to attach one end of the molecule to the tip, and then to modify the surface such that it will bind the other end of the molecule. This modification can be difficult and is seldom used for studying the stretching of individual polysaccharides. However, if you wish to study interactions between different molecules, then it is essential to attach one molecule to the tip and the other to the substrate.

An example of this type of study that has proved to be of interest in the food area, and which will be discussed in more detail later in this chapter, involves probing carbohydrate binding to lectins (Gunning *et al.*, 2009). In this case, the carbohydrate is bound directly through its reducing end to the tip, or to a spherical particle that is then glued onto the cantilever-tip assembly. The lectin is attached to a suitably derivatised surface, taking care to ensure that the binding site of the protein is not blocked during the attachment. Potential inter-molecular binding can be probed by raster scanning the sample surface relative to the modified tips. Binding events are revealed by adhesion and rupture events detected in the retract curve show that the area under the retract curve (Fig. 2.4), in the hysteresis region, is a measure of the energy required to extend the polysaccharide chain and then to break the carbohydrate-lectin bond, and the force required to break the linkage can be determined at the rupture point. The observed extension of the polysaccharide chain, prior to rupture, allows non-specific interactions that will

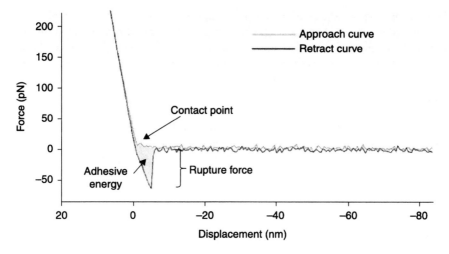

Fig. 2.4 Force spectroscopy data illustrating an interaction between an oligosaccharide bound directly onto the tip–cantilever assembly and a lectin attached to the substrate surface.

occur at the point of zero separation, to be distinguished from specific interactions, which will occur a distance characterised by the contour length of the polysaccharide.

For smaller oligosaccharides, it may be useful to add a spacer molecule, such as a polyethylene glycol (PEG) molecule of defined length, when attaching either the protein or the oligosaccharide to surfaces (Hinterdorfer *et al.*, 1996; 2002). These spacers assist the oligosaccharides or proteins to find and bind to each other. They also provide a defined extension region, allowing the specific interactions to be identified. Specific interactions can also be demonstrated through the addition of a suitable inhibitor molecule, which should compete for binding with the lectin, and thus eliminate or markedly reduce the level of carbohydrate-lectin binding. The studies can be used to demonstrate specific binding and also to characterise the nature of the binding site (Evans and Richie, 1997; 1999).

2.3.5   Colloidal particle interactions

If colloidal particles are attached to the tip and to a surface, then it becomes possible to probe interactions between these particles in a range of media. The effects on the forces of interaction of modifying the surfaces of the particles or the nature of the bulk medium can be probed in detail (Ralston *et al.*, 2005). Studies of such interactions between hard and more recently soft particles are now part of the emerging field of 'Soft Matter'. For the food industry, such studies allow an investigation of the behaviour of dispersions, emulsions and foams. For hard particles used as model systems for air bubbles or oil droplets, the point of contact between the particles can be determined, and thus the absolute forces and the separation between the surfaces can be defined. The main limiting factors in the use of AFM are difficulties in studying very low speed interactions, associated with problems due to piezoelectric drift and creep of the scanners, and difficulties associated with maintaining the high standard of cleanliness required when using small AFM liquid cells.

Most model studies have been made using hard particles but newer studies directly on deformable particles are emerging, which provide better models for soft particulates, food emulsions and foams (Dagastine *et al.*, 2004, 2006 *et al.*, 2004a; Gromer *et al.*, 2010; Manor *et al.*, 2008; Vakarelski *et al.*, 2010; Woodward *et al.*, 2010a). Some of the difficulties involved in these studies concern how to attach deformable particles to the AFM cantilevers and substrate surfaces, how to determine particle size and area of contact with the surfaces and, most importantly, the difficulty of defining the 'point of contact' between the surfaces. Despite these difficulties, deformable particles are better models for the adsorption and assembly of surface-active molecules at interfaces, and thus of foams and emulsions, because they take into account the effects of deformation of the particles on close approach, and the consequent effect on inter-particle interactions. Thus, despite the experimental problems, this field is growing and it has already led to new insights and findings relative to food research.

2.3.6   Coupling colloidal particles to cantilevers

Given the size of colloidal particles and the size of the cantilever, we might expect it to be extremely difficult to attach such particles to the cantilever. In practice, at least for hard particles, we can simply glue them to the tip-cantilever assembly (Morris *et al.*, 2010). One approach involves the use of fine wires, a micromanipulator and an optical microscope to apply glue and then press the particles onto the glue on the cantilever.

Another method involves dusting the particles (usually silica spheres) onto a hydrophobic surface (e.g. the lid of a plastic Petri dish) and then depositing a strip of a recently mixed slow-setting 2-part epoxy adhesive onto a glass microscope slide. The tip can be lowered down towards the glue and it will then suddenly 'snap into' the liquid glue. The tip plus a 'blob' of glue is retracted and can then be pressed down onto the particle, using a microscope to attach it near the end of the cantilever. The AFM will 'snap onto' the particle and then can be retracted with the particle glued to the cantilever. The tricky part of the operation is using just enough glue to attach the particle, but not too much such that the whole particle becomes engulfed with glue. Sometimes the particles will resist removal from the surface but, if the tip is moved parallel to the surface, then the small level of shear can dislodge the particle. It is useful to use techniques such as scanning electron microscopy (SEM) to examine the quality of the modified cantilevers, but this can only be done after completion of the experimental studies (Morris, 2012).

Attaching deformable particles (oil droplets or air bubbles) to cantilevers and surfaces is slightly more difficult. Oil droplets can be sprayed onto a 'cleaned' glass surface and ultrapure water carefully added to the slide without dislodging the droplets. Using the low power optical microscope of a hybrid optical/AFM, it is possible to position the AFM tip over a droplet attached to the surface and then to lower the tip down onto the droplet. The cantilever 'snaps onto' the droplet and can then be retracted, carrying the captured droplet away from the surface. Once the droplets are attached to the cantilever, the bulk composition of the aqueous media can be altered, allowing different surface-active molecules to adsorb and form structures on the surface of the droplets. Note that this needs to be done after attachment – once the droplets are coated they can become almost impossible to remove from the surface onto the cantilever (Morris *et al.*, 2010).

Similar approaches can be used to attach air bubbles to cantilevers. In this case, the air bubbles are produced ultrasonically on a glass surface, which has been modified to make it slightly hydrophobic. Rectangular tip-less cantilevers can be modified by forming a small gold-coated circular platform, which has been made hydrophobic by adsorbing a layer of decanethiol layer onto this region at the end of the cantilever. These modified tips can be made slightly more hydrophobic than the glass surface, and can then be used to capture an air bubble: the patch on the cantilever defines the contact area and the contact angle formed by the attached bubble (Vakarelski *et al.*, 2010).

2.3.7   Measurement and analysis

Colloidal force spectroscopy using AFM is now a fairly well established field, and the methodology and data analysis is described in recent reviews (Leite and Herrmann, 2005; Liang *et al.*, 2007; Ralston *et al.*, 2005). The interactions between charged particles can be modelled by DVLO theory and, depending on the nature of the surface of the particles, other types of interaction such as steric repulsion or bridging effects can be observed on close approach of the particles. In a fluid environment, hydrodynamic forces will also be important: as particles approach each other closely the liquid medium has to drain out from between the particles. Thus the form of the interaction between the particles on approach and retraction will depend on the size and approach speed of the particles, the rate of drainage of media from between the particles, and the viscosity of the bulk medium. If the fluid contains polymers, or small particles, then drainage will result in the exclusion of this material from the local region between the particles, giving rise to an attractive, depletion force. Thus this approach provides a simple 2-particle model system for studying colloidal interactions. Because of the nature of the dependence of the interactions between particles on particle size, it is usual practice to normalise the force data with respect to particle size, and the experimental data is usually plotted as force divided by particle radius versus distance curves.

A new and exciting aspect of this work is the study of deformable particles, in which the deformation of the droplets or bubbles can be taken into account in describing the interaction. Because the particles deform on approach, it becomes difficult to define the 'point of contact' and hence to determine the absolute separation of the surfaces. Dagastine's group (Dagastine *et al.*, 2004, 2006; Tabor *et al.*, 2012) in Australia have developed theoretical models to describe the interactions between deformable particles and one approach is to define the point of contact from a theoretical fit to the experimental data. Such analysis requires knowledge of contact angles, contact area and particle size, and the use of modified cantilevers and surfaces to control and measure these parameters is becoming important. An alternative and recent approach uses a combined AFM – confocal microscope to directly measure the separation of the particle (Tabor *et al.*, 2011a). Experimentally, the approach and retract speeds between particles can be varied in order to study the effects on the interactions between deformable particles. It is also possible to impose a dwell time after approach, before the particles are pulled apart, in order to probe transient effects and relaxation behaviour. This approach is yielding new insights into the behaviour of foams and emulsions and examples of this type of study will be discussed later in this chapter.

2.4   Machines and methods: optical tweezers and photonic microscopy

A complementary technique to AFM, both as a force transducer and as a microscope, is that of optical tweezers and their use as a microscopic tool. The original use of optical tweezers was as a nano-manipulator for collecting and positioning particles. This led to a natural extension to the study of particle–particle interactions. The lack of commercial apparatus has restricted use of this technique in food science. As commercial systems are now becoming available, it is likely that this approach will become more widely used, both for measuring forces and as a 3D microscope. The intention here is simply to describe the methodology although, at present, applications in food science are few and far between.

    The basis of the optical tweezers technique is the generation of an optical trap to capture and manipulate particles. The essential experimental requirements for an optical tweezers system (Fig. 2.5) are a suitable optical microscope with

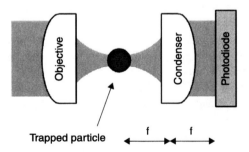

Fig. 2.5   The key features of the optical assembly in an optical tweezers apparatus. 'f' is the focal length of the condenser lens.

attached trapping laser(s) for generating optical trap(s) and a high numerical aperture (NA) microscope objective (Neuman and Nagy, 2008). Focusing the laser beam to a diffraction-limited spot using the objective lens creates the optical trap. The result is that, in the vicinity of this focused spot, there exists a 3D force, which acts to attract dielectric particles towards the focal point of the laser beam. The magnitude of the force will depend on the laser power and the polarisability of the particle. A requirement is that the laser used to produce the trap should have a Gaussian intensity profile: this is needed in order to achieve the smallest size focused spot and thus the largest optical gradient. If the trap is to be stable, then the NA of the objective lens should be at least 1.2, which requires the use of water or oil immersion objective lenses. Water objectives do not suffer from the spherical aberration effects experienced with oil immersion objectives and allow particles to be trapped deep within the solution. It is normal to use near infrared lasers to produce the traps to minimise any optical-induced damage to biological samples. Noise in the experimental data will arise from fluctuations in laser power, and any movement of the focused spot and thus these factors will influence the ultimate choice of laser: for biological samples, the laser of choice is usually a diode-pumped neodymium yttrium aluminium garnet (Nd:YAG) laser with a wavelength of 1.064 μm.

2.4.1   Modes of operation

The primary advantage of the use of optical tweezers as a force transducer is the ability to measure forces that are too small to measure by AFM (typically in the range 0.1–100 pN). By attaching molecules to carrier particles, it is possible to probe their interactions with derivatised surfaces or natural biological surfaces. Through the use of multiple traps, it should be possible to probe colloidal interactions between particles, or to monitor interactions between different molecules. A possible limitation of the use of this technique, certainly for colloidal interactions, might be difficulties in probing short-range interactions where adverse effects may arise due to the proximity of neighbouring optical traps. As yet there are no reported measurements of this type relevant to the food area. An advantage of the technique would be the ability to probe the forces acting on particles in complex environments, such as inside cells or complex food structures like gels (Gisler and Weitz, 1998; Furst, 2005), providing information on local viscosity or elasticity of the medium. Finally, the ability to 'feel' the interactions of probes with their environment makes it possible to generate pictures of bulk or interfacial structures containing the probe, offering applications as a 3D or 2D microscope (Tischer et al., 2001).

An optical arrangement positioned between the laser and the microscope can be used to locate the optical trap within the sample. The objective lens of the microscope is used to focus the laser beam and to view the trapped particle. In a passive mode, focusing the image of the trapped particle onto a four-quadrant photodiode can monitor the movement of trapped particles: the back focal plane of the microscope condenser lens is imaged on the photodiode. Motion of the particle will alter the interference pattern formed between the reflected and transmitted light on the photodiode. These changes provide information on the position of the particles in three dimensions. If particle displacement from an equilibrium position within the trap is small ($\leq$ 150 nm), the restoring force is linearly proportional to displacement. However, the restoring force is asymmetric. Restoring forces along the direction of the laser beam depend on the degree of focus of the laser beam, and thus the laser power and the NA of the objective lens. Within the image plane, perpendicular to the beam direction, the restoring force is determined by the Gaussian profile of the laser and is thus axially symmetric.

In order to calculate the forces acting on a particle as it probes its local environment, it is necessary to determine the 'spring constant' or stiffness of the system. The position detector can be calibrated by observing the motion of a bead over a known distance. Common methods for determining the stiffness involve either modelling the thermal motion of the particle within the trap, or using Stokes law to compare the trapping force with the viscous drag exerted on the particles undergoing motion within a given liquid environment. With optical tweezers, it is possible to measure forces in the range of 0.1 to 100 pN at a temporal resolution of 10^{-4}: however, the achievable resolution depends on the dielectric properties of the trapped particles and the characteristics of the optical trap.

Rather than simply monitoring changes in the position of the particles, it is also possible to control the position of the trapped particle in the specimen plane through the use of a feedback circuit. This makes it possible to maintain a constant force on the particle or to compensate for factors such as thermal drift. There are two ways to operate a feedback system, which result in different degrees of spatial and temporal resolution. Piezoelectric-controlled mirrors or acousto-optic deflectors can be used to determine the position of the trap within the sample chamber. This approach allows the trap to be moved very quickly (~10 µs), but limits the extent to which it can be displaced within a single axial plane to distances of about a few micrometres. An alternative is to use piezoelectric devices to move the trapping chamber relative to the trap. This permits a larger effective 3D spatial displacement ($\leq$100 µm) of the particles, but the temporal resolution is reduced (~10 ms).

2.5   Applications of the atomic force microscope as a microscope

There are no significant examples of the use of optical tweezers in the food area, although there is clearly potential for use either as a force transducer or as a microscope. Probe particles have been used to probe motion in interfacial structures and to monitor the internal structure of gels (Tischer et al., 2001). In contrast, the AFM has been used extensively as an imaging tool in food science and the use of AFM as force transducer is growing, both for studying molecular and colloidal interactions. There are reviews (Morris, 2012; Morris et al., 2010, 2011), which comprehensively cover these areas and provide a route to exploring the relevant scientific literature. As with most new microscopic methods, the application of AFM to food science has led to new structural information, new mechanisms and new insights, which can be used to understand and rationally design food structures for desired functions. Thus, rather than just listing the results of every study on food systems using AFM, the intention is to give examples where new information and new insights have been obtained using AFM, both as a microscope and as a force transducer.

2.5.1   Molecular heterogeneity and molecular complexes

The AFM is ideally suited for imaging food macromolecules deposited onto atomically flat substrates, such as freshly cleaved mica (Morris *et al.*, 2010). In the case of food polysaccharides, the technique provides information on the size and shape of the polymers, and allows us to investigate how they self-associate or interact with other biopolymers.

Polysaccharides exhibiting free rotation about the inter-sugar linkages adopt a spherical, random coil configuration in solution. When deposited onto mica, and even when imaged in air, the polysaccharides remain within a thin layer of solvent retained on the substrate. Thus they are able to access all configurations over periods of time that are rapid compared to the scan time used to image the molecules. The AFM images are of time-averaged structures with the molecules appearing as blurred spherical objects, the diameter of which increases with molecular weight.

Polysaccharides that show restricted motion about the inter-sugar linkages will exist as semi-flexible coils in solution (e.g. amylose, xyloglucans, arabinoxylans, alginates or pectin). These types of polymers can be deposited onto mica under appropriate conditions, which allow the polymer chains to be visualised: the images yield information on the contour length (molecular weight) distribution and the stiffness, through the calculation of the persistence length (Morris *et al.*, 2010; Morris, 2012).

A novel feature of the use of AFM has been the ability to obtain new information on the heterogeneity of polysaccharides. An example (Fig. 2.6) is the exploration of the branching of polysaccharides (Gunning *et al.*, 2003; Ikeda *et al.*, 2005; Kirby *et al.*, 2007; Pose *et al.*, 2012; Round *et al.*, 2001, 2010). Flexible short side chains, or side chains that are linked to backbones by flexible linkages, are difficult to visualise. However, even in these cases, the distribution of such side chains along the backbone of the molecule will give rise to local changes in properties such as the height or the stiffness of the polysaccharide. This type of variation has been observed for mucins (Round *et al.*, 2004) and could be used to analyse branching distributions for individual polysaccharide chains.

Another approach to mapping heterogeneity is to emphasise structural variation by observing the binding of molecules that recognise specific structural features. This approach has been used to visualise the heterogeneity of mucins and arabinoxylans through imaging and analysing the binding of antibodies (Round *et al.*, 2007) or inactivated enzymes (Adams *et al.*, 2004) to selected regions of the polymer chains. The presence and location of specific structural features can also

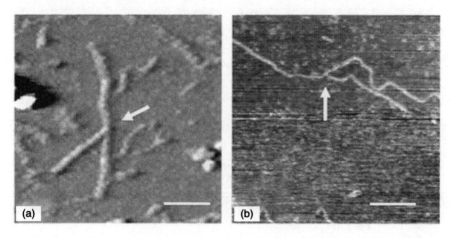

Fig. 2.6   AFM images showing branching of the backbone of polysaccharides. Branch points are indicated by the white arrows: (a) branched pectin molecule (scale bar 46 nm); (b) branched amylose molecule (scale bar 100 nm).

be determined through the use of selective enzymatic or chemical degradation: pectin extracts from plant cell walls are known to contain homopolymeric regions of polygalacturonic acid plus complex branched regions containing acid-labile neutral sugar side chains, although the detailed spatial distribution of these structural types within individual 'pectin' molecules is still ill-defined.

Most textbooks represent pectin as a composite linear structure with the branched regions located between homopolymer sequences. AFM images of pectin extracts reveals that the extract consists of homopolymers of polygalacturonic acid, plus aggregated structures containing the neutral sugars (Kirby *et al.*, 2007; Pose *et al.*, 2012; Round *et al.*, 2001). On acid hydrolysis, breakdown of the aggregates through cleavage of neutral sugar linkages released homopolymers. The homopolymers were not degraded on cleavage of neutral sugar linkages, suggesting that the branched regions are at the ends of the homopolymers or confined solely to the aggregated structures. Thus AFM provides new information on the potential assembly of pectin within plant cell walls (Round *et al.*, 2010).

AFM has also yielded new information on unsuspected and infrequent branching of the polysaccharide backbone of pectin homopolymers (Kirby *et al.*, 2007; Pose *et al.*, 2012; Round *et al.*, 2001) (Fig. 2.6a), the distribution of branches in water-soluble arabinoxylans (Adams *et al.*, 2003), and on small levels of branching for amylose molecules (Gunning *et al.*, 2003) (Fig. 2.6b). Ikeda *et al.* (2005) showed evidence for complex multi-branched structures for the surface-active soyabean polysaccharide. In all of these cases, the AFM can provide new information on the number of branches per molecule, the branch length distribution and the relative numbers of linear, branched or multi-branched molecules within a population (Pose *et al.*, 2012). In many cases the observed levels of branching are unexpected, because they are too low to detect by chemical analysis.

Knowledge of the molecular size, shape and branching is important for understanding the viscosity and viscoelastic behaviour of these molecules in solution. In addition to mapping polysaccharide structure, the binding of proteins can also be used to explain unusual functional behaviour, such as surface activity (Gromer *et al.*, 2009, 2010; Kirby *et al.*, 2006, 2007), or to facilitate understanding of molecular mechanisms of the activity of enzymes (Morris *et al.*, 2007).

2.5.2   Biopolymer gels

Many of the more familiar food polysaccharides are used to impart useful functional properties, such as thixotropy and gelation, and these properties arise because of the ability of these polysaccharides to adopt ordered conformations in solution. Such helix-forming polysaccharides include xanthan gum, gellan gum, agar and the carrageenans. For these materials, the AFM has been used to generate new information on the modes of association of these polysaccharides responsible for structuring food materials.

AFM imaging on sufficiently dilute solutions drop deposited onto suitable substrates such as mica provides information on the structure of individual polysaccharides. The use of higher concentrations gives rise to aggregation on drying and this can provide information on the self-association of polysaccharides yielding new insights into mechanisms of gelation (Morris, 2007a,b). Charged helix-forming polysaccharides, such as carrageenans and gellan, are examples of thermo-reversible gelling agents used in the food industry: they gel on cooling a hot sol and melt on re-heating. The gelation mechanisms for these helix-forming polysaccharides are similar and can be illustrated by studies on the model system gellan gum: gellan gum is a bacterial polysaccharide with a regular chemical repeat unit, unlike the carrageenans, which contain irregularities in their primary structure.

Extensive studies have revealed that in general gelation of thermo-reversible polysaccharides is triggered by helix formation on cooling from the sol state (Morris, 2007a), and involves two discrete and separable steps: helix formation and helix aggregation. The polysaccharides are charged, and hence increased ionic strength screens the charge, promoting helix formation, stabilising and raising the melting temperature of the helix. Gelation can be cation sensitive, with certain so-called 'gel-promoting cations' enhancing helix aggregation and gelation. Even in the absence of 'gel-promoting cations', some association will occur, but this leads to weak networks that break easily at low deformation.

Previous physical chemical studies have characterised the local, ordered structures (junction zones) formed in the gels, but say little about the long-range network structures that determine the mechanical properties of the gels. For gellan gum, AFM provided new information on the mechanism of gelation and the nature of the gel networks (Gunning *et al.*, 1996a; Morris, 2007a) through studies on the structure of gel precursors, on the network structures formed in aqueous films and, in certain cases, the actual structures within bulk gels (Fig. 2.7a–d). Association in the absence of 'gel-promoting cations' resulted in thin, branched fibrillar gel

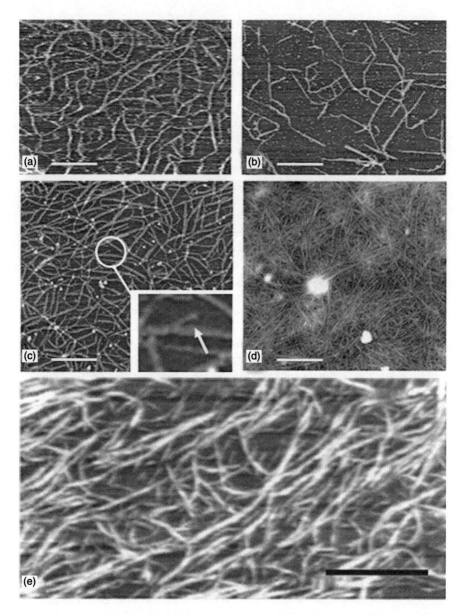

Fig. 2.7 AFM images of gel precursors, films and gels: (a) AFM topography image of gellan fibrils formed in the absence of gel-promoting cations (scale bar 200 nm); (b) topography image of gellan fibres formed in the presence of gel-promoting (potassium) cations (scale bar 160 nm); (c) topography image of a hydrated gellan film formed on mica (scale bar 160 nm). The inset shows a zoomed image of the ringed region of a fibre showing a fibril branch (arrow) that has not developed further into a thicker fibre; (d) topography image of the fibrous network structure at the surface of a hydrated bulk acid-set gellan gel (scale bar 400 nm); (e) AFM image of a fragment from a cold-setting calcium-induced pectin gel (scale bar 250 nm).

precursors (Fig. 2.7a). Gellan gels formed under these conditions show no hysteresis on setting and melting, indicating that the association involves just double helix formation, usually between gellan chains of unequal length. Addition of 'gel-promoting cations' led to the fibrils assembling into thicker branched fibres (Fig. 2.7b) attributed to cation-induced side-by-side association of the fibrils. Images of the gel precursors at sub-gelling concentrations suggest the form of association, which might be expected to occur at the higher concentrations that lead to gelation.

In general, it is difficult to image the network structures formed in bulk gels, as the tip tends to 'snap-in' to the gel leading to damage, or deformation of the network on scanning, resulting a blurred image. However, by depositing from still higher concentration sols, it is possible to form a thin hydrated film on the mica substrate. These types of networks are easier to image, because imaging effectively involves compressing the polymer network down onto the hard mica substrate, resulting in less deformation of the network.

For gellan, the long-range structure observed within such a film was a continuous branched fibrous network composed of aggregated gellan fibrils (helices) (Fig. 2.7c). As shown in the inset in Fig. 2.7c, it is possible to observe embryonic fibrils branching out from fibrous bundles. For gellan, it proved possible to form high-modulus gels that showed negligible distortion on scanning, and the AFM image (Fig. 2.7d) of the network structure present on the upper surface of the gel was seen to be equivalent to the branched fibrous structure previously observed in the hydrated films. Thus the AFM images showed that the gels are fibrous networks in which the 'junction zones' may be thought of as glue binding the helices (fibrils) into the fibres. This long-range structure is in contrast with previous models for thermo-reversible gels, which were generally pictured as rubber-like networks, with discrete extended junction zones linked by disordered polysaccharide chains. This type of fibrous structure for gels appears to be generic and has also been observed for cold-setting, thermally irreversible polysaccharides gels such as pectin gels (Fig. 2.7e), where once again the ordered junction zones appear to act as sticky patches linking the individual polysaccharides together into fibres (Morris et al., 2009).

2.5.3   Interfacial structures, foams and emulsions

Emulsions and foams are stabilised by the addition of surface-active agents. Surfactants absorb rapidly at the interface. Proteins tend to absorb more slowly but lead to structures that are stable over longer periods of time. Protein adsorption at air-water and oil-water interfaces results in the formation of 2D elastic networks, which is the important factor in promoting stable food foams and emulsions. The investigation of the structure of such interfacial networks is an area where AFM has made a large impact, mainly because it allowed imaging, for the first time, of these structures at the molecular level (Morris and Gunning, 2008).

In order to image the structures formed by proteins at interfaces, it is necessary to use a model system: the interfacial structures are formed on a Langmuir trough and Langmuir–Blodgett (LB) methods are used to sample the interfacial structures and to deposit them onto substrates, usually freshly-cleave mica, for imaging by AFM. Both air-water and oil-water interfaces can be studied and the interfacial structures can be formed by either spreading the molecules at the interface, or by allowing the proteins to adsorb from the bulk phase (Morris *et al.*, 2010).

The first and perhaps most important result of imaging interfacial proteins networks was the observation (Gunning *et al.*, 1996b) that the networks contained defects (holes). These holes occur because proteins partially unfold and associate on absorption at the interface. As they form networks the space available for further adsorption decreases and, eventually, there is insufficient room for further protein adsorption, leaving holes in the network. Such molecular heterogeneity proved to be crucial for understanding the mechanism of competitive displacement of proteins by surfactants: a process of practical importance in determining the stability and lifetime of food foams and emulsions.

Surfactants are more surface-active than individual proteins and they will, given sufficient time and sufficiently high bulk concentrations, eventually completely displace proteins from interfaces. However, the mechanism by which this occurs is not self-evident: individual proteins can be easily displaced but, when they are linked together into a network at the interface. they collectively resist displacement, and it is difficult to understand how such displacement does occur. The use of AFM revealed the answer to this question.

Representative AFM images of protein displacement by surfactant from an air-water interface are shown in Fig. 2.8: in this case the protein is the whey protein β-lactoglobulin and the displacer is the non-ionic surfactant Tween 20. The images reveal what appears to be a progressive colonisation of the interface by the surfactant. Initially small surfactant domains nucleate at the defects within the protein structure. This shows the importance of these defects because, without such holes, it is difficult to see how the surfactant could infiltrate onto the interface. With increased bulk concentration of surfactant or, if the surfactant concentration is sufficient, increased time, the surfactant domains grow larger (Fig. 2.8a–c). The AFM data shows that the colonisation of the interface by surfactant is heterogeneous rather than homogeneous, and explains why this is so (Mackie *et al.*, 1999).

AFM images are 3D profiles of the surface: the area occupied by protein and also the height of the protein network can be determined, allowing the protein volume (surface concentration) to be monitored during displacement. The key finding is that, as the surfactant domains grow, the protein concentration remains unchanged and then, suddenly, at a certain surface pressure (adsorbed surfactant concentration), the protein is expelled into the bulk phase. This happens because as more surfactant enters the holes, the domains expand compressing the protein network, causing refolding of individual proteins, and then folding and buckling of the network structure. The process, which involves the folding, buckling and eventual failure of the protein network, has thus been termed 'orogenic displacement' (Mackie *et al.*, 1999). The irregular shape of the surfactant domain boundaries indicates the heterogeneous nature of the protein network; different degrees of unfolding and association on adsorption make certain regions of the

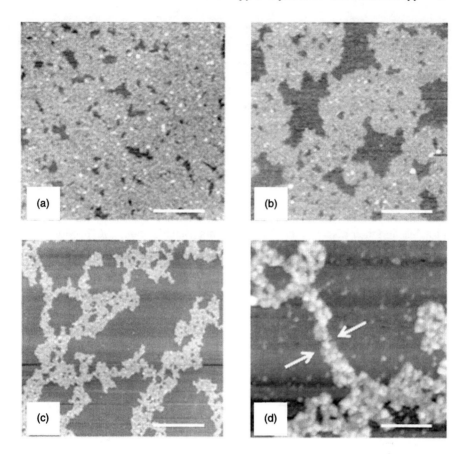

Fig. 2.8 AFM images showing orogenic displacement of the whey protein β-lactoglobulin (grey) from an air–water interface by the neutral surfactant Tween 20 (black domains) at increasing surface pressures of: (a) 18.6 mN/m, scale bar 400 nm; (b) 20.2 mN/m, scale bar 400 nm; (c) 22.5 mN/m, scale bar 640 nm; and (d) 22.5 mN/m, scale bar 150 nm. The AFM image (d) shows individual proteins (arrowed) in a stretched region of the network just prior to collapse. The surface pressure is a measure of the degree of colonisation of the interface by surfactant.

network weaker and easier to deform. At the higher surfactant concentrations, the elastic protein network is stretched (Fig. 2.8d) until the protein strands surrounding the surfactant domains eventually break. Once the network is broken then protein (or protein aggregates) can be displaced. Thus the holes in the protein network are crucial for displacement to occur.

The mechanism of displacement appears to be generic for all proteins studied and for different types of surfactants (neutral, charged, water-soluble and oil-soluble), at both air-water and oil-water interfaces: basically, if the proteins form networks at an interface, then the networks have to be broken in order to release and displace the protein (Morris and Gunning, 2008). Thus the generic nature of the displacement process means that interfaces can be stabilised either by strengthening the protein network or by preventing surfactants reaching the interface.

Early mechanistic studies were made on purified proteins and individual surfactants. In practice, commercial food products contain mixtures of proteins, surfactants and even possibly some surface-active polysaccharide extracts. Proteins used as commercial foam stabilisers or emulsifiers will be isolates rather than pure proteins, and thus the interfaces in real foods will be complicated structures, containing mixtures of proteins and other surface-active components. An important question is whether the simple models of protein displacement described above can be extended to accommodate this level of structural complexity? Through the use of AFM and fluorescence microscopy, it has been possible to examine the structures formed by binary mixtures of proteins, and their displacement by surfactants (Mackie et al., 2001b; Morris and Gunning, 2008).

These studies show that when mixtures of proteins adsorb at interfaces, they interact rapidly forming kinetically trapped, immobile networks, with little evidence for phase separation. Surfactant displacement occurs through an orogenic mechanism and, for mixed networks the important observation is that the final failure of the network is dominated by the protein components, which on their own would form the strongest network (Mackie et al., 2001b; Morris and Gunning, 2008; Woodward et al., 2004), even if this protein is only a minor component of the mixture (Woodward et al., 2009). Thus it is possible to predict the general behaviour expected for protein isolates if the composition is known, although the preparation and isolation of pure proteins and isolates may be different, which can modify further the quality of the network. The importance of the protein component in polysaccharide-protein complexes, such as sugar beet pectin (SBP) extracts, on emulsion stability has also been demonstrated through visualisation of individual complexes (Kirby et al., 2006, 2007), the structures formed at model interfaces and the effects of these interfacial structures on interactions between emulsion droplets (Gromer et al., 2009, 2010). These studies were consistent with the formation of a protein network at the interface further stabilised by the polysaccharide component.

The AFM studies on structures sampled from interfaces prepared on Langmuir troughs have been complemented by studies that demonstrate orogenic displacement from actual interfaces through Brewster angle microscopy (Mackie et al., 2001a) and studies on liquid lamellae as models for foam drainage (Clark et al., 1990, 1994; Wilde and Clark, 1993). Orogenic displacement has also been shown to occur in model emulsions through force spectroscopy studies of emulsion droplet deformation (Gunning et al., 2004; Woodward et al., 2010b), and in real emulsions through studies on protein displacement via direct colorimetric assays, or indirectly through measurements of surface charge (Woodward et al., 2010b). These studies have shown that orogenic displacement occurs from the curved surfaces of finite-sized oil droplets, and hence is applicable to the discussion of emulsion stability.

The techniques used to image interfaces have been used to investigate the physical chemical and enzymatic aspects of digestion on interfaces (Maldonado-Valderrama *et al.*, 2008; 2009; 2010a,b; 2012; Morris and Gunning, 2008). In this context, the orogenic mechanism, discovered through the use of AFM, has provided a rational basis for the modification of interfacial structures to reduce the rate of lipolysis for food emulsions, with a view to designing foods that induce satiety (Morris *et al.*, 2011; Woodward *et al.*, 2010c). The orogenic model provides new insights into the complex interactions between proteins and surfactants in food systems. The generic aspects of the model also apply to any protein-stabilised emulsion or foam. Non-food applications based on this model include the potential to develop novel anti-foaming agents for protein-stabilised foams (Christiano and Fey, 2003), nano-technological applications such as the printing of protein 'inks' (Jung *et al.*, 2004) and protein arrays (Deng *et al.*, 2006), and suggested use of surfactants for attenuation of gas embolism-induced thrombin production (Eckmann and Diamond, 2004).

2.6   Applications of atomic force microscopes as a force transducer

2.6.1   Intermolecular binding and bioactivity

Applications of force spectroscopy to food science are still in their infancy. However, the technique has yielded new information on the origins of the bioactivity (anti-cancer activity) of a food polysaccharide (modified pectin). Certain chemically or enzymatically-modified forms of commercial (citrus) pectin have been found to show a broad-spectrum action against the development and spread of a range of cancers (Glinsky and Raz, 2009; Maxwell *et al.*, 2012; Nangia-Makker *et al.*, 2002). A suggested molecular explanation for these effects is that the modification procedures generate fragments from the pectin molecules that bind to and inhibit the roles of a key regulatory protein galectin 3 (Gal3): Gal3 is mammalian lectin which controls key stages in the differentiation of normal cells and the development of primary and secondary tumours (Glinsky and Raz, 2009). It is possible to prepare the molecular fragments produced by the chemical or enzymatic modification of pectin and to test their ability to bind specifically to Gal3 using microscopy, flow cytometry and force spectroscopy (Gunning *et al.*, 2009).

The force spectroscopy studies show that the only fragment released from pectin that binds specifically into the carbohydrate recognition site of Gal3 is a pectin-derived galactan. Furthermore, it can be shown that the terminal disaccharide (β-galactobiose) at the non-reducing end of the linear galactans, which would be released from modified citrus pectin, will also bind specifically to Gal3 (Fig. 2.9). The measured rupture force is dependent on the rate at which the bond is ruptured, and analysis of this dependence provides information on the nature of the binding site and the lifetime of the interaction (Evans and Ritchie, 1997, 1999; Strunz *et al.*, 2000). These studies support the molecular hypothesis

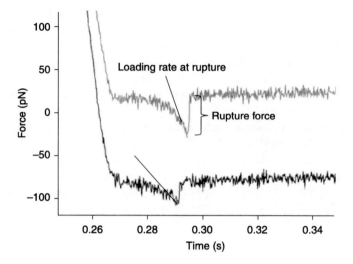

Fig. 2.9   Force spectroscopy data showing the specific binding of β-galactobiose to the mammalian lectin Gal3. The detachment events have been plotted as a function of time to illustrate that rupture of the bond occurs at a finite loading rate and the measured rupture force depends on the loading rate at rupture. Analysis of this dependence provides information on the nature of the binding site and the lifetime of the interaction.

for the role of modified pectin. Modified pectin is undergoing clinical trials as an anti-cancer drug, and there is clearly an interest in whether pectin fragments generated as digestion products of fruit and vegetable cell walls (dietary fibre) may be able to reduce the risk factors for the onset and progression of cancers. Elucidating the origins of bioactivity at the molecular level is crucial to establishing the basis for health claims for food products, and for defining foods or food supplements that can deliver protection against the onset and spread of chronic diseases such as cancers

Force spectroscopy can be used to probe specific interactions between carbohydrates on model systems at the molecular level (Touhami *et al.*, 2003; Gunning *et al.*, 2009), to probe interactions between molecules and receptors on cellular systems under physiological conditions (Gunning *et al.*, 2008), or through the use of suitable probes to map particular molecules, such as adhesins, on bacterial surfaces (Dupres *et al.*, 2005). The latter type of study promises an opportunity to probe the molecular interactions important in determining the interplay between friendly gut bacteria and pathogenic bacteria in colonising and maintaining a healthy gut.

2.6.2   Deformable colloids, foams and emulsions
A new and emerging area is the use of AFM-based colloidal force spectroscopy studies to probe interactions between deformable particles. This new area of soft

matter is already generating new insights into factors that control the stability of foams and emulsions.

In the simplest approach, hard spheres attached to the end of tip-less cantilevers can be used to probe the deformation of air bubbles and oil droplets. However, recently it has been shown that it is possible to attach air bubbles or oil droplets to AFM cantilevers, allowing their interaction with surfaces, or with other bubbles or droplets attached to suitable surfaces, to be probed in a liquid medium (Dagastine *et al.*, 2004, 2006; Gunning *et al.*, 2004; Gromer *et al.*, 2010; Manor *et al.*, 2008; Tabor *et al.*, 2011a,b; 2012; Vakarelski *et al.*, 2010; Woodward *et al.*, 2010a). These types of study provide simple 2-particle models for the interactions expected to occur in real foams and emulsions.

Studies of time-dependent interactions between two bubbles under controlled conditions in a liquid media have provided new direct information on the effects of surface deformation and hydrodynamic flow on the interactions between the bubbles and on coalescence (Vakarelsk *et al.*, 2010). Understanding the behaviour within thin films formed between interacting droplets is important for the modelling of film drainage and hence the stability of foam structures. Drainage behaviour between the surfaces of soft particles is also of interest in understanding lubrication behaviour between surfaces. The experimental data on the interactions of the bubbles can be modelled by newly developed theories, which describe the effects of hydrodynamic flow and particle deformation on the interactions between the particles (Dagastine *et al.*, 2006; Manor *et al.*, 2008; Tabor *et al.*, 2012). These experimental results and their analysis have revealed unexpected features that would not have been observed from model studies made using rigid particles: it has been shown that there are no specific ion effects observed at high ionic strength and also that thermal fluctuations are not required to trigger the onset of coalescence between bubbles.

Direct measurements of the deformability of oil droplets, when pressed against solid surfaces, and of the time-dependent interactions between oil droplets in an aqueous medium, has led to new and unexpected findings, important for understanding the stability of emulsions (Gunning *et al.*, 2004; Gromer *et al.*, 2010; Woodward *et al.*, 2010a). When a rigid particle attached to a cantilever is pressed against a hard surface, the slope of the force distance curve changes at the point of contact: once contact has occurred, the motion of the cantilever follows the advance or retraction of the surface as controlled by the piezoelectric transducer (Fig. 2.2). This part of the force-distance curve is called the constant compliance region.

For deformable oil droplets, the point of contact is difficult to define, but the slope of the force–distance curve in the region previously called the constant compliance region changes, and this change in slope reflects the deformability of the droplet (Fig. 2.10a). If two droplets are pressed together, then the effect on the slope of the curve is enhanced. Thus the change in slope can be used qualitatively to probe the effects of interfacial structure on the deformability of droplets, and droplet deformability can be used to monitor formation and modification of interfacial structures. Once 'bare' droplets are attached to the surfaces and the

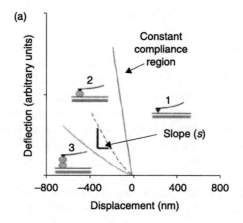

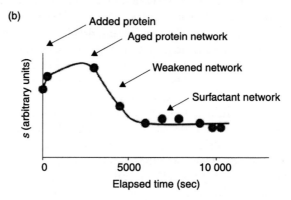

Fig. 2.10 Diagram illustrating droplet deformation. (a) In the region known as the constant compliance region for rigid particles (1) the slope of the curve (*s*) decreases when droplets are pressed against a solid surface (2) and decreases even further when two droplets are pressed together (3). The slope of the curve (*s*) provides a measure of droplet deformability, which can be used to monitor the effects of adsorption of material at the oil–water interface; (b) measurements of the slope following adsorption and ageing of the protein (*β*-lactoglobulin) network, which hardens the droplets, and the effect of subsequent addition of surfactant (Tween 20), which leads to weakening and collapse of the protein network.

cantilever, then introducing surface-active components into the bulk media, and allowing them to adsorb onto the interface, can be used to modify the interfacial structure. The expectation is that adsorption will lower the interfacial tension, thus reducing the Laplace pressure within the droplets: the droplets should become more deformable.

In practice, when protein adsorbs onto the interface, droplets harden despite a decrease in interfacial tension (Fig. 2.10b). This demonstrates that the behaviour of the droplets is dominated by the elasticity of the 'protein skin' formed on the surface of the droplets. The ageing of the protein network can lead to a further progressive hardening of the droplets. The addition of surfactant softens the droplets, and measurements of droplet deformability can be used to follow the surfactant–protein interactions at the interface. Such studies have demonstrated surfactant-induced 'orogenic' displacement of protein networks from the surface of finite-sized oil droplets in an aqueous medium (Gunning *et al.*, 2004; Woodward *et al.*, 2010a).

Time-dependent interactions between oil droplets can be employed as a model 2-particle system for investigating the effects of interfacial structure and composition on the forces between droplets that determine the stability of emulsions. Recent extensive studies of this type have been made to investigate the role of sugar beet pectin (SBP) as an emulsifier (Gromer *et al.*, 2009, 2010). AFM images of individual SBP components showed the presence of protein–pectin complexes (Kirby *et al.*, 2006, 2007) that are considered to be the principal determinant of the surface activity of SBP. The effects of bulk SBP concentration on interfacial structure was probed for model interfaces. These studies suggested that the protein component formed an elastic network shielded by the associated pectin chains. At low bulk concentrations the adsorbed networks were flat but, at the higher bulk concentrations, the interfaces became rougher, with the pectin chains extending out into the aqueous medium (Gromer *et al.*, 2009). These rough surfaces led to observable steric repulsion between droplets (Gromer *et al.*, 2010). By lowering the pH below the pK_a of the pectin chains, or through the addition of calcium ions, it was possible to induce favourable attractive interactions between pectin chains and bridging interactions between droplets (Gromer *et al.*, 2010).

Novel effects were observed for droplet interactions at low bulk SBP concentrations, where imaging of model interfaces suggested the presence of flat interfacial films (Gromer *et al.*, 2009). In this case (Gromer *et al.*, 2010), the observed effects at relatively slow approach speeds were characteristic of depletion interactions: on approach, the droplets flatten allowing bulk fluid to drain from between the droplets, with the consequent removal of SBP from the interfacial layer between the droplets generating an attractive interaction between the droplets.

Removal of SBP from the bulk phase by perfusion caused the interaction between the droplets to return to a reversible repulsive interaction between the droplets. Depletion was confirmed by showing that the effects were also observed in the presence of the non-adsorbing polymer polystyrene sulphonate (PSS). New and unexpected results were observed when the approach and retract speeds were increased (Fig. 2.11). For the slowest approach speeds, the droplets deform as they approach and then show a characteristic 'snap-in', indicating adhesion but not coalescence. Retraction reveals a hysteresis effect with a larger amount of energy required to pull the drops apart. Increasing the approach speed causes the snap-in event to move up along the approach curve, and then over and down the retract curve. At this point and beyond, the snap-in arises as the drops start to move apart. In addition, the energy required to pull the drops apart is observed to decrease as the approach speed is increased.

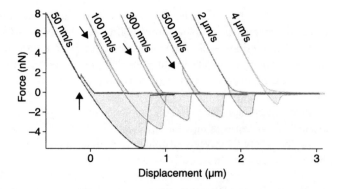

Fig. 2.11 Novel depletion effects of the type observed with sugar beet pectin extracts. The novel aspects of such effects are illustrated for the model system polystyrene sulphonate (PSS). The unexpected results are that, as the bulk polymer concentration and the loading rate increase, the 'snap-in' (arrowed) moves initially up the approach curve and then down the retract curve and, with increasing loading rate, the adhesive energy (shaded) and the rupture force decrease.

The theoretical models, developed to model the behaviour of surfactant-stabilised droplets in terms of droplet deformation and hydrodynamic drainage, do not predict these effects. Modification of such models by the addition of a depletion interaction accounts for certain generic aspects of the interactions but does not fully account for the experimental observations (Gromer *et al.*, 2010). Clearly this is a good example of how force spectroscopy can provide a means of examining the effects of interfacial structures on the interactions between deformable particles, which yields new insights into the mechanisms of stability.

The use of AFM to image interfacial structures offers a route to the rational design of such structures to generate required functionality. Force spectroscopy of deformable particles provides a further route to extend such rational design to a higher level of structural hierarchy, by providing a model 2-particle system for examining the effects of interfacial structure on interactions between particles in foams and emulsions.

2.7   Conclusion

The use of AFM as an imaging tool has matured and although still not routine it is now widely used to solve problems in food science. There is still a challenge in characterising the heterogeneity of individual biopolymers and mapping changes in these structures due to chemical, physical or enzymatic modification. Such studies will probably require the use of modelling and simulations to predict or explain the changes in molecular structures. Such approaches may prove to be important in studying important processes such as ripening, spoilage or the effects of genetic or natural mutations on natural food materials. The newer generation instruments are allowing the technique to be extended beyond studies of single molecules and molecular interactions to more complex structures, such as plant and mammalian tissue.

The newer generation instruments will also greatly assist in the development of force spectroscopy as a tool for solving problems in food science. Although still in its infancy, force spectroscopy offers the prospect of examining the molecular origins of processes such as adhesion, friction and lubrication. The extension of colloidal force spectroscopy to the study of deformable particles promises better models for foams, emulsions and the understanding of the general behaviour of dispersions of soft particulate materials. There is the prospect of characterising these materials through processing steps or ultimately of modelling the behaviour of these systems during complex processes such as digestion. The use of optical tweezers as a force transducer or as a microscopic tool has yet to be explored, but the availability of commercial instruments should allow such applications to be explored in the near future. Likely applications are in the exploration of microrheology in complex environments (cellular or plant tissue, mucin) or in complex food systems such as interfaces, concentrated dispersions or gels.

2.8   References

ADAMS, L. L., KROON, P., WILLIAMSON, G. and MORRIS, V. J. (2003), Characterisation of heterogeneous arabinoxylans by direct imaging of individual molecules by atomic force microscopy, *Carbohydrate Research*, **338**, 771–80.

ADAMS, E. L., KROON, P. A., WILLIAMSON, G., GILBERT, H. J. and MORRIS, V. J. (2004), Inactivated enzymes as probes of the structure of arabinoxylans as observed by atomic force microscopy, *Carbohydrate Research*, **339**, 579–90.

ALBRECHT, T. R. and QUATE, C. F. (1988), Atomic resolution with the atomic force microscope on conductors and nonconductors, *Journal of Vacuum Science & Technology*, **6**, 271–4.

BINNIG, G., QUATE, C. F. and GERBER, C. (1986), Atomic force microscope, *Physical Review Letters*, **56**, 930–3.

CHRISTIANO, S. P. and FEY, K. C. (2003), Silicone antifoam performance enhancement by nonionic surfactants in potato medium, *Journal of Industrial Microbiology & Biotechnology*, **30**, 13–21.

CLARK, D. C., COKE, M., MACKIE, A. R., PINDER, A. C. and WILSON, D. R. (1990), Molecular-diffusion and thickness measurements of protein-stabilised thin liquid-films, *Journal of Colloid and Interface Science*, **138**, 207–19.

CLARK, D. C., MACKIE, A. R., WILDE, P. J. and WILSON, D. R. (1994), Differences in the structure and dynamics of the adsorbed layers in protein-stabilised model foams and emulsions, *Faraday Discussions*, **98**, 253–62.

DAGASTINE, R. R., STEVENS, G. W., CHAN, D. Y. C. and GRIESER, F. (2004), Forces between two oil droplets in aqueous solution measured by AFM, *Journal of Colloid and Interface Science*, **273**, 339–42.

DAGASTINE, R. R., MANICA, R., CARNIE, S. L., CHAN, D. Y. C., STEVENS, G. W. and GRIESER, F. (2006), Dynamic forces between two deformable oil droplets in water, *Science*, **313**, 210–13.

DENG, Y., ZHU, X. Y., KIENIEN, T. and GUO, A. (2006), Transport at the air/water interface is the reason for rings in protein microarrays, *Journal of the American Chemical Society*, **128**, 2768–9.

DUPRES, V., MENOZZI, F. D., LOCHT, C., CLARE, B. H., ABBOTT, N. L. et al., (2005), Nanoscale mapping and functional analysis of individual adhesins on living bacteria, Nature Methods, 2, 515–20.

ECKMANN, D. M. and DIAMOND, S. L. (2004), Surfactants attenuate gas embolism-induced thrombin production, Anesthesiology, 100, 77–84.

EVANS, E. and RICHIE, K. (1997), Dynamic strength of molecular adhesion bonds, Biophysical Journal, 72, 1541–55.

EVANS, E. and RICHIE, K. (1999), Strength of a weak bond connecting flexible polymer chains, Biophysical Journal, 76, 2439–47.

FURST, E. M. (2005), Applications of laser tweezers in complex fluid rheology, Current Opinion Colloid & Interface Science, 10, 79–86.

GARCIA, R. and PEREZ, R. (2002), Dynamic atomic force microscopic methods, Surface Science Reports, 47, 197–301.

GISLER, T. and WEITZ, D. A. (1998), Tracer microrheology in complex fluids, Current Opinion Colloid & Interface Science, 3, 586–92.

GLINSKY, V. V. and RAZ, A. (2009), Modified citrus pectin anti-metastatic properties: one bullet, multiple targets, Carbohydrate Research, 344, 1788–99.

GROMER, A., KIRBY, A. R., GUNNING, A. P. and MORRIS, V. J. (2009), Interfacial structure of sugar beet pectin studied by atomic force microscopy, Langmuir, 25, 8012–18.

GROMER A., PENFOLD R., GUNNING A. P., KIRBY A. R. and MORRIS V. J. (2010), Molecular basis for the emulsifying properties of sugar beet pectin studied by atomic force microscopy and force spectroscopy, Soft Matter, 6, 3957–69.

GUNNING, A. P., KIRBY, A. R., RIDOUT. M. J., BROWNSEY, G. J. and MORRIS, V. J. (1996a), Investigation of gellan networks and gels by atomic force microscopy, Macromolecules, 29, 6791–6.

GUNNING, A. P., WILDE, P. J., CLARK, D. C., MORRIS, V. J., PARKER, M. L. and GUNNING, P. A. (1996b), Atomic force microscopy of interfacial protein films, Journal of Colloid and Interface Science, 183, 600–2.

GUNNING, A. P., GIARDINA, T. P., FAULDS, C. B., JUGE, N., RING, S. G. et al. (2003), Surfactant mediated solubilisation of amylose and visualisation by atomic force microscopy, Carbohydrate Polymers, 51, 177–82.

GUNNING, A. P., MACKIE, A. R., WILDE, P. J. and MORRIS, V. J. (2004), Atomic force microscopy of emulsion droplets: probing droplet–droplet interactions, Langmuir, 20, 116–22.

GUNNING, A. P., CHAMBERS, S., PIN, C., MAN, A. L., MORRIS, V. J. and NICOLETTI, C. (2008), Mapping specific adhesive interactions on living intestinal epithelial cells with atomic force microscopy, FASEB Journal, 22, 2331–9.

GUNNING, A. P., BONGAERTS, R. J. M. and MORRIS, V. J. (2009), Recognition of galactan components of pectin by galectin-3, FASEB Journal, 23, 415–24.

HIGGENS, M. J., PROKSCH, R, SADER, J. E., POLCIK, M., MCENDOOM, S. et al. (2006), Noninvasive determination of optical lever sensitivity in atomic force microscopy, Review of Scientific Instruments, 77, 013701.

HINTERDORFER, P., BAUMGARTNER, W., GRUBER, H. J., SCHILCHER, K. and SCHINDLER, H. (1996), Detection and localisation of individual antibody-antigen recognition events by atomic force microscopy, PNAS USA, 93, 3477–81.

HINTERDORFER, P., GRUBER, H. J., KIENBERGER, F., KADA, G., RIENER, C. et al. (2002), Surface attachment of ligands and receptors for molecular recognition force microscopy, Colloids and Surfaces B: Biointerfaces, 23, 115–23.

HOLSCHER. H. and SCHIRMEISEN, A. (2005), Dynamic force microscopy and spectroscopy, Advances in Imaging and Electron Physics, 135, 41–101.

HUMPHRIS, A. D. L., MILES, M. J. and HOBBS, J. K. (2005), A mechanical microscope: high-speed atomic force microscopy, Applied Physics Letters, 86(3), 034106.

IKEDA, S., FUNAMI, T. and ZHANG, G. Y. (2005), Visualizing surface active hydrocolloids by atomic force microscopy, Carbohydrate Polymers, 62, 192–6.

JUNG, H., DALAL, C. K., KUNTZ, S., SHAH, R. and COLLIER, C. P. (2004), Surfactant activated dip-pen nanolithography, Nanotechnology Letters, 4, 2171–7.

KIRBY, A. R., MACDOUGALL, A. J. and MORRIS, V. J. (2006), Sugar beet pectin – protein complexes, *Food Biophysics*, **1**, 51–6.

KIRBY, A. R., MACDOUGALL, A. J. and MORRIS, V. J. (2007), Atomic force microscopy of tomato and sugar beet pectin molecules, *Carbohydrate Polymers*, **71**, 640–7.

LEITE, F. L. and HERRMANN, P. S. P. (2005), Application of atomic force spectroscopy (AFS) to studies of adhesion phenomena: a review, *Journal of Adhesion Science and Technology*, **19**, 365–405.

LIANG, Y., HILAL, N., LANGSTON, P. and STAROV, V. (2007), Interaction forces between colloidal particles in liquid: theory and experiment, *Advances in Colloid and Interface Science*, **134–5**, 151–66.

MACKIE, A. R., GUNNING, A. P., WILDE, P. J. and MORRIS, V. J. (1999), The orogenic displacement of protein from the air/water interface by competitive adsorption, *Journal of Colloid and Interface Science*, **210**, 157–66.

MACKIE, A. R., GUNNING, A. P., RIDOUT, M. J, WILDE, P. J. and MORRIS, V. J. (2001a), Orogenic displacement in mixed β-lactoglobulin/β-casein films at the air/water interface, *Langmuir*, **17**, 6593–8.

MACKIE, A. R., GUNNING, A. P., RIDOUT, M. J., WILDE, P. J. and PATINO, J. R. (2001b), *In-situ* measurement of the displacement of protein films from the air/water interface by surfactant, *Biomacromolecules*, **2**, 1001–6.

MALDONADO-VALDERRAMA, J., WOODWARD, N. C., GUNNING, A. P., RIDOUT, M. J., HUSBAND, F. A. *et al.* (2008), Interfacial characterization of beta-lactoglobulin networks: Displacement by bile salts, *Langmuir*, **24**, 6759–67.

MALDONADO-VALDERRAMA, J, GUNNING, A. P, RIDOUT, M. J, WILDE, P. J. and MORRIS, V. J. (2009), The effect of physiological conditions on the surface structure of proteins: setting the scene for human digestion of emulsions, *The European Physical Journal E – Soft Matter and Biological Physics*, **30**, 165–74.

MALDONADO-VALDERRAMA, J., GUNNING, A. P., WILDE, P. J. and MORRIS, V. J. (2010a), *In-vitro* gastric digestion of interfacial protein structures: visualisation by AFM, *Soft Matter*, **6**, 4908–15.

MALDONADO-VALDERRAMA, J., MILLER, R., FAINERMAN, V. B., WILDE, P. J. and MORRIS, V. J. (2010b), The effect of gastric conditions on beta-lactoglobulin interfacial networks: influence of the oil phase on protein structure, *Langmuir*, **26**(20), 15901–8.

MALDONADO-VALDERRAMA, J., WILDE, P. J., MULHOLLAND, F. and MORRIS, V. J. (2012), Protein unfolding at fluid interfaces and its effect on proteolysis in the stomach, *Soft Matter*, **8**, 4402–14

MANOR, O., VAKARELSKI, I. U., STEVENS, G. W., GRIESER, F., DAGASTINE, R. R. and CHAN, D. Y. C. (2008), Dynamic forces between bubbles and surfaces and hydrodynamic boundary conditions, *Langmuir*, **24**, 11533–43.

MARSZALEK, P. E., OBERHAUSER, A. F., PANG, Y. P. and FERNANDES, J. M. (1998), Polysaccharide elasticity governed by chair-boat transitions of the glucopyranose ring, *Nature*, **396**, 661–4.

MARSZALEK, P. E., LI, H. and FERNANDES, J. M. (2001), Fingerprinting polysaccharides with single molecule atomic force microscopy, *Nature Biotechnology*, **19**, 258–62.

MAXWELL, E. G., BELSHAW, N. J., WALDRON, K. W. W. and MORRIS, V. J. (2012), Pectin – an emerging new bioactive food polysaccharide, *Trends in Food Science and Technology*, **24**, 64–73.

MEYER, E. (1992), Atomic force microscopy, *Progress in Surface Science*, **41**, 3–49.

MEYER, G. and AMER, N. M. (1988), Novel approach to atomic force microscopy, *Applied Physics Letters*, **53**, 1045–7.

MORRIS, V. J. (2007a). Gels, in, BELTON, P. S. (ed.), *The Chemical Physics of Food*, Oxford, Blackwell, Ch. 6, 151–98.

MORRIS, V. J. (2007b), Atomic force microscopy (AFM) techniques for characterising food structure, in, D. J. MCCLEMENTS (ed.), *Understanding and Controlling the Microstructure of Complex Foods*, New York, CRC Press, Ch. 8, 209–31.

MORRIS, V. J. (2012), Atomic force microscopy (AFM) and related tools for the imaging of foods and beverages on the nanoscale, in, Q. HUANG (ed.), *Nanotechnology in the Food, Beverage and Nutraceutical Industries*, Cambridge, UK, Woodhouse Publishing Ltd, Ch. 5, 99–148.

MORRIS, V. J. and GUNNING, A. P. (2008), Microscopy, microstructure and displacement of proteins from interfaces: implications for food quality and digestion, *Soft Matter*, **4**, 943–51.

MORRIS, V. J., GUNNING A. P., FAULDS C. B., WILLIAMSON, G. and SVENSSON, B. (2007), AFM images of complexes between amylose and Aspergillus niger glucoamylase mutants, native and mutant starch binding domains: a model for the action of glucoamylase, *Starch–Stärke*, **57**, 1–7.

MORRIS, V. J., GROMER, A. and KIRBY, A. R. (2009), Architecture of intracellular networks in plant matrices, *Structural Chemistry*, **20**, 255–61.

MORRIS, V. J., KIRBY, A. R. and GUNNING, A. P. (2010), *Atomic Force Microscopy for Biologists*, 2nd edition, London, Imperial College Press.

MORRIS, V. J., WOODWARD, N. C. and GUNNING, A. P. (2011), Atomic force microscopy as a nanoscience tool in rational food design, *Journal of the Science of Food and Agriculture*, **91**, 2117–25.

NANGIA-MAKKER, P., CONKLIN, J., HOGAN, V. and RAZ, A. (2002), Carbohydrate-binding proteins in cancer and their ligands as therapeutic agents, *Trends in Molecular Medicine*, **8**, 187–92.

NEUMAN, K. C. and NAGY, A. (2008), Single-molecule force spectroscopy: optical tweezers, magnetic tweezers and atomic force microscopy, *Nature Methods*, **5**, 491–505.

NG, S. P., RANDLES, L. G. and CLARKE, J. (2007), Single studies of protein folding using atomic force microscopy, *Methods in Molecular Biology*, **350**, 139–67.

POSE, S., KIRBY, A. R., MERCADO, J. A., MORRIS, V. J. and QUESADA, M. A. (2012), Structural characterization of cell wall pectin fractions in ripe strawberry fruits using AFM, *Carbohydrate Polymers*, **88**, 882–90.

RALSTON, J., LARSON, I., RUTLAND, M. V. V., FEILLER, A. A. and KLEIJN, M. (2005), Atomic force microscopy and direct surface force measurements – (IUPAC technical report), *Pure & Applied Chemistry*, **77**, 2149–70.

ROUND, A. N., RIGBY, N. M., RING, S. G. and MORRIS, V. J. (2001), Investigating the nature of branching in pectins by atomic force microscopy and carbohydrate analysis, *Carbohydrate Research*, **331**, 337–42.

ROUND, A. N., BERRY, M., MCMASTER, T. J., CORFIELD, A. P. and MILES, M. J. (2004), Glycopolymer charge density determines conformation in human ocular mucin gene products: an atomic force microscope study, *Journal of Structural Biology*, **145**, 246–53.

ROUND, A. N., MCMASTER, T. J., MILES, M. J., CORFIELD, A. P. and BERRY, M. (2007), The isolated MUC5AC gene product from human ocular mucin displays intramolecular conformational heterogeneity, *Glycobiology*, **17**, 578–85.

ROUND, A. N., RIGBY, N. M., MACDOUGALL, A. J. and MORRIS, V. J. (2010), A new view of pectin structure revealed by acid hydrolysis and atomic force microscopy, *Carbohydrate Research*, **345**, 487–97.

ROUSSEAU, D. (2007), The microstructure of chocolate, in D. J. MCCLEMENTS (ed.), *Understanding and Controlling the Microstructure of Complex Foods*, New York, CRC Press, Ch. 24, 648–90.

STRUNZ, T., OROSZLAN, K., SCHUMAKOVITCH, I., GÜNTHERODT H.-J. and HEGNER, M. (2000), Model energy landscapes and the force-induced dissociation of ligand-receptor bonds, *Biophysical Journal*, **79**, 1206–12.

TABOR, R. F., LOCKIE, H., MAIR, D., MANICA, R., CHAN, D. Y. C. *et al.* (2011a), Combined AFM-confocal microscopy of oil droplets: absolute separations and forces in nanofilms, *Journal of Physical Chemistry Letters*, **2**, 961–5.

TABOR, R. F., WU, C., LOCKIE, H., MANICA, R., CHAN, D. Y. C. *et al.* (2011b), Homo- and hetero-interactions between air bubbles and oil droplets measured by atomic force microscopy, *Soft Matter*, **7**, 8977–83.

TABOR, R. F., GRIESER, F., DAGASTINE, R. R. and CHAN, D. Y. (2012), Measurement and analysis of forces in bubble and droplet systems using AFM, *Journal of Colloid and Interface Science*, **371**, 1–14.

TISCHER, C., ALTMANN, S., FISINGER, S., HORBER, J. K. H., STELZER, E. H. K. and FLORIN, E. L. (2001), Three-dimensional thermal noise imaging, *Applied Physics Letters*, **79**, 3878–80.

TOUHAMI, A., HOFFMAN, B., VASELLA, A., DENIS, F. A. and DUFRENE, Y. F. (2003), Probing specific lectin-carbohydrate interactions using atomic force microscopy imaging and force measurements, *Langmuir*, **10**, 1745–51.

VAKARELSKI, I. U., MANICA, R., TANG, X., O'SHEA, S. J., STEVENS, G. W. *et al.* (2010), Dynamic interactions between microbubbles in water, *PNAS USA*, **107**, 11177–82.

VESENKA, J., MARSH, T., MILLER, R. and HENDERSON, E. (1996), Atomic force microscopy of G-wire DNA, *Journal of Vacuum Science and Technology*, **14**, 1413–17.

WILDE, P. J. and CLARK, D. C. (1993), The competitive displacement of beta-lactoglobulin by Tween 20 from oil-water and air-water interfaces, *Journal of Colloid and Interface Science*, **155**, 48–54.

WOODWARD, N. C., WILDE, P. J., MACKIE, A. R., GUNNING, A. P., GUNNING, P. A. and MORRIS, V. J. (2004), The effect of processing on the displacement of whey protein; applying the orogenic model to a real system, *Journal of Agricultural and Food Chemistry*, **52**, 1287–92.

WOODWARD, N. C., GUNNING, A. P., MACKIE, A. R., WILDE, P. J. and MORRIS, V. J. (2009), A comparison of the orogenic displacement of sodium caseinate with the caseins from the air-water interface by non-ionic surfactants, *Langmuir*, **25**, 6739–44.

WOODWARD, N. C., GUNNING, A. P., MALDONNA-VALDERRAMA, J., WILDE, P. J. and MORRIS, V. J. (2010a), Probing the *in situ* competitive displacement of protein by nonionic surfactant using atomic force microscopy, *Langmuir*, **26**, 12560–6.

WOODWARD, N. C., GUNNING, A. P., WILDE, P. J., CHU, B. S. and MORRIS, V. J. (2010b), Engineering interfacial structures to moderate satiety, in P. A. WILLIAMS and G. O. PHILLIPS (eds), *15th Gums & Stabilisers for the Food Industry Conference, 22nd–26th June 2009*, Cambridge, UK, Royal Society of Chemistry, special publication 325, 367–76.

3

Light microscopy: principles and applications to food microstructures

P. A. Gunning, Smith & Nephew Research Centre, UK

DOI: 10.1533/9780857098894.1.62

Abstract: History has shown light microscopy to be a ubiquitous and versatile tool in food science. This chapter discusses the history, usefulness and methods pertinent to the study of food microstructures. Setting up a microscope for best results, obtaining good specimen contrast and some examples of a variety of applications from basic transmitted light work to fluorescence and more specialised immuno-labelling methods and interfacial microscopy are discussed. Specimen preparation methods are described, including whole-mounts, hand-sectioning, resin and wax embedding. and the use of microtomes, vibratomes and cryo-microtomes is also described. The chapter concludes by considering recent and future developments of light microscopy.

Key words: light microscopy, physical contrast mechanisms, chemical contrast mechanisms, Köhler illumination.

3.1 Introduction

The importance of microscopic studies of foods has been demonstrated over several centuries, well before industrialised processed foods became commonplace. Much early work was closely linked to microbiological research. The individual often credited with the birth of scientific microscopy was a Dutchman named Antonie van Leeuwenhoek (1632–1723). Van Leeuwenhoek recorded detailed observations of microscopic organisms, which he termed 'animalcules', announcing his observations to the world in 1675. However, van Leeuwenhoek was inspired by the work of his contemporary Robert Hooke (1635–1703), an Englishman and Fellow of The Royal Society. Hooke compiled many detailed descriptions and illustrations of the microscopic world in his renowned book 'Micrographia' first published in 1665 (Hooke, 1665). Whilst van Leeuwenhoek used a single powerful glass lens in his small and eminently portable 'microscope', compound microscopes employing the use of multiple lenses as used by Robert Hooke were also available at the time, but their optics and illumination sources could not initially outperform Van Leeuwenhoek's simple and highly practical design.

The power of microscopy to contribute to human health was demonstrated when Louis Pasteur (1822–1895) later gained pivotal insights into the link between bacteria and food spoilage. Pasteur used microscopy to investigate the cause of souring of beet juice which was being fermented to produce alcohol, demonstrating that lactic acid bacteria were present in beet juice that produced sour liquor. Pasteur's work on microorganisms in fermentation led him to investigate the problem of wine spoilage. Pasteur was to provide a solution to extend the shelf-life of wine without affecting the quality and flavour. His solution of flash-heating the wine at the end of the fermentation process to inactivate all of the microorganisms present is the basis of the widespread process used to this day throughout the food and drinks industry that still bears his name (pasteurisation).

The microscope proved an invaluable tool in protecting the public at large from food-borne disease, as well as toxic ingredients or diluents that were sometimes used to adulterate foods by unscrupulous or ill-educated suppliers. During the nineteenth century, toxic lead, chromium, mercury, arsenic and copper salts were sometimes used as additives to foods to improve their appearance. Many organic adulterants were also added to foods, such as chicory, endives, burnt carrots or turnips in coffee, as well as inorganic substances such as sand or grit. An early pioneer in tackling this issue, Dr Arthur Hill Hassall (1817–1894) was a physician as well as an expert microscopist and public analyst. He dedicated his working life to researching and publishing the earliest and most complete body of work on food adulteration. Dr Hill Hassall came to public attention when he published his microscopic observations of drinking water in London in the 1850s. Following numerous publications in the medical journal *The Lancet*, his seminal work published in 1855 was the complete collection of his articles from *The Lancet* and was entitled *Food: Its Adulterations and the Methods for their Detection* (Hill Hassall, 1855). He described a multitude of different methods for the preparation and microscopic examination of foods, attracting such public interest that in 1860 his work directly led to the first Parliamentary legislation in the United Kingdom against food adulteration. In the context of food and drink at least, any notions that supposed microscopy to be merely a purely aesthetic/academic pursuit were comprehensively dispelled by the rigorous nature and tangible benefits of Dr Hill Hassall's methods.

Another pioneer, Henry George Greenish, published several books describing the microstructure of vegetable based foods and drugs in the early 1900s including, along with Eugène Collin, the first 'atlas' of vegetable microstructures replete with illustrations to aid the microscopist trying to identify whole, crushed or processed vegetable matter in different foods (Greenish and Collin, 1904). Several decades after his death, Hill Hassall's work was updated in 1929 by Edwyn Godwin Clayton who wrote *A Compendium of Food Microscopy with Sections on*

Drugs, Water and Tobacco (Clayton, 1929). The quality and detail of the descriptions and illustrations provided by these pioneers of food microscopy is such that any modern microscopist will instantly recognise the foodstuffs depicted and many of the methods that they developed are still routinely used. Although no longer in print, Hill Hassall's, Greenish and Collin's and Clayton's publications are freely available to read online within the 'Open Library' (*Openlibrary.org*). Each of these open access archived publications is informative and provides a truly fascinating insight into the extent of food adulteration in the nineteenth and early twentieth centuries.

Several other notable food microscopy books were written during the last century. Those that were more comprehensive in their scope included a series of books by Winton and Winton, published in four volumes from 1932 to 1939, providing detailed descriptions of the microstructure of most foods (Winton and Winton, 1932–1939). *Food Microscopy* edited by John Griffith Vaughan described microscopy of basic foodstuffs of meat, plants, fish, dairy and eggs (Vaughan, 1979), but the book contained little information regarding modern functional additives such as gums and stabilisers that are often found in today's manufactured foods. Sharing a similar title and having the advantage of still being in print, *Food Microscopy: a manual of practical methods, using optical microscopy* by Olga Flint is one of a number of titles in the Royal Microscopical Society handbook series (Flint, 1994). This excellent book covers basic foodstuffs as well as including methods for examining complex food mixtures that often include gums, thickeners and stabilisers. Written as a guidebook for microscopy of modern multi-component foods, it provides plenty of practical tips on how to use combinations of staining and optical contrast methods to help elucidate the composition of such foods, in addition to simply visualising their microstructures.

3.2   Fundamentals of light microscopy

3.2.1   Advantages and disadvantages of light microscopy

This chapter deals exclusively with light microscopy and it seems appropriate to address the question of the importance of the light microscope in an age where many microscope technologies have pushed the boundaries of spatial resolution many orders of magnitude beyond those obtained by conventional light microscopes that rely on visible light and an objective lens. The reasons that light microscopes are still such valuable analytical instruments are numerous. One advantage that remains unique to the use of a light microscope is that full colour information is preserved in the images obtained. This advantage opens up the possibilities of using colour to identify different substances, or indeed of staining or using other reagents to produce specific colouration of different substances, thereby aiding both interpretation and visualisation. Stains are used in other forms of microscopy (e.g. electron and X-ray microscopy), but the number of practically useful stains available are more limited compared to the plethora of stains

Table 3.1 Advantages and disadvantages of light microscopy

Advantages	Disadvantages/limitations
Modest capital and maintenance costs	
Convenience	Limited depth of focus
Preservation of full colour information	Diffraction limited spatial resolution
Fully hydrated/wet specimens	Difficulty eliminating optical aberrations
Ambient conditions (no vacuum requirements)	Fixed magnification lenses (compound microscopes)

available to the light microscopist. Light microscopy does of course have limitations and disadvantages, as well as many advantages; each is listed in Table 3.1.

Convenient 'workarounds' for some of the disadvantages are available to the modern microscopist, for example PC workstations and software are capable of post-processing methods such as 'z-stacking' (Section 3.7) to improve the depth of focus limitations of light microscopes. However, such methods are not well suited to following fast-moving dynamic processes, because it takes a significant length of time to acquire all of the images necessary to produce a z-stack.

3.2.2   Zoom stereomicroscope compared with compound microscope

Two major classes of light microscope are available to the food microscopist; compound microscopes and stereomicroscopes. The stereomicroscope is sometimes referred to as an 'inspection microscope' or 'dissection microscope', due to the low to intermediate magnifications commonly achieved. These microscopes provide truly stereoscopic image information and are best termed stereomicroscopes, because the image is obtained via two separate light paths inclined at a small angle with respect to the perpendicular to the specimen surface. Stereomicroscopes therefore possess two sets of essentially identical optics, although most modern stereomicroscopes use a design that employs a single 'common main objective' (CMO), allowing the resolving power of the final objective to be conveniently changed by swapping it with a different objective, supplementing it with an additional screw-on or bayonet mounted lens, or by mounting several objectives of different magnifications on a revolving turret/ nosepiece (nosepiece shown in Fig. 3.1), in a similar way that objectives are mounted on the revolving nosepiece of the compound microscope.

Research grade stereomicroscopes are most often equipped with twin zoom lens assemblies, one within each (left and right) optical path, allowing continuously variable magnification within a range of maximum and minimum dictated by the magnification/resolving power of the CMO lens in use. The dual optical paths and CMO lens configuration of most modern stereomicroscopes is shown in

Fig. 3.1   Nosepiece with objective lenses.

Fig. 3.2. Most stereomicroscopes can be configured to provide observation using reflected light, transmitted light (brightfield or darkfield) or epi-fluorescence, allowing the examination of both transparent/translucent and opaque specimens. Stereomicroscopes are the instruments of choice if low to moderate magnifications are required (typically capable of between 5× and 200×) and if access to the specimen is required during observations, as most designs incorporate an open specimen stage and the objective lens typically provides long working distances (working distance is the distance between the surface of the lens and the plane of sharp focus on or in the specimen). The lens design of zoom stereomicroscopes allow increased magnifications, whilst preserving the long working distance of the objective lens.

The other principal class of microscope is the compound microscope. The name is derived from the compound magnifications achieved by multiplying the magnification of the objective lens by one or several additional lenses, up to and including the eyepiece lens. The basic optical design of the compound microscope is shown in Fig. 3.3. Compound microscopes are the high resolution workhorse of the light microscopist, typically capable of achieving between 10× and 1000× magnifications. The trade-off for higher resolution performance is primarily that of much reduced working distances, which may be as short as one to two hundred microns for the highest resolution objectives. The working distance of each objective lens is usually different, with higher magnification objectives providing ever smaller working distances. This in turn means that access to the specimen is much more restricted during observations and specimens are usually

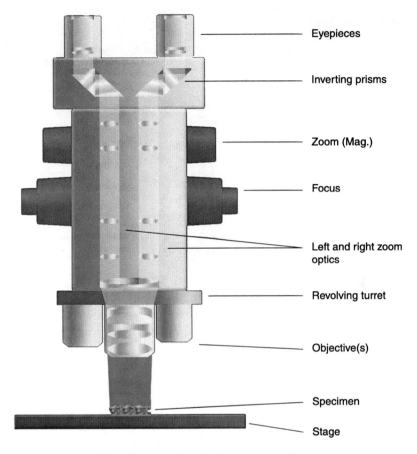

Eyepieces

Inverting prisms

Zoom (Mag.)

Focus

Left and right zoom optics

Revolving turret

Objective(s)

Specimen

Stage

Fig. 3.2   Diagram of a typical zoom stereomicroscope.

constrained between a flat glass slide and cover-slip, to improve light transmission, reduce scattering and to help ensure that the objective in use is protected from contamination by the specimen (with exception of special water-immersion lenses). The cover-slip also provides a means of mechanically flattening thicker deformable specimens (see 'smears' and 'squashes' in Section 3.3). Some compound microscope designs provide a degree of extra specimen accessibility by inverting the entire optical path by positioning the objective lens directly underneath the specimen stage and using a long working distance condenser lens above the specimen stage. Inverted microscopes allow better access to the specimen from above and they are commonly used by cell biologists observing *in-vitro* growth of cells in tissue-culture plates mounted directly on the microscope stage, but they are also useful if micromanipulators or tensile stages are to be used (Fig. 3.4).

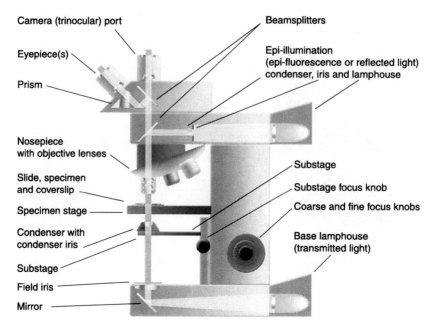

Camera (trinocular) port

Eyepiece(s)

Prism

Nosepiece
with objective lenses

Slide, specimen
and coverslip

Specimen stage

Condenser with
condenser iris

Substage

Field iris

Mirror

Beamsplitters

Epi-illumination
(epi-fluorescence or reflected light)
condenser, iris and lamphouse

Substage

Substage focus knob

Coarse and fine focus knobs

Base lamphouse
(transmitted light)

Fig. 3.3   Diagram of a compound microscope.

Fig. 3.4   Micro-tensile stage accessory.

Similar to stereomicroscopes, compound microscopes can be configured to provide observations using transmitted and reflected light as well as epifluorescence enabling transparent, translucent and opaque specimens to be examined. Although many compound microscopes provide dual ('binocular') eyepieces, the view afforded is a purely 2D mono view that is simply split equally into each eyepiece by means of a beam-splitter in the binocular head.

Whilst the compound microscope seems to provide much less practical convenience than a stereomicroscope, it remains the best light microscope to use for achieving magnifications of 100× or more. In addition, several useful optical contrast mechanisms can only be implemented on compound microscopes, because the long working distance and smaller numerical apertures of stereomicroscopes render such methods impractical (contrast enhancement methods such as differential interference contrast, phase contrast, etc. are described in Section 3.4).

3.2.3   Setting up the microscope

Time spent setting up a microscope correctly always pays dividends in the quality of the resulting images. However, this does not have to be too laborious; little more than 5 to 10 minutes should be needed to ensure that the optics and illumination are optimally configured to ensure that the best possible contrast and resolution are obtained.

Assuming a clean and serviceable instrument, stereomicroscopes require minimal work configuring the optics, other than ensuring that the eyepiece dioptre adjustment (if available) is correctly set for the person using it. Setting the correct dioptre adjustment is the same for both stereo and compound microscopes and merely involves bringing the specimen into sharp focus whilst viewing through the non-adjustable eyepiece only (this tends to be the left eyepiece by convention). Obscure the right eyepiece by covering it with dark paper, as opposed to closing your right eye, because keeping the eye tightly closed can affect the shape of both eyes causing them to focus slightly differently than they will when both eyes are kept open. After a sharply focused image is obtained in the left eyepiece, cover the left eyepiece, uncover the right eyepiece and adjust that dioptre adjustment collar (usually marked with +, 0, − symbols) until the image viewed through the (right) adjustable eyepiece is also sharp. Do not adjust the microscope focus between observing through left and right eyepieces, make any corrections necessary between each eyepiece only by means of the dioptre adjustment collar. Carrying out this step ensures that all of the subsequent observations are comfortably viewed without any need to wear spectacles for the correction of long or short sightedness. In theory, any individuals possessing perfect eyesight should be able to simply set any dioptre adjustable eyepieces to their zero (0) position to achieve comfortable viewing.

Compound microscopes often require adjustment of the sub-stage condenser to ensure flat, even illumination of the specimen and to optimise both image contrast and the resolving power of the objective lens in use. Most good-quality compound microscopes are equipped to provide 'Köhler illumination' by means of a properly adjusted and optically matched condenser lens assembly. Without going into a full description of optical theory, making these adjustments ensures that the full resolving power of each objective lens is realised by ensuring that the illumination provided evenly fills the entire numerical aperture of the objective in use. Köhler illumination is achieved in the following manner:

1. Mount a specimen on a glass slide and place a suitable cover-slip over it.
2. Bring the specimen into sharp focus as viewed through the eyepieces.
3. Bring the substage condenser upwards until it almost touches the underside of the glass slide.
4. Close down the field iris until it can be seen as a dark area superimposed over the image of the specimen, encroaching around the edges of the field of view as the field iris is closed further.
5. Re-adjust the substage condenser such that the edges of the illuminated field are brought into sharp focus (you should be able to see each of the iris 'blades' as a clean, sharp edge – on some microscopes they may exhibit some colour fringes around the edge (chromatic aberration)).
6. The brightly illuminated area must now be centred in the field of view; this can be accomplished by adjusting two little thumbscrews that are usually located at either side of the substage condenser assembly. Turn each of the thumbscrews until the bright area is as close to being central within the field of view as possible. Now open up the field iris again until it is just outside the field of view.
7. Adjust the condenser iris until the amount of illumination just begins to decrease; this is the point at which the condensed light no longer fills the entire numerical aperture of the objective lens in use) If you remain unsure about the condenser iris setting, remove one of the eyepieces and look down the eyepiece tube as you begin to close down the condenser iris. When the illuminated area in the eyepiece tube starts to diminish, stop closing the iris when seven-eighths of the total area remains illuminated. Closing the condenser iris further can produce better depth of focus, but it will also decrease the resolving power (optical quality) of the objective lens due to effectively reducing the numerical aperture of the optics and also due to the effects of diffraction; diffraction interference fringes can be seen to appear around the edges of features in the specimen image when this is severe.
8. Lastly, recheck the field iris each time you change to a different objective lens, as each objective has a different field of view; the higher the magnification of the objective, the smaller the field of view and the smaller the field iris can be. This will ensure that the contrast is optimal for each objective used. For critical work (e.g. photomicrography), it is worth rechecking most of the aforementioned adjustments, as they may be subtly altered after each change of objective, but can be re-optimised again very quickly once Köhler illumination has been set up at the start of the session.

The procedure for Köhler illumination on an inverted compound microscope is exactly the same as that for an upright, but the condenser assembly is located above the specimen and microscope stage rather than below it. A similar procedure is also used to set up epi-illumination for reflected light observations, but in this case the condenser adjustments are made to a condenser assembly that is usually located just in front of the epi-illumination lamp-house or just behind the microscope nosepiece, running from the nosepiece towards the epi-illumination lamp-house at the top and rear of the microscope body.

3.2.4   Accessories

Due to the relative convenience of the light microscope, many accessories are available to enable dynamic processes such as freezing, melting, tension, compression, etc, to be observed. Even dynamic biological processes such as cell division/proliferation/motility can be observed using *in-vitro* cell culture chambers with controllable temperature, humidity and gas chemistry. Thanks to the flexibility of modern digital cameras, high-quality photomicrography is accessible to most microscopists for relatively modest financial outlay. For brightfield work, even a consumer grade digital SLR camera with a suitable microscope adaptor can provide excellent micrographs. Many such cameras are capable of complex light metering, as well as offering facilities such as time-lapse and video functions. Some of the larger sensor (often termed 'full frame') digital SLR cameras also have good low light sensitivity and can be used for lower light applications such as DIC or fluorescence work. Dedicated, specialist microscopy cameras should provide excellent results in most situations and if low light fluorescence work is to be undertaken, a dedicated microscope camera with a cooled sensor (sometimes referred to as a 'cooled chip' or 'cooled CCD') should be capable of obtaining noise-free images of fluorescence so dim that it is barely visible through the eyepieces.

3.3   Specimen preparation

Having optimised the performance of the microscope, attention must be paid to preparing the specimen appropriately if the best-quality image information is to be obtained. The stereomicroscope scores highly for convenience again, in that it is capable of greater depth of focus than the compound microscope, so relatively large 3D specimens with appreciably high topographies can be observed directly at low magnifications with minimal specimen preparation necessary. The stereomicroscope is well suited to surface observations, but if internal microstructures of less transparent or opaque materials are to be investigated, the specimen must be cross-sectioned somehow in order to reveal this detail, or at least made thin enough that sufficient light can still be transmitted through the specimen to allow successful observations to be made. The requirement for a thin enough specimen is even more important for observations using a compound microscope, because a large amount of light scattering through a thick specimen will impair contrast and will also hamper some optical contrast enhancement methods such as DIC.

In addition, if large topographies are viewed using reflected light, the limited depth of focus of the compound microscope will preclude sharply focused observations of all but the smallest proportion of a rough surface. Liquids, melts, emulsions, dispersions or suspensions in liquids can simply be smeared out or squashed between the cover-slip and a glass slide if the mechanical disturbance imposed by such relatively robust treatment is acceptable. Any preparation of the specimen prior to microscopic observations (by any means; light, electrons, X-rays, ultrasound or scanning probe methods) must be well understood and carefully considered when interpreting the resulting images to avoid misrepresenting preparation-induced artefacts as truly representative information.

A good example of artefact creation from even a simple squash preparation is illustrated by considering observations of oil-in-water emulsions that may be subject to depletion flocculation effects. Depletion flocculation (aggregation) of oil droplets is easily disrupted by even very low shear forces, so squashing such flocs between a cover-slip and slide will often break them up, either partially or completely. Such emulsions may appear to be composed of individual droplets homogeneously dispersed across any given image field upon first observation (Fig. 3.5). If the specimen is allowed to rest for a significant length of time, and the droplet diameters are small enough not to be mechanically constrained between the cover-slip and the slide, the droplets may begin to flocculate again,

Fig. 3.5 Homogeneously dispersed emulsion.

Fig. 3.6 Flocculated emulsion.

reforming large flocs composed of many droplets and giving rise to a very different, heterogeneous appearance (Fig. 3.6). Thus the second look at such a sample may be more representative than the initial observation.

Smear mounts can be made of dry powdery foodstuffs and the addition of commonly available oils provides an ideal way of mounting such specimens if they are sensitive to water (where water may cause swelling, gelling or dissolution). Powders or granular materials such as flours, spray dried foods and water-soluble materials such as salt, sugar or vitrified sugar/carbohydrate glasses (often used as flavour carriers) can all be examined in oil. Fresh, clean oils such as vegetable oil, paraffin oil or even household lubricating oils can be used to good effect. Silicone oils have a high refractive index and may be useful for some materials, but may actually provide less contrast in some instances where the foodstuff possesses a similarly high refractive index (sugar crystals are a good example). The rule of thumb is that index matching equals zero contrast in optical microscopy! The use of other mounting media such as ethanol or other solvents is not recommended for most situations, as alcohols may begin to dehydrate and shrink moisture containing materials and other solvents may be less desirable from a health and safety perspective.

3.3.1   Hand-cut sections

If the specimen must be cross-sectioned, various methods can be employed, depending on the balance of speed/convenience and quality required. Simple hand sectioning, carried out by carefully slicing through the specimen with a good-quality razor blade can produce acceptable results with foodstuffs that have a firm but not hard and brittle texture. Foods such as many fresh and some cooked fruit/vegetable tissues, baked breads or pastries, firm gels, firm cooked or processed meat/fish products and some dairy products such as hard cheeses can be successfully hand sectioned. It is often easier to produce a thin 'wedge' of the foodstuff rather than trying to cut perfectly flat, plane-parallel sections when hand sectioning. The thin end of the wedge is usually suitably thin and most people become better at hand cutting sections with a little practice, taking care to avoid cutting fingers whilst doing so!

3.3.2   Vibratomes and microtomes

Hand-cut sections may not always be of sufficient quality to allow good observations of undisrupted microstructure and for some foods hand-cut sectioning is impractical. Precision instruments known as microtomes are available for producing better-quality cross-sections of controlled thickness. Vibrating microtomes (vibratomes) can cut some materials without the need to mechanically support the microstructure with any embedding mediums such as resins. Similar to hand-sectioning, vibratomes employ a razor blade to cut the cross-section, but the razor is mounted in a rapidly vibrating (frequencies typically between 0 and 100 Hz) knife holder that is advanced through the specimen material at a precisely controlled rate. Pre-defined section thicknesses can be obtained with great reproducibility.

The vibratome knife (razor) oscillates in a horizontal direction as it gently advances vertically through the section, keeping vertical forces on the specimen to a minimum. This works in much the same way as cutting a slice of bread from a loaf; 'sawing' the knife backward and forward across the loaf will produce a good slice, whilst just forcing the knife vertically through the bread is likely to deform the bread badly and result in a rather compressed slice. Many relatively soft materials such as unfixed vegetable or animal tissue can be sectioned by using a vibratome and, because the use of fixatives and embedding resins is not required, vibratomes can be helpful if living tissue or enzyme activity effects are to be studied, and also when immuno-histochemistry is to be carried out on the sections.

3.3.3   Embedding

However, some microstructures may be too delicate for successful hand-sections or vibratome sections to be cut, requiring mechanical support to be provided by chemically fixing the specimen (fixation helps to 'stiffen' delicate or relatively liquid regions within a specimen), followed by dehydration and infiltration of the specimen with a curable polymer resin. Once cured, the resin specimen 'block' is solidified and therefore far more robust. Resin embedded specimen blocks are capable of resisting the worst effects of the shear forces encountered when sections are cut. Chemical fixation is usually achieved using aldehydes such as formaldehyde or glutaraldehyde, which cross-link protein and lipoprotein/glycoprotein molecules. Kiernan published a good guide to the way in which aldehyde fixatives affect specimens (Kiernan, 2000). If lipids are to be preserved within the specimen microstructure, fixatives such as osmium tetroxide or ruthenium red can be used to cross-link the lipid molecules, thereby preventing their subsequent dissolution or removal during the infiltration of embedding resin. Both aldehydes and osmium/ruthenium compounds are volatile and extremely toxic and suitable protective measures must be taken to ensure personal safety when using fixatives.

Embedding procedure
Cut as small a piece of the specimen as possible before starting to fix and dehydrate it, too large a piece risks poor fixation/dehydration through the centre of the specimen, which will only become evident when the first sections are cut and examined – this is costly in time-considering the multiple steps which precede the cutting step. Most embedding resins will not polymerise in the presence of water, necessitating dehydration of the fixed specimen prior to infiltration with unpolymerised resin. Dehydration can be achieved by rinsing the fixative solution out of the specimen three times with fresh buffer, followed by exchanging the buffer solution for a series of aqueous ethanol solutions of progressively increasing ethanol concentration.

Progression through the ethanol concentrations must be gentle for delicate specimens, for example, starting at 10% v/v and increase in additional 10% increments all the way to absolute ethanol. More robust specimens (e.g. bone tissue, highly lignified vegetable tissue) can be progressed through ethanol gradients more rapidly, starting at 25% and increasing ethanol concentration by 25% for each solution exchange. Allowing a minimum of 20 minutes between exchanges of ethanol solutions is a good starting point, hard or dense specimens should be given more time between solution exchanges to ensure good dehydration through the entire specimen. Exchange the specimen three times in 100% (absolute) ethanol and if possible, allow equilibration in absolute ethanol overnight prior to commencing infiltration with embedding resin. Many resins are not miscible with ethanol, so it may be necessary to exchange the ethanol for three exchanges of acetone or propylene oxide before the embedding resin is used. Take care to keep the specimen submerged at all times during dehydration. Stoppered or sealed containers must be used, because ethanol, acetone and propylene oxide are hygroscopic and traces of water would otherwise be absorbed from moisture in the air. Alternatively, if large enough numbers of specimen dehydrations are likely to be required, it is possible to automate dehydration by investing in a 'tissue processor' machine. Most automated tissue processors allow the user to programme the desired dehydration sequence for gentle to more robust dehydration gradients under thermostatic control.

Resin embedding can be commenced once dehydration is completed. Embedding is achieved by slowly infiltrating the specimen with unpolymerised resin. Most resins are relatively low viscosity liquids with good penetration properties. The infiltration process generally follows the same routine of exchanging the dehydration solvent for resin several times to ensure complete exchange of the liquids and proper penetration of the resin through the entire specimen. Some resins achieve optimal infiltration when exchanged as mixtures of dehydration solvent plus resin, of increasing resin concentrations up to 100%; others can simply be exchanged into 100% resin directly and exchanged again for pure resin several times. The size and density of the specimen will dictate how many exchanges are required and how much time should be allowed between exchanges. Larger or dense specimens may require up to 24 hours between exchanges to ensure proper infiltration has taken place.

Some materials become more translucent as infiltration approaches completion and a quick visual check can be informative before committing the infiltrated specimen to final polymerisation. Different resins are polymerised using various methods; chemical catalysts/initiators, increased temperature and blue or UV light 'curing' methods are common. Selecting the most appropriate resin for the work to be carried out is also important. Try to match specimen hardness to resin hardness; a hard specimen is likely to be pushed by the microtome knife through a soft resin, and microstructures or enzymes/immunological epitopes sensitive to heating are best embedded in resins that do not polymerise exothermically and do not require much if any heating. Further information about embedding resins can be obtained in a review paper describing the various resins available to the light and electron microscopist and their development for differing applications over the last half of the twentieth century (Newman and Hobot, 1999).

Resin embedding also provides the flexibility of being suitable for cutting ultrathin sections for transmission electron microscopy (Chapter 2), if suitable fixatives were used during specimen preparation. Sections cut (using a rotary or ultramicrotome) from most resins can be stained or labelled directly, without the need for further processing. Sections are often cut onto water or 15% aqueous acetone (helps section spread onto surface of the solution) and then picked up onto a glass slide. If the sections do not stretch out well onto the water or acetone solution, try dabbing the torn edge of a small piece of filter paper into chloroform and then wafting the torn edge just above the section (get as close to the section as possible without actually touching it). Chloroform is extremely volatile and its vapour is much denser than air; the vapour will fall from the edge of the filter paper onto the floating section, causing it to stretch out flat, removing any wrinkles.

Wax embedding is not useful if sections are to be examined using electron microscopy as a follow-up examination to light microscopy, but it can be useful for preparing sections from soft tissue such as muscle, offal or softer vegetable tissues. The process is similar to resin embedding; starting with a fixative followed by rinsing and dehydration through an alcohol series. The process differs after completion of alcohol dehydration, when the specimen is immersed in a 'clearing agent' (solvent such as xylene, toluene, chloroform, cedar wood oil or 'Histo-Clear®' that is compatible with both alcohol and paraffin wax), which is usually exchanged several times prior to immersion in liquefied paraffin wax (between 50 °C and 70 °C, depending on wax used). The wax is usually exchanged for a minimum of two or three times before the specimen is transferred to a wax-filled mould and allowed to cool to produce the solidified specimen block.

Sections are cut using a rotary microtome and then floated out onto warm water. The warm water keeps the section supple and helps to prevent wrinkles and creases when the sectioned specimen is picked up on a glass slide. Unlike resin embedded sections, the wax must first be removed from the section before it can be stained. This is achieved by either placing the wax section on a slide in an oven or on a hot plate set at just below the melting point of the wax (wax gradually evaporates from the specimen), or by washing the section with a wax compatible solvent. Following wax removal, the sectioned specimen can be stained as required. Wax embedding does involve heating the specimen for molten wax and resin embedding sometimes involves exothermic resin polymerisation or thermal curing of the resin. Increased temperature can cause proteins or other epitopes to become denatured with a subsequent loss of antigenicity, so consider choosing a wax or resin that minimises exposure of the specimen to temperatures in excess of 65 °C to 70 °C if immuno-labelling (Section 3.5.3) is to be carried out.

Another alternative to supporting delicate microstructures by resin or wax embedding is to cut sections from frozen material, where the microstructure is mechanically supported by ice. Cryostats are essentially rotary microtomes within a sizeable chest freezer cabinet. The specimen is usually plunge frozen in liquid nitrogen or another cryogenic liquid to achieve as rapid a freeze as possible in order to encourage the formation of small ice crystals. Large ice crystals can damage microstructures, so keeping the specimen size as small as possible helps to ensure more rapid freezing, and hence much smaller ice crystals during plunge freezing. The specimen is mounted in the cryostat using cryogenic adhesive (commonly a viscous PVA solution). The specimen and the cryostat microtome knife blade are maintained at a constant and controllable low temperature, typically between −40 °C and −20 °C.

Frozen sections are cut in a similar manner to that used in a conventional rotary microtome and section thicknesses down to one micron can sometimes be obtained, although section thicknesses of 5 to 10 microns are more practicably achieved. Frozen sections can be placed on a glass slide, allowed to dry (with or without heating) and then stained as required. Cryostat sections provide a more rapid alternative to resin or wax embedded sections, because there is no need for lengthy dehydration or infiltration procedures, although cryostat sections are generally less robust and less well suited to subsequent long-term storage in a slide archive.

3.4   Specimen contrast enhancement: physical methods

Several physical methods are available for contrast enhancement of colourless or translucent specimens. Most animal-derived foods and some vegetable-based foods are relatively colourless when viewed in brightfield microscopy. Animal cells are particularly troublesome in this respect; owing to the relatively minor difference in refractive index between animal cells and water; animal cells are after-all composed chiefly of water enclosed within a phospholipid double membrane. If staining by chemical means is to be avoided for reasons of speed, convenience or practicality, optical methods may be employed to improve the contrast and visualisation of such specimens.

Differential Interference Contrast (DIC, sometimes referred to as 'Nomarski' or NIC after the Polish physicist Jerzy Nomarski) is perhaps the most versatile optical contrast enhancement method. DIC involves using several matched optical components (matched objectives and Wollaston/Nomarski prisms, along with a polariser and analyser). DIC optics are usually supplied as optional extras obtained at additional cost to a set of standard objectives, but if speed and convenience are important considerations, the additional cost is likely to prove worthwhile. DIC works by enhancing the contrast resulting from even very small changes in refractive index within a specimen. This produces higher edge contrast and gives the image an almost 3D quality, although the effect is purely qualitative and does not provide a true topographical representation of specific features within a specimen. Rotating the specimen on the microscope stage can change the nature of the contrast, such that apparent 'peaks' begin to look more like 'troughs' and vice-versa as the specimen is rotated through 180 degrees.

The optical design utilised in most DIC configurations is illustrated by Fig. 3.7. The optical theory underpinning DIC is rather complex and would require several

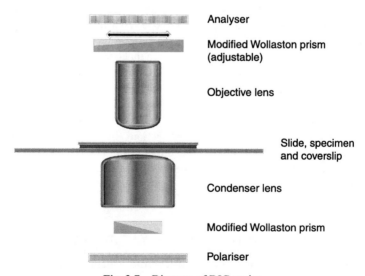

Analyser

Modified Wollaston prism (adjustable)

Objective lens

Slide, specimen and coverslip

Condenser lens

Modified Wollaston prism

Polariser

Fig. 3.7   Diagram of DIC optics.

pages of this book to describe it fully; however, a good description is provided by Ron Oldfield's '*Light Microscopy, an Illustrated Guide*' (Oldfield, 1994). DIC has a distinct advantage over several other optical contrast enhancement methods, in that the full numerical aperture of the objective/microscope can still be utilised and optimal resolution achieved. DIC also provides improved 'optical sectioning', resulting in sharp details in the plane of focus without too much confusing detail from defocused areas above and below. DIC can therefore be a good choice if 'z-stacks' are to be collected to produce extended depth of focus images via image processing software. However, there are some drawbacks of DIC, apart from the initial costs. DIC is not well suited to some types of image analysis work, because although edge enhancement is good, the contrast around the edge of an object changes with orientation and there remains little contrast across the middle of objects, making greyscale segmentation rather challenging. DIC does not always work well if the specimen itself is anisotropic, or if the specimen is observed through anisotropic plastics such as disposable petri dishes or tissue culture plastic multi-well plates.

Phase contrast microscopy is another method that achieves enhanced contrast by purely optical means. Dutch physicist Frits Zernicke (1888–1966) developed Phase Contrast microscopy in 1934. The new method revolutionised the microscopy of biological specimens to such an extent that his achievements were recognised with the award of the Nobel Prize in 1953. Phase contrast optics can be obtained at significantly lower cost than DIC optics, but their optical effect is similar to that of DIC, in that contrast is enhanced when light passes through the boundaries of materials, even where they possess only very small differences in refractive index, regardless of their optical absorbance characteristics. Although phase contrast is less costly than DIC, phase contrast images suffer the disadvantage that bright 'phase halos' can be seen around the boundaries of objects, and these can also be seen around the edges of objects that are significantly above and below the plane of focus, complicating the overall image.

Figure 3.8 shows the optical components of a typical phase contrast configuration. The sub-stage condenser contains an optically opaque annulus whose dimensions are matched to a corresponding annular ridge in a transparent 'phase plate' that is located inside the objective lens. The condenser annulus is often mounted within a rotatable turret within the condenser assembly, allowing the operator to select the annulus that matches the objective lens in use. The combination of annulus and phase plate work on the basis that changes in light/dark contrast (amplitude) that can be observed, correspond to phase change of half of one wavelength ($\lambda/2$, where λ is the wavelength of the light used to illuminate the specimen). The ridge in the phase plate inside the objective lens is coated with a transparent material that absorbs between 60 and 90% of the direct (undiffracted) light, whilst retarding the phase angle of the wave front by one-quarter of a wavelength ($\lambda/4$, where λ is assumed to be that of green light). The annulus and condenser produce a hollow cone shaped beam of light, some of which will pass through the specimen without encountering any objects.

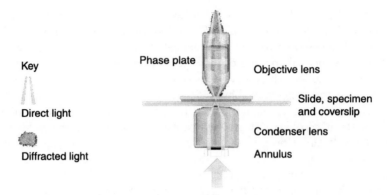

Key

Direct light

Diffracted light

Phase plate

Objective lens

Slide, specimen
and coverslip

Condenser lens

Annulus

Fig. 3.8   Diagram of phase contrast optics.

This direct undiffracted light will be collected by the objective lens into the ring-shaped phase plate ridge within the lens where the wavefront phase is deliberately retarded by $\lambda/4$. The light that does pass through the specimen will be diffracted (even if the change in refractive index is very small) as it crosses the boundaries between objects and their surroundings, and the phase of the diffracted light is also retarded by $\lambda/4$ as a consequence of the interaction. The diffracted light will spread across the entire area collected by the objective lens, including the areas within and outside the cone of illumination produced by the condenser annulus. When the diffracted light from specimen features reaches the phase plate within the objective, the difference in phase between light diffracted from specimen features and the undiffracted light from 'blank' areas within the specimen is half a wavelength ($\lambda/4 + \lambda/4 = \lambda/2$), thereby converting a change in phase to a change in amplitude (light/dark contrast), which can be visualised.

Darkfield illumination provides perhaps the simplest physical method of contrast enhancement. The optical configuration required is similar to that of phase contrast, but without the need for a phase plate within the objective lens. The condenser annulus used in phase contrast can be replaced by a simple disc, occluding the central part of the illumination to provide a 'cone' of light, projected through the specimen by the condenser. The light cone illuminates features within the specimen at a shallow angle, such that the objective lens collects light that has been reflected off the edges of specimen features. This produces images that are mostly dark, with high contrast bright edges delineating the microstructures being examined. In practice, many darkfield configurations utilise more refined darkfield condenser arrangements than a simple occluding disc, but the cost of a good darkfield attachment is likely to be quite modest.

Lastly, a relatively low-cost contrast enhancement method for light microscopy is that of polarising microscopy. Polarising microscopy provides great versatility in many applications for the food microscopist. Polarising attachments consist simply of two identical sheets of polarising material, generally termed a 'polariser' and 'analyser', depending on their position within the optical path of the microscope. The polariser is usually mounted just below the sub-stage condenser, such that it produces polarised illumination of the specimen. The analyser is usually located in the light tube above the objective lens/nosepiece assembly of the microscope and serves to filter the incoming polarised light from the specimen, extinguishing any illumination that is not subject to a change in polarisation angle as it interacts with any optically active material within the specimen. Microscopes

(a)

(b)

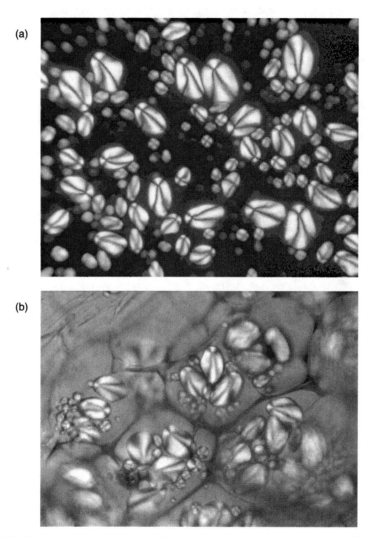

Fig. 3.9 Raw potato starch viewed through cross-polarisers (a) and partially cross-polarisers (b).

equipped with polarising attachments are usually also supplied with a 'strain-free' condenser lens to minimise any birefringence within the microscope optics themselves. The polariser and analyser are generally oriented such that their polarisation angles are perpendicular to each other (90-degree difference), causing complete extinction of unaltered light to maximise contrast from any birefringent materials within the specimen.

Most microscope designs enable rotation of either the polariser or the analyser to adjust the amount of birefringence contrast obtained in the image. This can be useful when viewing birefringent objects in context with any non-birefringent surrounding materials. However, birefringence is not exhibited by all crystalline materials, only by crystals with anisotropic symmetry. Perfectly symmetrical ordered crystals such as sodium chloride (a completely symmetrical cubic crystal) are not birefringent and therefore will not be visible when viewed using crossed polarisers. Other crystalline foods such as sugar, crystalline fats, liquid crystalline emulsifiers, etc. will be birefringent and thus visible under crossed polarisers. Highly ordered radially-aligned structures in starch granules are also beautifully visualised exhibiting a high contrast 'Maltese Cross' pattern when uncooked starch is viewed between crossed polarisers (Fig. 3.9a). Birefringent materials can be viewed in context with other components of a specimen by only partially crossing the polarisers, so that both brightfield details (including any light staining) and birefringence can be observed together, such as cell walls and starch granules from a hand-cut section of potato tissue (Fig. 3.9b).

3.5   Specimen contrast enhancement: chemical and biochemical methods

Whilst optical methods can provide a good range of options for improving contrast, chemical stains can enhance contrast, and provide selective contrast based on the nature of the chemical structure present within different parts of the specimen. The sheer breadth of chemical stains available to the microscopist provides a rich 'palette' with which to help identify different substances. This can be particularly useful in highly processed foods, where microstructure may have been lost, distorted or comminuted to such an extent that identification based on morphology alone becomes impractical. Many stains are relatively quick and easy to use, so there is little additional inconvenience required above simply mounting a specimen on a slide. The number of different available stains is so large that it is impractical to list all of them in this chapter, but it is possible to introduce and discuss some of the most versatile stains for use in food microscopy. Basic food groups such as polysaccharides, oils/fats and proteins can be stained independently of each other by using selected stains, and optical methods can often be used in conjunction with stains to help to identify different food components. The following paragraphs provide an overview of some of the most versatile stains used in food microscopy and, where relevant, the use of optical contrast (e.g. polarisation) is also suggested.

3.5.1   Brightfield stains for basic food groups

Polysaccharides starches, gums, thickeners and stabilisers
Iodine is one of the most commonly used chemicals for staining starches. Usually dissolved in a solution of potassium iodide, this staining solution is also known as Lugol's solution. A solution of 1% iodine in 2% potassium iodide gives a good working stock concentration that can be diluted further if less intense staining is desired. Iodine solutions are light sensitive and should be stored in dark glass (amber or brown) bottles or protected from light by wrapping the storage vessel in aluminium foil.

Raw starch is stained a dark blue/black, whilst gelatinised/partially gelatinised starch will be stained from a lighter blue/purple, through red/magenta to a light pink colour, depending on the state of the starch granules. Modified starches (usually partially hydrolysed by chemical action) stain from golden yellow to a deep brown colour. Iodine in potassium iodide solution can be used with whole mounts or with sections, although dry foods such as pastries or breads may become rather soggy and deformed when soaked, even causing some swelling or partial gelling of some starches. This problem can be overcome by staining with iodine vapour as an alternative to aqueous solution. The specimens are simply placed in a suitable gas-tight container (a sealed desiccator works well), into which some iodine crystals have been placed on a moistened piece of filter paper. Iodine is volatile at room temperature and the vapour diffuses into the specimen over a relatively short time period. The degree of staining required can be regulated by leaving the specimens in the iodine vapour for a shorter or longer period of time and the staining intensity is usually easier to control when using vapour.

Elemental iodine is toxic by ingestion and inhalation and the use of personal protective equipment and a fume cupboard is necessary when working with iodine crystals or vapour. Although iodine stains starches, it can also stain several other biologically derived materials such as amyloid proteins (possibly present in wound tissue in meat/skin) and will stain other proteins, albeit lightly to give a straw yellow colouration.

The transition from raw to gelatinised starch can also be visualised using crossed or partially crossed polarisers to observe the gradual disappearance of birefringence, as the ordered crystalline structure of the granules is disrupted by moisture and heat, or by chemical modification. If birefringence is to be examined, it is best not to allow dense staining of the granules and use of a heavily diluted iodine solution is recommended in such circumstances.

Other polysaccharides are also stained by iodine, but most apart from gum-tragacanth and agarose are not birefringent when viewed under polarisation. Guar gum, carob (locust bean gum), Irish moss and carrageenan will all be stained by iodine, although some of these gums will only appear stained in specific parts of their microstructure (e.g. gum-tragacanth will exhibit dark blue/black stained starch granules within unstained bodies (unstained cells/cell fragments), Guar gum exhibits unstained cells with yellow stained contents).

The polychromatic stain Toluidine Blue provides more versatile staining for most polysaccharide gums. Different gums can in a large part be unambiguously identified on the basis of their colouration and morphology. Acidifying the specimen can help to provide further identification when Toluidine Blue is used, because the colouration of the gum will change predictably for each different gum type, and unique combinations of acidified and unacidified colours can produce unambiguous identifications. Naturally acidic polysaccharides such as alginic acid, hyaluronic acid or chondroitin sulphate will be stained in shades of pink. Many different gums and their corresponding colour when stained with acidified or unacidified Toluidine Blue are described fully in Olga Flint's *Food Microscopy: a manual of practical methods, using optical microscopy* (Flint, 1994), with some accompanying full colour plates.

Toluidine Blue will also produce beautiful colour selective staining of specimens when heavily diluted in water or buffer. The specimen should be allowed to stain gradually by immersing in the diluted stain for periods of several hours, with proteins stained deep blue, phenolics and lignified tissue stained turquoise and cell walls and some polysaccharides stained magenta/pink.

Oils, fats and lipids (animal or vegetable, shortenings, spreads, waxes and mono- and di-glyceride emulsifiers)
Several different stains can be used to visualise oils, fats and lipids and they are usually relatively conveniently stained in whole mounts, smears or squashes. Thin sections can also be stained if simply cut by hand or cryo-sectioned, although resin or wax embedding for microtome sectioning can occasionally lead to dissolution of oils during the embedding process. Liquid oils are more susceptible to such loss than solid crystalline fats. Solid fats do not generally stain well, but they can be conveniently observed by examining the specimen through crossed polarisers, revealing the presence of any birefringent crystalline fats. Using partially crossed polarisers allows observation of both birefringent material and other materials within the specimen, so that the crystalline fat can be seen in context with its surroundings.

Oil Red O, Sudan Black B and Sudan IV (Scarlet R) stains can all be used to stain oils and fats for brightfield observations. Most stains for oil must be dissolved in aqueous alcohol solutions, as they are not soluble in water alone. Solutions of 60% iso-propanol or 70% ethanol are used to dissolve Oil Red O and the Sudan stains respectively. Saturated solutions of these stains (supersaturated in the case of Oil Red O) tend to be necessary to ensure adequate staining, but the solution often has to be filtered just prior to use to remove any precipitated stain. Indeed, stain may also precipitate out of solution during the staining process when using supersaturated solutions of Oil Red O. Precipitation during staining can be mitigated by adding 1% dextrin to the Oil Red O solution, but not all dextrins are effective: best results can be obtained using dextrin that is derived from mildly hydrolysed maize starch. Intensity of staining can also be improved by mixing Oil Red O and Sudan IV stains together to saturation in 70% ethanol.

Nile Red can also be used to stain oils and fats (neutral lipids), leading to a pale red/orange colouration when viewed using brightfield. Nile Red stained oils/ fats fluoresce brightly (bright yellow to bright red) when examined using epi-fluoresence illumination (485–540 nm excitation wavelengths). Nile Red can also be present in Nile Blue stain that has been stored over long time periods, because Nile Blue slowly oxidises to form Nile Red (Nile Blue Oxazone). Oils and fats can therefore also be stained using an aqueous solution of old Nile Blue, whereby the Nile Red component will preferentially partition into any liquid oils and fats (neutral lipids) in the specimen. The water-soluble blue component of the stain will additionally provide good staining of any cell nuclei within the specimen when observed using brightfield illumination.

In general, staining of liquid oils and fats has to be undertaken with great care to avoid mechanically disrupting the position and morphology of the oils within the specimen when this location is of importance (e.g. investigating the spatial distribution of lipids within bread dough). For materials such as emulsions, it may be acceptable to stir the staining solution into the specimen, but care must still be exercised that any emulsion droplets are not destabilised leading to coalescence, or induced to aggregate or flocculate by such processing. This can be checked by examination by brightfield or, better still DIC, before and after staining to investigate any gross changes to the size and shape of the emulsion droplets or flocs.

When sections are to be obtained from wax or resin embedded specimens, oils, fats or lipids may be removed or dissolved in the dehydration or embedding media used, making subsequent visualisation in the sectioned specimen difficult or impossible. Loss of oils can be avoided by fixing (cross-linking) any oils and fats with osmium tetroxide. Osmium cross-links the lipid to produce a dark/dull brown colouration, effectively ensuring that the oils/fats will not be solubilised during any subsequent dehydration or embedding process. Osmium tetroxide is supplied as an aqueous solution, usually packed in small sealed glass vials (typically 5 ml). Specimens can be either immersed in the solution or placed in a sealed container (again, a sealed desiccator is useful) above a broken vial of the solution, allowing osmium tetroxide vapour to stain any specimens within the container. By analogy with iodine vapour staining, the intensity of staining can be controlled by allowing the specimen to remain exposed to the vapour for a longer or shorter time.

Osmium staining also provides high contrast to fats and lipids if the specimen is to be examined using Transmission Electron Microscopy (TEM). Bulk lipid droplets and ultra-fine phospholipid plasma membranes in cells are stained and would otherwise possess very low contrast during TEM observations (Chapter 2). Unfortunately osmium tetroxide is both extremely volatile at room temperature and extremely toxic, so it must be handled with great care using a fume cupboard and suitable personal protective equipment. Osmium must never be handled outside a fume cupboard. Waste osmium tetroxide must be disposed of by first immersing it, and any fragments of the original glass vial, in a large excess of oil (inexpensive vegetable oil can be used) inside a suitable sealable container, thus helping to ensure that all unreacted osmium is cross-linked into the oil molecules and rendered less reactive. The sealed waste container must then be passed to a specialist chemical waste disposal contractor. It is strongly recommended that all other staining methods are attempted before resorting to the use of osmium tetroxide, because of the risks associated with the use of this chemical.

Proteins

Proteins can sometimes be identified based on their morphology alone, for instance striated collagen fibrils in muscle (meat) tissue are readily identifiable, particularly when viewed under polarisers revealing their birefringence. Other less ordered protein deposits often benefit from staining when viewed with simple brightfield. Toluidine Blue provides great versatility in staining both polysaccharides and proteins. Cell nuclei and nuclear proteins are stained dark blue, whilst most other proteins are stained a lighter shade of blue to purple to lilac. Vegetable proteins can also be stained by Toluidine Blue, producing dark blue cell nuclei and staining lighter shades of blue, purple and lilac for other proteins (often present as small, discrete 'protein bodies' in vegetable seeds and legumes). Alternatively iodine in potassium iodide can be used to stain proteins, generally producing a pale yellow colour or deeper golden yellow if allowed to stain for longer. Iodine also has the advantage of being able to vapour stain any moisture sensitive specimens. Staining of specific proteins for direct identification is possible using immuno-labelling methods and these will be discussed later in this chapter.

Moulds, fungi and bacteria

Moulds, fungi and bacteria may all be present in foods, either beneficially (in yoghurts, cheeses, yeast doughs, and other fermented foods and drinks), as contaminants or as the result of decay.

Fungi and moulds can both be stained using a solution of Trypan Blue in lactophenol (phenol and lactic acid). Methyl Blue (also called Cotton Blue) in lactophenol solution also stains the cell walls of fungi very well. Please note that phenol-containing materials should be handled with care and appropriate personal protection equipment (PPE) must be used when handling lactophenol based solutions for staining. Both of these stains work by binding to chitin in the fungal cell wall. Most moulds produce long thin structures known as 'hypae' that are easily recognisable from their shape, and fungal spores within fruiting bodies are also instantly recognisable. Individual spores can be trickier to recognise, especially if the specimen has lots of other small spheroidal features of similar dimensions. Selective staining using either Cotton Blue or Trypan Blue can therefore be used for confirmation of the presence or absence of fungal hyphae or spores.

Yeasts can also be observed using either of these stains because they also possess chitin in their cell walls. Yeasts can be distinguished from moulds by the presence of 'bud-scars', visible as ring shapes on the cell wall. Bud scars develop after new yeast cells have 'budded off' when the cells divide. For larger yeast cells, bud scars can be seen using DIC, but a good way of visualising them is to stain the yeasts using Calcofluor White (a fluorescent stain discussed in the next section). Both yeasts and fungal cells can have a wide variety of sizes depending on maturity and species, so they are not as easily identified as bacteria by simply considering their dimensions alone.

Most bacterial cells (with some rare exceptions) tend to have a limited range of sizes, typically being less than 10 microns and larger than 0.5 microns. Bacilli are easier to spot because of their rod-shaped cells. Coccoid bacteria have a spherical/spheroid morphology, which makes them easy to confuse with other small spherical features that may be present (e.g. some protein bodies in seeds, beans, etc.) in the sample. Toluidine Blue stains bacteria dark blue and is a good stain to use for quick brightfield work, as staining can be achieved in a single step. Gram stain is the classical microbiological method for staining bacteria, providing extra differentiation between 'Gram-positive' and 'Gram-negative' bacteria but, because it is a multi-step staining procedure, requiring several stains (Crystal Violet, Iodine and Saffranin or Basic Fuschin), it is not as convenient or quick to use as Toluidine Blue. Gram-positive bacteria are stained purple and Gram-negative bacteria are magenta/pink by Gram staining. The difference in staining is due to the presence or absence of a peptidoglycan layer in the bacterial cell wall (only present in Gram-positive bacteria). Not all bacteria stain predictably using Gram staining, but most common food-borne bacteria will stain as expected in terms of colouration.

3.5.2   Fluorescent stains and the use of auto-fluorescence

Fluorescence microscopy produces high image contrast, with features of interest generally well separated in brightness and colour (emission wavelength) from a dark background. This high contrast lends itself well to applications requiring image analysis or automated counting of individual objects or features. Fluorescent stains (sometimes referred to as 'fluorophores', 'labels' or 'probes') form the basis of most Flow Cytometry and FACS (Fluorescence Activated Cell Sorting) techniques and the majority of Confocal Scanning Laser Microscopy (CSLM) techniques make use of fluorescent stains or fluorescently tagged immuno-labels. CSLM is discussed in Chapter 4 of this book and Flow Cytometry and FACS methods are used primarily with cells as opposed to complex food matrices, therefore only the use of fluorescence methods for providing contrast in direct far-field microscopic observations of foods will be considered in this chapter.

Fat-soluble fluorescent stains (Oil Red O, Sudan IV, Nile Red) have already been mentioned in the previous section. ODAF (5-(N-octadecanoyl) aminofluorescein) is a useful lipid soluble stain that can be used to good effect for staining liposome structures. Liposomes are difficult to see using conventional brightfield illumination and, if DIC is not available, ODAF staining produces good results if epifluorescence capabilities are available. ODAF will also stain the phospholipid membranes of cells. Aqueous based fluorescent stains can be used for proteins, nuclear proteins and structural polysaccharides. Calcofluor White can be used to stain any β-linked sugars in the cellulose content of plant cell walls, and it also binds to chitin, present in the cell walls of fungi and the exoskeleton of insects. α-linked sugars, such as glucose and mannose, can be stained with fluorescent probes conjugated to the lectin ConA (Concanavalin A) as a good alternative to Calcofluor White (Chen *et al.*, 2007).

Some plant cell walls exhibit auto fluorescence in specific aqueous environments (different pH conditions), obviating the need to stain. A particularly striking example is that of Chinese water chestnut cell walls, which contain di-ferrulic acid that fluoresces bright blue in neutral pH under UV excitation, but will fluoresce a bright turquoise colour when in alkaline pH conditions (Parker and Waldron, 1995). Most plant cell wall autofluorescence is blue; the fluorescence intensity varies for different tissues and different species. Autofluorescence is usually stable during extended observations, whereas many fluorescent probes tend to suffer from 'photobleaching' effects. Photobleaching results in the gradual loss of fluorescence with increased illumination time and/or intensity, but the mechanism that leads to photobleaching is rather complex and a full description falls outside the scope of this book. Photobleaching can be quite pronounced with some stains such as Fluorescein isothiocyanate (FITC), but there are specimen mounting solutions that help mitigate gradual fading of the fluorescence intensity (Diaspro *et al.*, 2006). Many new probe molecules have now become available that are much more stable with respect to photobleaching effects, examples being the Alexa dyes (Molecular Probes, Invitrogen) that are both relatively photo-stable and available with many different excitation/emission characteristics.

Many fluorescent stains can be used for nuclear proteins (DNA, RNA), and ethidium bromide and propidium iodide are commonly used to stain cell nuclei. Acridine Orange is a versatile stain for DNA and RNA: it is water soluble and working solutions can be quickly made and conveniently stored in glass bottles. Working solutions of most fluorescent stains are best stored in cool dark conditions, and the use of amber or brown glass bottles or aluminium foil coverings around bottles is recommended. Acridine Orange is a good rapid staining method for bacteria, because it is not a multi-step staining procedure as required for classical gram staining, and also for the simple reason that it is relatively easy to spot brightly fluorescing bacteria in any specimen containing lots of sub-micron to several micron features when using a brightfield staining method. Acridine Orange has also been reported to provide some degree of differential staining between live (fluoresces green) and dead (fluoresces orange to red) mammalian cells, although it is not as unambiguous as may be obtained using commercially available Live-dead staining kits (Duffy and Sheridan, 1998) and it is of questionable validity for determining bacterial cell viability. Live-dead kits are based on Calcein and Ethidium stains; ethidium stain is prevented from entering cells whose plasma membrane is intact, but it can pass through damaged plasma membranes and will fluoresce red once bound to any nucleic acids. Calcein can pass through intact plasma membranes and fluoresces bright green due to the enzymatic action of esterases present within the cytoplasm. Both Acridine Orange and Live-dead staining kits can be used to stain eukaryotic cells as well as prokaryotes.

Care must be exercised when handling Ethidium bromide, propidium iodide or Acridine Orange, because these stains intercalate with DNA and RNA. They are potentially mutagenic/teratagenic/carcinogenic and appropriate PPE must be used. Hoechst stains (available as a variety of Bis-benzimide stains with different excitation/emission wavelengths) and DAPI (4',6-diamidino-2-phenylindole) are also used to stain cell nuclei and are believed to be less mutagenic but, because both stains still bind to DNA and RNA, it is advisable when handling them to wear gloves.

3.5.3   Immuno-labelling methods

Immuno-labelling is a potentially powerful tool for identifying the presence and spatial distribution of specific substances within a specimen. However, it is also somewhat less than straightforward to accomplish successfully at the first attempt, owing to the propensity of many labels to bind non-specifically across the specimen.

Non-specific labelling can be caused by the presence of unexpected epitopes in the specimen or even by simple physico-chemical effects, such as capillary action into grooves or fine pores in the specimen, or even simply electrostatic attraction towards charged groups in the specimen material or towards any charged mounting substrates. This is compounded by the critical nature of the optimal concentration of primary and labelled secondary antibodies needed to achieve just the right amount of labelling. Fortunately the possibility of non-specific binding can be reduced by carrying out a 'blocking' step in the labelling procedure. Blocking is achieved by incubating the specimen with a solution of protein such as bovine serum albumin or simply milk powder that will readily adsorb on the surface of the specimen, whilst leaving the epitopes of interest exposed.

Following the blocking step, the specimen is rinsed and then incubated with the primary antibody, which should bind to the exposed epitopes. Further rinsing steps are followed by incubation with a secondary antibody that has been conjugated to a fluorescent probe molecule, or to a gold nanoparticle (different diameters can be selected, typically between 5 nm and 25 nm). Further rinsing steps are carried out and, if secondary antibody/gold nanoparticle conjugates are used, it is usually necessary to carry out a 'silver enhancement' step, followed by further rinses. Silver enhancement simply involves incubating the labelled specimen with a strong solution of silver salts to deposit elemental silver as crystalline beads, nucleated around the gold nanoparticles, in order to increase their diameter towards sizes that may be more easily resolved in the light microscope. Incubation of the specimen is usually carried out at room or physiological temperatures, to ensure good antibody activity and a physiological buffer close to neutral pH must be used throughout to avoid denaturation of the antibodies.

Combining all of the steps is inadvisable, as primary and secondary antibodies would almost certainly bind to each other in solution before reaching any of the epitopes of interest on the specimen, and blocking proteins may also bind to the antibodies in solution, preventing them from subsequently binding to the specimen or to each other. Antibodies are themselves protein molecules and, as such are sensitive to pH, ionic strength, temperature and concentration. Care should even be taken to avoid 'foaming' antibody solutions by shaking them, as adsorption to air bubbles may cause conformational changes to the antibody molecules, affecting their ability to bind to their target molecules.

'Negative control' specimens should also be prepared to confirm that any antibody binding observed in the 'positive' specimen is specific. This is typically accomplished by running a duplicate specimen, prepared identically with the exception of omitting the primary antibody labelling step. Ideally another similar specimen that is known to be completely devoid of antigenic epitopes can also be run as a control.

Antibody labelling can be carried out on either whole-mounts (foods that will not dissolve away or be excessively deformed by immersion in aqueous solutions), or on resin or wax-embedded sections. If fixation can be avoided in the case of whole-mounts or vibratome sections, then it is best omitted, to preserve the antigenicity of the epitopes to be labelled. Fixation with aldehydes causes cross-linking and potential conformational changes to proteinaceous antigens and, if the specimen is over-fixed, the primary antibodies may not be able to bind properly. Unfortunately it is not possible to provide any single fixation protocol that may be best for all specimen types, mounts or embedding media other than following the advice provided by the supplier of the antibody, and reading around the literature specific to the specimen and antibody system to be used.

Choosing between gold nanoparticle labels and fluorescent labels may be dictated by the equipment (little choice if no epi-fluorescence attachments are available) or may be dictated by the nature of the specimen. If the specimen is a particularly dense material, 'large' nanoparticles may not penetrate the specimen well enough. This limitation can even apply to larger fluorescent probe molecules, hindering adequate diffusion of labels into the specimen. In many instances there may be little choice but to cut sections because of label penetration issues. If other fluorophores are being used to identify other substances and structures in the specimen, gold labels may be a better choice to avoid any spectral overlaps, therefore the choice of gold nanoparticles or fluorophore labels is best made pragmatically.

Despite all of the technical hurdles to be overcome, immuno-labelling is a useful tool for obtaining information that is both highly substance-specific and sensitive to even very small concentrations.

3.6   Interfacial microscopy

Many foods contain foams, emulsions or dispersions, where much of their physical behaviour during manufacture and subsequent storage may be influenced by the air–water, oil–water or liquid–solid interfaces that they contain. Studying the nature of these interfaces can provide valuable information for foods such as creams, sauces, batters, beverages, foamed products (cakes, mousses, beer foams, whipped creams, etc.), so it is worth considering some of the microscopic methods that are available to help characterise the interfaces in these food systems.

Perhaps one of the simplest ways to examine air–water interfaces (most pertinent to foams) is to set up a glass annulus much like a child's bubble blowing ring, under a microscope objective and observe the drainage behaviour of thin films of liquid (lamellae) as the film entrapped by the annulus gradually thins and drains liquid towards its edges. Placing the annulus with a model 'foam film' between crossed polarisers will enhance the interference patterns (Newton's Rings) in the film, such that different thicknesses of the film will have different colours. The shape of the interference pattern produced is characteristic of different interfacial rheology. Stiff, visco-elastic interfaces resist drainage for longer and produce regular concentric ring patterns, much like the Newton's rings that might be produced by a lens-shaped object. Lower viscosity interfaces allow much more rapid drainage, quickly showing featureless interference patterns that may be thicker (showing some colour difference) on only one side or one region of the edge of the annulus.

Whilst thin film microscopy is relatively simple to set up, it is essentially a qualitative method. Quantitative studies of the molecular dynamics of interfacial adsorbed layers can be obtained using a technique known as FRAP (Fluorescence Recovery After Photobleaching). FRAP turns the 'unhelpful' tendency of many fluorescent probes to photobleach when subjected to intense or prolonged exposure to light (usually of the correct excitation wavelength) into an advantage. FRAP can be used to study model lamellae contained within a glass annulus, as used in thin film studies, or to study the behaviour of other interfaces such as cell membranes or liposome structures by staining them with a limited amount of fluorescent probe; FITC is a good choice, as it is a small molecule and sensitive to photobleaching.

With fluorescent probe molecules incorporated into the test interface, a signal is excited by means of low power illumination by a laser beam of the appropriate wavelength. The fluorescence intensity across the entire area illuminated by the laser beam is monitored through a microscope objective and recorded as the 'baseline' intensity. The laser power is momentarily increased by several orders of magnitude in one spot, causing sudden, rapid photobleaching of all of the probe molecules within the targeted region. The fluorescence intensity drops almost to zero as the laser power is reduced again, but continues to be monitored and recorded as bleached probe molecules begin to diffuse out of the bleached area and 'fresh' unbleached probe molecules diffuse into the illuminated area. The rate of recovery of fluorescence intensity can be quantified and interpreted in terms of the nature of the interface. Stiff, highly visco-elastic interfaces will exhibit slow recovery of fluorescence, whereas highly mobile interfaces stabilised by small molecule surfactants will recover extremely quickly.

Both thin film microscopy and FRAP describe the mechanical nature of the interface, but cannot directly visualise any order/disorder/packing of the adsorbed/surface-active molecules at the interface. Indirect visualisation of adsorbed interfacial layers can be accomplished using 'Langmuir–Blodgett' dipping methods to collect surface-active molecules adsorbed to an air–water interface. The molecules of interest must first be irreversibly labelled with a fluorescent probe at a low enough concentration that their surface-activity must be unaffected. The interface is prepared in a specialised trough (known as a Langmuir trough) by allowing the surface active molecules to spread across the surface of surface-chemically clean water or buffer solution.

The surface pressure exerted by the presence of the adsorbed molecules is monitored using a Wilhelmy plate (usually a ground-glass plate touching the surface of the liquid, attached to a sensitive micro-balance), such that the surface pressure will plateau when spreading and molecular rearrangement at the surface is complete (including any folding/unfolding of globular protein molecules). The adsorbed molecules can be sampled onto a surface chemically clean glass cover-slip or a piece of freshly cleaved mica, by using a precision motorised dipping apparatus to dip the glass or mica vertically into the liquid contained in the trough and then slowly withdrawing it again. A barrier at one end of the trough can be swept forwards across the interface to compress it (increasing surface pressure) or backwards to stretch it (decreasing surface pressure). Distortion of the adsorbed layer is avoided during dipping by using feedback from the Wilhelmy plate monitor to advance or retract the interface barrier to maintain the surface pressure at precisely the same value throughout the dipping process. The retrieved cover-slip/mica with adsorbed molecules can be imaged using fluorescence microscopy to examine the spatial distribution of mixtures of differently labelled, or labelled and unlabelled molecules (Mackie, *et al.*, 2001).

Air–water interfaces can be directly visualised using a more specialised microscope; the Brewster Angle Microscope (BAM). BAMs work, as the name implies, by illuminating and observing the interface of interest at the 'Brewster Angle'; the angle of incidence of polarised illumination (usually laser) to a surface/interface at which no reflection takes place. The lack of reflections at the Brewster Angle result in enhanced visualisation of even very minute changes in refractive index due to any adsorbed layers that may be present at the interface. Monolayers or multilayers can be imaged and phase separation, domain shape and size and other heterogeneity can be monitored. Although the area coverage of different adsorbed domains can be measured, the thickness of adsorbed layers is not directly measureable from BAM observations. The thickness of thin adsorbed layers on solids can be measured with great precision using a non-contact optical method called 'ellipsometry'. The use of ellipsometry on liquid-liquid or air-liquid interfaces is not straightforward, but is being investigated by some workers and may prove to be a useful adjunct to BAM in future.

3.7   Recent and future developments

Although light microscopy is a well established technique that has been in use for centuries, new developments and refinements are still being made. Recent developments include the use of light emitting diode (LED) based illumination sources for fluorescence microscopy (less expensive, less hazardous and more energy efficient than mercury arc lamps). A host of improvements have been made possible by ever increasing computational power; extended depth of focus via z-stacking, automated tiling (montaging) of multiple image fields allowing large areas to be imaged at high resolution and various deconvolution algorithms (Wallace *et al.*, 2001; Shaw, 1998) that can be used to improve contrast and resolution. The effectiveness of deconvolution algorithms for improving contrast and resolution has even been compared to confocal scanning laser microscopy (Shaw, 1995).

Perhaps one of the more ambitious and striking examples of software-assisted microscopy is that of Ptychography (Rodenburg, 2008; Maiden *et al.*, 2010). Ptychography replaces the objective lens with a set of algorithms capable of reconstructing an image from the defocused illumination collected after coherent light from a laser interacts with (passes through) a specimen. The algorithms are based on knowledge of the optical principles governing the interaction of light (or other illumination) with materials and unsurprisingly, are computationally intensive. Once the defocused laser light containing both phase and intensity information is acquired from the specimen, a phase image and an intensity image are essentially 'back-calculated' iteratively from the diffraction pattern. The advantages of calculating the phase image are that any optical aberrations inherent in physical lenses are not present and there is no need to stain to enhance specimen contrast. The phase information retrieved is quantitative, allowing height information to be obtained, providing an accurate profilometry capability, free of the limitations imposed by vibrations or physical imperfections in a lens/ interferometric based profilometry system. Transmitted light images can be obtained at high resolutions without the need for short working distances, further enhancing the practical nature of Ptychography

There are some limitations to Ptychographic microscopy at present. Reconstructing the phase image from thicker specimens containing several overlaid features is not straightforward, but progress is being made and as the algorithms are developed further and computing power continues to increase, it will be interesting to watch the progress of such methods.

3.8   Conclusion

Light microscopy has played a pivotal part in the origins and development of food science and will continue to be an invaluable tool in this role in the future, despite the availability of increasingly diverse and advanced microscopy technologies. The reasons for the durability of the light microscope are varied, but it still enjoys a unique position in providing full colour information with, arguably, relatively rapid and convenient sample handling for initial microscopic examinations.

The optical principles at the core of the light microscope are well understood, allowing several specialist instruments to be developed, permitting long working distances for convenient observations of large samples or online process monitoring situations and even interfacial or thin-film observations capable of directly visualising molecular/macromolecular behaviour *en-masse*.

The future usefulness of light microscopy to the food scientist remains undiminished. The availability of powerful computers and advanced software have provided increased functionality and versatility to the light microscope and look set to facilitate further advances in convenience and performance in the future.

3.9   References

CHEN, M-Y., LEE, D-J., TAY, J-H. and SHOW, K-Y. (2007), Staining of extracellular polymeric substances and cells in bio-aggregates, *Appl. Microbiol. Biot.*, **75**, 467–74.

CLAYTON, E. G. (1909), *A Compendium of Food Microscopy with Sections on Drugs, Water and Tobacco*, London, Ballière Tindall and Cox. Available from: *http://www.archive.org/stream/compendiumoffood00clayrich#page/n1/mode/2up* [accessed July 2012].

DIASPRO, A., CHIRICO, G., USAI, C., RAMOINO, P. and DOBRUCKI, D. J. (2006), Photobleaching, in, J. B. Pawley (ed.), *Handbook of Biological Confocal Microscopy*, New York, Springer-Verlag, 690–99.

DUFFY, G. and SHERIDAN, J. (1998), Viability staining in a direct count rapid method for the determination of total viable counts on processed meats, *J. Microbiol. Meth.*, **31**, 167–74.

FLINT, O. (1994), *Food Microscopy: A Manual of Practical Methods, Using Optical Microscopy*, RMS Handbook 30, Oxford, BIOS Scientific Publishers Ltd.

GREENISH, H. G. and COLLIN, E. (1904), *An Anatomical Atlas of Vegetable Powders*, J. London, and A. Churchill. Available from: *http://archive.org/stream/anatomicalatlaso00greeuoft#page/n7/mode/2up* [accessed July 2012].

HILL HASSALL, A. (1855), *Food: Its Adulterations and the Methods for their Detection*, London, Longman Brown Green and Longmans. Available from: *http://www.archive.org/stream/fooditsadulterat00hassrich#page/n7/mode/2up* [accessed July 2012].

HOOKE, R. (1665), *Micrographia: orSome Physiological Descriptions of Minute Bodies Made by Magnifying Glasses with Some Observations and Inquiries Thereupon*, London, Martyn and Allestry.

KIERNAN, J. A. (2000), Formaldehyde, formalin, paraformaldehyde and glutaraldehyde: What they are and what they do, *Microscopy Today*, **00-1**(8), 8–12.

MACKIE, A. R., GUNNING, A. P., RIDOUT, M. J., WILDE, P. J. and MORRIS, V. J. (2001), Orogenic displacement in mixed *β*-lactoglobulin/*β*-casein films at the air/water interface, *Langmuir*, **17**, 6593–8.

MAIDEN, A. M., RODENBURG, J. M. and HUMPHRY, M. J. (2010), Optical ptychography: a practical implementation with useful resolution, *Opt Lett*, **35**, 2585–7.

NEWMAN, G. R. and HOBOT, J. A. (1999), Resins for combined light and electron microscopy: a half century of development, *Histochem. J.*, **31**(8), 495–505.

OLDFIELD, R. (1994), *Light Microscopy: An Illustrated Guide*, Aylesbury, UK, Wolfe Publishing/Elsevier Health Sciences.

PARKER, M. L. and WALDRON, K. W. (1995), Texture of Chinese water chestnut: involvement of cell wall phenolics, *J. Sci. Food Agric.* **68**, 337–46.

RODENBURG, J. M. (2008), Ptychography and related diffractive imaging methods, *Adv. Imag. Elect. Phys.*, **150**, 87–184.

SHAW, P. J. (1995), Comparison of wide-field/deconvolution and confocal microscopy for 3D imaging, in J. B. Pawley (ed.), *Handbook of Biological Confocal Microscopy*, 2nd edition, New York, Plenum Press, 373–87.

SHAW, P. J. (1998), Computational deblurring of fluorescence microscope images, in *Cell Biology: A Handbook*, 2nd edition, San Diego, Academic Press, 206–17.

VAUGHAN, J. G. (ed.) (1979), *Food Microscopy*, London, Academic Press.

WALLACE, W., SCHAEFER, L. and SWEDLOW, J. (2001), A working person's guide to deconvolution in light microscopy, *BioTechniques*, **31**(5), 1076–97.

WINTON, A. L. and WINTON, K. B. (1932–1939), *The Structure and Composition of Foods*, vols I–IV, Chichester, UK, John Wiley and Sons Ltd.

4

Confocal microscopy: principles and applications to food microstructures

M. A. E. Auty, Teagasc Food Research Centre, Ireland

DOI: 10.1533/9780857098894.1.96

Abstract: This chapter describes the principles and applications of confocal microscopy to food research. The advantages of the technique are that real food products can be studied directly and the microstructures of their constituents and their interactions characterised. A brief review of the literature highlights the growing use of confocal microscopy in a wide variety of food applications across all food sectors. Practical details of labelling techniques for different types of food are illustrated with examples. Dynamic microscopy techniques for visualising structure formation and deformation are presented and new instrumentation and measurement methodologies are also discussed.

Key words: confocal, fluorescence, food, microstructure, confocal scanning laser microscopy (CSLM), microscopy.

4.1   Introduction

Food research is the archetypal multidisciplinary subject, reflecting the need for a holistic scientific approach, combining biology, chemistry and materials science disciplines, among others. Microscopy generally, and confocal microscopy in particular, helps to relate food structure at the *meso* scale to physico-chemical properties, texture and sensory attributes. Since the first published food application of confocal microscopy (Heertje *et al.*, 1987), there has been an almost exponential increase in the use of confocal microscopy in food research. The confocal microscope is now seen by food scientists as a key tool for complementing, and often interpreting, more established physical and chemical food analyses. However, no single microscopy technique gives all the answers and where possible a correlative approach employing other microscopy techniques should be considered (Aguilera and Stanley, 1990).

This chapter will briefly describe the principle of confocal microscopy and review its applications to a wide range of food products. Newer techniques such as dynamic or time-dependent confocal imaging, hybrid configurations and recent technological developments will also be highlighted. As in all microscopy techniques, effective use of confocal microscopy is a skill to be learned and does not lend itself to the 'push button' approach of modern, computer-driven analytical instrumentation. It is very easy to misinterpret images, particularly when investigating new food systems or to fit some strongly held theory. The student and researcher must develop hands-on experience and a thorough knowledge of the physicochemical properties of the sample, the effect of sample preparation and its interaction with light, in addition to the correct operation of the instrument itself. This takes time and there is no substitute for experience. Consequently, this chapter will emphasise the practical use of confocal microscopy in real food products and is based on the author's trials and (frequent) errors, in using confocal microscopy to understand the behaviour of complex food systems.

4.2   Principle of confocal microscopy

4.2.1   Principles

Confocal microscopy is a form of optical microscopy and has much in common with an epifluorescence microscope. The essential feature of confocal imaging is that both the illumination and detection systems are focused on a single volume element in the specimen (Minsky, 1957). Conventional light microscopy employs wide-field illumination: the volume of sample above and below the plane of focus is uniformly and simultaneously illuminated. This usually requires a thin, relatively transparent, sample but often results in out-of-focus blur that reduces resolution and specimen contrast. In confocal scanning laser microscopy (CSLM), a diffraction-limited illuminated spot is detected by means of a small aperture (pinhole) placed in front of the emitted light detector, greatly reducing out-of-focus information (Brakenhoff *et al.*, 1988; Wright *et al.*, 1993). The illuminated spot is then scanned across the specimen (Fig. 4.1).

Incident light is absorbed by fluorophores within the sample and emitted at a longer wavelength, which is deflected via a dichroic mirror onto a photon detector such as a photomultiplier tube or CCD. The signal is then amplified and converted from photon electron intensity into pixels. The beam is scanned, raster fashion, and the size of the scan can be varied but is typically 512×512 or 1024×1025 pixels to give a 2D image. A stepping motor fitted to the sample stage enables the acquisition of consecutive x-y planes of focus through the z-plane, resulting in a 3D dataset (Fig. 4.2), which can then be recombined and rendered into a 3D projection. The resolution of a confocal microscope is governed by the same optical limitations affecting conventional light microscopy, but with some subtle differences. The theoretical resolution of an optical microscope is limited by diffraction (Abbe, 1884) and is given by:

$$\Delta x, \Delta y = \lambda/2n\sin\alpha \qquad\qquad [4.1]$$

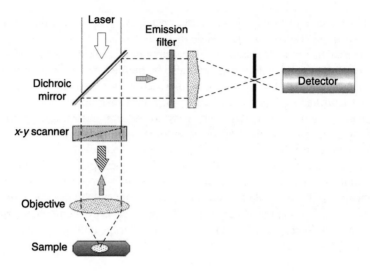

Fig. 4.1   Schematic of the principal components of a confocal scanning laser microscope.

where λ, n and α are the wavelength of the incident light, refractive index of the sample and semi-aperture of the objective lens, respectively. Along the optical axis, z-resolution, $\Delta z = 2\lambda / n \sin 2\ \alpha$, giving a maximum lateral resolution of 190 nm and axial resolution of more than 500 nm (Born and Wolf, 1980). By reducing out-of-focus blur, confocal microscopes under ideal conditions can approach these theoretical resolutions as predicted; however, in practice and with blue excitation (e.g. 488 nm), x-y resolution of approximately 250 nm and z resolution of approximately 750 nm for food materials is more realistic. The z-thickness of the optical sections can be adjusted by opening and closing the pinhole

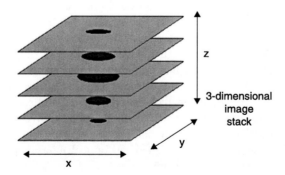

Fig. 4.2   Diagram showing sequential confocal optical sections through a sample (black).

aperture. For most modern instruments, the pinhole diameter will be automatically set to optimise resolution, depending on the objective magnification and numerical aperture (NA), typically equivalent to the diameter of 1 Airy disc, and this approximates to a minimum axial resolution of approximately 730 nm with a high-quality lens such as an X63, 1.4 NA oil-immersion apochromatic objective.

The advantages of CSLM over conventional light microscopy include:

* 3D imaging of bulk samples by 'optical' sectioning and digital reconstruction;
* minimal sample preparation and disturbance;
* improved resolution over conventional optical microscopy;
* sensitive detection of multiple fluorochrome probes;
* dynamic processes can be studied under controlled environmental conditions using appropriate sample stages and fast acquisition rates.

Disadvantages include:

* Food components usually require labelling with a suitable fluorochrome, which may involve solvents and/or multiple processing steps that could adversely affect the sample;
* The sample surface has to be flat to ensure an even illumination and emission signal across the field of view;
* Diffraction-limited lateral resolution is ~200 nm at best, making it impossible to properly resolve colloidal food systems such as casein micelles, or interfacial films.

Despite these disadvantages, the CSLM has become probably the most useful and versatile microscopy technique in food research; moreover, it is now regularly used by the food industry for product development and as a troubleshooting tool (Tamine *et al.*, 2011). The ability to 'optically' section a food product gives unique insight into its true 3D structure and the likely relationship between different ingredients through differential labelling. Furthermore, samples can be imaged under ambient conditions in real time, unlike most electron microscopy techniques that require the sample to be under a vacuum, relatively conductive or extremely thin.

4.2.2   Confocal microscope designs
There are three basic confocal microscope designs:

1. spinning (Nipkow) disk, also called tandem scanning confocal microscopes;
2. stage scanning; and
3. beam scanning.

The spinning disk-types can use conventional white light and have very fast acquisition rates. Recent disk-scanning instruments, such as the Yokogawa-type, have greatly improved laser thoughput and give real-time imaging of biological specimens in real colour. Beam (or point) scanning instruments, although relatively slower and more complicated (and therefore expensive), rely on detection of fluorescent light using laser excitation. Multiple lasers and separate emission detection channels facilitate specific labelling of different food components simultaneously. The stage scanning instrument is rarely used today and nearly all published food research studies use the point or beam scanning design. This technique employs lasers with defined emission wavelengths, typically in the range 488 to 647 nm but may also include blue 405 nm or UV lasers. Modern instruments, such as the Leica SP5 series, may be fitted with up to five separate confocal channels and acousto-optical tuneable filters can optimally excite a wide range of fluorochromes. In addition, acousto-optical tuneable filters and beam splitters may be used to replace conventional epifluorescence filters, greatly reducing emission spectral overlaps and improve bleed-through/crossover where multiple dyes are employed or autofluorescence is an issue.

4.3   Chemical contrast: identifying ingredients

4.3.1   Fluorescence basics

Most food ingredients are not naturally autofluorescent, so the main strategy to visualise them in the CSLM is to use a specific fluorescent label or probe. Fluorochromes molecules, being rich in chromophores, absorb photons of a particular wavelength or wavelength band, and emit them at longer wavelengths. The mechanism underlying fluorescence emission from the fluorescent molecules on excitation with photons from a light source is illustrated in Fig. 4.3a. Initially, the dye molecule is at a relaxed ground state, S_0. On excitation with photons, the dye molecule absorbs energy equivalent to E_V and attains an excited state energy level (S_1') for approximately 1 to 10 nanoseconds, after which the dye molecule dissipates some energy (E_M), reaching an intermediate molecular energy state, S_1. The dye molecule may then return to the relaxed energy level S_0 by further dissipation of energy E_C. This energy dissipation by the dye molecules (E_M, E_C) occurs as fluorescence energy, in the form of photons that are of lower energy than the photons used to excite the dye (E_V) (Fig. 4.3b). Thus fluorescent light emission occurs at a longer peak wavelength than the peak wavelength of the excitation laser light, a phenomenon known as Stokes' shift.

The choice of laser excitation light to be used for CLSM depends on the particular excitation wavelength of the fluorescent dye molecule and they should be matched as closely as possible. Laser diode arrangements containing semiconductor materials have recently been developed for obtaining laser excitation wavelengths in the range 300 to 460 nm; gas lasers, such as Argon ion laser or lasers containing a mixture of krypton and argon, are commonly used for obtaining laser excitation wavelengths in the range 450 to 565 nm, while helium neon lasers with excitations of 543 or 633 nm may be used. Laser technology is advancing rapidly and newer CSLMs are more increasingly being fitted with solid state lasers.

(a)

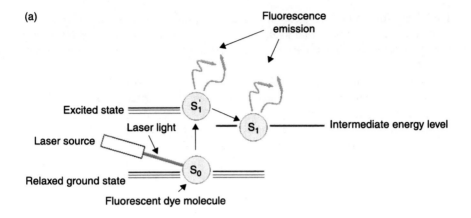

(b)

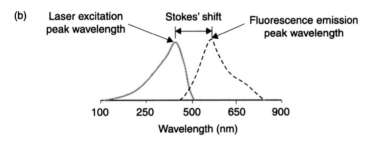

Fig. 4.3   Schematic representation of (a) the mechanism of fluorescence emission from a fluorescent dye molecule on excitation with laser light at peak wavelength; and (b) the laser excitation (continuous line) and fluorescence emission wavelength (dashed line) spectra of a fluorescent dye molecule illustrating the Stokes' shift phenomenon.

There are three main strategies for labelling food components with fluorochromes:

1. generic or non-specific labelling via electrostatic binding (proteins/ polysaccharides), or hydrophobic modification (lipids);
2. specific labelling of individual proteins (immuno-labelling) or polysaccharides (lectin binding);
3. covalent labelling of proteins or polysaccharides.

4.3.2   Generic labelling of fats and proteins

Fluorescent dye molecules used conventionally for CLSM are organic in nature and can attach non-covalently to the biopolymer of interest in a test specimen via ionic, hydrogen bonding, electrostatic or hydrophobic interactions; besides, dye molecules may also attach to a biopolymer due to covalent interactions between reactive functional groups present on the dye molecule and the biopolymer (Patonay et al., 2004). A wide range of organic fluorescent dyes having laser light excitation wavelengths in the range of 300 to 900 nm are now commercially available. More information can be found on the online version of the *Molecular Probes Handbook*, an invaluable resource for developing new fluorescent labelling strategies (*http://www.invitrogen.com/site/us/en/home/References/Molecular-Probes-The-Handbook.html*).

Fat or protein distribution in most food products is relatively straightforward to visualise, using either a single fluorochrome such as Nile Blue (Brooker, 1995) or a dual-labelling approach (Auty *et al.*, 2001a). Herbert *et al.* (1999) used Acid Fuchsin, Bodipy® 665/676 and DM-NERF to label proteins, lipids and whey, respectively, in a model dairy-based gel. Generic labelling with fluorescent probes generally relies on passive diffusion of dye molecules to the site of interest, thereby minimising sample disturbance. For oils and fats, Nile Red, an oxazone derivative of Nile Blue, is one of relatively few lipophilic probes that fluoresce when in contact with a liquid lipid phase and has been virtually the only fat dye available. Recently, a new fat dye, VO03–01136 (Dyomic, Jena, Germany) was successfully used to label fats in a mixed food biopolymer gel (Heilig *et al.*, 2009). This dye could be particularly useful as its excitation maximum is 654 nm, thus giving additional options for multiple dye applications.

For general electrostatic labelling of proteins there are many fluorochromes available and these include Rhodamine B, Fast Green FCF and Nile Blue. These are best applied as a single aqueous dye preparation; typically 10 µl of dye is added directly to the food ingredient or product. In some cases, such as dairy spreads, solid dye crystals can be directly sprinkled onto the product surface to allow the dye to slowly solubilise and migrate through the sample over time (Brooker, 1995). Labelling polysaccharides *in situ* is more problematic as there are few fluorochromes analogous to the generic protein dyes that can be used *in situ*. Fluorescent probes such as fluorescein isothiocyanate (FITC), Rhodamine B and Safranin O have been used for non-covalent labelling of starch (van de Velde *et al.*, 2002).

However, more research is needed to understand the mechanism of fluorescent binding, particularly for carbohydrates. Some care is needed when preparing the sample, as the CSLM requires a flat surface to ensure an even 2D emission signal. Incident beam penetration depth is strongly sample dependent, ranging from approximately 100 µm for transparent gels to less than 15 µm for optically dense, refractile material such as chocolate or hard cheeses. For solid or self-supporting visco-elastic materials such as cheese, hand-cut sections may be used, whilst for meat products, frozen cryostat sections give good results. For powders, small amounts of the powder are suspended in an excess of suitable dye formulation designed to prevent dissolution of the powder whilst facilitating diffusion of dye molecules to the site of interest (Auty *et al.*, 1999). Advantages of this one-step approach are that it is rapid with results being obtained within five minutes, and does not require multiple treatments, such as sequential labelling, washings etc., which may remove or distort unstable food constituents. The main disadvantage is that these dyes are relatively non-specific, for example, they cannot distinguish between different proteins. For more specific labelling, there are two main approaches:

1. immuno-labelling of proteins (and sometimes polysaccharides) or lectin binding of polysaccharides; and
2. covalent labelling of proteins or polysaccharides.

Immuno-labelling is more commonly used for proteins and requires a specific antibody to target the protein of interest, and methodology is usually adapted from its extensive use in cell biology (Tsien *et al.*, 2006).

4.3.3 Immuno-labelling and specific labelling of polysaccharides with lectins

Perhaps surprisingly, there have been few reports using immuno-labelling to identify specific proteins as biopolymers and food applications have tended to concentrate on localisation of food pathogens such as *Escherichia coli* O157:H7 on vegetable surfaces (Auty *et al.*, 2005a). Immuno-labelling depends heavily on the specificity of the primary antibody to the target epitope. In particular, the effect of heat denaturation on the immuno-reactivity of the target protein should be characterised and ideally antibodies raised against the denatured protein, rather than native protein, as appropriate. Two-step immuno-labelling, using a fluorescently labelled secondary antibody to bind to the primary antibody, is generally recommended, as this increases the signal-to-noise ratio.

Lectins are naturally occurring proteins and glycoproteins that have the ability to bind to specific carbohydrate monomers present in a biopolymer (Brooks *et al.*, 1997). One of the most commonly used lectins for carbohydrate recognition is Concanavalin A (ConA), which is isolated from the Jack Bean *Canavalia ensiformis* (Loris *et al.*, 1998). In the presence of selected metal ions, ConA has the ability to bind with the D-glucose and D-mannose residues at the non-reducing terminus of polysaccharides and glycoproteins. Arltoft *et al.* (2007) used enzyme-linked immunosorbent assay (ELISA) techniques to investigate the affinity of ConA with the sugar residues in different food polysaccharides and observed that ConA can bind with galactomannans, xanthans and bacterial exopolysaccharides (EPS). ConA has also been found to bind with the α-D-glucose residues in amylopectin (Matheson and Welsh, 1988). A commercially available ConA–fluorophore conjugate is often used as a label marker to localise sugar residues in biological cells using fluorescence microscopy techniques. In food microscopy, the lectin labelling technique has been used to localise bacterial ESP in dairy yoghurt gels (Arltoft *et al.*, 2007; Hassan *et al.*, 2002, 2003). The specificity of ConA may also allow localisation of Konjac glucomannan in a mixed biopolymer system using CLSM techniques. Arltoft *et al.* (2007) screened two lectins: Con A and wheat germ agglutinin (conjugated with Alexa Fluor 488 dyes) and one polyclonal anti-carageenan antibody, JIM7, as probes for specific localisation of a wide range of polysaccharides in foods. Despite the cross-reactivity of some of these probes, the authors were able to demonstrate good localisation of pectin and carrageenan in dairy products. For cell wall components including cellulose, glucans and glycoproteins, a range of monoclonal antibodies and carbohydrate binding molecules are now available (Hervé *et al.*, 2011).

4.3.4   Covalent labelling

Polysaccharide staining by the covalent labelling technique involves heating a mixture containing a polysaccharide and a fluorescent probe at high temperature in a non-aqueous environment followed by alcohol precipitation of the polysaccharide-fluorescent probe conjugate (Belder and Granath, 1973). Covalent labelling techniques have been used to localise polysaccharides such as carrageenan, dextrans and gellans in multi-component biopolymer systems (Tromp *et al.*, 2001). In a recent study, Abhyankar *et al.* (2011a) used covalently-labelled Konjac glucomannan to study phase behaviour in a mixed biopolymer system containing whey protein. Covalent conjugation of Konjac with FITC led to a shift in the absorbance spectrum peak of FITC to a lower wavelength and a decrease in the average molecular weight distribution of Konjac. Furthermore, covalently labelled Konjac showed reduced apparent viscosity compared to unlabelled Konjac. Van der Velde *et al.* (2003) covalently-labelled proteins and polysaccharides and added them to biopolymer mixtures containing starch and gelatine. The authors were able to successfully discriminate between starch, polysaccharides and proteins. Disadvantages of covalent labelling are that it can only really be used in model systems and not retrospectively to label real food products. There is also a concern that the rather harsh chemical labelling process, often involving high temperatures and pH adjustment, may modify the functionality of the target molecule (van de Velde *et al.*, 2003; Abhyankar *et al.*, 2011a). Kett *et al.* (2013) were able to localise covalently labelled milk protein fractions in biopolymer mixtures containing modified starch, suggesting that this technique could be useful for tracking specific proteins in heterogeneous food models.

Specific labelling of foods, in particular polysaccharide-based components, is not trivial and it is important to ensure that positive and negative controls are always included to reduce misinterpretation of images. Lectins and polyclonal antibody probes in particular are prone to non-specific binding, although the use of monoclonal antibodies can greatly reduce this. The reader is encouraged to visit Paul Knox's website at the University of Leeds (UK) (*http://www.personal.leeds. ac.uk/~~bmbjpk/*), for more information on how to obtain specific probes for plant-derived compounds. This research is very much at an early stage and further work is needed to develop specific labelling methods that can be applied to localise bioactive compounds in complex food matrices.

4.4   Confocal microscopy of food products: a brief review

Several early review papers highlighted applications of CSLM to food research (Heertje *et al.*, 1987; Brooker, 1991; Blonk and van Aalst, 1993; Vodovotz *et al.*, 1996; Ferrando and Spiess, 2000; Dürrenberger *et al.*, 2001) and more recently by

Loren *et al.* (2007). All major agri-food products have now been investigated by confocal microscopy, almost always employing CSLM. These include bakery, vegetable, confectionery, meat, dairy and beverage products and a summary of specific food-related applications of CSLM is given below.

4.4.1   Meat, fish and meat products

Although a primary application of CSLM has been in mammalian cell biology, there is relatively little published material on the use of the CSLM to study meat products. Velinov *et al.* (1990) used Nile blue to examine the distribution of fat in Frankfurters and acridine orange to identify bacteria in summer sausage. More recently, Bertram *et al.* (2006) used CSLM to study pressure-induced changes in the muscle structure of ham. Chattong *et al.* (2007) used a combination of FITC and Nile Red to show protein and fat, respectively, in an ostrich meat-based product. Meat emulsions and gels have also been studied using CSLM to help understand rheological or texture relationships (Brenner *et al.*, 2009; Klongdee *et al.*, 2012; Trespalcios and Pla, 2007). Microbiological applications of CSLM in meat studies include adhesion of bacteria to pork (Delaquis *et al.*, 1992) and chicken (Kim, *et al.*, 1996). An example of immuno-labelling is given in Plate I in the colour section between pages 222 and 223, which shows *E. coli* cells that have been specifically labelled using an FITC-labelled antibody conjugate within the beef connective tissue (Auty *et al.*, 2005a).

Meat and meat products such as sausages, burgers, etc. are best prepared by cutting frozen sections of the products in a cryostat (Flint, 1994). Fairly thick sections, 20 to 50 μm thick, may be cut to facilitate optical sectioning and reconstruction of 3D images. Sections may be stained using the generic protein and fat dyes such as Fast Green FCF and Nile Red described above (see also colour Plate II).

4.4.2   Fruit and vegetable products

Despite the application of CSLM to visualise intact plant cellular tissue, most CSLM applications have concentrated on isolated cell cultures or extracts (Verleben and Stickens, 1995). Until recently, there have been relatively few CSLM studies of fruit and vegetable tissues *in situ*. Lapsley *et al.* (1992) used CSLM with acridine orange staining to study the degree of cell adhesion in apple tissue. Alvarez *et al.* (2010) used CSLM observations of potato cells stained with acridine orange to support cutting energy data, while Ferrando and Speiss (2000) used CSLM with Congo Red staining to visualise fresh apple tissue, including cell walls and cytoplasm. It should be mentioned that the cited works required extensive sample preparation including chemical fixation, ethanol dehydration and embedding in a plastic resin prior to confocal examination. Attachment of bacteria to lettuce leaves (Takeuchi *et al.*, 2000) and apple cells (Burnett *et al.*, 2000) has been demonstrated by CSLM. Bouchon and Aguilera (2001) used CSLM to localise oil in fried potatoes.

Altan *et al.* (2011) used a spinning disk confocal for imaging changes in microstructure of almonds as a function of thermal processing. Staining with Nile Red and Calcofluor White was used to differentiate cell wall structures and oil bodies, respectively, within individual almond cells. Redgwell *et al.* (2008) used confocal microscopy to show that a shear-induced increase in viscosity cell wall extracts was accompanied by fragmentation of the kiwi fruit and tomato cell walls, which increased the available surface area for particle-particle and/or particle-solvent interaction. Roessle *et al.* (2010) used an *in situ* viability stain in conjunction with CSLM to localise probiotic bacteria in apple tissues, which can be labelled with Trypan Blue to visualise the cellulosic cell walls (see colour Plate III).

4.4.3   Cereal and bakery products

CSLM has been used to study the rising of dough at room temperature (Heertje *et al.*, 1987; Blonk and van Aalst, 1993). Lee *et al.* (2001) used CSLM with an alkaline FITC solution to visualise the protein matrix of wheat doughs at various stages of structural development. Dürrenberger *et al.* (2001) used CSLM to study yam parenchyma, wheat dough, bread and spaghetti. Samples were stained with Safranin O and Acid Fuchsin, enabling visualisation of starch and protein components, respectively. Starch granules before and during gelatinisation have been visualised by CSLM following labelling with rhodamine, FITC or safranin (van de Velde *et al.*, 2002). The microstructure of frozen dough incorporating sodium caseinate or whey protein concentrate has been visualised by CSLM with Nile Blue labelling, which revealed the gluten network (Kenny *et al.*, 2001). Nile Blue and FITC can be used to visualise partially gelatinised starch grains, the gluten network and lipid droplets using triple-channel imaging (see colour Plate IV). CSLM has also been used to study gluten in tortillas (Alviola *et al.*, 2008), starch hydrolysis (Apinan *et al.*, 2007; Bordoloi *et al.*, 2012) and gelatinisation (Chen *et al.*, 2009).

Particular interest has been shown in using CSLM to characterise the microstructure of cereal products when developing novel gluten-free products (Mariotti *et al.*, 2009; Shober *et al.*, 2008; Alvarez-Jubete *et al.*, 2010). A primary aim of this research is to develop new ingredients that can mimic the functional and microstructural properties of the gluten proteins.

4.4.4   Confectionery products

Brooker (1995) first used CSLM to identify components of chocolate. Fat was labelled with Nile Red and this enabled sugar crystals, milk protein and cocoa solids to be identified by negative contrast. To study the distribution of an added ingredient in the chocolate by CSLM, a hydrophilic polysaccharide was pre-labelled with the fluorescent dye, Cy5, which was then added to the chocolate. Little other published data is available on confectionery, possibly due to the commercial sensitivity of these products. Labelling confectionery products can be difficult as they are usually hard materials, often sugar-continuous; making the use of aqueous-based dyes problematic. Techniques are available though and it is possible to simultaneously label the protein and fat phases, for example in a caramel (toffee) by using a dye mixture dissolve in a low molecular weight polyethylene glycol, glycerol or propan1,2-diol (Auty *et al.*, 2001a). Two confectionery products, a milk-based caramel (toffee) and a milk chocolate, are illustrated in colour Plates V and VI, respectively.

4.4.5   Dairy: fat-based products

The non-invasive nature of confocal microscopy makes it a useful technique for examining shear-sensitive samples such as fat spreads and mayonnaise (Blonk and van Aalst, 1993; Langton *et al.*, 1999). Care is required during sample preparation in order to maintain structural integrity, for example to preserve delicate fat crystal networks. Heertje *et al.* (1987) used a plunger-like sampling tube to extract a sample of spread for CSLM. The spread sample was then pushed out of the tube for sectioning prior to staining with Nile Blue. Brooker (1995) suggested applying solid crystals of fluorochrome to the spread surface to allow slow solubilisation and diffusion of the stain into the product and this approach can work well. The use of a cooling stage (Peltier-type) can be useful to characterise the fat crystal network at refrigeration/chilled food temperatures (4–10 °C). The fat phase may be stained using a lipid stain such as Nile Red or Nile Blue, whereas the aqueous phase can be stained by FITC or Rhodamine (Heertje *et al.*, 1987; Blonk and van Aalst, 1993). The microstructure of low fat spreads containing sodium caseinate has been studied by CSLM (Clegg *et al.*, 1996); the fat phase was stained with 0.1%, w/v, Nile Red in polyethylene glycol. Fat droplet and egg yolk protein aggregate sizes were measured from CSLM images of mayonnaises stained with Nile Blue (Langton *et al.*, 1999; see colour Plate VII). van Dalen (2002) used CSLM to quantify water droplet size distributions in fat spreads labelled with Nile Red (see colour Plate VII). One useful effect of labelling with fat dyes in CSLM is that fat crystals can be seen in negative contrast within the stained liquid fat phase, as shown in the confocal micrograph of butter (see colour Plate VIII), which shows characteristic rounded clusters of butterfat crystals. This can be useful when comparing fat crystal networks of churned spreads, such as butter, with margarine-type spreads which has been through a scrape-surface heat exchanger.

CSLM is also useful for characterising interfaces in shear-sensitive food systems such as food foams. An example of this is whipped cream, where dual-labelling reveals that the air bubble interface is stabilised by a mixture of partially coalesced fat globules and protein aggregates (see colour Plate IX).

4.4.6   Dairy: fermented milks

Hassan *et al.* (1995a,b) used reflected light confocal microscopy to visualise protein aggregation in directly acidified and fermented milks. In addition, a pH-sensitive probe, CL-NERF, was used to visualise pH gradients around encapsulated bacteria. The effect of ropy yoghurt cultures on the structure of yoghurt gels was studied by CSLM using Rhodamine B staining and image analysis to quantify pore size and protein aggregates. Bacterial EPS have recently been localised in fermented milk and Feta cheese using fluorescently labelled wheat germ agglutinin and concanavalin A lectins (Hassan, *et al.*, 2002, 2003). The effect of high shear milk processing on sensory and rheological properties yoghurts has been studied by CSLM (Ciron *et al.*, 2012).

4.4.7   Dairy: cheese

Most cheese varieties comprise a visco-elastic protein-continuous phase with dispersed fat globules. CSLM imaging of these two phases is relatively straightforward using a dual-labelling approach (see colour Plate X). Different cheese varieties can have very different microstructures. Processed cheeses tend to have discrete spherical fat globules in a homogeneous protein continuum; Cheddar (and Gouda) type cheeses have a protein continuous phase with curd junctions and irregular, fairly large fat globules or pools; stretched curd cheeses such as Mozzarella have protein fibres with interstitial fat and cream cheeses contain small fat droplets clustered within an aggregated protein matrix (Auty *et al.*, 2001a).

Several CSLM studies at various key stages in cheese manufacture have been published. The microstructure of rennet curd made from non-fat dried milk, standardised to 40 g fat/l, has been studied by CSLM (Hassan and Frank, 1997). The protein phase was visualised by reflectance and the fat was labelled with Nile Red. Several cheese varieties have been studied by CSLM employing dyes added to the cheese surface: these include Gouda (Heertje *et al.*, 1987; Blonk and van Aalst, 1993); Cheddar (Brooker, 1991; Everett *et al.*, 1995; Gunasekaran and Ding, 1999; Guinee *et al.*, 1999, 2000a,b; Everett and Olson, 2003); Feta (Hassan, *et al.*, 2002); Mozzarella (Guinee, *et al.*, 2002; Rowney *et al.*, 2003a,b); and processed cheese (Sutheerawattanononda *et al.*, 1997). Studies using CLSM have examined the fat globule structure in cheese (Gunasekaran and Ding, 1999) milk gelation and cheese melting (Auty *et al.*, 1999), permeability of rennet casein gels (Zhong *et al.*, 2004), the effect of the pasta filata process on fat globule coalescence in Mozzarella cheese (Rowney, *et al.*, 2003b) and Cheddar cheese (Everett and Olson, 2003; Everett *et al.*, 1995), localisation and viability assessment of probiotic bacterial cells (Auty *et al.*, 2001b) and starter cells in cheese (Hannon *et al.*, 2006), location of EPS in cheese (Hassan *et al.*, 2002), and correlation with sensory data of acid milk gels (Pereira *et al.*, 2006). O'Reilly *et al.* (2002) studied the effect of high pressure treatment on Mozzarella cheese. CSLM proved a useful technique to show the increased hydration and swelling of the protein phase. Three-dimensional reconstructions from CSLM image stacks (Fig. 4.4) confirmed the more swollen nature of the protein phase in the 1-day-old HP-treated sample (Fig. 4.4(b)) compared to the control at 1 day (Fig. 4.4(a)).

More recently, Ong *et al.* (2011) used dual-labelling of the fat and protein phases to study the effect of processing on microstructural changes at gel formation, curd and cheddaring stages, and compared CSLM results with cryo-scanning electron micrographs.

4.4.8   Dairy: powders

Confocal microscopy was used to localise fat and phospholipid in whole milk powder (McKenna, 1997). To visualise the fat distribution, milk powder particles were immersed in various mounting media containing Nile blue. Glycerol prevented dissolution of powder particles and enabled localisation of fat globules. Lecithin, pre-labelled with BODIPY 3806, was visualised by CSLM and was located on the surfaces of instantised milk powder agglomerates. By careful formulation of staining mixtures, it is possible to simultaneously visualise both the fat and protein phases of spray dried milk powders *in situ* (Fig. 4.5, Auty *et al.*, 1999).

An alternative to post-labelling powders is to pre-label one of the phases prior to spray drying. Although impractical for large-scale pilot or commercial dryers, this approach has been used to spray dry microcapsules containing fish oils pre-labelled with Nile Red on a laboratory scale using a bench-top dryer (Drusch and Berg, 2008). Recently, the effect of protein content and homogenisation on emulsion droplet size in spray dried infant milk formula was studied using *in situ* powder labelling (McCarthy *et al.*, 2012).

4.5   Model food systems

Whilst the main aim of food microstructure studies is to understand the behaviour of real foods, much published research has, unsurprisingly, been focused on model systems. Model systems have the advantages of being well-defined and that individual components can be specifically labelled prior to mixing. Using model systems and in combination with other rheological and textural analyses, CSLM has become a key tool for understanding and directly visualising structure-function properties such as gelation, phase-separation, foaming and emulsion stability.

4.5.1   Protein gel systems

Bremer *et al.* (1993) used CSLM to demonstrate the fractal nature of particulate casein gels. Sodium caseinate was labelled with FITC or Rhodamine B prior to acidification by glucono-delta-lactone. Schorsch *et al.* (2001) pre-labelled casein-whey mixtures with Rhodamine B prior to the study of gelation by CSLM. Herbert *et al.* (1999) employed multiple labelling using acid Fuschin, Bodipy 665/676 and DM-NERF to label milk proteins, lipid and the aqueous phase, respectively, in acid-coagulated raw milk prior to CSLM imaging. Post-labelling of milk protein

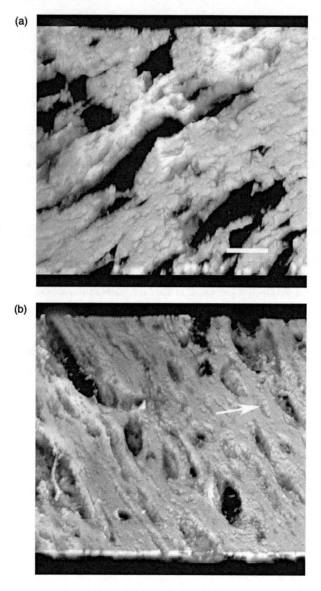

Fig. 4.4 Three-dimensional reconstructions of confocal scanning laser micrographs of control (a) and high-pressure treated (b) 1-day-old low moisture Mozzarella cheese labelled with Fast Green FCF to show protein phase. Note concave impressions of fat globules in the HP-treated sample (arrow). Bar = 25 μm.

gels has also been performed: β-Lactoglobulin gels have been imaged by CSLM following immersion of heat-induced gels in 0.001 wt% FITC (Hagiwara *et al.*, 1997), and Olsson *et al.* (2002) used 0.01% Texas Red to visualise the protein phase of β-lactoglobulin/amylopectin gels by CSLM.

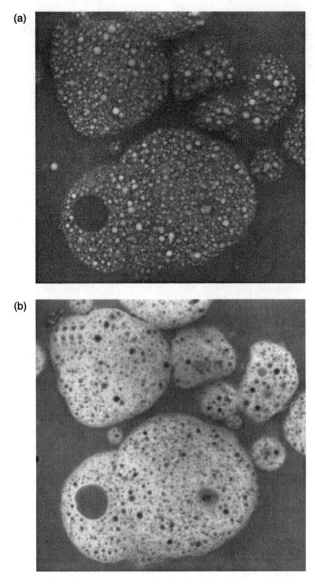

Fig. 4.5 Confocal micrographs of spray dried whole milk powder showing dual fluorescent labelling with Nile Red and Fast Green FCF: (a) 488 nm excitation, fat phase; and (b) 633 nm excitation, protein phase. Bar = 25 μm.

4.5.2   Phase separation studies

Phase separation has been studied by CSLM in several mixed biopolymer systems, including sodium caseinate/sodium alginate (Blonk *et al.*, 1995), dextran/micellar casein and guar gum/micellar casein (Garnier *et al.*, 1998), milk protein/locust bean gum/guar gum (Schorsch *et al.*, 1999; Goff *et al.*, 1999; Bourriot *et al.*,

1999), and colloidal casein/amylopectin (de Bont *et al.*, 2002). In all reported cases, the protein component was labelled with an aqueous solution of fluorochrome. Polysaccharides require covalent labelling (Blonk *et al.*, 1995), which involves treating the sample with organic solvents, dehydration and heating, resulting in some depolymerisation of the polysaccharide molecule (Garnier *et al.*, 1998) (see colour Plate XI).

4.5.3   Emulsion stability
Emulsions are common structural features of many foods and have been the subject of many confocal microscopy studies. Even in the early days, CSLM was used to study emulsifier displacement at the oil-water interface (Heertje *et al.*, 1990, 1996). The displacement of sodium caseinate, covalently labelled with FITC, by monoacylglycerols, was observed and quantified by measuring emitted fluorescence intensity.

4.5.4   Human-food interface
An exciting new area of microstructural research is concerned with the human-food interface. In particular, research is focusing on the matrix effect of food on nutrient bioavailability and the microstructural design of new delivery vehicles, generally based on encapsulation technology and innovative biopolymer design (McClements *et al.*, 2007; Jones *et al.*, 2010). The interaction of foods with the palate, tongue and the transformations that occur during gastric transit is now a 'hot topic' and the confocal microscope is becoming a key tool to visualise these dynamic and complex events. The role of saliva in emulsion breakdown and the effect of oral tissue on foods have been studied using CSLM (Dresselhuis *et al.*, 2008a,b). Li *et al.* (2012) studied the digestion of various emulsions as potential delivery systems both *in vitro* and *in vivo* using CSLM. The interaction of milk stabilised emulsions with bile salts has also been studied (Sarkar *et al.*, 2010).

4.6   Reflectance confocal microscopy
Reflectance confocal microscopy is a rarely used imaging mode whereby the emission wavelength is set to overlap the excitation beam wavelength and a reflected laser signal is detected. The advantage of this approach is that no fluorescent dyes are needed, although optical interference effects can occur, requiring careful interpretation. Reflected confocal has been used to visualise casein in fermented milks (Hassan *et al.*, 1995a). Reflectance mode is particularly useful for characterising the surface of solid foods and has been used to study texture differences in sliced ham, cheese and salami (Sheen *et al.*, 2008), fat bloom in chocolate (Pedreschi *et al.*, 2002), and highlighting defects in food packaging material and food contact surfaces (Flores *et al.*, 2006).

4.7   Image processing and analysis

Single confocal images are effectively 2D with high contrast, making them amenable to image analysis. Basic measurements such as relative phase area, particle size and shape are fairly straightforward using basic image analysis routines (see below). Emulsion droplet size is an important factor controlling the quality and stability of a range of food products including dairy spreads, spray dried milk powders and cheese. Generic fat and/or protein labelling results in clear edge discrimination of phases that are suitable for measuring. An example is shown in Fig. 4.6. The primary image (also see colour Plate XII) is a dual-labelled Cheddar cheese sample showing fat globules (green) and the continuous protein phase (red). The green channel is separated and a simple thresholding procedure gives a binary image (Fig. 4.6(b)). After binary inversion (Fig. 4.6(c)), erosion and dilation filters help to separate touching objects and help reduce noise (Fig. 4.6(d)). Once the fat globules have been discriminated, the equivalent circular diameter of all the fat globules in the image can be measured and plotted as a histogram (Fig. 4.6(e)).

It should be noted that image analysis using this approach is number-based, whereas particle size measurement by laser scattering is volume-based, plotted on a logarithmic scale. Although this method gives a good approximation and is useful for comparative purposes, another factor to consider is that the optical plane may not pass through the centre of a fat globule. Everett and Olson (2003) used CSLM with image analysis to measure fat globule diameter and circularity in Cheddar cheese containing fat globules stabilised with different proteins. These researchers calculated the true globule diameter by applying a correction factor of $1/\sqrt{3}$ to the apparent diameter of fat globules as imaged by CSLM, to account for the fact that a random section through the plane of a sphere may not pass through its equator and thus the apparent size will be smaller than the true diameter. The correction factor assumes that all fat globules are spheres and that the images are 2D. In reality, fat globules in cheese are rarely perfectly circular and also CSLM optical sections have a minimum axial resolution of 1 μm, due to the point spread function of the beam.

Another consideration is that a random section will contain a relatively greater number of larger droplets than smaller ones (compared to the true distribution), because the section plane will more frequently intersect the former. Thus, the apparent size will be greater than the true diameter. Ko and Gunasekaran (2007) have thoroughly evaluated the various errors that can occur when quantifying microstructure of food with CSLM. An alternative approach to the measurement of fat globule size is to use a statistical method, such as shape unfolding, otherwise called stereology, which describes the size distribution of many particles rather than measurement of the apparent diameter of individual particles. Van Dalen (2002) used a modification of the Wicksell transform, which is a shape unfolding numerical method characterising the distribution of planar intercepts of zero thickness through a sphere. A simpler stereological approach is to use the line intercept measurement of 2D binary images to calculate the volume-weighted

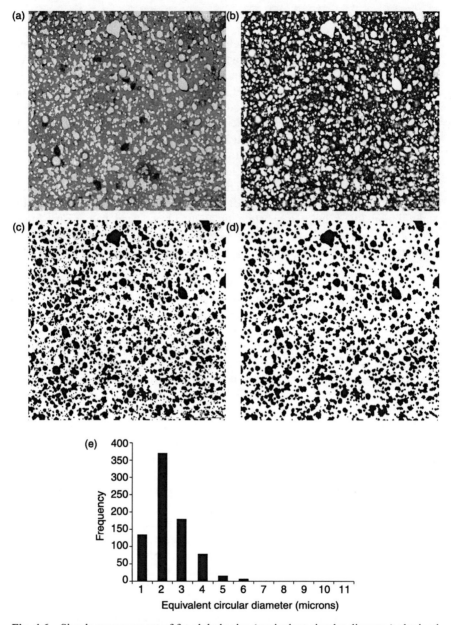

Fig. 4.6   Simple measurement of fat globule size (equivalent circular diameter) obtained from confocal micrograph of Cheddar cheese dual-labelled to show fat (green) and protein (red). Panel (a) appears in colour as Plate XII in the colour section between pages 222 and 223. Typical image processing steps are: (b) binarization of the green channel; (c) binary inversion (black objects will be measured); (d) opening filter (erosion and dilation) to separate touching droplets and remove very small objects; and (e) histogram of equivalent circular diameter size distribution.

mean volume of particles (Russ, 2006). This technique assumes that samples are isotropic, uniform and randomly sampled (a big assumption!), but does not require particles to be spherical, is statistically rigorous and has been used to measure whey protein particle size and pore star volume from TEM images (Langton and Hermansson, 1996) and star volume of pores in acidified casein gels (Auty *et al.*, 2005b). More complex structural parameters such as texture require more sophisticated approaches and these have been reviewed by Zheng *et al.* (2006).

High-order statistical analysis (co-occurrance and run length matrices) and grey-scale mathematical morphology were used to characterise texture from confocal micrographs of cream cheeses (Fenoul *et al.*, 2008). Another image processing approach is that of deconvolution. This is based on the fact that the emission signal from a sub-resolution point is blurred or 'convolved' by the various components of the optical train. Image restoration can be achieved by dividing the Fourier transform of the image by that of the point spread function, which can be calculated from the NA of the objective, incident and emitted wavelengths and refractive index of the medium (Cannell *et al.*, 2006). Since single photon CSLMs have a relatively poor axial resolution, this technique is particularly useful for restoring 3D image stacks. Although not yet applied to food systems, this approach could be useful for resolving fine detail of protein network systems or interfacial films.

For more details on some of these techniques, as well as practical help on how to get meaningful data from images of food materials, the reader is referred to Russ (2005).

4.8   Time dependent studies: dynamic confocal microscopy

4.8.1   General comments

Food manufacture is not a static process and many foods undergo several transformations during manufacture, storage and consumption. These processes often include mixing, heating and pH adjustment, as well as complex biochemical transformations such as fermentation. There is a need to effectively model these transformations to give insight into how food structures are formed and deformed. It should be remembered that foods are designed to be unstable! Environmental effects such as temperature can be effectively modelled using temperature-controlled heating and cooling stages that can be purchased from specialist suppliers such as Linkam Scientific (Surrey, UK). Heuer *et al.* (2007) used a pressure stage to study bubble size distribution and coalescence by direct CSLM imaging. In order to understand how the different food processing stages affect the microstructure and ultimately behaviour of a food product, confocal images taken at key process steps can be useful, and combining CSLM with rheological measurement is a common research approach (Auty *et al.*, 2005b; Manski *et al.*, 2007). However, these 'snapshots' cannot show what is happening during a fast-moving process such as acid gelation or during breakdown of a product under shear, and dynamic imaging techniques are needed. This has led to the development of new hybrid configurations combining shear/oscillatory rheometry and biaxial compression with a confocal microscope to facilitate real-time imaging during small and large deformation measurement (Nicolas *et al.*, 2003). Two examples of dynamic CSLM are given below.

4.8.2   Structure formation: CSLM monitoring of acid gelation in milk

An early example of dynamic CSLM was the direct monitoring of acid gelation milk using time lapse CSLM imaging (Auty *et al.*, 1999). Using this technique, it was possible to follow the particle movement, aggregation and subsequent network formation of casein micelles during real-time acidification by glucono-delta-lactone at a controlled temperature (Fig. 4.7).

The progress of gelation could be followed rheologically using low amplitude oscillatory rheometry and images mapped to changes in the storage modulus (Fig. 4.8) CSLM analysis of skim milk showed that quiescent acidification resulted in three main stages:

1. initiation of aggregation of casein micelles at pH 5.58 accompanied by an increase in viscosity;
2. the onset of gelation as reflected by a reduction in protein mobility at pH 5.48 and increase in storage modulus;
3. gradual formation and consolidation of a 3D gel network as the pH decreased below 5.48 and a linear increase in storage modulus.

From 1 to 20 minutes (pH 6.51–5.64) after GDL addition, small particles of protein, less than 1 μm in diameter, were observed to move rapidly in random directions, presumably due to Brownian motion and/or small, localised convection currents. After 30 minutes, the pH had dropped to 5.58 and aggregation of milk

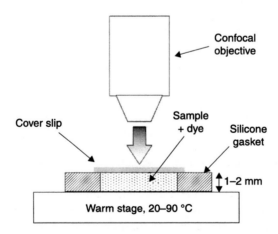

Fig. 4.7   Diagram showing warm-stage modified for dynamic CSLM studies of milk gelation.

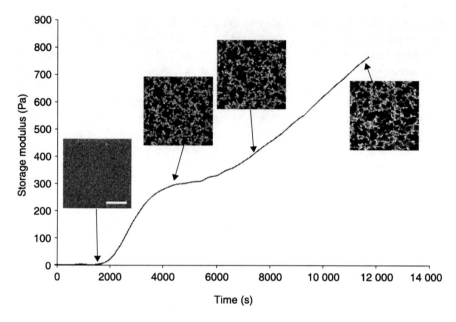

Fig. 4.8   Acid-induced gelation of skim milk, showing storage modulus (G′) as a function of time. Confocal images of the protein-labelled milk highlight the onset of gelation and subsequent network formation and consolidation. Bar = 25 μm.

proteins into small (<2 μm) particles was evident. The onset of gelation and the formation of a visible protein network occurred at a pH of approximately 5.4, at which point much of the stained protein became immobilised. Time-lapse animation of the CSLM image sequence enabled dynamic visualisation of the gelation process. Further reduction in pH to 5.00 was accompanied by an increase in the staining intensity of the network. Simultaneously, the aqueous phase became darker, indicating a decrease in staining intensity, suggesting further consolidation of proteins onto the initial network. Whilst most of the protein appeared to be immobile at a pH of less than 5.4, although time-lapse animation revealed that some protein strands attached to the network appeared to flex (Fig. 4.8).

4.8.3   Structure deformation: fracture behaviour of filled gels
In addition to structure formation described above, it should be remembered that food is designed to break down in the mouth and the fracture behaviour of food has been little studied using CSLM. Microtensile stages are now commercially available from Deben Ltd (UK) and these can be useful for studying deformation of solid food materials at the microscopic scale. Brink *et al.* (2007) used a tensile stage in combination with a confocal laser scanning microscope to study the large deformation and fracture properties of whey protein/polysaccharide gels. Uniaxial

compression and fracture analysis of food composites and their relationship with sensory perception has also been studied by van den Berg *et al.* (2007, 2008).

Tensile testing using fluorescent particles embedded in zein films has also been monitored dynamically with CSLM (Emmambux and Stading, 2007). Abhyankar *et al.* (2011b) employed a standard failure analysis technique called notch propagation to monitor the fracture of filled whey protein gels. A microtensile stage (Deben Ltd, UK) was fitted to an upright confocal microscope and the progress of a fracture through a heat-induced whey protein gel filled with sunflower oil droplets was monitored in real time both visually and through a force transducer (2N load cell). The effect of the gel type (coarse and particular vs fine stranded) on the fracture properties and release of fat was studied. Selected images from the CLSM image time series taken during tensile stretching of the filled gel prepared at pH 7.0 and at 5.4, all containing 16% WP in the aqueous phase and 7% sunflower oil in the overall emulsion filled gel, are shown in Plate XIII (in colour section between pages 222 and 223). On stretching the gel prepared at pH 5.4, the oil droplets were disconnected from the protein matrix (light blue arrows in images).

Panels (a), (b) and (c) in Plate XIII show disconnected oil droplets, and formation of oil pools was observed due to oil droplet coalescence/breakage, while the notch propagated in a zig-zag manner due to the particulate nature of the gel matrix (panels (c) and (d)). It was found that the fracture plane in the coarse gel went around fat droplets leaving them intact. By contrast, the fracture plane in the fine gel (pH 7.0) propagated through the fat droplets, since they were tightly bonded to the continuous protein phase. Oil droplets in the gel prepared at pH 7.0 deformed from a spherical to an elliptical shape on stretching (panels (e) and (f)), indicating that the oil droplets were bound to the protein gel matrix and behaved as an integral part of the matrix. Notch propagation in the emulsion filled gel occurred cleanly either through the gel matrix (f) or through the oil droplet (panels (g) and (h)). Time-lapse animations graphically show how the gel breaks and the mobility of the fat phase. This type of study highlights the importance of microstructure on the fat release properties of food and how new reduced-fat products can be designed with defined fat release and sensory properties.

During stretching, the stress in the gels increased linearly with increasing tensile strain and at a critical strain the stress dropped drastically, indicating the fracture of the gels through the notch. The WP emulsion filled gel prepared at pH 5.4 had a higher Young's modulus (higher slope of the stress vs. strain curve), higher critical stress at fracture and a lower fracture strain than that of the WP emulsion filled gel prepared at pH 7.0 (Fig. 4.9).

4.9   Future trends

A significant development in confocal technology has been the introduction of dual- and multi-photon excitation (Denk *et al.*, 1995). In two-photon fluorescence

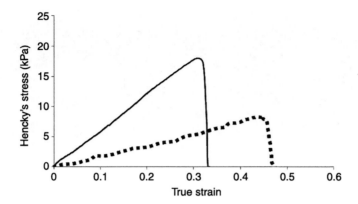

Fig. 4.9   Hencky's strain dependence of the true stress on tensile stretching at a rate of 1.5 mm.min^{-1} of emulsion-filled whey protein gels prepared at pH 5.4 (---) and pH 7.0 (—) (Abhyankar and Auty, unpublished data).

microscopy, dye molecules are excited by the simultaneous absorption of two photons of longer wavelength. Advantages of multi-photon technology are that signal penetration depth increases considerably (>500 µm) and photobleaching is significantly reduced (Sheppard and Shotton, 1997). However, the lower energy of the longer wavelength excitation beam renders it prone to scattering by concentrated or refractile structures, which unfortunately includes many food materials, limiting its applications. Other developments, such as use of synchrotrons as possible light sources for ultra-violet high resolution imaging (van der Oord *et al.*, 1996) or infra-red microspectroscopy (Murdock *et al.*, 2010) have been developed for biological systems and may yet find an application in foods.

4.9.1   Molecular diffusion measurement: FRAP techniques

The study of molecular diffusion is opening up a new area of research for confocal microscopy. The need to protect sensitive bioactives during processing and the migration of solutes and water mobility are important factors affecting functionality. Fluorescence recovery after photobleaching (FRAP) and related techniques such as fluorescence resonance energy transfer (FRET) are useful chemometric techniques used in the life sciences, which could be more widely applied to food systems. FRAP can be a useful tool for measuring local diffusion in biopolymer systems. A high intensity excitation beam is focused on a spot or region of a fluorescently labelled sample and this inactivates the fluorescent molecules – this is called photobleaching. However, after bleaching, unbleached molecules from the surrounding area begin to diffuse into the bleached region and bleached molecules diffuse out, thereby increasing the fluorescence intensity of the previously bleached region (see colour Plate XIV and Fig. 4.10). By measuring the rate of fluorescence recovery, the diffusion constant of a localised area of a

(a)

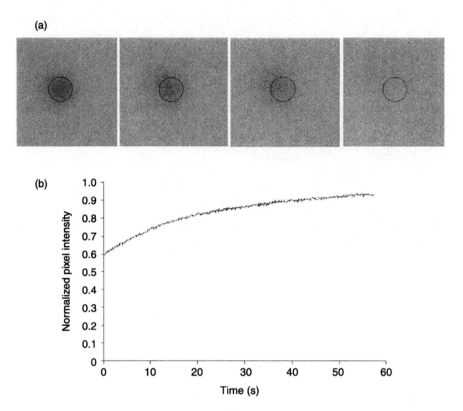

Fig. 4.10 Fluorescence recovery after photobleaching (FRAP). (a) Confocal images of bleached (Ar laser, 10 s, 100 mW) region of interest (ROI, dark circle), in a whey protein gel containing 4 kDa FITC-dextrans, showing gradual fluorescence recovery: panel (a) appears in colour as Plate XIV in the colour section between pages 222 and 223; and (b) fluorescence recovery curve of bleached area (ROI), courtesy of Mamdhou El-Bakry, Teagasc.

heterogeneous biomaterial can be calculated (Loren *et al.*, 2009). This approach has recently been applied to a model cheese system using FITC-dextrans of known molecular weights to calculate the diffusion coefficient of the protein gel (Floury *et al.*, 2012).

4.9.2   New lasers

Laser technology is rapidly developing and there is a move away from gas lasers towards using solid state lasers, as these have more stable emission outputs and longer lifetimes, whilst being relatively trouble free. The rapid development of non-laser light sources, for example with new inexpensive light emitting diodes with higher light output and configured as microarrays, could eventually be used in all fluorescence applications (Nolte *et al.*, 2006). The increasing number of laser excitation lines and newer fluorochromes from companies such as the Invitrogen Corporation (formerly *Molecular Probes*) now offer food scientists a wide range of options for multiple labelling of complex and heterogeneous food materials.

4.9.3   Breaking the diffraction barrier

There have been several successful attempts to break the diffraction limit of optical microscopes and these include linear methods such as 4Pi, I5 and structured illumination microscopy, and non-linear approaches such as stimulated emission depletion (STED) and saturated structured illumination. Localisation approaches (e.g. PALM and STORM), near-field methods and total internal refraction microscopy have also been developed for biological systems (for a short review, see Heinztmann and Ficz, 2012). Few of these techniques have yet been applied to actual food products and it seems likely that most will be used initially to study food models. One of the most promising of these developments is STED microscopy. This involves the concept of reversible saturable optical fluorescence transitions, developed by Hell and Wichman (1994). STED microscopy involves the use of an 'excitation laser' and a 'stimulation laser' surrounding the excitation laser beam for microscopic imaging. The excitation laser excites the fluorescent dye molecules at a focused point in the test specimen, while the stimulation laser quenches the excited dye molecules in the periphery of the focused point in the test specimen to their ground state energy level using the phenomenon of stimulated emission.

Quenching of the excited dye molecules using the stimulation laser reduces the unwanted background fluorescence during microscopic imaging by stimulating the excited dye molecules to release photons of similar wavelength to that of a stimulation laser, which can be separated from the in-focus fluorescence using spectral filters. This reduction in the background fluorescence results in an axial resolution of 30 to 40 nm resolution, well beyond the diffraction limit of optical microscopes.

4.9.4   Confocal Raman: the future of chemical mapping?

Spectroscopic analysis of microstructures *in situ* has been the ideal for many years and there have been several attempts to combine spectrometers with conventional optical microscopes, notably recently where Wellner *et al.* (2011) developed an *in situ* Raman microscope to study starch granule structure. This approach is non-destructive and has the great advantage of not requiring staining and the problems associated with it. The most promising new technique is Raman confocal microscopy. This was first applied to multiphase food materials by Pudney *et al.* (2004), where a Raman spectrometer at 785 nm excitation wavelength was coupled to a light microscope fitted with a pinhole to allow the confocal signal to be collected as a confocal plane. This instrument had a notional spatial resolution of 1.9 µm in the x-y plane and 2.9 µm in the z-plane, although this was seldom achieved for real food samples. Phase separated gellan/κ-carrageenan mixtures were distinguished based on their Raman scattering spectra using a statistical self-modelling curve. The Raman spectra of other biopolymers, such as β-lactoglobulin and pectin, as well as fat, were also characterised. Early problems included slow acquisition speeds (up to 100 sec per image) and relatively poor spatial resolution (>5 µm), although these have been improved upon. Confocal Raman microscopy has been used to analyse the composition of milk fat droplets. Gallier *et al.* (2011) demonstrated that is was possible to distinguish between triglycerides and carotenoids, as well as the milk fat globule membrane composition, based on the Raman peaks. Recent improvements in spectral acquisition rates, data processing and resolution have resulted in new commercial confocal Raman systems such as the Witech Alpha 300R, which shows great promise for chemical mapping of complex multiphase foods and emulsions (see Plate XV in colour section between pages 222 and 223.

4.10   Conclusion

Confocal microscopy has become an essential tool for understanding the structure–function relationships of food materials. A key advantage of the technique is that real food products can be examined directly and the distribution of the main ingredients quickly determined. New methods for monitoring dynamic events and breakthroughs in technology, particularly confocal spectroscopy and nanoscale resolution, ensure that food microstructure research has an exciting future.

4.11   Sources of further information and advice

4.11.1   Buying a confocal microscope for food applications

For the fortunate new purchaser, there are several well-established manufacturers making excellent confocal microscopes; these include Leica Microsystems (Mannheim), Carl Zeiss, Olympus and Nikon. Even basic configurations from any of these suppliers can be used effectively to study food materials. When purchasing a new confocal for food research, the main applications must be identified and flexibility should be prioritised. Features such as the number of lasers, separate confocal channels and fast acquisition rates may be useful, but will add significantly to the cost, particularly if fitted retrospectively. As always, buyers should insist on using their own samples for instrument demonstrations to enable clear comparisons between different models. For most food applications, a single photon confocal with a minimum of two detection channels and three lasers to include blue, green and red emission lines is perfectly adequate. Acquisition speed can be increased by reducing image size and increasing scan rate and there is always the option of line or point scanning. If speed is the most important criterion, then a tandem scanning or Yokogawa disk-scanning instrument should be considered. It should also be remembered that optical microscopes exist in two basic configurations: upright or inverted. Most confocal microscopes in biology laboratories are inverted, that is, the objective is underneath the sample. Inverted microscopes can be used perfectly well for food analysis, although upright configurations are more flexible and allow for heating, tensile or shearing stages to be fitted easily onto the existing microscope stage assembly.

4.11.2   Books and software

For more in-depth study of confocal microscopy, the *Handbook of Biological Confocal Microscopy*, 4th edition by J. B. Pawley is essential reading and covers much of the theoretical basis of confocal imaging, optimising its use and hardware innovations, as well as chapters on more advanced labelling techniques and 3D image reconstruction and analysis. For image analysis of food materials, *Image Analysis of Food Microstructure* by John Russ is indispensable. This book has many examples of how to derive numerical data from a wide range of real food products and is based on the same author's definitive work for image processing: *The Image Processing Handbook*. Regarding software for image analysis, there is a wide range from commercial high end software such as Image Pro Plus or Imaris through to freely available image analysis algorithms, the most popular for biological applications being ImageJ (*http://rsb.info.nih.gov/ij/*). Advantages of the commercial programs are that they are easier to use for non-programmers and repetitive processing routines can be automated relatively easily. The advantage of the Java-based ImageJ is that it is free, has an active user community and is constantly updated with new plug-ins. However, ImageJ does require some programming or at least scripting skills for routine handling of large datasets. A cost-effective compromise would be Fovea Pro (Reindeer Graphics, *www.reindeergraphics.com*), which are a set of image processing and analysis plug-ins for common graphics packages such as Adobe Photoshop.

4.12   References

ABBE, E. (1884), Note on the proper definition of the masgnifying power of a lens or a lens-system, *Journal of the Royal Microscopical Society*, **4**, 348–51.
ABHYANKAR, A. R., MULVIHILL, D. M., CHAURIN, V. and AUTY, M. A. E. (2011a), Techniques for localization of Konjac glucomannan in model milk protein-polysaccharide mixed systems: physico-chemical and microscopic investigations, *Food Chemistry*, **129**, 1362–8.

ABHYANKAR, A. R., MULVIHILL, D. M., and AUTY, M. A. E. (2011b), Combined microscopic and dynamic rheological methods for studying the structural breakdown properties of whey protein gels and emulsion filled gels, *Food Hydrocolloids*, **25**, 275–82.

AGUILERA, J. M. and STANLEY, D. W. (1990), *Microstructural Principles of Food Processing and Engineering*, Essex, UK, Elsevier Science, Ltd., 47.

ALTAN, A., MCCARTHY, K. L., TIKEKAR, R., MCCARTHY, M. J. and NITIN, N. (2011), Image analysis of microstructural changes in almond cotyledon as a result of processing, *Journal of Food Science*, **76**, 212–21.

ALVAREZ-JUBETE, L., AUTY, M. A. E., ARENDT, K. and GALLAGHER, E. (2010), Baking properties and microstructure of pseudocereal flours in gluten-free bread formulations, *European Food Research and Technology*, **230**, 437–45.

ALVIOLA, J. N., WANISKA, R. D. and ROONEY, L. W. (2008), Role of gluten in flour tortilla staling, *Cereal Chemistry*, **85**, 295–300.

APINAN, S., YUJIRO, I., HIDEFUMI, Y., TAKESHI, F., MYLLAERINEN, P. *et al.* (2007), Visual observation of hydrolyzed potato starch granules by alpha-amylase with confocal laser scanning microscopy, *Starch–Starke*, **59**, 543–8.

ARLTOFT, D., MADSEN, F. and IPSEN, R. (2007), Screening of probes for specific localisation of polysaccharides, *Food Hydrocolloids*, **21**, 1062–71.

AUTY, M. A. E., FENELON, M. A., GUINEE, T. P., MULLINS, C. and MULVIHILL, D. M. (1999), Dynamic confocal scanning laser microscopy methods for studying milk protein gelation and cheese melting, *Scanning*, **21**, 299–304.

AUTY, M. A. E., TWOMEY, M., GUINEE, T. P. and MULVIHILL, D. M. (2001a), Development and applications of confocal scanning laser microscopy methods for studying the distribution of fat and protein in selected food products, *Journal of Dairy Research*, **68**, 417–27.

AUTY, M. A. E., GARDINER, G., MCBREARTY, S., O'SULLIVAN, E., MULVIHILL, D. M. *et al.* (2001b), Direct *in situ* viability assessment of bacteria in probiotic products using viability staining in conjunction with confocal scanning laser microscopy, *Applied and Environmental Microbiology*, **67**, 420–5.

AUTY, M. A. E., DUFFY, G., O'BEIRNE, D., MCGOVERN, A., GLEESON, E. and JORDAN, K. (2005a), *In situ* localisation of *Escherichia coli* O157:H7 in food by confocal scanning laser microscopy, *Journal of Food Protection*, **68**: 482–6.

AUTY, M. A. E., O'KENNEDY, B. T., ALLAN-WOJTAS, P. and MULVIHILL, D. M. (2005b), The application of microscopy and rheology to study the effect of milk salt concentration on the structure of acidified micellar casein systems, *Food Hydrocolloids*, **19**, 101–9.

BELDER, A. N. D. and GRANATH, K. (1973), Preparation and properties of fluorescein-labeled dextrans, *Carbohydrate Research*, **30**, 375–8.

BERTRAM, H. C., WU, Z., STRAADT, I. K., AAGAARD, M. and AASLYNG, M. D. (2006), Effects of pressurization on structure, water distribution, and sensory attributes of cured ham: can pressurization reduce the crucial sodium content? *Journal of Agricultural and Food Chemistry*, **54**, 9912–17.

BLONK, J. C. G. and VAN AALST, H. (1993), Confocal scanning light microscopy in food research, *Food Research International*, **26**, 297–311.

BLONK, J. C. G., VAN EENDENBURG, J., KONING, M. M. G., WEISENBORN, P. C. M. and WINKEL, C. (1995), A new CSLM-based method for determination of the phase behaviour of aqueous mixtures of biopolymers, *Carbohydrate Polymers*, **28**, 287–95.

BORDOLOI, A., SINGH, J. and KAUR, L. (2012), *In vitro* digestibility of starch in cooked potatoes as affected by guar gum: microstructural and rheological characteristics, *Food Chemistry*, **133**, 1206–13.

BORN, M. and WOLF, E. (1980), *Principles of Optics*, 6th edition, Oxford, Pergamon Press.

BOUCHON, P. and AGUILERA, J. M. (2001), Microstructural analysis of frying potatoes, *International Journal of Food Science and Technology*, **36**, 1–8.

BOURRIOT, S., GARNIER, C. and DOUBLIER, J-L. (1999), Phase separation, rheology and structure of micellar casein-galactomannan mixtures, *International Dairy Journal*, **9**, 353–7.

BRAKENHOFF, G. J., VAN DER VOORT, H. T. M., VAN SPRONSEN, E. A. and NANNINGA, N. (1988), Three-dimensional imaging of biological structures by high resolution confocal scanning laser microscopy, *Scanning Microscopy*, **2**, 33–40.

BREMER, L. G. B., BIJSTERBOSCH, B. H., SCHRIJVERS, R., VAN VLIET, T. and WALSTRA, P. (1993), Formation, properties and fractal nature of particle gels, *Advances in Colloid and Interface Science*, **46**, 117–28.

BRENNER, T., NICOLAI, T. and JOHANNSSON, R. (2009), Rheology of thermo-reversible fish protein isolate gels, *Food Research International*, **42**, 915–24.

BRINK, J., LANGTON, M., STADING, M. and HERMANSSON, A-M. (2007), Simultaneous analysis of the structural and mechanical changes during large deformation of whey protein isolate/ gelatin gels at the macro and micro levels, *Food Hydrocolloids*, **21**, 409–19.

BROOKER, B. E. (1991), The study of food systems using confocal laser scanning microscopy, *Microscopy and Analysis*, **27**, 13–15.

BROOKER, B. E. (1995), Imaging food systems by confocal scanning laser microscopy, in, *New Physico-chemical Techniques for the Characterization of Complex Food Systems*, E. Dickinson (ed.), London, Blackie Academic and Professional, 53–68.

BROOKS, S. A., LEATHEM, A. J. C. and SCHUMACHER, U. (1997), *Lectin Histochemistry*, Oxford, BIOS Scientific Publishers Limited.

BURNETT, S. L., CHEN, J. and BEUCHAT, L. R. (2000), Attachment of *Escherichia coli* O157:H7 to the surfaces and internal structures of apples as detected by confocal scanning laser microscopy, *Applied and Environmental Microbiology*, **66**, 4679–87.

CANNELL, M. B., MCMORLAND, A. and SOELLER, C. (2006), Image enhancement by deconvolution, in, *Handbook of Biological Confocal Microscopy*, 3rd edition, J. B. Pawley (ed.), New York, Springer Science+Business Media, 488–500.

CHATTONG, U., APICHARTSRANGKOON, A. and BELL, A. E. (2007). Effects of hydrocolloid addition and high pressure processing on the rheological properties and microstructure of a commercial ostrich meat product 'Yor' (Thai sausage), *Meat Science*, **76**, 548–54.

CHEN, P., YU, L., SIMON, G., PETINAKIS, E., DEAN, K. and CHEN, L. (2009), Morphologies and microstructures of cornstarches with different amylose-amylopectin ratios studied by confocal laser scanning microscope, *Journal of Cereal Science*, **50**, 241–7.

CIRON, C. I. E., KELLY, A. L. and AUTY, M. A. E. (2012), Modifying the microstructure of low-fat yoghurt by microfluidization of milk under different pressures to enhance rheological and sensory properties, *Food Chemistry*, **130**, 510–19.

CLEGG, S. M., MOORE, A. K. and JONES, S. A. (1996), Low-fat margarine spreads as affected by aqueous phase hydrocolloids, *Journal of Food Science*, **61**, 1073–9.

DE BONT, P. W., VAN KEMPEN, G. M. P. and VREEKER, R. (2002), Phase separation in milk protein and amylopectin mixtures, *Food Hydrocolloids*, **16**, 127–38.

DELAQUIS, P. J., GARIÉPY, C. and MONTPETIT, D. (1992), Confocal scanning laser microscopy of porcine muscle colonized by meat spoilage bacteria, *Food Microbiology*, **9**, 147–53.

DENK, W., PISTON, D. W. and WEBB, W. W. (1995), Two-photon molecular excitation in laser-scanning microscopy, in, *Handbook of Biological Confocal Microscopy*, 2nd edition, J. B. Pawley (ed.), New York, Plenum Press, 445–58.

DRESSELHUIS, D. M., DE HOOG, E. H. A., COHEN, S. M. A. and VAN AKEN, G. A. (2008a), Application of oral tissue in tribological measurements in an emulsion perception context, *Food Hydrocolloids*, **22**, 323–35.

DRESSELHUIS, D. M., STUART, M. A. C., VAN AKEN, G. A., SCHIPPER, R. G. and DE HOOG, E. H. A. (2008b), Fat retention at the tongue and the role of saliva: adhesion and spreading of 'protein-poor' versus 'protein-rich' emulsions, *Journal of Colloid and Interface Science*, **321**, 21–9.

DRUSCH, S. and BERG, S. (2008), Extractable oil in microcapsules prepared by spray-drying: localisation, determination and impact on oxidative stability, *Food Chemistry*, **109**, 17–24.

DÜRRENGBERGER, M. B., HANDSCHIN, S., CONDE-PETIT, B. and ESCHER, F. (2001), Visualisation of food structure by confocal scanning laser microsopy, *Lebensmittel Wissenschaft und Technologie*, **34**, 11–17.

EMMAMBUX, M. N. and STADING, M. (2007), *In situ* tensile deformation of zein films with plasticizers and filler materials, *Food Hydrocolloids*, **21**, 1245–55.

EVERETT, D. W. and OLSON, N. F. (2003), Free oil and rheology of Cheddar cheese containing fat globules stabilised with different proteins, *Journal of Dairy Science*, **86**, 755–63.

EVERETT, D. W., DING, K., OLSON, N. F. and GUNASEKARAN, S. (1995), Applications of confocal microscopy to fat globule structure in cheese, in, *Chemistry of Structure-Function Relationships in Cheese*, E. L. Malin and M. H. Tunick (eds), New York, Plenum Press, 321–30.

FENOUL, F., LE DENMAT, M., HAMDI, F., CUVELIER, G. and MICHON, C. (2008), Technical note: Confocal scanning laser microscopy and quantitative image analysis: application to cream cheese microstructure investigation, *Journal of Dairy Science*, **91**, 1325–33.

FERRANDO, M. and SPEISS, W. E. L. (2000), Review: Confocal scanning laser microscopy: a powerful tool in food science, *Food Science and Technology International*, **6**, 267–84.

FLINT, O. (1994), *Food Microscopy*, Oxford, Bios Scientific Publishers Ltd.

FLORES, R. A., TAMPLIN, M. L., MARMER, B. S., PHILLIPS, J. G. and COOKE, P. H. (2006), Transfer coefficient models for *Escherichia coli* O157:H7 on contacts between beef tissue and high density polyethylene surfaces, *Journal of Food Protection*, **69**, 1248–55.

FLOURY, J., MADEC, M-N., WAHARTE, F., JEANSON, S. and LORTAL, S. (2012), First assessment of diffusion coefficients in model cheese by fluorescence recovery after photobleaching (FRAP), *Food Chemistry*, **133**, 551–6.

GALLIER, S., GORDON, K. C., JIMENEZ-FLORES, R. and EVERETT, D. W. (2011), Composition of bovine milk fat globules by confocal Raman microscopy, *International Dairy Journal*, **21**, 402–12.

GARNIER, C., BOURRIOT, S. and DOUBLIER, J-L. (1998), The use of confocal laser scanning microscopy in studying mixed biopolymer systems, in, *Gums and Stabilisers in the Food Industry 9; 9th Conference*, Cambridge, UK, Royal Society of Chemistry, 247–56.

GOFF, H. D., FERDINANDO, D. and SCHORSCH, C. (1999), Fluorescence microscopy to study galactomannan structure in frozen sucrose and milk protein solutions, *Food Hydrocolloids*, **13**, 353–62.

GUINEE, T. P., AUTY, M. A. E. and MULLINS, C. (1999), Observations on the microstructure and heat-induced changes in the viscoelasticity of commercial cheeses, *Australian Journal of Dairy Technology*, **54**, 84–9.

GUINEE, T. P., AUTY, M. A. E. and FENELON, M. A. (2000a), The effect of fat content on the rheology, microstructure and heat-induced functional characteristics of Cheddar cheese, *International Dairy Journal*, **10**, 277–88.

GUINEE, T. P., AUTY, M. A. E., MULLINS, C., CORCORAN, M. O. and MULHOLLAND, E. O. (2000b), Preliminary observations on the effects of fat content and degree of emulsification on the structure-functional relationship of Cheddar cheese, *Journal of Texture Studies*, **31**, 645–63.

GUINEE, T. P., FEENEY, E. P., AUTY, M. A. E. and FOX, P. F. (2002), Effect of pH and calcium concentration on some textural and functional properties of Mozzarella cheese, *Journal of Dairy Science*, **85**, 1–15.

GUNASEKARAN, S. and DING, K. (1999), Three-dimensional characteristics of fat globules in Cheddar cheese, *Journal of Dairy Science*, **82**, 1890–6.

HAGIWARA, T., KUMAGAI, H., MATSUNAGA, T. and NAKAMURA, K. (1997), Analysis of aggregate structure in food protein gels with the concept of fractal, *Bioscience, Biotechnology, Biochemistry*, **61**, 1663–7.

HANNON, J. A., LOPEZ, C., MADEC, M-N. and LORTAL, S. (2006), Altering pH of renneting changes cell distribution, microstructure and level of lysis of *Lactococcus lactis* AM2 in cheese made from milk concentrated by ultrafiltration (UF-cheese), *Journal of Dairy Science*, **89**, 812–23.

HASSAN, A. N. and FRANK, J. F. (1997), Modification of microstructure and texture of rennet curd by using a capsule-forming non-ropy lactic culture, *Journal of Dairy Research*, **64**, 115–21.

HASSAN, A. N., FRANK, J. F., FARMER, M. A., SCHMIDT, K. A. and SHALABI, S. I. (1995a), Observation of encapsulated lactic acid bacteria using confocal scanning laser microscopy, *Journal of Dairy Science*, **78**, 2624–8.

HASSAN, A. N., FRANK, J. F., FARMER, M. A., SCHMIDT, K. A. and SHALABI, S. I. (1995b), Formation of yoghurt microstructure and three-dimensional visualisation as determined by confocal scanning laser microscopy, *Journal of Dairy Science*, **78**, 2629–36.

HASSAN, A. N., FRANK, J. F. and QVIST, K. B. (2002), Direct observation of bacterial exopolysaccharides in dairy products using confocal scanning laser microscopy, *Journal of Dairy Science*, **85**, 1705–8.

HASSAN, A. N., IPSEN, R., JANZEN, T. and QVIST, K. B. (2003), Microstructure and rheology of yoghurt made with cultures differing only in their ability to produce exopolysaccharides, *Journal of Dairy Science*, **86**, 1632–8.

HEERTJE, I., VAN DER VLIST, P., BLONK, J. C. G., HENDRICKX, H. A. C. and BRACKENHOF, G. J. (1987), Confocal scanning laser microscopy in food research: some observations, *Food Microstructure*, **6**, 115–20.

HEERTJE, I., NEDELOF, J., KENDRICKX, H. A. C. M. and LUCASSEN-REYNDERS, E. (1990), The observation of the displacement of emulsifiers by confocal scanning laser microscopy, *Food Structure*, **9**, 305–16.

HEERTJE, I., VAN AALST, H., BLONK, J. C. G., NEDERLOF, D. J. and LUCASSESN-REYNDERS, E. H. (1996), Observations on emulsifiers at the interface between oil and water by confocal scanning light microscopy, *Lebensmittel Wissenschaft und Technologie*, **29**, 217–26.

HEILIG, A., GOEGGERLE, A. and JOERG, H. (2009), Multiphase visualisation of fat containing beta-lactoglobulin-kappa-carrageenan gels by confocal scanning laser microscopy, using a novel dye, V03–01136, for fat staining, *LWT-Food Science and Technology*, **42**, 646–53.

HEINZTMANN, R and FICZ, G. (2012), Breaking the resolution limit in light microscopy, *Briefings in Functional Genomics and Proteomics*, **5**, 289–301.

HELL, S. W. and WICHMANN, J. (1994), Breaking the diffraction resolution limit by stimulated emission: stimulated emission depletion microscopy, *Optics Letters*, **19**, 780–2.

HERBERT, S., BOUCHET, B., RIAUBLANC, A., DUFOUR, E. and GALLANT, D. J. (1999), Multiple fluorescence labelling of proteins, lipids and whey in dairy products using confocal microscopy, *Lait*, **79**, 567–75.

HERVÉ, C., MARCUS, S. E. and KNOX, J. P. (2011), Monoclonal antibodies, carbohydrate-binding modules, and the detection of polysaccharides in plant cell walls, in, *The Plant Cell Wall: Methods and Protocols, Methods in Molecular Biology*, vol. 715, Z. A. Popper (ed.), New York, Springer/Humana Press, 103–13.

HEUER, A., COX, A. R., SINGLETON, S., BARIGOU, M. and GINKEL, M-V. (2007), Visualisation of foam microstructure when subject to pressure change, *Colloids and Surfaces A: Physicochemical and Engineering Aspects*, **311**, 112–23.

JONES, O. G., LESMES, U. DUBIN, P. and MCCLEMENTS, D. J. (2010), Fabrication and characterization of filled hydrogel particles based on sequential segregative and aggregative biopolymer phase separation, *Food Hydrocolloids*, **24**, 689–701.

KENNY, S., WEHRLE, K, AUTY, M. and ARENDT, E-K. (2001), Influence of sodium caseinate and whey protein on baking properties and rheology of frozen dough, *Cereal Chemistry*, **78**, 458–63.

KETT, A. P., CHAURIN, V., FITZSIMONS, S. M., MORRIS, E. R., O'MAHONY, J. A. and FENELON, M. A. (2013), Influence of milk proteins on the pasting behaviour and microstructural characteristics of waxy maize starch, *Food Hydrocolloids*, **30**, 661–71.

KIM, K. Y., FRANK, J. F. and CRAVEN, S. E. (1996), Three-dimensional visualisation of *Salmonella* attachment to poultry skin using confocal scanning laser microscopy, *Letters in Applied Microbiology*, **22**, 280–2.

KLONGDEE, S., THONGNGAM, M. and KLINKESORN, U. (2012), Rheology and microstructure of lecithin-stabilized tuna oil emulsions containing chitosan of varying concentration and molecular size, *Food Biophysics*, **7**, 155–62.

KO, S. and GUNASEKARAN, S. (2007), Error correction of confocal microscopy images for in situ food microstructure evaluation, *Journal of Food Engineering*, **79**, 935–44.

LANGTON, M. and HERMANSSON, A-H. (1996), Image analysis of particulate whey protein gels, *Food Hydrocolloids*, **10**, 179–91.

LANGTON, M., JORDANSSON, ELVY., ALTSKÄR, A., SØRENSEN, C. and HERMANSSON, A-H. (1999), Microstructure and image analysis of mayonnaises, *Food Hydrocolloids*, **13**, 113–25.

LAPSLEY, K. G., ESCHER, F. E. and HOEHN, E. (1992), The cellular structure of selected apple varieties, *Food Structure*, **11**, 339–49.

LEE, L., NG, P. K. W., WHALLON, J. H. and STEFFE, J. F. (2001), Relationship between rheological properties and microstructural characteristics of nondeveloped, partially developed and developed doughs, *Cereal Chemistry*, **78**, 447–52.

LI, Y., KIM, J., PARK, Y. and MCCLEMENTS, D. J. (2012), Modulation of lipid digestibility using structured emulsion-based delivery systems: comparison of *in vivo* and *in vitro* measurements, *Food & Function*, **3**, 528–36.

LOREN, N., LANGTON, M. and HERMANSSON, A. M. (2007), Confocal fluorescence microscopy (CLSM) for food structure characterisation, in, *Understanding and Controlling the Microstructure of Complex Foods*, D. J. McClements (ed.), 232–60.

LOREN, N., NYDEN, M. and HERMANSSON, A. M. (2009), Determination of local diffusion properties in heterogenous biomaterials, *Advances in Colloid and Interface Science*, **150**, 5–15.

LORIS, R., HAMELRYCK, T., BOUCKAERT, J. and WYNS, L. (1998), Legume lectin structure, *Biochimica Et Biophysica Acta-Protein Structure and Molecular Enzymology*, **1383**, 9–36.

MANSKI, J. M., VAN DER GOOT, A. J. and BOOM, R. M. (2007), Influence of shear during enzymitic gelation of caseinate-water and caseinate-water-fat systems, *Journal of Food Engineering*, **79**, 706–17.

MARIOTTI, M., LUCISANO, M., AMBROGINA, P. M. and NG, P. K. W. (2009), The role of corn starch, amaranth flour, pea isolate, and Psyllium flour on the rheological properties and the ultrastructure of gluten-free doughs, *Food Research International*, **42**, 963–75.

MATHESON, N. K. and WELSH, L. A. (1988), Estimation and fractionation of the essentially unbranched (amylose) and branched (amylopectin) components of starches with concanavalin A, *Carbohydrate Research*, **180**, 301–13.

MCCARTHY, N. A., GEE, V L., HICKEY, D. K., KELLY, A. L., O'MAHONY, J. A. and FENELON, M. A. (2012), Effect of protein content on the physical stability and microstructure of a model infant formula, *International Dairy Journal*, doi: 10.1016/j.idairyj.2012.10.004.

MCCLEMENTS, D. J., DECKER, E. A. and WEISS, J. (2007), Emulsion-based delivery systems for lipophilic bioactive components, *Journal of Food Science*, **72**, 109–24.

MCKENNA, A. B. (1997), Examination of whole milk powder by confocal scanning laser microscopy, *Journal of Dairy Research*, **64**, 423–32.

MINSKY, M. (1957), Microscopy Apparatus, US Patent No. 3013467.

MURDOCK, J. N., DODDS, W. K., REFFNER, J. A. and WETZEL, D. L. (2010), Measuring cellular-scale nutrient distribution in algal biofilms with synchrotron confocal infrared microspectroscopy, *Spectroscopy*, **25**, 32–41.

NICOLAS, Y., PAQUES, M., VAN DEN ENDE, D., DHONT, J. K. G., VAN POLANEN, R. C. *et al.* (2003), Microrheology: new methods to approach the functional properties of food, *Food Hydrocolloids*, **17**, 907–13.

NOLTE, A., PAWLEY, J. B and HÖRING, L. (2006), Non-laser light sources for three-dimensional microscopy, in, *Handbook of Biological Confocal Microscopy*, 3rd edition, J. B. Pawley (ed.), New York, Springer Science+Business Media, 338–48.

OLSSON, C., LANGTON, M. and HERMANSSON, A-M. (2002), Microstructures of β-lactoglobulin/amylopectin gels on different length scales and their significance for rheological properties, *Food Hydrocolloids*, **16**, 111–26.

ONG, L., DAGASTINE, R. R., AUTY, M. A. E., KENTISH, S. E. and GRAS, S. L. (2011), Influence of coagulation temperature on the microstructure and composition of full fat Cheddar cheese, *Dairy Science & Technology*, **91**, 739–58.

O'REILLY, C. E., MURPHY, P. M., KELLY, A. L., GUINEE, T. P., AUTY, M. A. E. and BERESFORD, T. P. (2002), The effect of high-pressure treatment on the functional and rheological properties of Mozzarella cheese, *Innovative Food Science & Emerging Technologies*, **3**, 3–9.

PATONAY, G., SALON, J., SOWELL, J. and STREKOWSKI, L. (2004), Noncovalent labeling of biomolecules with red and near-infrared dyes, *Molecules*, **9**, 40–9.

PAWLEY, J. B (ed.) (2006), *Handbook of Biological Confocal Microscopy*, 3rd editon, New York, Springer Science+Business Media, 985.

PEDRESCHI, F., AGUILERA, J. M. and BROWN, C. A. (2002), Characterization of the surface properties of chocolate using scale-sensitive fractal analysis, *International Journal of Food Properties*, **5**, 523–35.

PEREIRA, R., MATIA-MERINO, L. JONES, V. and SINGH, H. (2006), Influence of fat on the perceived texture of set acid milk gels: a sensory perspective, *Food Hydrocolloids*, **20**, 305–13.

PUDNEY, P. D. A., HANCEWICZ, T. M., CUNNINGHAM, D. G and BROWN, M. C. (2004), Quantifying the microstructures of soft solid materials by confocal Raman spectroscopy, *Vibrational Spectroscopy*, **34**, 123–35.

REDGWELL, R. J., CURTI, D. and GEHIN-DELVAL, C. (2008), Physicochemical properties of cell wall materials from apple, kiwifruit and tomato, *European Food Research and Technology*, **227**, 607–18.

ROESSLE, C., AUTY M. A. E., BRUNTON, N., GORMLEY, R. T. and BUTLER, F. (2010), Evaluation of fresh-cut apple slices enriched with probiotic bacteria, *Innovative Food Science and Emerging Technologies*, **11**, 203–9.

ROWNEY, M. K., ROUPAS, P., HICKEY, M. W. and EVERETT, D. W. (2003a), The effect of compression, stretching and cooking temperature on free oil formation in Mozzarella curd, *Journal of Dairy Science*, **86**, 449–56.

ROWNEY, M. K., ROUPAS, P., HICKEY, M. W. and EVERETT, D. W. (2003b), The effect of homogenisation and milk fat fractions on the functionality of Mozzarella cheese, *Journal of Dairy Science*, **86**, 712–18.

RUSS, J. C. (2005), *Image Analysis of Food Microstructure*, Boca Raton, FL, CRC Press, 369.

RUSS, J. C. (2007), *The Image Processing Handbook*, 5th edition, Boca Raton, FL, CRC Press, 817.

SARKAR, A., HORNE, D. S. and SINGH, H. (2010), Interactions of milk protein-stabilized oil-in-water emulsions with bile salts in a simulated upper intestinal model, *Food Hydrocolloids*, **24**, 142–51.

SCHOBER, T. J., BEAN, S. R., BOYLE, D. L. and PARK, S-H. (2008) Improved viscoelastic zein-starch doughs for leavened gluten-free breads: their rheology and microstructure, *Journal of Cereal Science*, **48**, 755–67.

SCHORSCH, C., JONES, M. G. and NORTON, I. T. (1999), Thermodynamic incompatibility and microstructure of milk protein/locust bean gum/sucrose systems, *Food Hydrocolloids*, **13**, 89–99.

SCHORSCH, C., WILKINS, D. K., JONES, M. G. and NORTON, I. T. (2001), Gelation of casein-whey mixtures: effects of heating whey proteins alone or in the presence of casein micelles, *Journal of Dairy Research*, **68**, 471–81.

SHEEN, S., BAO, G. and COOKE, P. (2008), Food surface texture measurement using reflective confocal laser scanning microscopy, *Journal of Food Science*, **73**, 227–34.

SHEPPARD, C. J. R. and SHOTTON, D. M. (1997), *Confocal Laser Scanning Microscopy*, Oxford, Bios Scientific Publishers, 2.

SUTHEERAWATTANANONDA, M., FULCHER, R. G., MARTIN, F. B. and BASTIAN, E. D. (1997), Fluorescence image analysis of process cheese manufactured with trisodium citrate and sodium chloride, *Journal of Dairy Science*, **80**, 620–7.

TAKEUCHI, K., MATUTE, C., HASSAN, A. F. and FRANK, J. F. (2000), Comparison of the attachment of *Escherichia coli* 0157:H7, *Listeria monocytogenes*, *Salmonella typhimurium* and *Pseudomonas fluorescens* to lettuce leaves, *Journal of Food Protection*, **63**, 1433–7.

TAMIME, A. Y., MUIR, D. D., WSZOLEK, M., DOMAGALA, J., METZGER, L. *et al.* (2011), Quality control in processed cheese manufacture, in A. Y. Tamine (ed.), *Processed Cheese and Analogues*, Oxford, West Sussex, UK, Blackwell Publishing Ltd, 245–332.

TRESPALACIOS, P. and PLA, R. (2007), Simultaneous application of transglutaminase and high pressure to improve functional properties of chicken meat gels, *Food Chemistry*, **100**, 264–72.

TROMP, R. H., VAN DE VELDE, F., VAN RIEL, J. and PAQUES, M. (2001), Confocal scanning light microscopy (CSLM) on mixtures of gelatine and polysaccharides, *Food Research International*, **34**, 931–8.

TSIEN, R. Y., ERNST, L. and WAGGONER, A. (2006), Fluorophores for confocal microscopy: photophysics and photochemistry, in, *Handbook of Biological Confocal Microscopy*, 3rd edition, J. B. Pawley (ed.), New York, Springer Science+Business Ledia, 338–48.

VAN DALEN, G. (2002), Determination of the water droplet size distribution of fat spreads using confocal scanning laser microscopy, *Journal of Microscopy*, **208**, 116–33.

VAN DEN BERG, L., VAN VLIET, T., VAN DER LINDEN, E., VAN BOEKEL, M. A. J. S. and VAN DE VELDE, F. (2007), Breakdown properties and sensory perception of whey proteins/ polysaccharide mixed gels as a function of microstructure, *Food Hydrocolloids*, **21**, 961–76.

VAN DEN BERG, L., KLOK, H. J., VAN VLIET, T., VAN DER LINDEN, E., VAN BOEKEL, M. A. J. S. and VAN DE VELDE, F. (2008), Quantification of a 3D structural evolution of food composites under large deformations using microrheology, *Food Hydrocolloids*, **22**, 1574–83.

VAN DER OORD, C. J. R., JONES, G. R., SHAW, D. A., MUNRO, I. H., LEVINE, Y. K. and GERRITSEN, H. C. (1996), High-resolution confocal imaging using synchrotron radiation, *Journal of Microscopy*, **182**, 217–24.

VAN DE VELDE, F., VAN RIEL, J and TROMP, R. H. (2002), Visualisation of starch granule methodologies using confocal scanning laser microscopy (CSLM), *Journal of the Science of Food and Agriculture*, **82**, 1528–36.

VAN DE VELDE, F., WEINBRECK, F., EDELMAN, M. W., VAN DER LINDEN, E. and TROMP, R. H. (2003), Visualisation of biopolymer mixtures using confocal scanning laser microscopy (CSLM) and covalent labelling techniques, *Colloids and Surfaces B: Biointerfaces*, **31**, 159–68.

VELINOV, P. D., CASSENS, R. G., GREASER, M. L. and FRITZ, J. D. (1990), Confocal scanning optical microscopy of meat products, *Journal of Food Science*, **55**, 1751–2.

VERLEBEN, J. P. and STICKENS, D. (1995), *In vivo* determination of fibril orientation in plant cell walls with polarization CSLM, *Journal of Microscopy*, **177**, 1–6.

VODOVOTZ, E., VITTADINI, E., COUPLAND, J., MCCLEMENTS, D. J. and CHINACHOTI, P. (1996), Bridging the gap: the use of confocal microscopy in food research, *Food Technology*, **50**, 74–82.

WELLNER, N., GEORGET, D. M. R., PARKER, M. L. and MORRIS, V. J. (2011), *In situ* Raman microscopy of starch granule structure in wild type and *ae* mutant maize kernals, *Starch–Stärke*, **63**, 128–38.

WRIGHT, S. J., CENTONZE, V. E., STRICKER, S. A., DEVRIES, P. J., PADDOCK, S. W. and SCHATTEN, G. (1993), Introduction to confocal microscopy and three-dimensional reconstruction, *Methods in Cell Biology* **38**, 1–45.

ZHENG, C., SUN, D. W. and ZHENG, L. (2006), Recent applications of image texture for evaluation of food qualities – a review, *Trends in Food Science and Technology*, **17**, 113–28.

ZHONG, Q., DAUBERT, C. R. and VELEV, O. D. (2004), Cooling effects on a model rennet casein gel system – Part II: Permeability and microscopy, *Langmuir*, **20**, 7406–11.

5

Optical coherence tomography (OCT), space-resolved reflectance spectroscopy (SRS) and time-resolved reflectance spectroscopy (TRS): principles and applications to food microstructures

A. Torricelli, Politecnico di Milano, Italy, L. Spinelli, IFN-CNR, Italy, M. Vanoli, CRA-IAA, Italy, M. Leitner and A. Nemeth, RECENDT GmbH, Austria and N. N. D. Trong, B. Nicolaï and W. Saeys, KU Leuven, Belgium

DOI: 10.1533/9780857098894.1.132

Abstract: This chapter presents the recent developments in advanced optical methods for exploring food microstructure. The chapter first discusses the basics of light propagation in food and the main limitations of classical approaches (e.g. continuous wave near infrared (NIR), colorimetry) for the measurement of the optical properties of food. It then describes the physical principles, the technological solutions and the advantages of optical coherence tomography, and of space- and time-resolved reflectance spectroscopy. The chapter includes examples of applications and an overview of future prospects.

Key words: optical coherence tomography, space-resolved reflectance spectroscopy, time-resolved reflectance spectroscopy.

5.1   Introduction

Whilst the final judgement on food quality definitely involves taste and smell, it is a typical everyday experience (from time immemorial) that we choose food primarily on the basis of visual and tactile attributes: a glossy and coloured apple is often preferred to a dull and pale one. Similarly nobody would pick up a withered and irregular shaped apple in preference to a stiff and round one. Only in cases where the choice is among foods that look just the same to the eye and feel the same to the fingertips, do we bring them close to our nose and smell them in order to make a final decision.

It is therefore not surprising that, together with mechanical methods, optical techniques have been playing a primary role in food research aimed at assessing food quality. Vision systems have been introduced into automatic grading lines in order to monitor external colour and shape by means of visible light (Yud-Ren et al., 2002), whilst ultraviolet light has been used for inspection of external defects (Saito, 2009).

When it was understood that food microstructure could affect food quality, the use of light microscopy became crucial in supporting food research and applications (Gunning, 2012; Lorén, 2012; Wellner, 2012). Unfortunately, light microscopy (both classical and sophisticated recent variants such as confocal and multi-photon microscopy) is a destructive technique that, in many cases, requires preparation and manipulation of samples and thus cannot be used for continuous monitoring of products.

Visible (VIS) light and near infra-red (NIR) radiation have lately been introduced as tools for monitoring internal food quality: they have mostly been focused on estimating the concentration of internal constituents (e.g. water, sugar, antioxidants) based on changes in light absorbance. VIS-NIR spectroscopy has the advantage of being easily implemented in the grading lines, but the use has been based mainly on statistical (multivariate) analysis rather than on a direct link to phenomenological properties of food microstructure.

This chapter will focus on advanced optical techniques that have been developed within the last couple of decades, principally in the biomedical optics community for diagnostic applications, and more recently introduced into the food sector. Optical coherence tomography (OCT), space-resolved and time-resolved reflectance spectroscopy (SRS and TRS, respectively) offer the possibility for non-destructive investigation of food structure and microstructure, therefore complementing the cornucopia of optical techniques available for food research and applications.

To properly understand the basics of OCT, SRS and TRS, we need to recall the principles of light–matter interaction. Light (and more generally optical radiation) is propagated by means of electromagnetic waves. The interaction with matter may vary considerably, depending on the frequency of the incident radiation. Therefore it is usual to consider the electromagnetic spectrum as divided in different partially overlapping regions: radio (10^2–10^9 Hz), microwave (10^9–10^{11} Hz), infrared (IR, 3×10^{11}–4×10^{14} Hz), visible (VIS, 4×10^{14}–8×10^{14} Hz), ultraviolet (UV, 8×10^{14}–10^{17} Hz), X-ray (10^{17}–5×10^{19} Hz), and γ ray (5×10^{19}–10^{21} Hz) regions. The term 'optical' generally refers to electromagnetic radiation in the IR to UV regions. More specifically in this chapter we use the term optical to include VIS and near infrared (NIR, 10^{14}–4×10^{14} Hz) radiation. We can use the wavelength to identify these regions: 0.37 to 0.75 μm for the VIS and 0.75 to 2.50 μm for the NIR (Born and Wolf, 1999).

Reflection and refraction of light are the main phenomena occurring at the interface between two isotropic media (typically air and the sample under investigation): these phenomena depend on sample geometry and on sample refractive index (the ratio of the speed of light in vacuum and in the sample). Snell's laws are the formulae used to describe the relationship between the angles of incidence and refraction, whilst Fresnel's equations are typically used when polarization effects have to be taken into account (Born and Wolf, 1999). The refractive index strongly depends on sample microstructure and composition. In the food sector this is illustrated by the use of refractometers as an instrument that exploits light refraction to estimate the degree BRIX (soluble solids content) in aqueous solutions (e.g. fruit juices): sugars are known to affect the refractive index of the sample (ICUMSA, 2009).

The wave nature of the electromagnetic radiation is fully revealed in the phenomenon of light interference. Interference is driven by wave coherence; that is, the characteristic of maintaining a constant phase difference between superimposing waves during the measurement time in the region of observation. The use of alternating layers with different refractive indices is a typical way of producing constructive and destructive interference patterns (Born and Wolf, 1999). Usually, spatial differences in the refractive index can introduce phase shift in the wave, thus contributing to interference, and this is the basis underpinning OCT.

Whilst reflection, refraction and interference of light determine the overall redistribution of energy in the medium (i.e. with no energy loss), more generally the refractive index can also account for light absorption in the medium. In the classical electromagnetic theory (Lorentz model), light absorption is related to frequency changes in the complex refractive index (Bohren and Huffman, 1983). Materials preferentially absorb light when the incident frequency falls within the absorption bands of the material. Moreover, the Lambert–Beer law links the energy lost due to light attenuation to the concentration of the constituents within the medium, and this forms the basic principle of modern spectrophotometry (Sumpf, 2009).

5.1.1   Advances in optical measurements of food microstructure

Many foods are visually opaque media that not only absorb, but also scatter optical radiation (this includes VIS, but also NIR radiation). The physical origin of light scattering lies in microscopic spatial variations of the refractive index. In a simple (classical) picture of light–matter interaction, when a particle with a different refractive index with respect to the surrounding medium is illuminated by light, the electric charges (dipoles) contained within the particle are excited into oscillation. These excited dipoles re-radiate electromagnetic waves at the same frequency as that of the incident light. The overall distribution and intensity of the scattered radiation is determined by the coherent superposition of the waves radiated by all electrical dipoles within the particle, and is determined by particle size (as compared to the wavelength), shape and orientation with respect to the incident light (Bohren and Huffman, 1983).

Electromagnetic theory (Kong, 1986) can provide a complete description of this phenomenon in the single scattering regime, that is for a single particle, or for a collection of particles when particle density in the medium is low. In the single scattering regime, the intensity of the light scattered by every particle is small compared to the incident field, and the overall light distribution can be expressed as the superposition of the scattering from individual single particles.

When the particle density is high and the distribution in space is random, then the single scattering regime is violated and multiple scattering events need to be considered (Bohren and Huffman, 1983). No workable solutions for this regime can be derived through the application of electromagnetic theory: unfortunately this is the typical situation for many foods. Light scattering overwhelms light absorption in the VIS and NIR spectral region and we refer to food as a diffusive or turbid medium, in order to differentiate such materials from clear (transparent) media, where light attenuation is determined uniquely by light absorption.

To overcome the complexity of the electromagnetic theory and the lack of knowledge on the random microscopic properties of the investigated sample, light propagation in a diffusive medium is typically described by the Radiative Transfer Theory (RTT). In this framework, a heuristic description of light as a stream of particles (photons) is used, and energy balance is used to derive workable solutions in the diffusive regime (Ishimaru, 1978, 1989; Martelli *et al.*, 2009).

Although the RTT model is simpler in implementation than the Electromagnetic Theory, it is still too complicated to be employed for modelling light propagation in biological materials in practical applications. Therefore, for highly-scattering media, in which the diffuse intensity tends to be isotropic, a simplified form of the RTT has been derived by considering the light propagation as a diffusion process (Martelli *et al.*, 2009).

Despite the availability of physical models that can accurately describe light propagation in diffusive media, such as biological tissue and microstructured foods, their use is at present confined mainly to the biomedical community (Tuchin, 2007). In the food sector, the common approach to reduce the disturbing influences of scattering is to apply indices that yield spectral readings of light attenuation, normalized by readings at specific wavelengths where food ingredients do not absorb photons, and therefore causing all variations to appear due to differences in scattering (Zude, 2003; Ziosi *et al.*, 2008; Noferini *et al.*, 2009). Alternatively, adapted data pre-processing in whole-spectra analyses can be applied, such as multiplicative scatter correction, standard normal variates, or working on derivatives (Naes *et al.*, 2002). Whilst having the advantage of being an easy way to deal with the problem, these approaches do not in themselves characterize the phenomenon.

The approach pursued by the advanced optical techniques described in this chapter – OCT, SRS and TRS – is to rely on a sound theoretical basis provided by proper physical methods, for example, the radiative transfer equation and diffusion approximation for SRS and TRS, and electromagnetic theory for OCT.

5.2   Optical coherence tomography (OCT)

5.2.1   Introduction

OCT (Huang *et al.*, 1991) is an emerging non-destructive and contactless imaging technique, which allows for real-time and *in situ* acquisition of high-resolution

1-, 2-, and 3D images. OCT is based on the physical phenomenon of white light interferometry and therefore well suited to image layered structures. The image contrast is due to inhomogeneity in the refractive index of the sample material. Thus OCT provides complementary information to other high-resolution imaging techniques such as X-ray computed tomography (CT) and magnetic resonance imaging (MRI, Barreiro, 2012). Nowadays, with state-of-the-art light sources, it is possible to achieve depth (axial) resolutions of as high as one micrometre (Drexler *et al.*, 1999).

OCT can provide both quantitative and qualitative information. The former includes, for example, the thickness of layers, or the size and distribution of pores or cells. With the aid of state-of-the-art image processing tools, it is possible to automatically analyse sample features. Qualitative information is obtained through the visual inspection of the acquired images. In the biomedical area, features such as surfaces, impurities or cells can easily be detected giving rapid indication on the condition of the sample, and providing important information to physicians or technicians.

So far, the research in the field of OCT has mainly been directed towards biomedical applications such as ophthalmology (Huang *et al.*, 1991), dermatology (Podoleanu *et al.*, 2000), dentistry (Colston *et al.*, 1998), or developmental biology (Davis *et al.*, 2007). However, over the last few years, the number of OCT applications outside the field of biomedicine has increased substantially (Liang *et al.*, 2005; Wiesauer *et al.*, 2005, 2007; Stifter, 2007; Targowski *et al.*, 2009). In the field of food research, little use has been made of this method at present, with most studies focusing on the characterisation of onions (Meglinski *et al.*, 2010; Ford *et al.*, 2011; Landahl *et al.*, 2012) and apples (Verboven *et al.*, 2013).

5.2.2   Background

Variants of OCT

Optical coherence tomography is an extension of low coherence interferometry. The simplest approach includes a modified Michelson interferometer (Born and Wolf, 1999), where the mirror in one interferometer arm is replaced by the sample. The probe beam is focused onto the object and photons are back-scattered from structures such as interfaces, impurities, pores, or cells at different depths within the sample. By comparing the arrival times of the scattered photons with those from a reference light beam it is possible to obtain a depth scan. Reconstruction of depth-resolved cross-sections (2D images) or volumes (3D images) is performed by scanning the probing beam laterally across the sample with the aid of galvanometer mirrors, and subsequent acquisition of depth scans at adjacent lateral positions.

The depth-resolved OCT signal can be acquired either in the time-domain or the Fourier-domain, with the latter approach offering advantages in terms of imaging speed and sensitivity (Leitgeb *et al.*, 2003), enabling video rate imaging (van der Jeught *et al.*, 2010) and on-line applications.

In time-domain OCT, the length of the reference arm in the interferometer is scanned over several millimetres while the sample is kept static. Due to the interferometric detection scheme, and the short coherence length of the light sources, a signal is detected only if the photons reflected from both interferometer arms have travelled the same optical distance to the photodetector; otherwise only noise is detected. Obviously, the mechanical movement of the reference mirror is rather time consuming and may lead to mechanical instabilities and noise.

One way to speed up the image acquisition process is to use the so-called Fourier-domain OCT approach. Here the reference mirror is fixed and the light coming from both the reference and the object is detected in a spectrally resolved way. This can be done either in parallel (spectral-domain OCT) by using a dispersing element and a CCD or CMOS camera, or sequentially by scanning a narrow laser line across a broad spectral region (swept-source OCT). Whatever the approach, in Fourier-domain OCT the depth information is encoded in a sinusoidal modulation of the acquired spectrum, and can be accessed by applying an inverse Fourier transform. With state-of-the-art cameras or swept laser sources, it is possible to achieve frame rates of several hundred kHz.

Figure 5.1 shows a schematic of a spectral-domain OCT system (Fig. 5.1(a)), and a photograph of an industrial OCT system, as developed in the laboratories at RECENDT (Fig. 5.1(b)).

Resolution
One of the most important features of OCT is the outstanding depth (axial) resolution, which can be as good as one micrometre. The axial resolution Δz is defined as one half of the coherence length l_c of the light source, which is a function of the square of the central wavelength over the bandwidth of the source:

$$\Delta z = l_c/2 = K\lambda_c^2/\Delta\lambda \tag{5.1}$$

Here K denotes a constant factor (0.44 for an optical source with a Gaussian spectral distribution), and λ_c and $\Delta\lambda$ describe respectively the source's central wavelength and full width at half maximum (FWHM) spectral bandwidth.

One particular feature of OCT, in comparison with other high-resolution microscopic approaches such as confocal microscopy, is that the axial and lateral resolutions are decoupled. Therefore the imaging optics can be located several centimetres away from the sample without penalising the axial resolution.

The lateral resolution (Δx) is determined mainly by the probe optics and can be calculated from the equation:

$$\Delta x = 1.22\lambda f/(nD) \tag{5.2}$$

where f is the focal length of the lens, D is the beam diameter, and n is the refractive index of the sample material.

Light sources
The choice of the correct light source is crucial in OCT since it determines, on the one hand the axial resolution (as described in Equation 5.1), and on the other hand

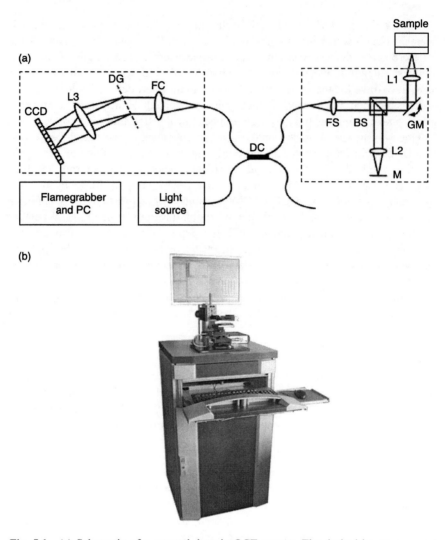

Fig. 5.1   (a) Schematic of a spectral-domain OCT system. The dashed boxes represent portable and independent modules: DC – directional coupler; FC – fibre coupler; BS – beam splitter; (G)M – (galvanometer) mirror; Lx – lens; DG – diffraction grating. (b) Photograph of a spectral-domain OCT system developed at RECENDT.

the penetration depth. Light sources used in OCT should have very high spatial but very low temporal coherence (i.e. a large spectral bandwidth) and provide a smooth Gaussian shaped spectrum. In this way it is possible to achieve axial resolutions in the range of one micrometre.

The intensity measured at the detector is determined by the attenuation characteristics of the sample, which can be described by the Lambert law:

$$I_D(\lambda, z) = I_0 R_z \exp\{-2\int_0^z \mu_{att}(\lambda, z')dz'\}$$ [5.3]

where I_D describes the intensity measured at the detector, I_0 the initial intensity, R_z the sample reflectivity at depth z, and μ_{att} is the attenuation coefficient. The attenuation coefficient μ_{att} is a sum of the scattering and the absorption coefficients, μ_s and μ_a, respectively, which are both functions of wavelength. Consequently, the penetration depth is strongly dependent on the wavelength range used for the imaging process, since this determines the absorption and scattering of the light. With the right choice of the light source it is possible to achieve penetration depths of several millimetres.

Historically the most important application for OCT is ophthalmology. Therefore most of the commercially available OCT systems are designed for this specific application and comprise light sources centred around 800 nm, where absorption due to water is low. For many non-ophthalmic applications it has been shown that it might be advantageous to image at different central wavelengths (Stifter, 2007). Promising spectral regions are located around 1300 nm and 1550 nm. Here scattering is reduced for many samples and the development of high-power and broadband light sources triggered by the telecommunications industry is leading to rapid development of relatively low-cost light sources.

The three types of light sources mostly applied in OCT are superluminescence diodes, femtosecond lasers and supercontinuum laser sources. Since superluminescence diodes are low-cost, provide nicely shaped Gaussian spectra, and are available at many different central wavelengths, they are the most popular kind of light source for OCT. However, the FWHM bandwidth of superluminescence diodes is limited to $\approx$100 nm, which confines the axial resolution to several micrometres. Femtosecond laser sources offer higher bandwidths (up to 200 nm) and increased optical power, but they are relatively expensive. Supercontinuum laser sources are priced in between superluminescence diodes and femtosecond lasers, and can cover a spectral range from around 400 nm to 2400 nm, allowing for the acquisition of ultra-high resolution images.

Additional sample information
Besides providing images in which the contrast is based on different reflectivities, OCT also allows for the study of sample characteristics such as attenuation properties (e.g. differential absorption OCT; DA-OCT) (Pircher *et al.*, 2003), the flow direction and velocity of fluids (Doppler OCT) (Chen *et al.*, 1997) and birefringence due to strain or anisotropy (polarization sensitive OCT; PS-OCT) (Wiesauer *et al.*, 2006).

5.2.3   Applications

OCT on apples
Fruit such as apples are often stored for several months, and the quality and thickness of the wax layer is an important quality parameter throughout storage

(Veraverbeke *et al.*, 2001). With the aid of OCT, it is possible to analyse and evaluate the wax layer thickness quickly and in a fully non-destructive way. Figure 5.2(a) shows an OCT cross-section image of a Braeburn apple acquired with a spectral-domain (SD)-OCT system. Several layers of the paring can clearly be distinguished.

Another interesting feature for the storage life of apples is the lenticels, which act as a bypass medium for the exchange of gases between the fruit flesh and the environment. However, bacteria and fungi can also penetrate into the fruit through the lenticels. Figure 5.2(b) shows an OCT cross-section of a lenticel, as acquired with a time-domain ultrahigh-resolution (TD-UHR)-OCT set-up. The lenticel is clearly visible in the lateral centre of the image. Figure 5.2(c) depicts an *en-face* scan of the sub-surface pore structure of a Braeburn apple. The fragmented appearance of the *en-face* OCT image is caused by the high axial resolution of the imaging system.

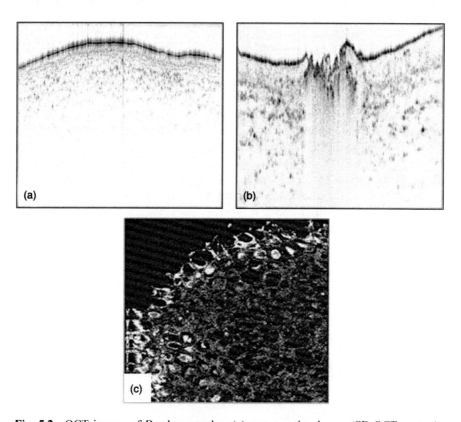

Fig. 5.2   OCT images of Braeburn apples: (a) cross-section image (SD-OCT system), image size: $4 \times 1.25 \, \text{mm}^2$; (b) cross-section image of a lenticel (TD-UHR-OCT system), image size: $3 \times 0.3 \, \text{mm}^2$; and (c) *en-face* OCT image of subsurface pores (TD-UHR-OCT system), image size $3 \times 3 \, \text{mm}^2$.

OCT on extruded breakfast cereals

In the production of extruded breakfast cereals it is common practice to apply sugar coatings onto the product. In this way it is possible to control taste, rehydration properties, and the crispiness and crunchiness of the extruded cereals. Figure 5.3 displays the rehydration process of extruded breakfast cereals (NESTEC) in semi-skimmed milk (20°C), as imaged with an SD-OCT system. The extruded cereals were fixed on the bottom of a recipient vessel, which was

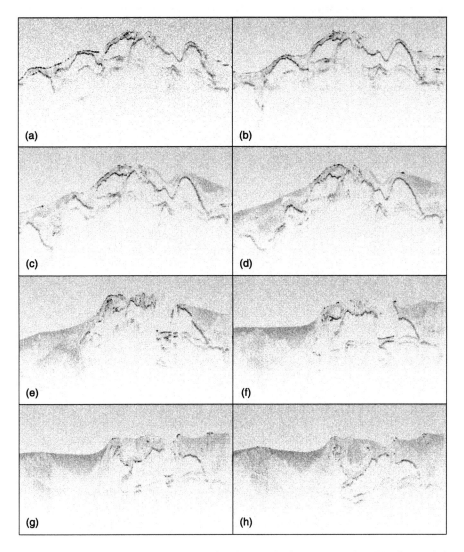

Fig. 5.3   Sequence of OCT images showing the rehydration process of NESTEC extruded cereals in semi-skimmed milk. The images are ordered from (a) to (h), according to the progress in time; image size: $4 \times 1.5\,\mathrm{mm}^2$.

subsequently filled with milk up to approximately 80% of the height of the cereal balls. OCT cross-section images were constantly acquired from the highest point of the cereals throughout the whole rehydration process. The images in Fig. 5.3 are ordered from (a) to (h) according to the progress in time during the experiment. Panel (a) illustrates the surface of the cereals with no milk visible at the surface. In panel (b), some milk has appeared at the left-hand side of the image discernible as a grey shaded feature. In panels (c) to (h), more and more milk appears along the surface of the cereal ball, resulting in a reorganisation of the surface and the pore structure. Also shrinkage of the height of the cereal becomes evident, as the structure of the extruded cereal ball collapses when immersed in milk.

5.2.4   Conclusions

OCT is a fully non-destructive and contactless imaging technique. Its application in the field of food research provides information on quality attributes such as the thickness of the wax layer of apples. Furthermore, it allows a real-time study of dynamic processes such as the rehydration of extruded breakfast cereals in milk.

5.3   Space-resolved reflectance spectroscopy (SRS)

5.3.1   Instrumentation for SRS

In the steady-state SRS approach, the diffusely reflected light is collected at multiple distances from the illumination point. A spatially resolved (or space-resolved) diffuse reflectance profile is then constructed by plotting the diffuse reflectance values as a function of the source–detector distance (Farrell et al., 1992). If the illuminating beam is monochromatic (of a single wavelength), a spatially resolved diffuse reflectance profile of that wavelength is obtained. Similarly, if a broadband beam is used for illumination, a diffuse reflectance spectrum can be acquired at each source–detector distance, and spatially resolved diffuse reflectance profiles obtained at all wavelengths. In this case, the method is sometimes called steady-state continuous wave (cw) SRS. The term 'steady-state' indicates that the intensity of the illumination beam is constant over time, which means that the intensity, or the number of photons at any position inside the illuminated sample, or leaving the sample surface as diffuse reflectance, theoretically remains constant over time. This constraint or controlled condition allows some simplifications to be made in the light propagation models needed for estimation of the optical properties.

Figure 5.4 illustrates the concept of SRS. The longest arrow represents the vertical illumination beam with an illumination intensity I_0, whilst the shorter arrows represent the diffuse reflectances captured at increasing source–detector distances, with intensities $I_1, I_2, \ldots, I_N$ (shorter arrows correspond to smaller intensities). The shaded banana-like shapes are simplified visualisations of the paths travelled by the photons through the sample from the illumination point to the corresponding detector at a given source–detector distance. Longer and wider shapes correspond to longer path lengths. A closer look at these simplified photon

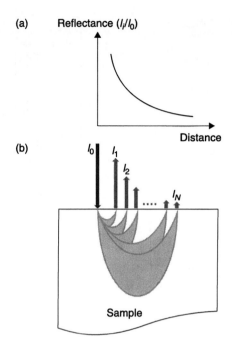

Fig. 5.4 Principle of a cw SRS measurement: a typical spatially resolved diffuse reflectance profile (a) and an illustration of the corresponding illumination and detection positions on the sample and the light paths in the sample (b): I_0 is the illumination intensity; I_1, I_2,...,I_N are the diffusely reflected intensities captured at different source–detector distances.

paths shows that photons acquired at a larger source–detector distance have travelled a longer path and consequently have had a higher chance of being scattered or absorbed when compared to those detected at smaller source–detector distances. As a result of this, the intensity of the diffusely reflected light, or the number of photons exiting the sample, decreases with increasing source–detector distance, as observed in the plot in Fig. 5.4(a). The decrease in intensity with increasing path length is caused by scattering and absorption, hence the chance of a scattering or absorption event per infinitesimal step is described respectively by the scattering coefficient μ_s and the absorption coefficient μ_a. The anisotropy factor g then indicates whether the scattering is mostly forward ($g > 0$), backward ($g < 0$) or isotropic ($g = 0$). The scattering coefficient μ_s and the anisotropy factor g are often combined in the reduced or transport scattering coefficient μ_s', calculated as $\mu_s (1 - g)$.

Different approaches and set-ups are available for acquiring spatially resolved reflectance profiles from a sample. In contact SRS measurement, optical fibres are used to guide the light from the light source to the sample, and to collect the diffusely reflected light and guide it to the detectors. The acquisition at multiple source–detector distances is either done sequentially by precisely positioning one detection fibre at different positions through use of a translation stage (Fig. 5.5(a)),

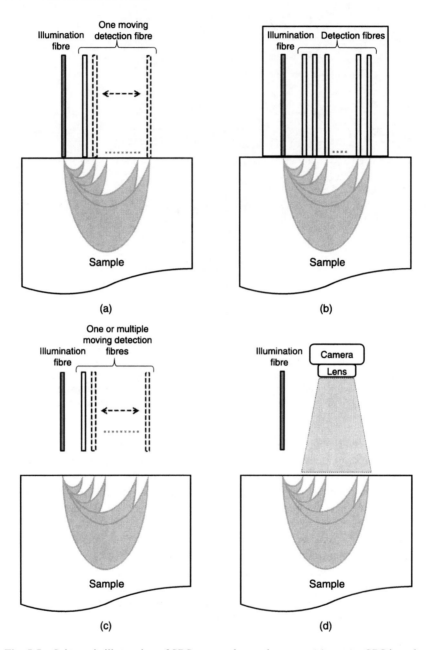

Fig. 5.5   Schematic illustration of SRS approaches and set-ups: (a) contact SRS based on a translation stage; or (b) a fibre-optics probe; (c) contactless SRS based on optical fibres; or (d) a camera (hyperspectral scatter imaging).

or simultaneously with a probe containing an integral array of several optical fibres at different distances from the illumination fibre (Fig. 5.5(b)). Contactless SRS measurement can be obtained by placing one or more detection fibres above the sample surface (Fig. 5.5(c)) or by imaging the surface with a (spectral) camera (Fig. 5.5(d)).

5.3.2   Physical models for SRS

Monte Carlo (MC) methods for describing light propagation
As anticipated in Sections 5.1 and 5.2, the description of light propagation in food can be conveniently tackled by adopting the diffusion approximation of the RTE. However, it should be noted that the diffusion approximation only gives an adequate description of the diffuse reflectance for radial distances larger than 1 mfp (transport scattering mean free path). This is the average distance between two scattering events. As short source–detector distances are typically used in SRS in order to acquire sufficiently large reflected intensities, this requirement might not be fulfilled for a given sample. In these cases the diffusion approximation might lead to considerable errors in the simulated SRS profiles and thus should be replaced by a more accurate light propagation model, such as an MC simulation.

MC simulations are used widely for modelling light propagation in tissue or organs in biomedical research (Wilson and Adam, 1983; Keijzer *et al.*, 1989; Wang *et al.*, 1995; Fang, 2010; Shen and Wang, 2011). It is considered to be a precise approach for modelling light propagation in turbid media. Recently this method has also been utilised in the food and agriculture domain to model the light distribution in fruit (Fraser *et al.*, 2003; Qin and Lu, 2009). The MC method is fundamentally a numerical approach for solving the steady-state RTE, based on random movements of the photon particles in the medium under the effects of scattering and absorption. Because of its stochastic approach, a large number of photon packets (10^6–10^7) have to be launched into the tissue and tracked to obtain statistically meaningful results. The main advantages of MC simulations are the high accuracy, universal applicability for any medium geometry, and validity even for source–detector distances smaller than 1 mfp'. However, the major drawback lies in the high computation time needed to track the large numbers of photons required to obtain an acceptable accuracy in the final results. Due to recent advances in computer hardware, the computation time required for MC simulations has been reduced significantly, such that the method can now be used even for complex microstructures (Alerstam *et al.*, 2008; Fang and Boas, 2009; Ren *et al.*, 2010; Watté *et al.*, 2012).

Estimation of optical properties
The optical properties (absorption and scattering or reduced scattering coefficients) of a sample can be derived from the acquired spatially resolved diffuse reflectance profiles by inverting the light propagation models. When the diffusion model is applied, the reduced scattering coefficient μ_s' and absorption coefficient μ_a can be

estimated by iterating from an initial guess, until the spatially resolved diffuse reflectance profile calculated from the model fits to the acquired profile. Due to the large computation time required for one simulation, such an iterative approach is not feasible in combination with MC simulations. Therefore typically a library with simulated profiles is created for a dense grid of scattering coefficient μ_s, absorption coefficient μ_a, and anisotropy factor g values. To estimate the optical properties of a given sample based on the acquired profile, either this library is then used as a look-up table (Hjalmarsson and Thennadil, 2007) or a meta-model is trained on this library (Kienle *et al.*, 1996; Dam *et al.*, 1998; Hjalmarsson and Thennadil, 2008).

Optical phantoms
Both the measurement performance and the procedure employed to estimate the optical properties must be validated using reference samples with known optical properties. For this purpose, model systems are made that are usually called 'phantoms'. In general, the phantoms are made by mixing a substance that induces the scattering effects (scatterers) with a substance that exhibits absorption in the targeted wavelength range (absorbers). Two types of phantoms have been used in practice: liquid phantoms and solid phantoms. A popular type of liquid phantom used in biomedical research consists of aqueous mixtures of a fat emulsion known as Intralipid (scatterer) and black ink (absorber) (Martelli and Zaccanti, 2007). Solid phantoms are often made by dispersing TiO_2 as a scatterer and black ink as an absorber in an epoxy resin base matrix (Pifferi *et al.*, 2005). Polystyrene spherical particles with known particle size distribution are also often used as scatterers, whilst food colouring agents or other chromophores have been used to obtain absorption at specific wavelengths. Because the method used for validation should provide confidence in the estimated optical properties, it is important that the optical properties of the phantoms cover the anticipated ranges of optical properties of the measured samples.

There are two steps of validation: *forward* and *inverse validation*. Forward validation evaluates how well the employed light propagation model is able to simulate the measured profiles. This is achieved by investigating the differences between the measured and simulated spatially resolved diffuse reflectance profiles for the optical phantoms with known optical properties. It should be noted here that, due to practical difficulties in measuring absolute diffuse reflectance values, a scaling factor is typically needed to scale the simulated profiles. Once the forward validation step is successful, inverse validation is performed to determine the accuracy of the procedure used for the estimation of optical properties, by comparing the estimated and known values for the optical phantoms. Once the methodology for estimation of the optical properties has been properly validated, it can be applied to samples for which the optical properties are unknown.

Food microstructure: scattering property relation
The absorption of light by a food product is related to the presence of molecular bonds that can be excited by photons of the right wavelength. Scattering of

the photons happens at each interface between two materials with a different refractive index. The scattering coefficient thus provides information on the sample microstructure. One common approach to model food microstructure is to consider it as an ensemble of discrete particles. Depending on the shapes, inter-distances and sizes of the particles relative to the propagating wavelengths, many physical models could be used to describe the light scattering (Mie scattering, Rayleigh scattering, etc.). Since the Vis/NIR/SWIR wavelength ranges (400–2500 nm) are commonly used in applications in food and agriculture, and the fact that biological media basically consist of organelles such as nuclei, mitochondria, vesicles, membranes and cell walls with sizes in the range 0.1 to 10 µm, comparable to the wavelengths of the scattered light, Mie theory is usually applied (Bohren and Huffman, 1983; Born and Wolf, 1999; Saeys *et al.*, 2010; Aernouts *et al.*, 2012) to relate scattering properties to the microstructure of biological tissue.

5.3.3   Application: differentiation of microstructure of sugar foams

Sugar foams are used as a model system for a wide range of foods, including candy (marshmallows), foamed dairy products, mousses etc. They can be prepared by aerating agar-fructose gels made of agar-agar powder, fructose, albumin and water. Air bubbles can be created by means of a mixer, by whipping the mixture of ingredients obtained after completely dissolving fructose in a heated agar solution. As the refractive index of air ($n = 1$) is considerably lower than that of the agar-fructose medium ($n \approx 1.4$), these air bubbles act as scatterers. The light scattered by the air bubbles is responsible for the typical white colour of these sugar foams. Different microstructures (air bubble size distributions) can be introduced by changing the whipping time.

These foams were measured by an SRS set-up based on a fibre optics probe containing an integrated array of 1 illumination and 6 detection fibres in the range 500 to 1000 nm (Nguyen Do Trong *et al.*, 2011, 2012; Verhoelst *et al.*, 2011). An example of the spatially resolved diffuse reflectance profiles of the foam obtained after whipping the sample for 10 minutes is illustrated in Fig. 5.6.

In Fig. 5.6, a clear decrease of the diffuse reflectance with increasing source–detector distance can be observed at each wavelength. This can be explained by the fact that light exiting the foam, at a greater distance from the incident light beam, has travelled a longer path through the sample, and thus has had more chance to be absorbed or scattered. This is actually the spatially resolved diffuse reflectance profile of the sugar foam at that wavelength. A flat spectrum before 900 nm observed at each fibre position is a good indication that the sugar foam absorbs more or less evenly and little at all wavelengths in this region. A small 'valley' appears in the wavelength region close to 970 nm at each source–detector distance, indicating the presence of water in the sugar foams.

The estimated reduced scattering coefficients μ'_s and absorption coefficients μ_a, acquired by fitting the light propagation model based on the diffusion equation

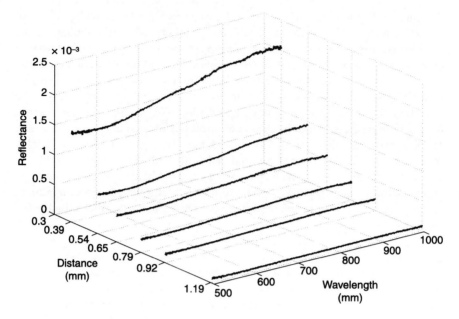

Fig. 5.6   Spatially resolved diffuse reflectance profiles of the sugar foam with 10-minute whipping time.

(Farrell *et al.*, 1992) to the measured spatially resolved diffuse reflectance profiles at all wavelengths, by means of a non-linear least squares method, are shown in Fig. 5.7.

In Fig. 5.7, the reduced scattering coefficients μ_s' of the three sugar foams increase with longer whipping times, which indicates that the average bubble sizes in these foams were getting smaller during whipping. This also agrees well with the expectation that the mechanical energy supplied by the mixer or stirrer, during the foaming process, was being transferred to the sugar foam medium to break down larger air bubbles into smaller ones, and to maintain the stability of these newly created bubbles. All reduced scattering profiles are flat, which indicates that, according to the Mie theory for light scattering, large distributions of the bubble sizes in the range 500 to 1000 nm actually exist, rather than a homogeneous bubble size of specific or narrow range of diameters.

The absorption coefficients μ_a of the foams are low in the range 500 to 750 nm, which might be expected from the white colour of the foam. Because of the presence of water in the foams, the absorption values start to increase from 800 nm and peak at around 970 nm. For the three foams, the absorption coefficients μ_a are almost identical, which would be expected for three foams made from the same mixture of ingredients.

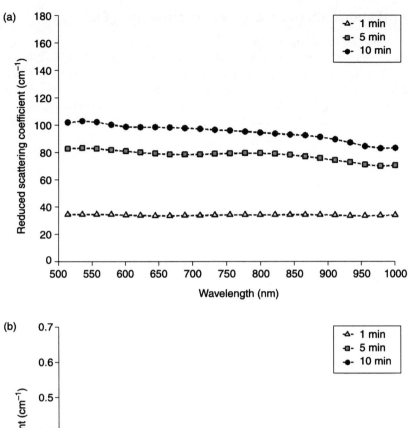

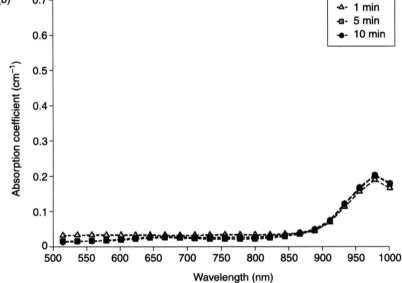

Fig. 5.7   Scattering (a) and absorption (b) coefficient spectra acquired for the three sugar foams.

5.4   Time-resolved reflectance spectroscopy (TRS)

5.4.1   Instrumentation for TRS

TRS relies on the ability to measure the distribution of photon time-of-flight (DTOF) in a diffusive medium. Following the injection of the light pulse within a turbid medium, the DTOF measured at a fixed distance from the injection point (typically in the range 1–4 cm) will be delayed, broadened and attenuated. The delay is a consequence of the finite time light takes to travel the distance between the source and detector; broadening is mainly due to the different paths that photons undergo because of multiple scattering; attenuation appears because absorption reduces the probability of detecting a photon; and diffusion into other directions within the medium decreases the number of detected photons in the considered direction. Increasing the source–detector distance yields an increased delay of the DTOF and decreases the number of detected photons. Similar behaviour is observed when the scattering increases. Finally, absorption affects both the signal intensity and the trailing edge (i.e. slope of the tail) of the DTOF, whilst leaving the temporal position of the DTOF substantially unchanged (Jacques, 1989; Patterson *et al.*, 1989; Wilson and Jacques, 1990; Cubeddu *et al.*, 1994).

Typical values of the optical parameters in the VIS and NIR spectral regions set the timescale of TRS measurements in the range 1 to 10 ns, and fix the ratio of detected to injected power at about −80 dB. The two key points in the design of a TRS system are therefore temporal resolution and sensitivity. The overall temporal resolution is affected by the duration of the light pulse and by the temporal response of the detection apparatus. Nowadays pulsed lasers, which produce short (<100 ps) light pulses with repetition frequency up to 100 MHz, and detectors with temporal resolution in the range of less than 150 ps are widely available. TRS laboratory set-ups, based on the use of mode-locked solid state lasers and streak camera detection, can reach temporal resolution of less than 1 ps, while compact and portable TRS systems are typically based on pulsed semiconductor (or fibre laser) and Time-Correlated Single Photon Counting (TCSPC) (Becker, 2006) detection with less than 200 ps resolution.

Two different systems for TRS measurements of food quality have been developed and used at Politecnico di Milano: a broadband TRS system for multi-wavelength measurements (TRS_MW), and a compact TRS system, working at a single discrete wavelength (TRS_SW). Both systems are based on the TCSPC and exploit the expertise of the research group in the biomedical field. A detailed description of the two systems can be found in Pifferi *et al.* (2007) and in Cubeddu *et al.* (2001a), respectively. A brief description of the two systems is reported below.

TRS_MW system

A schematic overview of the TRS_MW system is illustrated in Fig. 5.8. The light source consists of a fibre laser (SC450, Fianium, UK) providing white-light pulses shorter than 10 ps, at a repetition rate of 20 MHz in the 465–750 nm bandwidth,

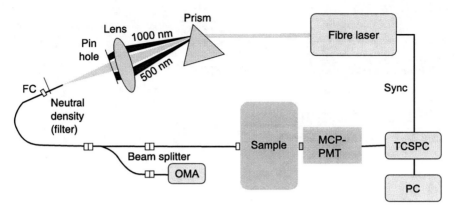

Fig. 5.8 Schematic of the TRS_MW system. FC, fiber connector; OMA, optical multichannel analyzer.

with a total power of 2.6 W. The supercontinuum is spectrally dispersed by an F2-glass prism (PS854, Thorlabs, Germany) and then focused by an achromatic doublet with focal length of 75 mm (AC508-075-B, Thorlabs, Germany) onto a 50 μm core graded-index fibre. The fibre – FC connectorized – is mounted with a 2-axis tilt on a 3D translational stage for precise alignment. The prism is mounted on a motorized rotational stage in order to sequentially tune the light source: it is rotated in 120 steps of 1.1 mrad to cover the 600 to 1100 nm bandwidth. A 1 mm pinhole is placed after the prism to limit the incident power at the fibre head and to avoid dangerous back reflections to the laser. The power can be adjusted by means of a motorized circularly variable neutral-density filter placed in front of the fibre. A small part of the light (3%) is split by using a fused splitter (Phoenix, UK) and delivered to a spectrometer (USB2000, Ocean Optics, USA) for the purpose of monitoring the light signal. The remaining light (97%) is guided to the sample. Light diffusely transmitted through the sample is detected using a 1 mm plastic-glass fibre guide (APCS1000, Fiberguide, USA) coupled to a micro-channel plate photomultiplier (MCP-PMT, R1564U, Hamamatsu Photonics, Japan). The signal from the detector is delivered to a time-correlated single-photon counting board (TCSPC in Fig. 5.8) (SPC-130, Becker & Hickl, Germany); synchronization is provided directly from the laser. The temporal resolution of the overall system is about 70 ps, limited by the chromatic dispersion in the optical fibres and the MCP-PMT temporal resolution. The acquisition of the time-resolved curves and the synchronous movement of the prism and the variable filter are automatically controlled by a personal computer.

TRS_SW system
This is a compact prototype for a TRS at a discrete wavelength (Fig. 5.9). The light source is a pulsed laser diode (mod. PDL800, PicoQuant GmbH, Germany) working at 670 nm, with an 80 MHz repetition frequency, 100 ps duration and

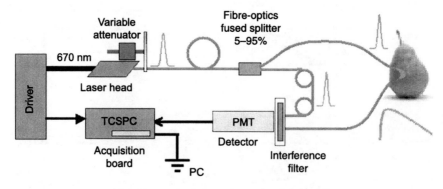

Fig. 5.9   Schematic of the TRS_SW system.

1 mW average power. A compact photomultiplier (mod. R5900U-L16, Hamamatsu Photonics, Japan) and an integrated PC board for time-correlated single photon counting (mod. SPC130, Becker&Hickl GmbH, Germany) are used to detect the TRS data. Typical acquisition time is 1 s per point. A couple of 1 mm plastic fibres (Mod. ESKA GK4001, Mitsubishi, Japan) deliver light into the sample and collect the emitted photons. A bandpass filter, tuned at the laser wavelength, is used to cut off the fluorescence signal due to chlorophyll or inks. Overall, the instrumental response function (IRF) duration is less than 180 ps.

5.4.2   Physical models for TRS
Fitting the experimental DTOF with a standard solution of the diffusion approximation to the RTE yields the reduced scattering coefficient μ'_s and the μ_a absorption coefficient. A complete description of available solutions for different geometries (e.g. semi-infinite, slab, parallelepiped, layered medium and sphere) can be found in Martelli *et al.* (2009).

The reflectance $R(r,t)$, that is the photon probability to be re-emitted from the tissue at a time t and a distance r from the injection point, can be expressed as:

$$R(r, t) = A\, t^{-5/2}\, e^{-\mu_a vt}\, e^{-\frac{r^2}{4Dvt}} \left(z_o e^{-\frac{z_o^2}{4Dvt}} - z_p e^{-\frac{z_p^2}{4Dvt}} \right)$$ [5.4]

where A is a normalization constant, $v = c/n$ is the speed of light in the medium, $z_o = 1\mu'_s$ is the scattering mean free path, while z_p derives from the extrapolated boundary conditions and depends on the refractive index of the tissue. The diffusion coefficient $D = 1/(3\mu'_s)$ was taken to be independent of the absorption properties of the medium.

The theoretical curve is convoluted with the instrumental response function (IRF) and normalized to the area of the experimental curve. The fitting range includes all points with a number of counts higher than 10% of the peak value on the rising edge of the curve and 1% on the tail. The best fit is reached with a

Levenberg–Marquardt algorithm (Press *et al.*, 2007) by varying both μ_s' and μ_a in order to minimize the reduced chi-square.

In the range of measured values of the optical coefficients, with the set-up and the theoretical model used, the accuracy in the absolute estimate of both μ_s' and μ_a is usually better than 10%. However, the error made in the assessment of the absorption line shape is definitely smaller (<2%) (Cubeddu *et al.*, 1996).

Figure 5.10 shows a typical TRS curve, the IRF and the fitting with the model for photon diffusion. When TRS measurements are performed at multiple wavelengths, the absorption spectrum and the scattering spectrum of the diffusive medium are obtained by iterating the fitting procedure for all wavelengths. A more sophisticated and robust procedure is the so-called spectrally constrained approach (D'Andrea *et al.*, 2006). The spectral dependence of the optical properties is inserted into the analytical solution for light transport in diffusive media and a best fit is obtained by simultaneously considering TRS data at multiple wavelengths, and using structure parameters (e.g. density and size of scatterers) and chromophore concentrations (e.g. chlorophyll and water) as free parameters.

The relationship between the absorption coefficient and tissue constituents is given by the Beer Law:

$$\mu_a(\lambda) = \sum_i c_i \varepsilon_i(\lambda) = c_{CHL}\varepsilon_{CHL}(\lambda) + c_{H_2O}\varepsilon_{H_2O}(\lambda) + bkg \qquad [5.5]$$

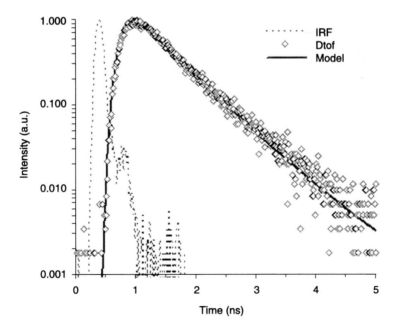

Fig. 5.10   Typical TRS curve, IRF and the fitting with the model for photon diffusion.

where c_{CHL} and c_{H_2O} are the chlorophyll and water concentration, respectively, ε_{CHL} and ε_{H_2O} are the chlorophyll and water specific absorption coefficients, respectively, and *bkg* is a constant value to account for the contribution of other absorbers.

An approximation to the Mie theory is used to relate the reduced scattering coefficient to the structural properties of the measured diffusive sample:

$$\mu_s'(\lambda) = a(\lambda/\lambda_0)^{-b}$$

[5.6]

where *a* is the scattering coefficient at wavelength λ_0 (in this case $\lambda_0 = 600\,nm$) and *b* is a parameter related to the size of the scatterers.

5.4.3   Applications

TRS has been applied mainly to the non-destructive assessment of the bulk optical properties of several foods (Cubeddu *et al.*, 2001b). Since TRS relies on the broadening of light pulses in a diffusive medium, as a rule-of-thumb TRS would require a sample volume larger than $10\,cm^3$. For this reason, TRS studies have mostly focused on fruits (e.g. apple, pear, plum, melon and nectarine) and vegetables (e.g. potato). Indeed, it is the combination of optical properties and sample geometry that influences the applicability of TRS: a small but strongly diffusive sample produces a large broadening of laser pulses (e.g. this is the case for cherries). For the same reasons, TRS cannot provide robust results if applied to samples with weak diffusion (e.g. gels) or the presence of voids (e.g. cereal flakes). Application of TRS to milk and milk-based products (e.g. cheese and dairy cream) should be straightforward. In many research laboratories, milk is in fact used as a calibration phantom for TRS set-ups, as a cheaper alternative to the more expensive Intralipid (Di Ninni *et al.*, 2011).

In recent years, TRS has been applied mainly in postharvest studies on fruits for estimating fruit maturity at harvest (Eccher Zerbini *et al.*, 2006, 2011; Tijskens *et al.*, 2007), for the detection of internal defects in fruits and vegetables (Eccher Zerbini *et al.*, 2002, Vanoli *et al.*, 2010, Lurie *et al.*, 2011), and the study of structural changes in apples during storage. This latter application is briefly reviewed below in the next sub-section, and a detailed description can be found in Vanoli *et al.* (2011).

Monitoring changes in apple structure during storage by TRS
Sixty Pink Lady® apples, picked in Laimburg (Bolzano, Italy), were measured by TRS in the spectral range 600 to 1100 nm on two opposite positions around the equator region, at harvest and after 91 days of storage at 1°C in normal atmosphere.

Overall, at harvest the absorption spectra showed a maximum at 670 nm (chlorophyll-a) and at 980 nm (water), whilst the scattering coefficient slowly decreased with increasing wavelength (Fig. 5.11). Comparison of the TRS spectra measured at harvest, with those measured on the same fruit during storage, showed that the absorption coefficient at 670 nm values was lower in apples stored for

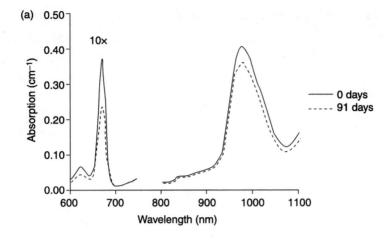

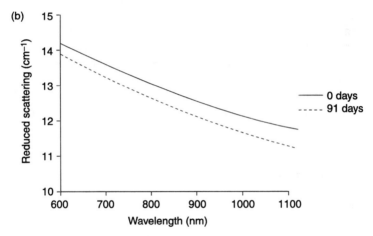

Fig. 5.11 Absorption (a) and scattering (b) spectra of 'Pink Lady®' apples at harvest (solid line) and after 91 days storage (dashed line).

91 days than in the same fruit measured at harvest. Differences in the absorption values at 980 nm were also evident. As for the scattering behaviour, the parameter a reported in Equation 5.6 is proportional to the density of the scattering centres and determines the absolute value of μ'_s, while the parameter b depends on the size of the scatterers, and determines the slope of the scattering spectra. Apples stored for 91 days showed values of both a and b that were lower than those of the same apples at harvest.

The decrease of the absorption coefficient at 670 nm and 980 nm was due respectively to a decrease in the chlorophyll and water contents (Table 5.1), the first related to fruit ripening and the latter to water loss. In fact, during cold storage, apples ripened as they lost firmness and hydrolyzed starch, as shown

Table 5.1 Chlorophyll content, water content and scattering parameters of 'Pink Lady®' apples during storage

	Days at 1 °C	
	0	91
Chlorophyll content (μM^{-1})	0.43±0.11	0.27±0.08
Water content (%)	81.8±1.4	71.8±1.4
a (cm^{-1})	14.2±1.1	13.9±1.4
b (−)	0.31±0.04	0.35±0.08

above, and the ripening process was accompanied by a progressive decrease in chlorophyll content (Vanoli et al., 2005; Rizzolo et al., 2010). In agreement with the very low μ_a670 values found in 'Pink Lady®' apples, the estimated chlorophyll content was also lower than that found by Cubeddu et al. (2001b): chlorophyll in 'Pink Lady®' apples was about half to one-third lower than that of 'Golden Delicious' and 'Granny Smith' apples, respectively.

The changes in scattering parameters reflected the changes in the pulp structure during storage. The decrease of the parameter a after 91 days storage could be linked to apple softening; this process is accompanied by the enzymatic breakdown of the cell wall, and thus the density of the scattering particles in the fruit flesh decreased, as found by Qin and Lu (2008) and Bobelyn et al. (2010). As a consequence of this phenomenon, the light scattering at the cell interfaces is reduced, leading to less scattering events in the tissue. At the same time, the intercellular space volume increased and some water was lost due to fruit transpiration, so the cells in the pulp tissue became smaller with more air-filled pores. The values of the parameter b decreased but scattering values increased, as there was a higher refractive index mismatch, leading to more and stronger scattering events. Actually the situation in the pulp tissue is much more complex than just described, as the scattering centres of a fruit are not expected to be homogeneous spheres. These parameters do not assess the real size of scattering centres in the tissue; rather they are average equivalent parameters that could be related to physical characteristics of fruit.

5.5   Conclusion and future trends

The use of optical techniques for investigation of food microstructure is an emerging field. As in the biomedical field, the optical techniques can complement other methodologies (e.g. X-ray, ultrasound, nuclear magnetic resonance) so as to allow data fusion strategies. A characteristic feature of optical methods is the opportunity to develop compact and portable devices (even hand-held), and this holds true also for OCT, SRS and TRS.

Due to the rapid rise and widespread use of OCT in the biomedical field, commercial OCT devices are now more readily available with compact and affordable systems found in the biomedical field and thus extension and use in the food sector should be straightforward. A hand-held and cost-effective SRS device could be made for certain practical applications by selecting one or more signature wavelengths, which could provide comparable results to those obtained using the full wavelength range. The number of detection fibres could also be optimised at some specific distances. This would allow use of either one or a few photodiodes, or one photodiode combined with a multiplexer for sequentially acquiring data from each detection fibre. An integrated chip would then serve as the data processing unit to produce output results. Recent advancements in optoelectronics devices could also allow the production of a hand-held SRS set-up, by using the combination of a camera chip with an on-chip grating, or on-chip waveguides, to guide the light to the detectors. The earliest TRSs occupied a space equivalent to a laboratory room, whereas nowadays we can easily assemble a compact tabletop or rack-based system. The use of semiconductor lasers and detectors (e.g. single photon avalanche diode, SPAD) to studies are under way to further scale down TRSs to the level of hand-held devices.

In the food sector, it is crucial to develop on-line applications of these measuring techniques. OCT is a fully non-destructive and contactless imaging technique; therefore, in principle, it should be simple to develop its use for in-line/on-line applications. Detailed implementation (e.g. choice of light sources, acquisition scheme) should be tailored to the specific application. Contact SRS based on a fibre-optics probe has the potential for on-line applications for liquids or food pastes. For solid foods transported on production lines, a contactless SRS approach is needed. For this purpose, the illuminated samples can be imaged by a hyperspectral camera, as is used in hyperspectral scatter imaging. The considerations required for SRS hold also for TRS, with the disadvantage that nowadays TRS technology (picosecond laser, fast detectors and readout electronics) is definitely more expensive.

5.6   Acknowledgements

This publication has been produced with the financial support of the European Union (project FP7-226783 – InsideFood – Integrated sensing and imaging devices for designing, monitoring and controlling microstructure of foods). The opinions expressed in this document do not necessarily reflect the official opinion of the European Union or its representatives.

AN and ML further acknowledge support from the European Regional Development Fund (EFRE) in the framework of the EU-program REGIO 13, and the federal state of Upper Austria.

WS is funded as a postdoctoral fellow of the Research Foundation-Flanders (FWO). NNDT acknowledges the Flemish Interuniversity Council (VLIR) for the financial support.

AT further acknowledges partial support from the ICT-AGRI Project ID 95 – 3D Mosaic.

5.7   References

AERNOUTS, B., WATTÉ, R., LAMMERTYN, J. and SAEYS, W. (2012), A flexible tool for simulating the bulk optical properties of polydisperse suspensions of spherical particles in an absorbing host medium, *SPIE Photonics Europe 2012*, Brussels, Belgium, Paper No. 8429-28.

ALERSTAM, E., SVENSSON, T. and ANDERSSON-ENGELS, S. (2008), Parallel computing with graphics processing units for high-speed Monte Carlo simulation of photon migration, *J. Biomed. Opt.*, **13**, 060504.

BARREIRO, P. (2012), MRI, in, V., Morris and K. Groves, *Food Microstructures: Microscopy, Measurement and Modelling*, Cambridge, UK, Woodhead Publishing Ltd.

BECKER, W. (2006), *Advanced TCSPC*, Berlin, Springer.

BOBELYN, E., SERBAN, A. S., NICU, M., LAMMERTYN, J., NICOLAÏ, B. M. and SAEYS, W. (2010), Post-harvest quality of apple predicted by NIR-spectroscopy: study on the effect of biological variability on spectra and model performance, *Postharvest Biol. Technol.*, **55**, 133–43.

BOHREN, C. F. and HUFFMAN, D. R. (1983), *Absorption and Scattering of Light by Small Particles*, New York, Wiley.

BORN, M. and WOLF, E. (1999), *Principles of Optics*, 7th edition, Cambridge, UK, Cambridge University Press, 40.

CHEN, Z., MILNER, Z., DAVE, D. and NELSON, J. (1997), Optical Doppler tomographic imaging of fluid flow velocity in highly scattering media, *Opt. Lett.*, **22**, 64–6.

COLSTON, B., SATHYAM, U., DASILVA, L., EVERETT, M., STROEVE, P. and OTIS, L. (1998), Dental OCT, *Opt Express*, 3, 230–238.

CUBEDDU, R., MUSOLINO, M., PIFFERI, A., TARONI, P. and VALENTINI, G. (1994), Time-resolved reflectance: a systematic study for application to the optical characterization of tissues, *IEEE J. Quantum Electron.*, **30**, 2421–30.

CUBEDDU, R., PIFFERI, A., TARONI, P., TORRICELLI, A. and VALENTINI, G. (1996), Experimental test of theoretical models for time-resolved reflectance, *Med. Phys.*, **23**, 1625–33.

CUBEDDU, R. D'ANDREA, C. PIFFERI, A., TARONI, P., TORRICELLI, A. *et al.* (2001a), Non-destructive measurements of the optical properties of apples by means of time-resolved reflectance spectroscopy, *Appl. Spectr.*, **55**, 1368–74.

CUBEDDU R., D'ANDREA C., PIFFERI A., TARONI P., TORRICELLI A. *et al.* (2001b), Nondestructive quantification of chemical and physical properties of fruits by time-resolved reflectance spectroscopy in the wavelength range 650–1000 nm, *Appl. Opt.*, **40**, 538–43.

DAM, J. S., ANDERSEN, P. E., DALGAARD, T. and FABRICIUS, P. E. (1998), Determination of tissue optical properties from diffuse reflectance profiles by multivariate calibration, *Appl. Opt.*, **37**, 772–8.

D'ANDREA, C., SPINELLI, L., BASSI, A., GIUSTO, A., CONTINI, D. *et al.* (2006), Time-resolved spectrally constrained method for the quantification of chromophore concentrations and scattering parameters in diffusing media, *Opt. Express*, **14**, 1888–98.

DAVIS, A., BOPPART, S. A., ROTHENBERG, F. and IZATT, J. (2007), OCT applications in developmental biology, in W. Drexler and J. G. Fujimoto, *Optical Coherence Tomography: Technology and Applications*, Berlin, Springer, 919–60.

DI NINNI, P., MARTELLI, F. and ZACCANTI, G. (2011). 'Intralipid: towards a diffusive reference standard for optical tissue phantoms, *Phys. Med. Biol.*, **56**, N21–8.

DREXLER, W., MORGNER, U., KÄRTNER, F. X., PITRIS, C., BOPPART, S. A. *et al.* (1999), *In-vivo* ultrahigh-resolution optical coherence tomography, *Opt. Lett.*, **24**, 1221–3.

ECCHER ZERBINI, P., GRASSI, M., CUBEDDU, R., PIFFERI, A. and TORRICELLI, A. (2002), Nondestructive detection of brown heart in pears by time-resolved reflectance spectroscopy, *Postharvest Biol. Tech.*, **25**, 87–99.

ECCHER ZERBINI, P., VANOLI, M., GRASSI, M., RIZZOLO, A., FIBIANI, M. *et al.* (2006), A model for the softening of nectarines based on sorting fruit at harvest by time-resolved reflectance spectroscopy, *Postharvest Biol. Tech.*, **39**, 223–32.

ECCHER ZERBINI, P., VANOLI, M., LOVATI, F., SPINELLI, L., TORRICELLI, A. *et al.* (2011), Maturity assessment at harvest and prediction of softening in a late maturing nectarine cultivar after cold storage, *Postharvest Biol. Tech.*, **62**, 275–81.

FANG, Q. (2010), Mesh-based Monte Carlo method using fast ray-tracing in Plücker coordinates, *Biomed. Opt. Express*, **1**, 165–75.

FANG, Q. and BOAS, D. A. (2009), Monte Carlo simulation of photon migration in 3D turbid media accelerated by Graphics Processing Units, *Opt. Express*, **17**, 20178–90.

FARRELL, T. J., PATTERSON, M. S. and WILSON, B. (1992), A diffusion theory model of spatially-resolved, steady-state diffuse reflectance for the non-invasive determination of tissue optical properties *in vivo*, *Med. Phys.*, **19**, 879–88.

FORD, H. D., TATAM, R. P., LANDAHL, S. and TERRY, L. A. (2012), Investigation of disease in stored onions using optical coherence tomography, *Acta Hort*, **945**, 247–54.

FRASER, D. G., JORDAN, R. B., KÜNNEMEYER, R. and MCGLONE, V. A. (2003), Light distribution inside mandarin fruit during internal quality assessment by NIR spectroscopy, *Postharvest Biol. Technol.*, **27**, 185–96.

GUNNING, P. (2012), Light microscopy, in V. Morris and K. Groves, *Food Microstructures: Microscopy, Measurement and Modelling*, Cambridge, UK, Woodhead Publishing Ltd.

HJALMARSSON, P. and THENNADIL, S. N. (2007), Spatially resolved *in vivo* measurement system for estimating the optical properties of tissue in the wavelength range 1000–1700 nm, *Proc. SPIE*, **6628**, 662805.

HJALMARSSON, P. and THENNADIL, S. N. (2008), Determination of glucose concentration in tissue-like material using spatially resolved steady-state diffuse reflectance spectroscopy, *Proc. SPIE*, **6855**, 685508.

HUANG, D., SWANSON, E. A., LIN, C. P., SCHUMAN, J. S., STINSON, W. G. *et al.* (1991), Optical Coherence Tomography, *Science*, **254**, 1178–81.

icumsa (2009), *The Determination of Refractometric Dry Substance (RDS %) of Molasses – Accepted and Very Pure Syrups (Liquid Sugars), Thick Juice and Run-off Syrups – Official*, Method GS4/3/8-13, London, International Commission for Uniform Methods of Sugar Analysis.

ISHIMARU, A. (1978), *Wave Propagation and Scattering in Random Media*, New York, Academic Press.

ISHIMARU, A. (1989), Diffusion of light in turbid material, *Appl. Opt.*, **28**, 2210–15.

JACQUES, S. L. (1989), Time-resolved reflectance spectroscopy in turbid tissues, *IEEE Trans. Biomed. Eng.*, **36**, 1155–61.

KEIJZER, M., JACQUES, S. L., PRAHL, S. A. and WELCH, A. J. (1989), Light distributions in artery tissue: Monte Carlo simulations for finite diameter laser beams, *Lasers Surg. Med.*, **9**, 148–54.

KIENLE, A., LILGE, L., PATTERSON, M. S., HIBST, R., STEINER, R. and WILSON, B. C. (1996), Spatially resolved absolute diffuse reflectance measurements for non-invasive determination of the optical scattering and absorption coefficients of biological tissue, *Appl. Opt.*, **35**, 2304–14.

KONG, J. A. (1986), *Electromagnetic Wave Theory*, New York, Wiley.

LANDAHL, S., TERRY, L. A. and FORD, H. D. (2012), investigation of diseased-onion bulbs using data processing of optical coherence tomography images, *Acta Hort.*, **969**, 269–70.

LEITGEB, R., HITZENBERGER, C. and FERCHER, A. (2003), Performance of Fourier domain vs. time domain optical coherence tomography, *Opt. Express*, **11**, 889–94.

LIANG, H., CID, M. G., CUCU, R. G., DOBRE, G. M., PODOLEANU, A. G. *et al.* (2005), En-face optical coherence tomography – a novel application of non-invasive imaging to art conservation, *Opt. Express*, **13**, 6133–44.

LORÉN, N. (2012), Confocal microscopy, in V. Morris and K. Groves, *Food Microstructures: Microscopy, Measurement and Modelling*, Cambridge, UK, Woodhead Publishing Limited.

LURIE, S., VANOLI, M., DAGAR, A., WEKSLER, A., LOVATI, F. *et al.* (2011), Chilling injury in stored nectarines and its detection by time-resolved reflectance spectroscopy, *Postharvest Biol. Tech.*, **59**, 211–18.

MARTELLI, F. and ZACCANTI, G. (2007), Calibration of absorption and scattering properties of a liquid diffusive medium at NIR wavelengths. CW method, *Opt. Express*, **15**, 486–500.

MARTELLI, F., DEL BIANCO, S., ISMAELLI, A. and ZACCANTI, G. (2009), *Light Propagation through Biological Tissue and Other Diffusive Media: Theory, Solutions, and Software*, Washington, DC, SPIE Press.

MEGLINSKI, V., BURANACHAI, C. and TERRY, L. A. (2010), Plant photonics: application of optical coherence tomography to monitor defects and rots in onion, *Laser Phys. Lett.*, **7**, 307–10.

NAES, T., ISAKSSON, T., FEARN, T. and DAVIS, T. (2002), *A User-friendly Guide to Multivariate Calibration and Classification*, Chichester, UK, NIR Publications.

NGUYEN DO TRONG, N., WATTÉ, R., AERNOUTS, B., HERREMANS, E., VERHOELST, E. *et al.* (2011), Spatially resolved spectroscopy for non-destructive quality inspection of foods, *2011 ASABE Annual International Meeting*, Louisville, KY, Paper No. 1111381.

NGUYEN DO TRONG, N., WATTÉ, R., AERNOUTS, B., VERHOELST, E., TSUTA, M. *et al.* (2012), Differentiation of microstructures of sugar foams by means of spatially resolved spectroscopy, *SPIE Photonics Europe 2012*, Brussels, Paper No. 8439–40.

NOFERINI, M., FIORI, G., CIOUS, V., GOTTARDI, F., BRASINA, M. *et al.* (2009), DA-Meter. easier control of fruit quality from farm to distribution, *J. Fruit Hort.*, **71**, 74–80.

PATTERSON, M. S., CHANCE, B. and WILSON, B. C. (1989), Time resolved reflectance and transmittance for the non-invasive measurement of tissue optical properties, *Appl. Opt.*, **28**, 2331–6.

PIFFERI, A., TORRICELLI, A., BASSI, A., TARONI, P., CUBEDDU, R. *et al.* (2005), Performance assessment of photon migration instruments: the MEDPHOT protocol, *Appl. Opt.*, **44**, 2104–14.

PIFFERI, A., TORRICELLI, A., TARONI, P., COMELLI, D., BASSI, A. and CUBEDDU, R. (2007), Fully automated time domain spectrometer for the absorption and scattering characterization of diffusive media, *Rev. Sci. Instrum.*, **78**, 053103.

PIRCHER, M., GÖTZINGER, E., LEITGEB, R., FERCHER, A. and HITZENBERGER, C. (2003), Measurement and imaging of water concentration in human cornea with differential absorption optical coherence tomography, *Opt. Express*, **11**, 2190–7.

PODOLEANU, A., ROGERS, J., JACKSON, D. and DUNNE, S. (2000), Three-dimensional OCT images from retina and skin, *Opt. Express*, **7**, 292–8.

PRESS, W. H., TEUKOLSKY, S. A., VETTERLING, W. T. and FLANNERY, B. P. (2007), *Numerical Recipes: The Art of Scientific Computing*, 3rd edition, Cambridge, UK, Cambridge University Press.

QIN, J. and LU, R. (2008), Measurement of the optical properties of fruits and vegetables using spatially resolved hyperspectral diffuse reflectance imaging technique, *Postharvest Biol. Tech.*, **49**, 355–65.

QIN, J. and LU, R. (2009), Monte Carlo simulation for quantification of light transport features in apples, *Comput. Electron. Agric.*, **68**, 44–51.

REN, N., LIANG, J., QU, X., LI, J., LU, B. and TIAN, J. (2010), GPU-based Monte Carlo simulation for light propagation in complex heterogeneous tissues, *Opt. Express*, **18**, 6811–23.

RIZZOLO, A., VANOLI, M., SPINELLI, L. and TORRICELLI, A. (2010), Sensory characteristics, quality and optical properties measured by time-resolved reflectance spectroscopy in stored apples, *Postharvest Biol. Tech.*, **58**, 1–12.

SAEYS, W. THENNADIL, S.N., RAMON, H. and NICOLAÏ, B. (2010a), Optical characterization of apple tissue: a multiscale approach, in, *Proceedings of the 14th International Conference on Near Infrared Spectroscopy*, 7–16, November 2009, Bangkok, Paper No. O-077.

SAEYS, W., NGUYEN DO TRONG, N., WATTÉ, R., TSUTA, M. *et al.* (2010b), Optical characterization of biological material: a multiscale approach, *Presented at the 2010 ASABE Annual International Meeting*, 20–23 June 2010, Pittsburgh, PA, ASABE, Paper No. 1000032.

SAITO, Y. (2009), Monitoring raw material by laser induced fluorescence spectroscopy in the production, in M. Zude, *Optical Monitoring of Fresh and Processed Agricultural Crops*, London, CRC Press, Taylor & Francis Group.

SHEN, H. and WANG, G. (2011), A study on tetrahedron-based inhomogeneous Monte Carlo optical simulation, *Biomed. Opt. Express*, **2**, 44–57.

STIFTER, D. (2007), Beyond biomedicine: a review of alternative applications and developments for optical coherence tomography, *Appl. Phy. B.*, **88**, 337–57.

SUMPF, B. (2009), Spectrophotometer technology, in M. Zude, *Optical Monitoring of Fresh and Processed Agricultural Crops*, London, CRC Press, Taylor & Francis Group.

TARGOWSKI, P., IWANICKA, M., TYMINLASKA-WIDMER, L., SYLWESTRZAK, M. and KWIATKOWSKA, E. A. (2009), Structural examination of easel paintings with Optical Coherence Tomography, *Acc. Chem. Res.*, **43**, 826–36.

TIJSKENS, L. M. M. P., ECCHER ZERBINI, P., SCHOUTEN, R., VANOLI, M., JACOB, S. *et al.* (2007), Assessing harvest maturity in nectarines, *Postharvest Biol. Tech.*, **45**, 204–13.

TUCHIN, V. (2007), *Tissue Optics – Light scattering methods and instruments for medical diagnosis*, 2nd edition, Washington, DC, SPIE Press.

VAN DER JEUGHT, S., BRADU, A. and PODOLEANU, A. (2010), Real-time resampling in Fourier domain optical coherence tomography using a graphics processing board, *J. Biomed. Opt.*, **15**, 030511.

VANOLI, M., ECCHER ZERBINI, P., GRASSI M., RIZZOLO, A., FIBIANI, M. *et al.* (2005) The quality and storability of apples cv 'Jonagored' elected at harvest by time-resolved reflectance spectroscopy, *Acta Hort.*, **682**, 1481–8.

VANOLI, M., RIZZOLO, A., ECCHER ZERBINI, P., SPINELLI, L. and TORRICELLI, A. (2010), Non-destructive detection of internal defects in apple fruit by time-resolved reflectance spectroscopy, in, C. Nune, *Proceedings of the International Conference Environmentally Friendly and Safe Technologies for Quality of Fruit and Vegetables*, Universidade do Algarve, Faro, Portugal, 14–16 January, 2009.

VANOLI, M., RIZZOLO, A., GRASSI, M., FARINA, A., PIFFERI, A. *et al.* (2011), Time-resolved reflectance spectroscopy non-destructively reveals structural changes in 'Pink Lady[®]' apples during storage, *Procedia Food Science*, **1**, 81–9.

VERAVERBEKE, E. A., VAN BRUAENE, N., VAN OOSTVELDT, P. and NICOLAÏ, B. M. (2001), Non-destructive analysis of the wax layer of apple (*Malus domestica* Borkh.) by means of confocal laser scanning microscopy, *Planta*, **213**, 525–33.

VERBOVEN, P., NEMETH, A., ABERA, M. A., BONGAERS, E., DAELEMANS, D., ESTRADE, P., HERREMANS, E., HERTOG, M., SAEYS, W., VERLINDEN, B., LFITNER, M., and NICOLAI, B. (2013), Optical coherence tomography visualizes microstructure of apple peel, *Postharvest Biol. Tech.*, **78**, 123–32.

VERHOELST, E., BAMELIS, F., KETELAERE, B., NGUYEN DO TRONG, N., DE BAERDEMAEKER, J. *et al.* (2011), The potential of spatially resolved spectroscopy for monitoring angiogenesis in the chorioallantoic membrane, *Biotechnol. Prog.*, **27**, 1785–92.

WANG, L., JACQUES, S. L. and ZHENG, L. (1995), MCML – Monte Carlo modeling of light transport in multi-layered tissues, *Comput. Methods Programs Biomed.*, **47**, 131–46.

WATTÉ, R., AERNOUTS, B. and SAEYS, W. (2012), A multi-layer Monte Carlo method with free phase function choice, *SPIE Photonics Europe*, Brussels, Paper No. 8429-27.

WELLNER, N. (2012), FTIR, NIR and Raman Microscopy, in, V. Morris and K. Groves, *Food Microstructures: Microscopy, Measurement and Modelling*, Cambridge, UK, Woodhead Publishing Ltd.

WIESAUER, K., PIRCHER, M, GOTZINGER, E., BAUER, S., ENGELKE, R. *et al.* (2005), En-face scanning optical coherence tomography with ultra-high resolution for material investigation, *Opt. Express*, **13**, 1015–24.

WIESAUER, K., PIRCHER, M., GOETZINGER, E., HITZENBERGER, C., ENGELKE, R. *et al.* (2006), Transversal ultra-high-resolution polarizationsensitive optical coherence tomography for strain mapping in materials, *Opt. Express*, **14**, 5945–53.

WIESAUER, K., PIRCHER, M., GÖTZINGER, E., HITZENBERGER, C. K., OSTER, R. and STIFTER, D. (2007), Investigation of glass-fibre reinforced polymers by polarisation-sensitive, ultra-high resolution optical coherence tomography: Internal structures, defects and stress, *Composites Science and Technology*, **67**, 3051–8.

WILSON, B. C. and ADAM, G. (1983), A Monte Carlo model for the absorption and flux distributions of light in tissue, *Med. Phys.*, **10**, 824–30.

WILSON, B. C. and JACQUES, S. L. (1990), Optical reflectance and transmittance of tissues: principles and applications, *IEEE J. Quantum Electron*, **26**, 2186–99.

YUD-REN, C., KUANGLIN, C. and MOON, S. K. (2002), Machine vision technology for agricultural applications, *Comput. Electron. Agric.*, **36**, 173–91.

ZIOSI, V., NOFERINI, M., FIORI, G., TADIELLO, A., TRAINOTTI, L. *et al.* (2008), A new index based on vis spectroscopy to characterise the progression of ripening in peach fruit, *Postharvest Biol. Tech.*, **49**, 319–29.

ZUDE, M. (2003), Comparison of indices and multivariate models to non-destructively predict the fruit chlorophyll by means of visible spectrometry in apples, *Anal. Chim. Acta*, **481**, 119–26.

6

Fourier transform infrared (FTIR) and Raman microscopy: principles and applications to food microstructures

N. Wellner, Institute of Food Research, UK

DOI: 10.1533/9780857098894.1.163

Abstract: Vibrational spectroscopies in combination with microscopy techniques can be used for analysing the chemical composition of microscopic domains in heterogeneous food materials. This chapter characterises the two most widely applied methods – Fourier-transform infrared (FTIR) and Raman microscopy. The resulting chemical maps give information about the different domains in heterogeneous food materials such as emulsions and gels, plant tissues and mixed food biopolymer systems. After discussing technical principles and applications, FTIR and Raman microscopy are compared with respect to their relative advantages and disadvantages. Finally, new advances in technology are presented, which aim to overcome physical and technical limitations.

Key words: infrared, Fourier-transform infrared (FTIR), Raman, microscopy, hyperspectral data, chemical imaging.

6.1   Introduction

Vibrational spectroscopy is often used in preference to exhaustive chemical analysis methods, especially in industrial settings where time is of the essence. In addition, samples can be examined *in situ*, avoiding the loss of structural information due to chemical and/or physical separation steps. The basic principle is the interaction of samples with light that can excite vibrational transitions in the molecules (Fig. 6.1). A molecule with N atoms has 3N-6 different vibrational modes, and the fundamental transitions in these modes (generally from the ground state to the first excited state), involve energies that correspond to that of infrared (IR) light (4000–400 cm^{-1}, 2.5–25 micrometres wavelength). The excitation of first and second overtones and combination bands requires higher energies, corresponding to light with shorter wavelengths, from 900 to 2000 nm in the near

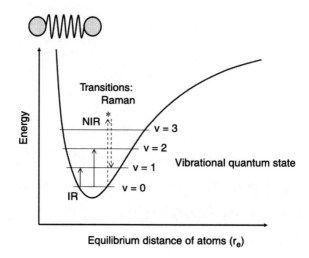

Fig. 6.1 Schematic drawing of a vibrational potential function and the energy level transitions measured by IR, NIR and Raman spectroscopy, * denotes a virtual excited state.

infrared (NIR), and is much less likely. Strictly speaking, the vibrational selection rules only allow ± 1 transitions; higher transitions are only possible because the asymmetry of the potential function of the normal modes gives them a small probability. IR spectra are mostly obtained directly by measuring the amount of IR light absorbed by a sample, but they can also be obtained in other ways. Raman spectroscopy utilises the Raman effect, that is, shifts in the wavelength of inelastically scattered visible light (400–1100 nm) that correspond to the vibrational transitions in the mid- and far IR. The Raman scattering is very inefficient (only 1 in 10^6 photons is scattered inelastically), but the shorter wavelength of the employed light has distinct advantages.

Because the extinction coefficients at each wavelength are determined by the geometry of the molecule and the strength of the bonds between its atoms, each compound has a characteristic absorption spectrum. The band pattern, especially in the 'fingerprint' region from 1500 to 400 cm^{-1}, allows the identification of compounds by IR and Raman spectroscopy. In contrast, the complex overtones observed in the NIR are much less characteristic and usually only relate to whole classes of molecules (Fig. 6.2).

The intensity of the observed spectrum depends on the number of molecules (or absorbing centres) in the sample. The well-known Beer–Lambert law states that the absorption (A) of a light beam passing through a sample of thickness d is directly proportional to the concentration (c):

$$A = -\log_{10} I/I_0 = \varepsilon cd \qquad [6.1]$$

where I and I_0 are the transmitted and incident intensities, respectively.

This relationship allows quantitative concentration measurements with a suitable calibration. A similar relationship also exists for scattered (diffusely

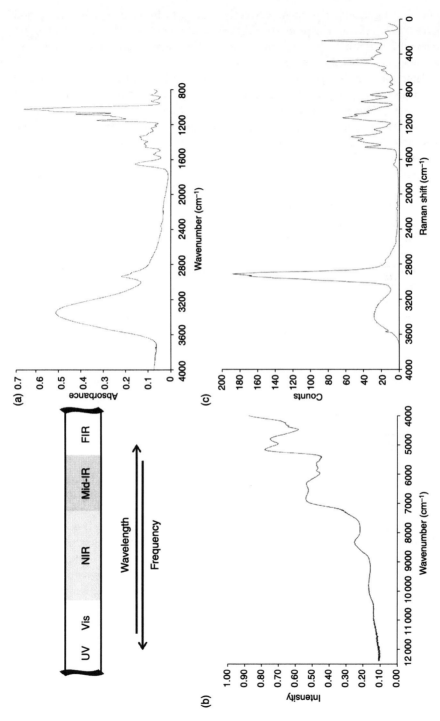

Fig. 6.2 Interaction of matter with light from different regions in the electromagnetic spectrum: comparison of the FT-IR (a), NIR (b) and Raman spectra (c) of maize endosperm.

reflected) light, although some corrections have to be made for scattering and particle size effects. The proportionality generally holds for absorption values up to about 2, but it becomes non-linear for very strong absorption corresponding to less than 1% light transmitted.

Thus, vibrational spectroscopic techniques can be used for rapid qualitative and quantitative analyses. This is relatively straightforward for bulk samples, but with appropriate optics these measurements can also be made on small parts of the samples, as long as we take into account that all these techniques have finite sampling volumes, determined by the spot size of the incident light, and its penetration depth in the sample. Spectra from distinct spatial domains (phases, components) can only be obtained when these domains are larger than the sampling volume, otherwise the experiment will return average spectra from several domains. Therefore the key step for moving from bulk analyses to the investigation of food microstructures on a microscopic scale is the reduction of the sampling size. The first IR microspectroscopy was reported as long ago as 1949 (Barer et al., 1949).

For IR and Raman microscopy, this is achieved by combining a spectrometer with microscopes, where the light beam can be tightly focused and the transmitted or scattered light collected from a small sample area. The dimension in the z-direction can be controlled by sample thickness or penetration depth. NIR is traditionally used for bulk analysis because of its relatively large penetration depth, but recently special NIR imaging systems have been introduced that focus reflected light from the sample surface onto an IR-sensitive digital camera and thus allow chemical imaging in analogy to the more widely established FTIR and Raman imaging systems.

6.2   Instrumentation

6.2.1   FTIR microscopy

An IR microscopy system consists of an IR microscope attached to an FTIR spectrometer. The main difference from a normal confocal microscope is that it must not contain any glass lenses, because glass would absorb all the IR light. Therefore the optical elements are gold- or aluminium-coated mirrors and some IR transparent windows. Focusing the beam on the sample is achieved by Cassegrain reflector - style condensers and collectors, and both transmission and reflection modes are available (Fig. 6.3). These mirror optics extend the usable frequency range into the IR region. In the visible light range (400–800 nm), the microscope operates as normal: the sample on the microscope stage is illuminated with white light from a small halogen bulb and an image is observed in the eyepiece or projected onto a digital camera.

When the optics are switched to IR light, the microscope takes on the role of a beam condenser in an IR spectrometer. IR light is passed through a set of apertures into a Michelson interferometer, where it is modulated with a moving mirror. The modulated light beam is then directed to the microscope

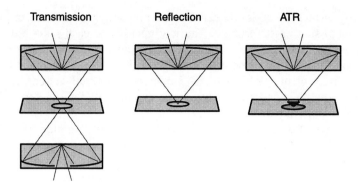

Fig. 6.3   Schematic comparison of FTIR microspectroscopy sampling configurations.

optics that focuses it onto the sample. The transmitted (or reflected) IR light is collected and focused onto the IR detector, where the light intensity is converted into an electrical signal. The measured intensity, which is a function of the mirror position, is converted into a single-beam spectrum with a fast Fourier transform, and then the instrument signal is removed with a division by a background spectrum. It is important to note that the finished spectrum is the ratio between two single beam spectra, that is, with and without sample, respectively, and their intensities have a strong impact on the signal-to-noise ratio.

The sampling area on the stage is determined by the magnification of the condenser and collector 'lenses', which is typically 15x, thus the spot size on the sample has a diameter of about 0.8 mm. The area can be reduced further by setting an aperture in one of the focal planes of the IR beam. However, reducing the aperture size cuts down the intensity of the beam and also enhances diffraction, therefore this approach becomes impractical for areas smaller than around 10×10 micrometres.[2]

Bench top FTIR spectrometers work with IR light from a glowing ceramic source, which is very divergent and not easy to focus onto a small spot. It is also not very bright, and using a small aperture drastically reduces the available intensity. In contrast, the light from a synchrotron is much more intense and has a low divergence, which means that sample spots of 3 to 10 micrometres can be easily measured with an excellent signal-to-noise ratio (Yu *et al.*, 2003). Most synchrotron sources (i.e. Diamond in the UK, Soleil in France, Brookhaven in the USA) have dedicated IR microscopy beam lines for high-resolution imaging and spectroscopy. The main drawback, of course, is the cost and accessibility of the synchrotron light source.

This is particularly valid because in recent years the major instrument manufacturers have been improving their IR microscope systems so that they can achieve comparable spatial resolution. The crucial step has been the introduction of focal plane imaging detectors. Instead of measuring spectra from one apertured sample spot at a time, these detectors act like digital cameras in the focal plane, with many elements working in parallel, each recording a spectrum originating from a small spot on the sample. The low intensity at each pixel reduces the signal-to-noise ratio and may require longer acquisition times, but the spatial resolution of these focal plane detector optics is basically diffraction-limited by the wavelength of the IR light (2.5–15 micrometres, 4000–800 cm^{-1}). Various models of point, line and array detectors, in combination with motorised stages are now available commercially.

The interaction of IR light with polar molecules is very strong – mid-IR light is strongly absorbed by materials with many polar groups such as OH, NH, C=O, etc. This concerns both the slides used as sample substrates, as well as the food materials that contain proteins, carbohydrates, and is especially true for water. Therefore the transmission measurements require thin sections, preferably dry, mounted on an IR transparent support. Care has to be taken when cutting and mounting the sections. The preferred method would be dry cutting or cryo-sectioning. Fixing and embedding of delicate samples in resins, paraffin or polyethylene glycol (PEG) should be avoided, because there would be band overlaps from the embedding/fixing material. It has been reported (Richter *et al.* 2011) that embedding of plant tissues can be done in PEG, which is washed out after the sectioning, but that would entail the loss of soluble constituents and therefore this approach is only suitable for studying insoluble structures, such as cell walls.

The thickness of the sections depends on the water content of the original material – some samples, especially plant tissues, with a high water content can be cut thick (40–70 micrometres), as the water will be lost on drying and will leave a much thinner dry section. However, dense samples need to be cut quite thin, perhaps 1 to 5 micrometres, in order to have a reasonable light throughput. The aim is to have a thin section with an IR absorption value in the range of 0.1 to 1.0. This can be challenging with samples where the dry matter content varies – such as in plant sections that contain dense vascular bundles or epidermal layers and much less dense parenchyma. The result is sometimes an uneven section where either the thick parts over-absorb, or which has large holes in the less dense areas.

As glass slides cannot be used for IR microscopy, the sections have to be mounted on an IR-transparent support. The most commonly used materials are CaF$_2$ or BaF$_2$ crystals, which are IR transparent down to 700 to 800 cm^{-1}, and, more importantly, water-resistant. However, because of their relatively high cost, only small disks are used (typically 13 mm diameter, 1–2 mm thick). They can be re-used, but are fragile and easily scratched. A much more economical alternative for large sample numbers is to mount the sample sections on mirror slides. These can be either normal metal mirrors (gold, aluminium) or glass slides coated with a special layer that renders them IR reflective, but still lets through normal visible light, so that they can also be used for normal optical microscopy. When using mirror supports, we have to bear in mind that the light in this set-up will pass twice (Fig. 6.3) through the sample, therefore thinner sections must be cut than for transmission.

The sections must adhere to the disk/support. Many materials do stick to glass or metal, but not all. Another problem with samples that contain a lot of water is that they shrink during drying and tend to shrivel and curl up. The most obvious solution would be to sandwich the sections between two disks, but this often has the drawback of creating interference fringes from multiple reflections at parallel crystal surfaces.

Liquid or gel samples can be investigated sandwiched between two BaF_2 disks, as long as the layer is very thin. Samples with large water content usually have to be no more than 5 to 10 micrometres thick, since otherwise the water absorption becomes too strong. In some cases, D_2O can be used to shift the strong water bands away from amide band region, but even this only works in thin layers. An advantage of working in aqueous system is that the refractive indices of the sample and the IR windows are similar, thus reducing the interference fringing from multiple internal reflections.

Preparing thin layers can be problematic for a number of reasons. To avoid these problems, it is possible to use a technique that has become the method of choice for IR spectroscopy when it comes to awkward, highly absorbing samples – Attenuated Total Reflection (ATR). In this technique, the IR light is bounced inside a crystal and only the evanescent wave is absorbed by material in contact with the surface. This equates to a reproducible and a very thin penetration depth of between 0.5 to 2 micrometres depending on wavelength and refractive index, so that the actual sample thickness is irrelevant.

The method has been adapted for IR microscopy by placing a small conical ATR crystal in the focus of the microscope. The IR light is reflected inside the crystal and only interacts with the sample where the crystal touches the sample. With a single point detector, the spatial resolution is determined by the size of the tip of the ATR crystal. Recently, it has even become possible to combine ATR tips with imaging detectors, which sub-sample the image of the tip. Then the sampling size becomes smaller, determined by detector pixel size and optical parameters such as magnification, refractive indices and diffraction. Samples that are measured with this technique can be sections, but are more often solid slices or blocks, or powder particles. The only requirement is a good contact between the sample and crystal. This requires some gentle pressing of the crystal on the sample, but care must be taken not to damage the tip. The workflow of IR microscopy measurements in the different configurations is compared in Fig. 6.4.

6.2.2   Raman microscopy

In contrast to the IR microscopy, Raman microscopy is carried out with conventional light microscopes, which have high-power lasers and Raman detectors attached, either directly or via optical fibres. The sample position and focus are selected in normal white light, then the system is switched over to the Raman mode (confocal, but cannot be used simultaneously for obvious laser safety reasons!). A range of diode-pumped solid state lasers and tuneable

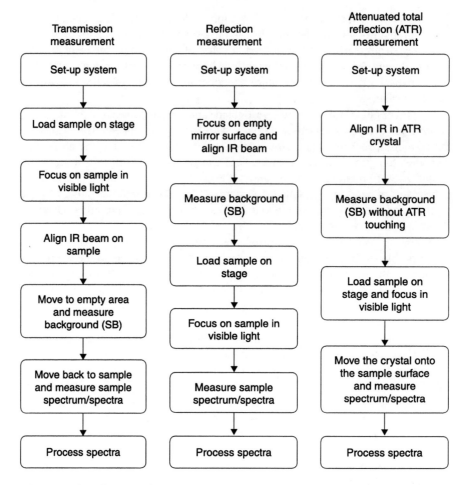

Fig. 6.4   Flow diagram of the measurement with IR microscopes in different sampling configurations.

diode-lasers in the yellow/blue/green/red (400–1100 nm) can be used for Raman excitation. The laser light is passed through the microscope onto the sample, and scattered light collected in the objective. After passing through narrow bandpass filters that block out the elastically scattered light at the excitation frequency, the Raman scattered (frequency shifted) light is spread out by a grating and collected with a highly sensitive multi-pixel photodiode detector. The spot size is determined by the magnification of the objective (10×–100×) and the laser optics configuration. Most instruments can reach a resolution of 1 micrometre or better, and resolution limits of 200 nm have been claimed (Witec, JobinYvon). Since the optical path of the microscope is fixed, mapping is performed with a moving stage. By using a high optical magnification and small step-size for the stage, samples can be mapped at sub-micrometre resolution.

It is often said that Raman and IR are complementary techniques. Raman scattering depends on electron polarisability and obeys different selection rules than IR absorption. Thus, Raman bands tend to be strongest for non-polar bonds such as C–C or C–H, whereas OH or NH bonds give only a weak signal. This leads to less abundant compounds (unsaturated, aromatic compounds, etc.) showing strong bands, while the most abundant food compounds (proteins, polysaccharides, etc.) have rather weak scattering. Water as the most polar molecule does not give rise to strong Raman bands, and neither does glass. Therefore samples can be mounted on normal glass slides, even with a cover-slip if required, and can be measured in water (but note that water does dampen scattering!)

Most Raman microscopy is performed in reflection mode – Raman spectra are usually collected in a 180-degree backscattering configuration. Therefore a wide range of sample forms can be used – thin sections as well as thick blocks, powders, gels and emulsions. The sampling volume is determined by the laser spot size (focus area) and the penetration depth, which can be variable depending on the optical properties of the sample. The easiest way to prepare samples is to cut a solid block with a flat surface. The z-dimension can be constrained by using thin sections. The drawback is that this reduces the amount of sample in the focus and so the observed signal may be weak. For cutting sections, the same considerations as for the FTIR apply and, in particular, we have to be careful to avoid contamination from embedding material. Gels and emulsions may be investigated sandwiched between a glass slide and a cover-slip. The sampling plane is determined by the z-focus, if transparent enough, perhaps using a refractive-index matched solvent.

A particular problem with Raman spectroscopy is fluorescence, because of the higher energy of the visible light compared to IR. Many aromatic compounds have low-lying electronic transitions that can give rise to strong fluorescence. This is a particular problem with phenolic compounds, which are frequently found in plant tissues. Also, some types of glass can have fluorescent additives.

The sample is positioned on the stage and focused in white light. With the measurement point/area selected, the system is switched over to Raman mode and the spectra are collected. Since Raman spectroscopy measures the scattered light, it is mostly not necessary to collect a background spectrum. Special care has to be taken to avoid overheating the sample with the concentrated laser beam!

6.2.3   Near infrared (NIR) imaging

NIR is traditionally used for bulk samples, because the weak interaction of NIR light with matter allows a relatively deep penetration of several millimetres into the sample. NIR light is therefore collected from a wide sample volume. However, recently systems have been developed which use an NIR sensitive focal plane camera in combination with tuneable frequency filters. These instruments acquire a series of images at various NIR frequencies, with the spatial resolution determined mainly by the optics and the pixel number of the detector. Stacking these images yields data cubes analogous to the IR and Raman data. This technique has not quite the spatial resolution of FTIR or Raman, but possible resolutions down to several 10s of micrometres have been claimed by the manufacturers. Despite the rather broad bands in the NIR spectrum, in combination with powerful chemometric procedures and image analysis tools, the technique gives fast maps of chemical composition. NIR imaging is mostly used for whole fruits or seeds, and so far little microscopy has been done in the food area, but the technique has found a ready use in pharmaceuticals where it is desirable to image the distribution and form of constituents (particle sizes and packing, crystal phases) in tablets either for understanding drug delivery characteristics or for detecting counterfeit products.

NIR imaging uses non-contact scattered light and therefore does not require any sample preparation. Samples are mostly solid blocks (tablets) placed in the focus of the camera system.

6.3   Data analysis

Each compound has a characteristic IR spectrum – that is, the absorption at each wavelength is determined by the geometry of the molecule – the masses and relative positions of all the atoms and the strength of the bonds between them. The band pattern enables easy identification of compounds, and the observed signal depends on the number of molecules in the sample. Together, this allows qualitative and quantitative analysis of samples. With the microscope set-up, all the generally used analysis procedures can be applied to small sample spots. Single spectra can be useful, for instance for the analysis of small sample or tissue areas or the identification of small inclusions or contaminants in particular samples. However, nowadays IR or Raman microscopy is routinely used as an imaging tool – spectra are obtained at hundreds or thousands of points in a grid pattern overlaid on the sample. The result of such an experiment is a hyperspectral data cube (Fig. 6.5), with one spectrum for each x,y data point. Thus, the spectral analysis has gained spatial dimensions.

6.3.1   Composition

Identification of chemical compounds involves matching specific features or bands with reference spectra. The general spectral features of the major food components – water, protein, oil/fat and carbohydrates – are well known, and each has one or more strong signature bands (Fig. 6.6). Thus, for instance, a map of the amide II band intensity at around $1550\,cm^{-1}$ allows the visualisation of the distribution of protein even in the presence of water, which has a strong OH deformation band at $1636\,cm^{-1}$. However, it is usually not possible to separate individual proteins in a food matrix with many different compounds, because their spectra are similar. Similarly, carbohydrate bands are strongly overlapping, although their spectral patterns in the fingerprint region do allow the distinction

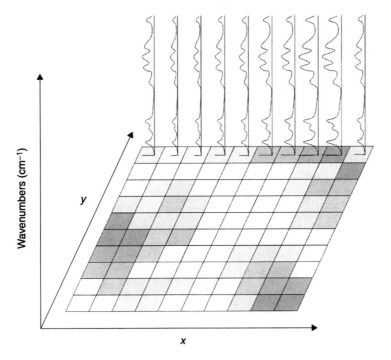

Fig. 6.5   Schematic drawing of a hyperspectral data cube with two spatial dimensions (x,y) and one spectral dimension.

between different types to a certain degree. However, 'foreign' materials such as plastics or solvents, etc. have different spectra and their distinct bands enable the identification of inclusions or contaminants.

A further piece of information that is crucial to understanding many food systems is the physical state of the components. The IR and Raman spectra of a chemical compound in solid, liquid and gas forms have marked differences because of the different molecular environments – gas spectra show rotation/vibration bands, but in the condensed phase the rotation fine structure is lost as the bands are broadened by collisions. In liquids, solutions and amorphous solids, the bands are deformed and shifted by molecular interactions. In the crystalline state, the molecules are fixed in regular conformations and the bands are much narrower. Furthermore, the spectra are no longer purely of individual molecules. The defined positions and intermolecular interactions in the crystal lattice can cause band splitting and give rise to many additional bands, especially in the fingerprint region (Fig. 6.7). This kind of information is useful, for instance, for food where the texture may be influenced by the phase behaviour of fats (lipids), the crystallinity of carbohydrates, especially starch or sugars, and the structures of proteins in foodstuffs such as bread, cakes, crisps or chocolate.

The quantitative determination of the amount of a compound present in a sample makes use of the linear relationship between absorbance and concentration.

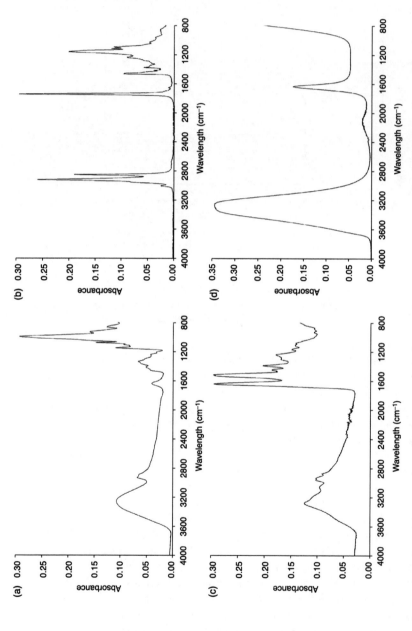

Fig. 6.6 Examples of FTIR–ATR spectra of major food components: (a) carbohydrates (wheat starch), (b) lipids (olive oil), (c) protein (bovine serum albumin), and (d) water. All samples were measured with ATR.

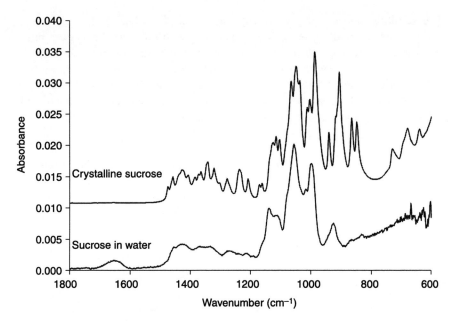

Fig. 6.7   Influence of hydration and phase on FTIR spectra – comparison of sucrose crystals and sucrose dissolved in water (spectra measured with ATR).

When the compound of interest has one or more unobstructed bands, then it is straightforward to generate a calibration curve by plotting the absorption as a function of the known concentrations for a range of reference samples. The largest bands from the main component can be treated this way. Thus it may be possible to approximately determine protein, water, carbohydrate, fat, etc. in selected, simple systems (e.g. where essentially only one component is varied). However, in many cases, the system is more complex – the samples contain many different proteins and different carbohydrates, in addition to other compounds (water, fat, nucleic acids, etc.) in variable amounts. This makes it difficult to assign the bands to individual proteins or carbohydrates. Furthermore, less abundant compounds may be largely obscured by the major constituents of the food matrix.

These systems can be analysed successfully using chemometric methods such as Principal Components Analysis (PCA), Partial Least Squares (PLS), clustering, etc. (Wold, 1995). Quantitative analysis by chemometrics methods is based on the two assumptions that each compound has a unique spectrum and its intensity is a function of the concentration. Therefore the observed spectrum of a mixture (M) of n components is the linear combination of the spectra (S_i) of its individual (i) components weighted by their respective concentrations (c_i):

$$M = c_1 * S_1 + c_2 * S_2 + \ldots + c_n * S_n \qquad [6.2]$$

The resulting linear equation system can be easily solved if the number of components is small, if each has a distinct spectrum, and the concentration dependence is strong. However, in practice, the number of components can be large, not all of the component spectra are available (or even known), and the concentrations may vary only over a limited range of a few percent. Moreover, all experiments contain some amount of noise, from instrument instabilities, variable sampling, etc. which adds another factor (E) to the equation:

$$M = c_1 * S_1 + c_2 * S_2 + \ldots + c_n * S_n + E \qquad [6.3]$$

Of course, some kinds of variability can be eliminated from the spectra by appropriate pre-processing, for example baseline correction or area normalisation, but some noise still remains. As a result, the equation does not have an exact solution. Instead it is usually approximated by some kind of fitting routine, which minimises E.

There are several methods used for the analysis of hyperspectral datasets. Since IR spectra are usually broad and have many bands, there is a lot of redundancy in the data. Therefore the first step is usually a principal component analysis (PCA), which rotates the n-dimensional spectral data (n = number of spectral wavelengths) into a new set of coordinates, which are aligned with the different sources of variance in the dataset. Since most of the variance will be explained by the first few principal components, this reduces the data considerably. The variation of composition can be visualised by plotting a few component scores for each spectrum. If appropriate, the point spreads in these (2D, 3D) plots can be used to assign the samples to different groups/classes. (Discriminant analysis, Mahalanobis distances, etc.). Of course, there are many other techniques for pattern recognition that can be applied for classifying spectra, from K-means clustering to neural networks. (Hopke, 2003).

In order to produce a quantitative model, a similar approach to PCA is used. PLS (partial least squares) regression produces a coordinate transformation to the spectra of a calibration set, which is aligned with their known concentrations.

The advantage of these techniques is that once a chemometrics model is made from a suitable calibration sample set, it can be applied to unknown spectra very quickly. This makes it an ideal method to analyse all the spectra in large imaging datasets.

6.3.2   Chemical imaging

Hyperspectral data cubes contain a spectrum for each x,y data point. The final step is converting the extracted chemical information – such as the concentration or crystallinity of a particular component – into a colour value and plotting it as a function of the x,y coordinates, and thus forming a picture of the chemical composition of the sample, a chemical image.

The spectra contain information about many different compounds and with suitable calibrations different chemical images can be overlaid onto the visual images to show the spatial distribution of various compounds. The beauty of the method is that it uses the information in the spectra, and therefore no staining of particular compounds is required.

6.4   Applications

Originally specific structure- or phase-dependent band shifts in the IR and Raman spectra of food materials were used to deduct molecular structures, but without physical separation, heterogeneous samples only provide average spectra in the bulk phase. Even food matrices that appear smooth to the eye, such as emulsions or gels, can nevertheless be heterogeneous at the microscopic level. The physical properties and textures of these systems may vary if their components are not mixed evenly or form discrete particles. The use of microscope optics make it possible to measure spectra from small sample spots, and the compositions and properties of individual phases can be investigated, as long as the physical domains are larger than the spatial resolution limit.

6.4.1   Mixed gels

One of the first food applications of FTIR microscopy was to study different domains in a phase-separated gelatin/amylopectin system, which forms thermo-reversible gels when a hot mixed solution is cooled to room temperature (Cameron et al., 1994; Durrani et al., 1993; Durrani and Donald, 1994; Prystupa and Donald, 1996; Sun et al., 1996a,b). A gel made up of 10% (w/w) amylopectin/7% (w/w) gelatin/83% (w/w) D_2O, sandwiched in a 12.5 micrometre thin layer between two AgCl disks, was studied with a FTIR microscope. In this system, the separated phase domains were large enough to allow mapping of the gelatin/amylopectin distribution, using a 40×40 micrometres^2 aperture and single point detector. Semi-quantitative maps of the chemical composition were generated, by plotting the spatial distribution of the band areas of the individual components (e.g. the amide I band). Quantitative determination of the gelatin, amylopectin and D_2O wt% was obtained by applying a PLS model using spectra obtained from bulk calibration samples. This allowed the direct comparison of the compositions of the microscopic domains within bulk phase-separated solutions above the gel temperature and showed that they had the same phase concentrations. The authors noted a spatial resolution limit of 30 micrometres – thus this approach works for systems with large domains (typically already phase-separated at high temperature), but not for systems where phase-separation on cooling generates small domains (Durrani and Donald, 1994, 1995).

The analysis of such systems was also carried out using the higher spatial resolution of Raman mapping (Pudney et al., 2002, 2003, 2004; Pudney and Hancewicz, 1999). Two different microstructures were mapped using a Raman microscope with a 785 nm laser at a spatial resolution of 1.9 micrometres in the x,y plane and 2.9 micrometres in the z-plane: a bulk phase separated system (gelatin/dextran), and a gelled microstructure of two carbohydrate polymers (gellan/kappa-carrageenan), which is phase separated on the micrometre scale. In the latter system, the highly overlapping spectra of the two similar polysaccharides were separated using multivariate curve resolution (MCR), to give quantitative maps of the two polymers in the gels (Pudney *et al.*, 2003). Such measurements allowed the determination of tie lines for different micro-phases, and the approach can be useful for other types of analyses, for example the measured local concentration of the included phase in a composite gel, allowed the modelling of physical properties of the gel, such as the kinetic evolution of the elasticity of the sample (Normand *et al.*, 2000). In principle, Raman imaging could also be applied to emulsion systems; however, the scanning does take several hours and the sample needs to be stable for the whole time (Andrew *et al.*, 1998).

6.4.2   Biopolymer films fracture

Understanding the organoleptic properties, as well as their processing and storage properties, of foods often requires the knowledge of other properties besides the concentrations of components. For instance, texture and mouth-feel of dried food systems depend on their fracture behaviour. Starch and gelatin are examples of biopolymers widely used as food ingredients or for producing capsules for pharmaceuticals. It is well known that most biopolymers are strongly affected by hydration, but the details of the connection between chemical structures and the visco-elastic behaviour and fracture of dried protein and starch systems are complex. FTIR microscopy maps were used to visualise the structural homogeneity and cohesivity of three different biopolymer materials, gelatin, hydroxypropyl cellulose (HPC) and cassava starch, in thin films that were mechanically deformed and fractured at different hydration levels (Paes *et al.*, 2010; Yakimets *et al.*, 2005, 2007). Figure 6.8 shows an example of the extensive deformation zone created around the propagating crack tip in a fractured HPC film.

These studies utilised the fact that the IR spectra can yield information beyond that of just the chemical concentration. The band shapes of the amide bands in proteins are indicative of their secondary structures (Byler and Susi, 1986). Besides depending on the molecular conformation, the IR spectra (or at least some of their bands) are also sensitive to the environment of the molecules. The most significant effect for many biopolymer systems is formation of hydrogen bonds with either other polymers or water. This causes shifts in the positions of both donor (NH, OH) and acceptor groups ($C=O$, $C–O$, $C–N$). By monitoring these shifts, as well as the intensity increases due to the water bands, the changes in structure and intermolecular interactions can be studied directly as a function of hydration.

Mechanical stress can also induce changes in the spectra. Spectral band positions can be affected by mechanical force directly, as for example when the stretching of large polymer molecules leads to widening of some bond angles. However, more often the applied force leads to changes in the structures and orientation of polymers. The latter can be observed with the use of polarised light, because the probability of a molecule absorbing an IR photon varies with $\sin(\theta)$, θ being the angle between the direction of the dipole moment of any given

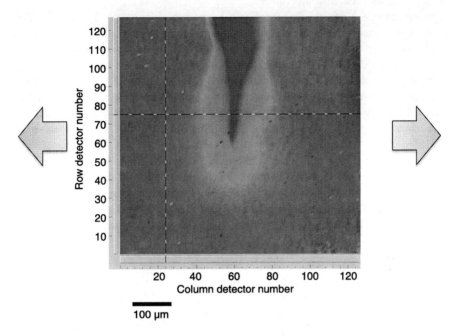

Fig. 6.8   IR map of the deformation zone around the progressing fracture tip in a stretched HPC film. The arrows show the stretching direction. The grayscale intensity in the plot indicates the extent of the film thinning and molecular orientation changes.

vibrational mode and the plane of the electric vector of the incident light. Not all vibration modes show a large dichroic effect, but in materials with orientated polymers, there are usually some bands that are sensitive enough to determine the preferential orientation in the material.

Crystalline cellulose with its well-orientated linear structure is a particularly striking example, but many other polysaccharide chains can have some bands showing a dichroic effect, either from the glycosidic linkage lying in the direction of the chain, or from side groups orientated perpendicular to the chain. In the particular studies discussed below, the authors used a polarised FTIR microscope with a Focal Plane Array (FPA) detector with 5×5 micrometres^2 nominal pixel resolution to map these changes in stretched hydrated polymer films close to fracture edges and in the deformation zone around a propagating crack tip, where the local stress is highest. Even at high humidity, gelatin films showed brittle fractures with little or no changes away from the fracture edge, whereas the much more ductile HPC had a large zone with elastic deformation and induced polymer orientation in the stretching direction. Cassava starch films showed a ragged break with some plasticity, but uneven deformation due to structural inhomogeneities (incomplete granule gelatinisation, ghosts) (Paes *et al.*, 2010; Yakimets *et al.*, 2005, 2007).

6.4.3   Plant cell wall mechanics

Similar studies of elastic mechanical deformation with polarised FTIR microspectroscopy have been made on plant cell walls and cell wall models, in order to understand the structural origins of the texture of vegetable tissues.

Onion epidermis, which consists of a single layer of cells, was used to study the response of cell walls to mechanical stress. Dichroic spectra (parallel/perpendicular polarised) were collected from individual epidermis cells with the help of a rectangular aperture (Fig. 6.9). Cellulose, hemicelluloses and pectin of the plant cell walls showed little orientation initially, but mechanical stress induced orientation of the polymers in the stretching direction. This was partially reversible, indicating both elastic and viscous deformation, especially at high strain rates. Two-dimensional dynamic FTIR spectra showed that in native, fully hydrated cell walls there was a difference in the elastic re-orientation kinetics of pectin and cellulose/hemicelluloses (Wilson *et al.*, 2000). This indication of a degree of independence between these components was confirmed with three

(a)

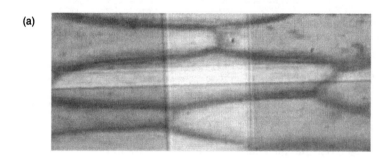

(b)

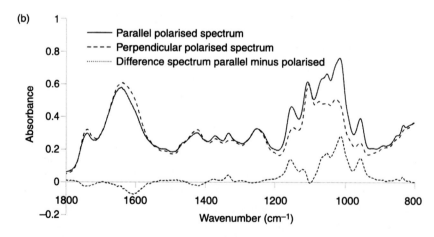

Fig. 6.9 Cell wall polymer orientation in stretched onion epidermis: (a) light micrograph and (b) FTIR microscopy spectra measured with a $50 \times 10\,\mu\text{m}^2$ aperture in parallel and perpendicular orientation.

Acetobacter cellulose composites: cellulose (C), cellulose/pectin (CP) and cellulose/xyloglucan (CXG), which showed no interaction between cellulose and pectin in CP, while cellulose and xyloglucan in CXG were uniformly strained. The results depended strongly on hydration, because at lower water content much of the flexibility was lost and the whole system behaved as a single network. (Kacurakova *et al.*, 2000, 2002; Kacurakova and Wilson, 2001; Wilson *et al.*, 2000).

6.4.4   Grains

To date, most food-related applications of FTIR and Raman microscopy have been on plant grain materials, because wheat, barley, oats, maize and rice play such an essential role both in human diet and animal feeds. The properties of isolated protein, starch, fibre, etc. are relatively well understood, but wherever whole grains are consumed, the nutritional value is influenced by their microstructure – both the accessibility and the digestibility of the components. Different cultivars can be easily distinguished with NIR spectroscopy, and this method is widely used for bulk and even single kernel analysis; however, NIR spectra measure the total material, and discriminations can be due to variations in any component – for example, protein, starch, fibre, oil and moisture.

FTIR and Raman microscopy are used to study the *in situ* structures at the cellular level. It is this ability to map the natural chemical variability *in situ* that is so important to understand the origins of the distinctions between these systems.

Cutting open the grains reveals the internal tissue structure. Embryo, endosperm, pericarp and aleurone can be easily distinguished visually, and not surprisingly, they have different spectra. Wetzel *et al.* (2003) have imaged 8 μm thick dry-cut cross-sections of wheat kernels with an FTIR microscope using a 64×64 pixel Focal plane detector, and could distinguish these different tissue types by the FTIR band patterns (Marcott *et al.*, 1999; Wetzel and LeVine, 1999). The endosperm consists of a mass of starch granules embedded in a protein matrix rich in prolamin storage proteins (Fig. 6.10). The individual starch granules can easily be seen under the light microscope, but are too small to be resolved with the standard FTIR microscope. By using the synchrotron light source at Brookhaven with a 3 to 5 micrometres beam spot (confocal 5 μm spot size mapping data with 5 μm step size in a 5×5 grid pattern and 3 μm single masking in the same step size and grid pattern), it was possible to increase the spatial resolution of the maps enough to study the protein and starch separately. Pixels with pure protein spectra were selected to avoid artefacts potentially created by artificially separating protein and starch spectra. From those pure protein spectra, they could estimate the relative amounts of protein secondary structures by amide I band deconvolution and fitting, and found relative differences in the α-helix content between hard and soft wheats, corroborating similar Raman microspectroscopy results (Bonwell *et al.*, 2008; Wetzel *et al.*, 2003).

The heterogeneous protein/starch ratio in a number of crops, and in particular the protein secondary structures, have been linked to nutritional (feed) quality.

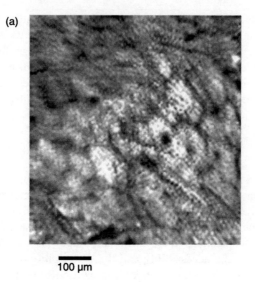

(a)

100 μm

(b)

Fig. 6.10   (a) FTIR map of protein/starch distribution in the endosperm of a maize kernel. Bright areas show starch, dark areas are protein-rich. The 2 μm thick section was mounted on BaF_2 and imaged in transmission with a spatial resolution of 5 μm; (b) ratio of the starch peak at $1022 \, cm^{-1}$ to the amide I band at $1654 \, cm^{-1}$, marked with arrows in the spectra. *ae* maize, isogenic *ae* mutant of maize; WT, wild-type.

By using appropriate bands (3350 cm^{-1} (OH and NH stretching), 2929 and 2885 cm^{-1} (CH stretching), 1736 cm^{-1} (carbonyl ester), 1650 cm^{-1} (amide I), 1550 cm^{-1} (amide II), 1510 cm^{-1} (phenolics), 1246 and 1160 cm^{-1} (cellulosic material) and 1150, 1080, 929, 860 cm^{-1} (carbohydrate)), it was possible to map the chemical composition of wheat, barley, corn and sorghum seeds. It was shown that a feed barley variety had a lower starch/protein ratio than a malting variety. There was also a wide variation in the protein structure in different grains, with barley containing about 17% β-sheet and 71% α-helix; oats about 2% β-sheet and 92% α-helix; and wheat about 42% β-sheet and 50% α-helix. In forage seeds such as flax seed and winterfat, which are used for animal feeds, wide variation in the secondary structure of the proteins were found, which were linked to the feed quality. The authors of the study postulated that the higher the native β-sheet structure in feeds, the more difficult it was to digest the feed, although this correlation is tenuous (Yu *et al.*, 2003, 2004a,b,c, 2005a,b; Yu 2004, 2005; 2011).

Besides the protein content, the presence of starch is also of great significance. Many grain seeds contain starches with A-type crystalline structures, with starch polysaccharide compositions of about 30% amylose and 70% amylopectin. Mutants with altered starch structure differ from their wild types in their nutritional and functional properties. Waxy (low-amylose) starches provide some advantageous properties for processing. High-amylose starch mutants are of interest, because their increased gelatinisation temperature alters their digestion (increases their resistant starch content).

Chemical imaging using PLS and PCA factor analysis of synchrotron IR maps was shown to objectively discriminate between normal wild-type parent and triple null waxy wheat, but could not discriminate single null cross-products (Dogan *et al.*, 2008).

The structure of starch granules in seeds of high-amylose mutants was compared with the wild type by high-resolution Raman microscopy (Wellner *et al.*, 2011). In the isogenic *ae* mutant of maize, the activity of the starch branching enzyme IIb is inhibited, which gives rise to a high-amylose starch. The granule structures in the wild-type samples were homogeneous, whereas those in the *ae* mutant were grossly heterogeneous. Wild-type samples had large, well-formed granules embedded in a protein matrix consisting mainly of zein, whereas the mutant seeds contained a whole range of granule shapes and sizes, usually smaller, and sometimes surrounded by a different polysaccharide as well as the protein.

By using thin sections (1–2 μm) with a high power 100× objective, Raman mapping at high resolution revealed information on the internal structure of the starch granules (Fig. 6.11). The laser polarisation also revealed the orientation of the starch helices. The normal wild-type granules showed uniform A-type starch crystal structures with a clearly defined radial orientation. In contrast, in the mutant seeds the starch crystallinity was more B-type and also variable within the seed. The inhibition of the branching enzyme should reduce the relative amounts of branched (1→4,6) glucose residues compared to linear (1→4) linked residues and, indeed in the mutant starches, the level of branching was smaller. Another feature of granule heterogeneity was that radial orientation of ordered structures was restricted to

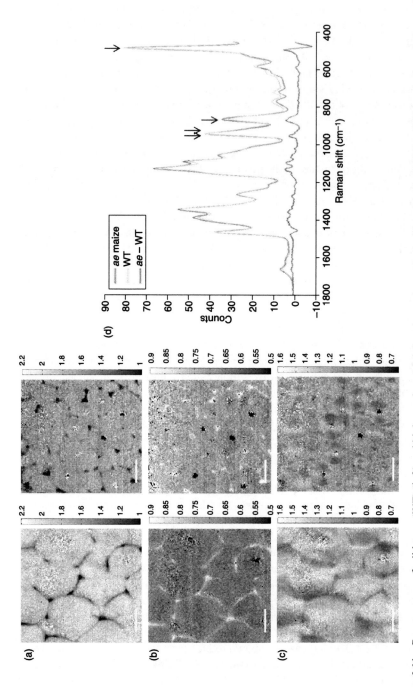

Fig. 6.11 Raman maps of wild-type (WT) (left) and high-amylose (right) maize endosperm sections: (a) crystallinity (WT has A-type, darker *ae* has B-type); (b) branching ratio (brighter = more linear); (c) helix orientation in starch granules (bright areas are aligned parallel to laser polarisation, darker areas perpendicular). The scale bars are 10 μm; (d) the arrows in the Raman spectra indicate the bands used to calculate these maps.

localised regions within certain granules (Fig. 6.11). The high-resolution mapping of granule structure revealed that a simple mutation in biosynthesis resulted in a grossly heterogeneous population of granule structures and that this heterogeneity is important in understanding their novel functional properties.

Another component in the cereal grains of interest are the endosperm cell walls. In a number of studies it was shown that in wheat, oat and barley seeds the presence of arabinoxylans (AX) and β-glucan polysaccharides varies during development and also that environmental factors influence this composition. AX and β-glucans have been shown to impart particular viscosity characteristics, important for processing (e.g. in milling or brewing) as well as influencing the digestion of resultant food products. These types of cell wall structures are difficult to study because in grain kernel sections the starch totally dominates the whole spectrum. Likewise, in flour samples, most of the non-starch carbohydrate material originates from the aleurone/bran. In order to study these cell walls *in situ*, 50 to 70 µm thick sections were cut with a vibratome in 70% ethanol to avoid solubilising too many carbohydrates, and gently sonicated to remove the starch granules. The remaining cell walls in these sections could then be imaged with a FTIR microscope (Fig. 6.12). The spectra were then analysed for the presence of characteristic patterns for AX, determined from model compounds, which had specific signatures for arabinose in several spectral regions (i.e. 400–600, 800–950 and 1030–1100 cm^{-1}). Mapping these spectra features revealed differing

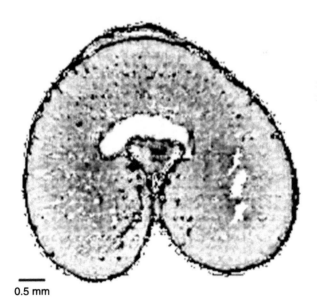

0.5 mm

Fig. 6.12   FTIR map of the 1079/1041 cm^{-1} band ratio in a wheat kernel section imaged after removal of starch. The greyscale gradient shows the different arabinoxylan structures in the endosperm cell walls. Bran layers have very different polysaccharide composition and appear in black.

changes in AX structure during grain development under cool/wet and hot/dry growing conditions, for differing cultivars (Mills *et al.*, 2005; Toole *et al.*, 2007, 2009). The AX consist of a xylan backbone with arabinose sidechains and the structure of the AX in the endosperm cell walls was found to change from a highly branched form to a less branched form, and the rate of restructuring during development was faster when the plants were grown at higher temperature with restricted water availability from 14 days after anthesis.

In developing grain, a clear difference in the composition of the endosperm cell walls of hard and soft wheat cultivars was observed as early as 15 days after anthesis. The soft wheats contained more of a polymer resembling water-soluble (branched) AX in the outer part of the grain.

Raman microspectroscopy was also used to determine the arabinose:xylose (A/X) ratio in the cell wall AX, as well as the amount of ferulic acid and related phenolic acids. The degree of esterification of the endosperm cell walls with ferulic acid was lower when the grains were grown in hot/dry conditions. Feruloylation of AX increased during the grain-filling stage, especially in the aleurone, where it was highly esterified with phenolic acids (Barron *et al.*, 2005, 2006; Barron and Rouau, 2008; Barron, 2011; Guillon *et al.*, 2011; Philippe *et al.*, 2006a,b; Robert *et al.*, 2005, 2011).

6.4.5   Other applications
The mapping of specific components can be extended to other nutrients in vegetables and fruits. By analogy with the analysis of ferulic acids in grain, other aromatic compounds can be visualised and mapped with Raman microscopy. Raman spectra of lycopene and β-carotene have been obtained from sampling oil droplets and plant cell structures, which suggested that the carotenoids were non-crystalline in these environments (Lopez-Sanchez *et al.*, 2011), and confocal Raman microspectroscopy was used to follow changes in the physical state of carotenoids – lycopene, β-carotene and lutein, in tomatoes (Pudney *et al.*, 2011).

Comparatively few studies have been made on meat or meat products, either because the cells are smaller or the tissue heterogeneity does not pose similar problems. Perhaps the reason is that food structure in this context comes a distant second to medical imaging, where IR and Raman maps of bone and soft tissues are widely used to study and diagnose diseases.

6.5   Conclusion and future trends

FTIR and Raman microscopy are both powerful techniques for studying the microscopic structures of food materials. In order to compare the techniques, we have to take into account a number of factors: first and foremost the material and the intended information, as well as the technical differences:

* IR is more suited for polar materials, OH, NH, water, give strong signals, whereas Raman is more suited for non-polar groups.

- Raman suffers from fluorescence and there is a high risk of sample heating from a focused laser.
- Quantitative analysis is well-established with IR; Raman intensity is more difficult because the scattering is influenced by experimental parameters.
- Qualitative analysis benefits from many reference spectra existing for IR, although Raman is catching up.
- Rastering a sample at high resolution is time-consuming. Raman microscopes can only do one spot at a time, while multi-pixel IR and NIR imaging detectors can acquire several thousand spectra simultaneously.
- IR needs special microscopes and sample slides/carriers, whilst Raman works with a normal optical microscope setup
- Spectral resolution in FTIR spectrometers is determined by the scan speed – in Raman systems with gratings and photodiodes, there is a trade-off between spectral bandwidth and resolution
- Spatial resolution is higher with Raman because of the shorter wavelength ($10\,\mu m$ vs $< 1\,\mu m$)

Recent technical developments have made these techniques both more powerful and easier to perform. Some of the most significant technical developments are better and faster IR imaging detectors (large array detectors run in normal rapid-scan mode instead of step-scan), cheap, powerful Raman laser modules, better fibre-optics links and photodiode detectors, and powerful computers that allow the handling of massive data files and run highly integrated microscope/ spectroscopy control software.

Both microspectroscopy techniques have some advantages and disadvantages. However, there are also some fundamental limits. In recent years, there have been a number of developments to overcome these limits:

- Both techniques are essentially measuring thin layers – IR because of high absorbance, Raman because of scattering. Both can handle sample blocks, but this is easier with Raman. IR needs micro-ATR to deal with non-transparent samples. To expand the measurements into 3D tomography, with IR we would need to image a stack of consecutive sections. Raman microscopes can image z-stacks at different focal planes; however, this is limited by the light penetration and scattering properties of the sample.
- Both techniques are diffraction-limited. Spatial resolution is higher with Raman because of the shorter wavelength ($\sim 10\,\mu m$ vs $<1\,\mu m$). Many interesting systems would need sub-micrometre resolution (small cell organelles, plant cell walls, individual domains in phase-separated systems, nanoparticles). This requires a fundamental change of the technique, using combined AFM/ spectroscopy set-ups instead of standard microscopes. Methods that have been employed are scanning near-field IR (using fine optical fibres instead of an AFM tip), optothermal AFM/IR (using the AFM tip as the detector) and tip-enhanced Raman microscopy.

However the imaging is done, the result is usually a large dataset. Besides the technical advances, the main other development is the arrival of software packages dedicated to measuring and analysing hyperspectral data. Many functions, such as advanced chemometrics/classification/etc. that originally had to be purpose-coded in Matlab (Mathworks Inc. USA), are now implemented in general spectrometer software, or can be interactively performed in program packages such as ENVI (EXELIS Visual Information Solutions Inc., USA).

Looking to the future, it is this field of automated analysis that will see the largest practical improvements – providing easier and more powerful tools to extract the information from the datasets.

6.6   Sources of further information and advice

The websites of spectrometer manufacturers contain various bits of up-to-date technical information and some application notes. For an excellent general overview of IR spectroscopy, the reader is referred to the *Handbook of Vibrational Spectroscopy*, edited by J. M. Chalmers and P. R. Griffiths (Wiley and Sons, 2002).

6.7   References

ANDREW, J. J., BROWNE, M. A., CLARK, I. E., HANCEWICZ, T. M. and MILLICHOPE, A. J. (1998), Raman imaging of emulsion systems, *Applied Spectroscopy*, **52**(6), 790–6.

BARER, R., COLE, A. R. H. and THOMPSON, H. W. (1949), Infra-red spectroscopy with the reflecting microscope in physics, chemistry and biology, *Nature*, **163**, 198–201.

BARRON, C. (2011), Prediction of relative tissue proportions in wheat mill streams by Fourier Transform Mid-infrared Spectroscopy, *Journal of Agricultural and Food Chemistry*, **59**(19), 10442–7.

BARRON, C. and ROUAU, X. (2008), FTIR and Raman signatures of wheat grain peripheral tissues, *Cereal Chemistry*, **85**(5), 619–25.

BARRON, C., PARKER, M. L., MILLS, E. N. C., ROUAU, X. and WILSON, R. H. (2005), FTIR imaging of wheat endosperm cell walls *in situ* reveals compositional and architectural heterogeneity related to grain hardness, *Planta*, **220**(5), 667–77.

BARRON, C., ROBERT, P., GUILLON, F., SAULNIER, L. and ROUAU, X. (2006), Structural heterogeneity of wheat arabinoxylans revealed by Raman spectroscopy, *Carbohydrate Research*, **341**(9), 1186–91.

BONWELL, E. S., FISHER, T. L., FRITZ, A. K. and WETZEL, D. L. (2008), Determination of endosperm protein secondary structure in hard wheat breeding lines using synchrotron infrared microspectroscopy, *Vibrational Spectroscopy*, **48**(1), 76–81.

BYLER, D. M. and SUSI, H. (1986), Examination of the secondary structure of proteins by deconvoluted FTIR spectra, *Biopolymers*, **25**(3), 469–87.

CAMERON, R. E., JALIL, M. A. and DONALD, A. M. (1994), Diffusion of bovine serum-albumin in amylopectin gels measured using Fourier-Transform Infrared Microspectroscopy, *Macromolecules*, **27**(10), 2708–13.

DOGAN, H., SMAIL, V. W. and WETZEL, D. L. (2008), Discrimination of isogenic wheat by InSb focal plane array chemical imaging, *Vibrational Spectroscopy*, **48**(2), 189–95.

DURRANI, C. M. and DONALD, A. M. (1994), Fourier-Transform Infrared Microspectroscopy of phase-separated mixed biopolymer gels, *Macromolecules*, **27**(1), 110–19.

DURRANI, C. M. and DONALD, A. M. (1995), Compositional mapping of mixed gels using FTIR microspectroscopy, *Carbohydrate Polymers*, **28**(4), 297–303.

DURRANI, C. M., PRYSTUPA, D. A., DONALD, A. M. and CLARK, A. H. (1993), Phase-diagram of mixtures of polymers in aqueous-solution using Fourier-Transform Infrared-Spectroscopy, *Macromolecules*, **26**(5), 981–7.

GUILLON, F., BOUCHET, B., JAMME, F., ROBERT, P., QUEMENER, B. *et al.*, (2011), Brachypodium distachyon grain: characterization of endosperm cell walls, *Journal of Experimental Botany*, **62**(3), 1001–15.

HOPKE, P. K. (2003), The evolution of chemometrics, *Analytica Chimica Acta*, **500**(1–2), 365–77.

KACURAKOVA, M. and WILSON, R. H. (2001), Developments in mid-infrared FT-IR spectroscopy of selected carbohydrates, *Carbohydrate Polymers*, **44**(4), 291–303.

KACURAKOVA, M., CAPEK, P., SASINKOVÁ, V., WELLNER, N. and EBRINGEROVÁ, A. (2000), FT-IR study of plant cell wall model compounds: pectic polysaccharides and hemicelluloses, *Carbohydrate Polymers*, **43**(2), 195–203.

KACURAKOVA, M., SMITH, A. C., GIDLEY, M. J. and WILSON, R. H. (2002), Molecular interactions in bacterial cellulose composites studied by 1D FT-IR and dynamic 2D FT-IR spectroscopy, *Carbohydrate Research*, **337**(12), 1145–53.

LOPEZ-SANCHEZ, P., SCHUMM, S., PUDNEY, P. D. A. and HAZEKAMP, J. 2011. Carotene location in processed food samples measured by Cryo In-SEM Raman, *Analyst*, **136**(18), 3694–7.

MARCOTT, C., REEDER, R. C., SWEAT, J. A., PANZER, D. D. and WETZEL, D. L. (1999), FT-IR spectroscopic imaging microscopy of wheat kernels using a Mercury-Cadmium-Telluride focal-plane array detector, *Vibrational Spectroscopy*, **19**(1), 123–9.

MILLS, E. N. C., PARKER, M. L., WELLNER, N., TOOLE, G., FEENEY, K. and SHEWRY, P. R. (2005), Chemical imaging: the distribution of ions and molecules in developing and mature wheat grain, *Journal of Cereal Science*, **41**(2), 193–201.

NORMAND, V., PUDNEY, P. D. A., AYMARD, P. and NORTON, I. T. (2000), Weighted-average isostrain and isostress model to describe the kinetic evolution of the mechanical properties of a composite gel: Application to the system gelatin: maltodextrin. *Journal of Applied Polymer Science*, **77**(7), 1465–77.

PAES, S. S., YAKIMETS, I., WELLNER, N., HILL, S. E., WILSON, R. H. and MITCHELL, J. R. (2010), Fracture mechanisms in biopolymer films using coupling of mechanical analysis and high speed visualization technique, *European Polymer Journal*, **46**(12), 2300–9.

PHILIPPE, S., BARRON, C., ROBERT, P., DEVAUX, M. F., SAULNIER, L. and GUILLON, F. (2006a), Characterization using Raman microspectroscopy of arabinoxylans in the walls of different cell types during the development of wheat endosperm, *Journal of Agricultural and Food Chemistry*, **54**(14), 5113–19.

PHILIPPE, S., ROBERT, P., BARRON, C., SAULNIER, L. and GUILLON, F. (2006b), Deposition of cell wall polysaccharides in wheat endosperm during grain development: Fourier transform-infrared microspectroscopy study, *Journal of Agricultural and Food Chemistry*, **54**(6), 2303–8.

PRYSTUPA, D. A. and DONALD, A. M. (1996), Infrared study of gelatin conformations in the gel and sol states, *Polymer Gels and Networks*, **4**(2), 87–110.

PUDNEY, P. D. A. and HANCEWICZ, T. M. (1999), Confocal Raman microspectroscopic study of phase separating mixed biopolymers: Concentration mapping by multivariate curve resolution, in J. Greve, G. J. Puppels and C. Otto (eds), *Spectroscopy of Biological Molecules: New Directions*, Kluwer, 615–16.

PUDNEY, P. D. A., GAMBELLI, L. and GIDLEY, M. J. (2011), Confocal Raman microspectroscopic study of the molecular status of casotenoids in tomato fruits and foods. *Applied Spectroscopy*, **65**(2), 127–34.

PUDNEY, P. D. A., HANCEWICZ, T. M. and CUNNINGHAM, D. G. (2002), The use of confocal Raman spectroscopy to characterise the microstructure of complex biomaterials: foods. *Spectroscopy: An International Journal*, **16**(3–4), 217–25.

PUDNEY, P. D. A., HANCEWICZ, T. M., CUNNINGHAM, D. G. and GRAY, C. (2003), A novel method for measuring concentrations of phase separated biopolymers: the use of confocal

Raman spectroscopy with self-modelling curve resolution, *Food Hydrocolloids*, **17**(3), 345–53.

PUDNEY, P. D. A., HANCEWICZ, T. M., CUNNINGHAM, D. G. and BROWN, M. C. (2004), Quantifying the microstructures of soft solid materials by confocal Raman spectroscopy, *Vibrational Spectroscopy*, **34**(1), 123–35.

RICHTER, S., MUSSIG, J. and GIERLINGER, N. (2011), Functional plant cell wall design revealed by the Raman imaging approach, *Planta*, **233**(4), 763–72.

ROBERT, P., MARQUIS, M., BARRON, C., GUILLON, F. and SAULNIER, L. (2005), FT-IR investigation of cell wall polysaccharides from cereal grains. Arabinoxylan infrared assignment, *Journal of Agricultural and Food Chemistry*, **53**(18), 7014–18.

ROBERT, P., JAMME, F., BARRON, C., BOUCHET, B., SAULNIER, L. *et al.* (2011), Change in wall composition of transfer and aleurone cells during wheat grain development, *Planta*, **233**(2), 3406.

SUN, L., DURRANI, C. M. and DONALD, A. M. (1996a), FTIR methods for the study of protein-polysaccharide mixtures, *Gums and Stabilisers for the Food Industry*, **8**, 423–32.

SUN, L., DURRANI, C. M., DONALD, A. M., FILLERYTRAVIS, A. J. and LENEY, J. (1996b), Diffusion of mixed micelles of bile salt-lecithin in amylopectin gels: A Fourier transform infrared microspectroscopy approach, *Biophysical Chemistry*, **61**(2–3), 143–50.

TOOLE, G. A., WILSON, R. H., PARKER, M. L., WELLNER, N. K., WHEELER, T. R. *et al.* (2007), The effect of environment on endosperm cell-wall development in *Triticum aestivum* during grain filling: an infrared spectroscopic imaging study, *Planta*, **225**(6), 1393–403.

TOOLE, G A., BARRON, C., GALL, G., COLQUHOUN, I., SHEWRY, P. and MILLS, E. (2009), Remodelling of arabinoxylan in wheat (*Triticum aestivum*) endosperm cell walls during grain filling, *Planta*, **229**(3), 667–80.

WELLNER, N., GEORGET, D. M. R., PARKER, M. L. and MORRIS, V. J. (2011), *In-situ* Raman microscopy of starch granule structures in wild type and ae mutant maize kernels, *Starch-Starke*, **63**(3), 128–38.

WETZEL, D. L. and LEVINE, S. M. (1999), Microspectroscopy – Imaging molecular chemistry with infrared microscopy, *Science*, **285**(5431), 1224–5.

WETZEL, D. L., SRIVARIN, P. and FINNEY, J. R. (2003), Revealing protein infrared spectral detail in a heterogeneous matrix dominated by starch, *Vibrational Spectroscopy*, **31**(1), 109–14.

WILSON, R. H., SMITH, A. C., KACURAKOVA, M., SAUNDERS, P. K., WELLNER, N. *et al.* (2000), The mechanical properties and molecular dynamics of plant cell wall polysaccharides studied by Fourier-transform infrared spectroscopy, *Plant Physiology*, **124**(1), 397–405.

WOLD, S. (1995), Chemometrics: What do we mean with it, and what do we want from it? *Chemometrics and Intelligent Laboratory Systems*, **30**(1), 109–15.

YAKIMETS, I., WELLNER, N., SMITH, A. C., WILSON, R. H., FARHAT, I. and MITCHELL, J. (2005), Mechanical properties with respect to water content of gelatin films in glassy state, *Polymer*, **46**(26), 12577–85.

YAKIMETS, I., PAES, S. S., WELLNER, N., SMITH, A. C., WILSON, R. H. and MITCHELL, J. R. (2007), Effect of water content on the structural reorganization and elastic properties of biopolymer films: a comparative study, *Biomacromolecules*, **8**(5), 1710–22.

YU, P. (2004), Application of advanced synchrotron radiation-based Fourier transform infrared (SR-FTIR) microspectroscopy to animal nutrition and feed science: a novel approach, *British Journal of Nutrition*, **92**(6), 869–85.

YU, P. (2005), Molecular chemistry imaging to reveal structural features of various plant feed tissues, *Journal of Structural Biology*, **150**(1), 81–9.

YU, P., CHRISTENSEN, D. A., CHRISTENSEN, C. R., DREW, M. D., ROSSNAGEL, B. G. and MCKINNON, J. J. (2004a), Use of synchrotron FTIR microspectroscopy to identify chemical differences in barley endosperm tissue in relation to rumen degradation characteristics, *Canadian Journal of Animal Science*, **84**(3), 523–7.

YU, P., MCKINNON, J. J., SOITA, H. W., CHRISTENSEN, C. R. and CHRISTENSEN, D. A. (2005a), Use of synchrotron-based FTIR microspectroscopy to determine protein secondary structures of raw and heat-treated brown and golden flaxseeds: a novel approach, *Canadian Journal of Animal Science*, **85**(4), 437–48.

YU, P., WANG, R. and BAI, Y. (2005b), Reveal protein molecular structural-chemical differences between two types of winterfat (forage) seeds with physiological differences in low temperature tolerance using synchrotron-based Fourier transform infrared microspectroscopy, *Journal of Agricultural and Food Chemistry*, **53**(24), 9297–303.

YU, P. Q. (2011), Microprobing the molecular spatial distribution and structural architecture of feed-type sorghum seed tissue (*Sorghum bicolor* L.) using the synchrotron radiation infrared microspectroscopy technique, *Journal of Synchrotron Radiation*, **18**, 790–801.

YU, P. Q., MCKINNON, J. J., CHRISTENSEN, C. R., CHRISTENSEN, D. A., MARINKOVIC, N. S. and MILLER, L. M. (2003), Chemical imaging of microstructures of plant tissues within cellular dimension using synchrotron infrared microspectroscopy, *Journal of Agricultural and Food Chemistry*, **51**(20), 6062–7.

YU, P. Q., MCKINNON, J. J., CHRISTENSEN, C. R. and CHRISTENSEN, D. A. (2004b), Imaging molecular chemistry of pioneer corn, *Journal of Agricultural and Food Chemistry*, **52**(24), 7345–52.

YU, P. Q., MCKINNON, J. J., CHRISTENSEN, C R. and CHRISTENSEN, D. A. (2004c), Using synchrotron transmission FTIR microspectroscopy as a rapid, direct, and nondestructive analytical technique to reveal molecular microstructural-chemical features within tissue in grain barley. *Journal of Agricultural and Food Chemistry*, **52**(6), 1484–94.

7

Ultrasonic and acoustic microscopy: principles and applications to food microstructures

M. J. W. Povey and N. Watson, Leeds University, UK and N. G. Parker, Newcastle University, UK

DOI: 10.1533/9780857098894.1.192

Abstract: The underlying principles of acoustic scanning/acoustic microscopy are reviewed, underpinning a discussion of the operation and application of acoustic microscopy and scanning to food materials. The construction of a versatile scanning acoustic platform (VSAP) capable of imaging over a wide range of scales, from centimetres to micrometres, is described in detail. Applications of the technique to the study of meat, food emulsion stability, and *in-vivo* study of plant cells are described. Future developments in conjunction with optical techniques (acousto-optics) are also reviewed.

Key words: ultrasound, food, tomography, imaging, microscopy.

7.1   Introduction

Acoustic imaging and characterization techniques in general respond to a different set of physical properties to those that affect electromagnetic propagation. Thermal conductivity, heat capacity, viscosity, density, elastic modulus (shear and compression) and acoustic attenuation in all of the components of the studied material will affect acoustic propagation.

In summary, the general benefits of acoustic techniques include (Maev, 2008; Parker *et al.*, 2010):

- they can measure *both* phase and signal amplitude simultaneously. This increases measurement versatility in comparison to optical techniques;
- they penetrate many optically opaque materials, providing 3D information. This ability has been widely exploited in medical imaging (Duck *et al.*, 1998),

non-destructive testing of materials (Krautkrämer and Krautkrämer, 1990; Kundu, 2004) and industrial fluid characterisation (Povey, 1997; Dukhin and Goetz, 2002);

- they are highly scalable. Sound can be employed from global scales, for example, in ocean and seismic tomography, down to the micro- and nano-scale. This latter regime is the realm of the scanning acoustic microscope, a precision device that uses tightly focused ultrasound to probe a sample (Yu and Boseck, 1995; Briggs and Kolosov, 2010);
- interfacial reflections can provide mechanical information at a microscopic level, an almost unique aspect of acoustic microscopy. Boundaries of acoustically-distinct components of the sample generate reflections whose time of flight/phase shift details the sample's surface and sub-surface structure. Furthermore, the amplitude and phase of the returning signal carries mechanical information, which can reveal elastic moduli, compressibility, stress (Drescher-Krasica and Willis, 1996), adhesion properties, thermophysical properties and phonon transport (Foster, 1984);
- they are complementary to optical methods. Take, for example, oil-in-water emulsions. Whilst the oil-water interface generally possesses a small refractive index difference, there is a large acoustic contrast due to thermal scattering. This is valuable, for example, in studies of oil crystallization in emulsions where the acoustic contrast is very sensitive to the liquid-solid phase transition (Povey, 2000).

Perhaps the greatest challenge in acoustic techniques is data analysis. Because acoustic propagation in heterogeneous materials is complex, good theoretical and practical mathematical models are necessary for accurate retrieval of sample properties.

The particular acoustic technique considered here is acoustic microscopy. This technique is analogous to well-known confocal optical microscopy, but with the light source and optical lens effectively replaced by a source of sound waves (an electro-mechanical device known as a sound transducer) and an acoustic lens. Sound waves are focused onto or into the sample and we 'listen' to the sound that is reflected or transmitted into the sample. This information can provide acoustic images of the sample. Moreover, if the acoustic data is merged with acoustic models of the material, we can characterise the material properties on a microscopic scale. Such material properties may be, for example, density and compressibility or, for a multi-phase system, volume fraction and solid fat content.

The first acoustic microscope was demonstrated by Lemons and Quate (1974a,b) with a resolution of approximately 10 microns. Following this, the resolution was progressively reduced, through optical (Jipson and Quate, 1978) and sub-optical (Hadimioglu and Quate, 1983) resolution, to the current record of 15 nanometres (150 Angström) (Muha et al., 1990). The latter case was performed in a cryogenic environment of super-fluid helium. In the more convenient environment of water, a resolution of below 200 nm has been achieved at an operating frequency of 4.4 GHz and an ambient temperature of 60 °C (Hadimioglou and Quate, 1983). The main application of acoustic microscopy to date has been in non-destructive flaw detection of hardware, although more recently there have been applications to studying biological matter, including live cells (Drescher-Krasica and Willis, 1996; Zinin and Weise, 2004).

In this chapter we will begin by reviewing acoustic microscopy, including its theoretical basis and a typical modern realization of such a system. Then, by presenting various case studies, we will highlight the potential of acoustic microscopy for revealing new levels of structural, mechanical and physico-chemical information in samples relevant to the food (and related) sector.

7.2   Theories of ultrasound propagation

The history of the physics of ultrasound propagation extends back together with optics, to Lord Rayleigh (Strutt, Baron Rayleigh, 1872, 1877, 1896; Dukhin and Goetz, 2010; Povey, 1997). Comprehensive discussions of acoustic microscopy, from the underlying theory to example applications in non-destructive testing and biological studies, can be found in Briggs and Kolosov (2010), Maev (2008), Kundu (2004) and Yu and Boseck (1995). A comprehensive discussion of the ultrasound characterization of colloids, together with an account of scattering theory is found in Challis, *et al.* (2005) and Pinfield and Challis (2011), an introduction to ultrasound techniques is found in Cheeke (2002) and an introduction to ultrasound techniques for fluids characterization is found in Povey (1997). One of the first papers to consider ultrasound scattering in emulsions was published in 1989 (McClements and Povey, 1989). Scattering is introduced here because of its importance in biological materials and its impact on sound velocity and attenuation. It is also important to recognise, in the context of water-containing materials such as many foods, that the large temperature coefficient of the velocity of sound in water at ambient temperature introduces a significant requirement for temperature control, which is less severe in many of the conventional applications of acoustic microscopy, due to the lower temperature coefficient of the velocity of sound in harder materials, such as metals and ceramics.

Acoustics, sound and ultrasound all involve motion of particles and masses which require elasticity, a combination of attractive and repulsive forces that causes oscillatory, repetitive or self-similar motion. Small displacement, linear elastic mathematical solutions of the acoustic propagation problem involve conserving momentum, energy and mass, ignoring relaxation and accounting for both shear and bulk viscosity (in the case of fluids) or the retarded integral-elastic stress-strain tensor in the case of soft (and hard) solids. These solutions are quasi-steady state, as changes in the mean temperature and pressure over time are neglected; small perturbations of velocity and pressure are considered that permits restricting the solutions to first- and second-order terms in the elastic tensor only. This also includes neglecting the variability of coefficients of viscosity and heat conduction with temperature, treating them as constants.

Theoretical, physical explanations of acoustic propagation make presumptions about the nature of the mechanical waves being measured and the physical properties that are measured. This cannot be analysed in the abstract, as the detailed process is important. The great majority of acoustic/ultrasound transducers detect sound pressure and convert that pressure to a voltage. This process takes place over the surface of the transducer and if the pressure varies across the surface, then so will the voltage. This voltage is integrated by the wire that conveys the voltage signal to the electronic detection system. If the pressure is positive on one side of the transducer and negative on the other, then it is possible for the voltage to sum to zero, even though there is a pressure wave present. This means that most ultrasound transducers, unlike their optical equivalents, are phase sensitive. However, this is an advantage, because phase can be measured. This is also a disadvantage, because it makes analysis of the acoustic pressure much more complicated. In particular, diffraction effects are much more important in acoustics than light and this makes the design of acoustical systems more sensitive to alignment and mechanical details than might at first be apparent (Povey, 1997).

Not all acoustic waves are oscillatory; it is possible for a pressure wave to propagate as a pulse. However, in our discussion, we consider only oscillatory motion that can be explained on the basis that stress and strain/strain rate are linearly related.

The most complicated elastic relations are found in so-called 'soft solids', which exhibit behaviour combining elements of both liquid and solid behaviour. Elastic behaviour in solids (Landau and Lifschitz, 1970; Love, 1944; Timoshenko and Goodier, 1970) is described by the relationship between stress and strain, in general a fourth-order tensor with 36 elements, called the elastic tensor. Stress is not uniformly distributed over the cross-section of a material and consequently has to be described in terms of the direction in which the force is applied and the direction of the area over which it is applied, so there are three orthogonal normal stresses and six orthogonal shear stresses. This indicates that sound may propagate in different modes, most commonly in solids as compression, shear and surface waves and in fluids as compression and surface waves. In fluids, this simplifies because the shear components cannot propagate over significant distances and are called 'non-propagational', and the three orthogonal normal stresses reduce to one isotropic compressional stress, otherwise known as pressure. In soft solids, processes such as polymer entanglement lead to a 'retardation' of the strain propagation, so that we must take account of the history of the stress-strain process. The stress tensor may be related to the strain tensor in a way that connects current stresses to the deformation history (Charlier and Crowet, 1986; Tschoegl, 1989; Povey, 1997).

7.2.1   Harmonic description of wave propagation

In the harmonic description, the pressure fluctuation is assumed to be sinusoidal with a radial frequency ω. The phase (ζ) of a wave defines its position in space, relative to some arbitrary point. Phase is therefore only meaningful when measured relative to a defined point:

$$\Delta p = \Delta p_0 exp\{i(\omega t - \zeta + i\alpha x)\} = \Delta p_0 exp(i\omega t)exp(-i\zeta)exp(-\alpha x) \qquad [7.1]$$

where

$$\zeta = k'_c x,$$

k'_c is the real part of the acoustic wave number:

$$k'_c = \frac{2\pi}{\lambda} = \frac{\omega}{v_p} = \frac{2\pi f}{v_p}, \qquad [7.2]$$

$$k_c = k'_c + ik''_c = k'_c + i\alpha \qquad [7.3]$$

Here, Δp and Δp_0 are the instantaneous pressure deviation and the maximum pressure deviation, respectively; $i = \sqrt{-1}$; k' and k'' are the real and imaginary parts of the complex wave vector $\mathbf{k}$; v_p is the phase velocity of the wave; λ is wavelength; f is frequency; and α is the attenuation coefficient.

Figure 7.1 assumes that only a single cycle exists. A true sinusoidal wave would repeat this single cycle infinitely. A similar assumption is made in Fig. 7.2 for a shear pulse. Figure 7.3 plots the field distribution for the normal type of piston transducer, illustrating the differences between the near field, the focal region and the far field. It is absolutely necessary to carefully locate where within

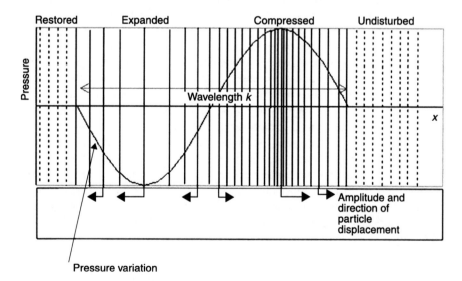

Fig. 7.1 Particle position, displacement and spatial pressure variation plotted against position (x) for a single cycle, sinusoidal, plane-travelling wave in a fluid medium (adapted from Pierce, 1981).

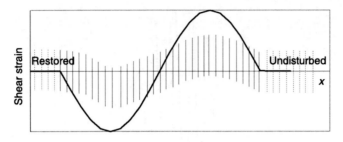

Fig. 7.2   A shear pulse (adapted from Povey, 1997).

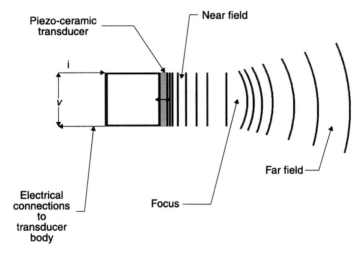

Fig. 7.3   Pressure field distribution in front of a piston transducer (Povey, 1997).

the field distribution measurements are being made. For instance, in the far field it is necessary to correct for the diffraction effects of the radiating field. Directly adjacent to the transducer face is called the 'region of confusion'; in this region behaviour is chaotic.

The group velocity v of a pulse such as that shown in Fig. 7.4 is the velocity of the envelope of the wave, which embraces all cycles in the pulse. This is given by

$$v = \frac{d\omega}{dk} \qquad\qquad [7.4]$$

whereas the phase velocity v_p represented in Equation 7.1 is the velocity of a given frequency component within the wave, of frequency ω given by

$$v_p = \frac{\omega}{k} = f\lambda \qquad\qquad [7.5]$$

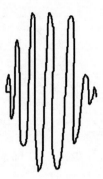

Fig. 7.4   A seven cycle pulse, also called a tone burst.

Pressure is related to amplitude or displacement ξ through

$$p \sim \xi^2 \qquad\qquad [7.6]$$

We can then define the attenuation in terms of measureable pressure differences as

$$-\alpha = \frac{1}{x} \ln \frac{\xi}{\xi_0} \text{Neper m}^{-1} \qquad\qquad [7.7]$$

or

$$-\alpha = \frac{1}{x} 20 log \frac{\xi}{\xi_0} \text{dB m}^{-1} \qquad\qquad [7.7]$$

The decibel (dB) and Neper (Np) are called dimensionless units, because they are ratios. Note that the attenuation is normally expressed in terms of pressure rather than displacement and, in this case:

$$-\alpha [\text{dBm}^{-1}] = \frac{1}{x} 10 log \frac{p}{p_0} = \frac{1}{x} 20 log \frac{\xi}{\xi_0} \qquad\qquad [7.8]$$

In ultrasound spectrometry, the attenuation α is measured as a function of frequency, producing a so-called attenuation spectrum. The phase velocity may also be measured as a function of frequency, producing a velocity spectrum. The measured attenuation spectrum will contain contributions from a number of phenomena, not just from scattering. Since it is the scattering that gives us information about the size of particles dispersed in a fluid (this may be any fluid, including a gas, not just water), it is necessary to subtract from the spectrum all non-scattering contributions. These generally comprise (a) the viscosity of the continuous phase and (b) relaxation effects in the dispersed phase. These effects may be measured by determining the attenuation spectrum for the pure continuous (α_c) phase and then subtracting that from the measured attenuation spectrum for the dispersion:

$$\alpha_s = \alpha_{tot} - \alpha_c \qquad\qquad\qquad [7.9]$$

The spectrum $\alpha(f)$ is then 'inverted'; a process of comparing measurement with a modelled spectrum, based on a postulated size distribution. An iteration procedure is then applied until the modelled spectrum most closely matches the measured one. The model is generally restricted to one or two log normal distributions, since an exploration of all possible distributions is extremely time-consuming and need not necessarily produce a more accurate result. This is then the determined particle size distribution. If the dispersed phase exhibits relaxation, as is the case for protein molecules (Povey *et al.*, 2011), then it is necessary to carry out a more elaborate procedure to pre-determine the relaxation spectrum in the dispersed phase, which then needs to be subtracted in order to obtain the scattering spectrum.

7.2.2   Sonic beam formation

Acoustic contrast

Reflection-mode acoustic microscopy detects sound waves that have been reflected and back-scattered. These phenomena both arise from boundaries in the elastic distribution, which is usually parameterised through the characteristic acoustic impedance $Z=\rho v$, where ρ is the density. Broadly speaking, the interaction is a reflection when the length scale of the boundary/bounded object l is much greater than the sound wavelength λ. Meanwhile, when $l \leq \lambda$, the feature acts as an inhomogeneity and scatters the sound. Scattering itself has a spectrum of behaviour, ranging from the mid-frequency regime ($l \sim \lambda$), where the scattering is sensitive to shape and size resonances (Uberall, 1992), up to the far limit of Rayleigh scattering where the scattering becomes insensitive to shape ($l \gg \lambda$) (Morse and Ingard, 1968).

Consider plane sound waves in medium I (Z_I, v_I and ρ_I) at normal incidence to an interface with medium II (Z_{II}, v_{II} and ρ_{II}). The reflection coefficient R, the ratio of the reflected pressure amplitude p_R to the incidence pressure amplitude p_I, (Krautkrämer and Krautkrämer, 1990) is given by

$$R = \frac{Z_{II} - Z_I}{Z_{II} + Z_I} \qquad\qquad\qquad [7.10]$$

This equation provides important intuition; reflection increases with the size of the impedance mismatch. However, strictly, R is a function of incident angle θ and can change markedly with θ, for example due to critical angles at which complete internal reflection or surface wave generation can occur. Indeed, the tightly focused beams employed in acoustic microscopes can include incident angles of up to 60 degrees. The true reflected signal thus requires a non-trivial integration of the reflection function over this angular distribution (Atalar, 1978).

Focusing aberrations

In chromatic aberration, different frequency components in the beam are refracted by the lens to differing degrees, leading to a spectral spread in focal positions.

However, acoustic microscopes are usually sufficiently dominated by a single frequency, such that chromatic aberration is not a significant effect (Briggs and Kolosov, 2010).

In an ideal spherical lens under monochromatic sound all paraxial incoming rays will be refracted to a common point at a focal distance F. In reality, rays at different radii from the lens axis have different focal distances, causing spherical aberration. Third-order theory reveals that the deviation of focal distance scales as $(v_{cf}/v_l)^2$ (Briggs and Kolosov, 2010), where v_{cf} and v_l are the speeds of sound in the coupling fluid and lens. Since the speed mismatch is typically large (e.g. for a quartz-water boundary $v_{cf}/v_l \sim 0.25$), this deviation can often be negligible in acoustic microscopy.

At ultra-high frequency and for wide-aperture lenses aberrations can become considerable, for example due to the differential attenuation experienced by the different beam path lengths. These complex effects are now well-understood (Maev, 2008).

Resolution
As described above, lens aberrations in acoustic microscopes can often be sufficiently small, such that the imaging resolution is close to the diffraction limit. Consider a plane wave diffracting through a circular aperture of diameter D (the lens). According to far-field wave theory, the pressure field $p(r)$ in the focal plane varies with radial position r according to

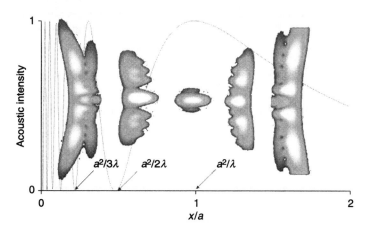

Fig. 7.5   Acoustic intensity distribution for the piston transducer with a radius a, along the x-axis (Povey, 1997). The focal point is at $x/a = 1$. The point spread function of a pulsed, focused transducer, depicting the lateral distribution of the field intensity at five points in the field, is superimposed in greyscale (darker is more intense) on top of the intensity plot. The point spread function is calculated computationally by first determining the spatial impulse response of the transducer, performed by integrating the analytical far-field contribution of each area element of the transducer face, and convolving this with a typical excitation pulse (performed using Field II (Tupholme, 1969; Stepanishen, 1971). Note that the focus is line-shaped, rather than a point.

$$p(r) = p_0 \frac{J_1(\pi Dr \,/\, F\lambda)}{\pi Dr \,/\, F\lambda} \qquad [7.11]$$

where $J_1(x)$ is the Bessel function of the first kind and F is the focal length. The first node of this pressure distribution lies at position $d = 1.22F\lambda/D$ (Fig. 7.5). According to the well-known Rayleigh criterion, two objects are just resolvable when their separation equals this value. For the emitter/receiver system considered here, the result become slightly modified to become (Kino, 1980)

$$d = 1.02 \frac{F\lambda}{D} \qquad [7.12]$$

Note that the definition of resolution is somewhat arbitrary and best determined experimentally.

For a given lens (fixed F and D), resolution is enhanced at greater frequencies. Figure 7.6 shows the resolution of the focused sound beam in water at 30 °C (assuming $F \approx D$, which is typically valid). At kHz frequencies, the resolution is of the order of a metre, whilst at GHz frequencies the resolution is less than a micron. The resolution can also be improved by a factor of $\sqrt{2}$, by operating in the nonlinear regime (Rugar, 1984).

Thermal aberrations

Distances are inferred from the time of flight of the returning echoes and the speed of sound of the coupling fluid. However, the speed of sound varies with temperature at up to 3 ms^{-1} per °C in water (Fig. 7.7). This can introduce considerable thermal aberration in the image and so temperature stabilisation is essential. Moreover, the effect of temperature variations can be reduced by operating at temperatures with the minimal gradient, for example around 70 °C for water.

Attenuation

The plane wave solution (Equation 7.1 and rewritten slightly differently below) is an ideal solution to the propagation problem. Real fluids possess viscosity and finite thermal conduction that dampen the beam during propagation and result in an attenuated form:

$$p(x) = p_0 \exp[-\alpha x] \, \exp[i(kx - 2\pi ft)] \qquad [7.13]$$

For almost all fluids at the frequencies of interest, the attenuation coefficient α has a quadratic dependence on frequency and can be expressed as $\alpha = \alpha' f^2$.

After a certain propagation distance the beam will be too weak to be detected. We define a propagation distance based on an arbitrary 100-fold (40 dB) decrease in beam amplitude, which is given by

$$L_{0.01} = -\frac{\ln(0.01)}{\alpha' f^2} \qquad [7.14]$$

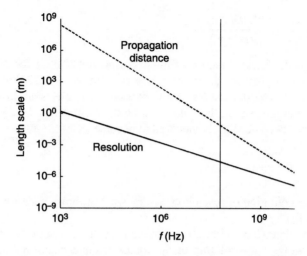

Fig. 7.6 Resolution (Equation 7.4, assuming $F = D$) and propagation distance (Equation 7.6) of a scanning acoustic microscope (SAM) in distilled water at 30 °C ($v = 1509\,\mathrm{ms}^{-1}$), as a function of frequency f. The vertical line indicates our operating frequency.

Figure 7.6 presents $L_{0.01}$ as a function of frequency for water at 30 °C ($\alpha' = 18 \times 10^{-15}$ $\mathrm{Np\,m}^{-1}\,\mathrm{Hz}^{-2}$). The propagation distance decreases more rapidly than the resolution and at a certain frequency becomes restrictively small. For example, at 1 GHz in water at 30 °C, the maximum propagation distance is limited to around 50 μm. The presence and nature of any intervening interfaces will further reduce the signal amplitude and maximum propagation distance.

The attenuation coefficient changes with temperature and this can be exploited to generate conditions with reduced beam loss. For water (Fig. 7.7, dashed line),

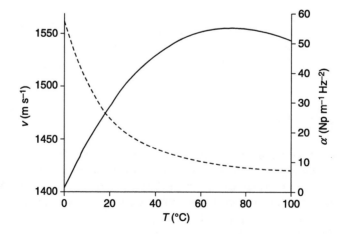

Fig. 7.7 Speed of sound v (continuous line) and attenuation coefficient (dashed line) in distilled water as a function of temperature (Kaye and Laby, 1995). See also Bilaniuk and Wong (1993, 1996) and Del Grosso and Mader (1972).

it is preferable to operate at raised temperatures, for example α' is halved from 20 °C to 50 °C.

7.2.3   Scanning methods

Acoustic microscopy can be either conducted in transmission mode or reflection mode. In the former, sound waves are focused onto the sample by one transducer, and waves that are forward transmitted through the sample are detected by an additional transducer in the opposing position. In the latter, the emitting transducer is also used as the receiver, and thus the detected sound waves are those that are reflected or back-scattered within the sample. An immediate implication of these set-ups is that reflection-mode systems are trivial to align and more economical, requiring only one transducer.

Transmission-mode acoustic microscopy may be operated with either pulsed or continuously-excited sound waves. Reflection-mode acoustic microscopy is restricted to pulsed excitation only, due to the requirement to temporally separate the emission and received signals in the single transducer. The choice of operational mode is often determined by the acoustic thickness of the sample. Clearly, if the sample highly scatters or attenuates the beam, negligible field strength may transmit through the sample, thus prohibiting the use of transmission mode. Conversely, for samples that allow sound transmission, transmission mode can often be preferable, due to its capacity to be driven continuously, thus enabling more rapid data acquisition. Overall, reflection-mode acoustic microscopes are more common than their transmission-mode counterparts, probably due to their greater versatility for a range of sample types. For this reason, we will focus on reflection-mode acoustic microscopy for the remainder of this chapter.

The operating principle of reflection-mode acoustic microscopy is illustrated schematically in Fig. 7.8. Sound waves generated by a piezoelectric element

Fig. 7.8   Schematic of the reflection-mode scanning acoustic microscope. A transducer unit generates and focuses an ultrasound pulse through a coupling fluid onto a sample. Reflected/ scattered waves detected by the transducer unit reveal the acoustic properties of the sample.

within the transducer unit are conveyed along a buffer rod to a spherical lens. The lens focuses the waves onto the sample through a coupling fluid, and the subsequent reflections are detected. Note that spherical transducer elements, which negate the need for a spherical lens, have also been employed (Liang, 1985). The transducer unit or sample is then scanned in space to form a spatial image of the acoustic contrast of the sample.

7.3   Construction of an acoustic microscope

Here we describe an acoustic microscope constructed in our laboratory. An image of our microscope is presented in Fig. 7.9. We will discuss the key components in turn.

7.3.1   Transducer unit

The transducer unit is a commercial high-frequency focused unit (Panametrics V3534) with a quoted fundamental frequency $f = 100$ MHz. A fused quartz delay rod serves to temporally separate the reverberations in the unit. A spherical lens of diameter $D = 6$ mm ground into the end of the delay rod focuses the pulse at a quoted distance $F = 5$ mm. Due to the large reflection at the fluid-lens interface, a quarter-wave layer is employed to enhance transmission.

7.3.2   External controls

A pulse-receiver unit (UTEX-320) generates a square wave pulse of 50 ns duration and 300 V amplitude to excite the transducer and receive the returning signal. The signal is digitised and averaged through a digital oscilloscope. A computer is

Fig. 7.9   Image of the SAM with key components highlighted. The sample unit is shown in further detail in Fig. 7.10 (Parker *et al.*, 2010).

employed to synchronise the data recording with the position system and visualise the signal.

7.3.3 Sample unit

An integrated sample unit, composed of aluminium, provides both an inner well and a surrounding temperature bath (Fig. 7.10). The inner well contains the sample and coupling fluid (in this case Millipore water), and the transducer is immersed in the coupling fluid from above. Water from an external temperature bath (Haake DC50 circulator and Haake B5 bath) circulates through the outer annular well to regulate the temperature of the sample unit, with an embedded thermometer providing feedback to the external unit. Thermal insulation is provided by a layer of foam at the top of the annular well and flexible sealing film stretched over the top of the whole unit (through which the transducer penetrates).

7.3.4 Optical access

High-quality glass windows of 3 cm diameter are embedded into opposing walls of the sample unit (not illustrated in Fig. 7.10). This allows optical access into the sample unit and is exploited to perform simultaneous optical imaging of the sample, using a CCD camera mounted on a Leica Monozoom® 7 parfocal lens, which offers a field of view ranging from 20 mm to 2 mm, and an optical resolution of down to 10 microns.

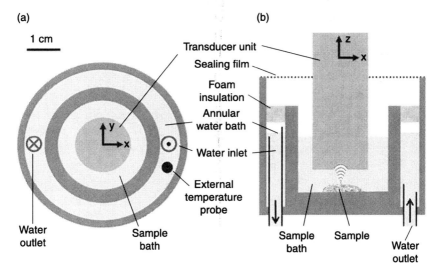

Fig. 7.10 Details of sample unit. Schematic views of the sample unit from the top (a) and side (cross-section) (b). Light grey denotes water, intermediate grey denotes the transducer unit, and dark grey denotes the aluminium housing. In the top view (a), the foam insulation and jacket are not shown.

7.3.5   Positioning system

To provide versatility to image over a range of length scales, we require a positioning system that combines high spatial precision with a large and configurable range of motion. We have constructed a bespoke device that offers a spatial resolution of the order of a micron and a range of several centimetres, all within a single positioning system. Our positioning system, constructed of arms and rotational joints, is based on arcular motion rather than the more conventional *x-y-z* motion. This design offers a simplicity that minimises the need for commercial parts and enhances the economy of the system.

7.4   Operation and calibration of an acoustic microscope

7.4.1   Transducer/beam characteristics

The properties of a high-frequency transducer can often deviate from its ideal, quoted properties. As such, it is essential to characterise the transducer experimentally. Further details can be found in published articles (Shiloh *et al.*, 1991; Lee and Bond, 1993, 1994), which detail the characterisation of transducers of similar construction and frequency.

Transducer electrical impedance

As an electro-mechanical device, we can characterise the transducer electrically. A plot of the electrical impedance of the transducer as a function of frequency is shown in Fig. 7.11. This is obtained by connecting the transducer to a network analyser (Agilent, E5062A, range 300 kHz-3 GHz). Impedance contributions from the connecting cables are observed to be very small in comparison. During measurement, the transducer is immersed in water to simulate its working conditions. The impedance response resembles that of an LC (Inductor/Capacitor)

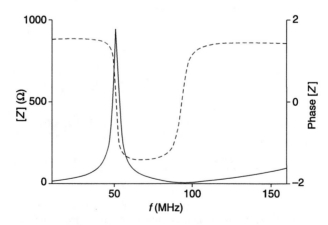

Fig. 7.11   Complex electrical impedance (solid line) and phase (dashed line) as a function of frequency for our transducer (Parker *et al.*, 2010).

resonator circuit, with a clear resonance at approximately 50 MHz and anti-resonance at approximately 95 MHz. The UTEX system is a low impedance system of approximately 50 Ω, hence maximum power transmission into the transducer can be expected when the electromechanical impedance of the transducers is of this order, that is at around 60 MHz and 100 MHz, the two main output frequencies of the transducer.

Transducer emission
Insight into the transducer emission can be obtained from the first acoustic reverberation within the transducer. This pulse is shown by the grey line in Fig. 7.12(a). Its Fourier spectrum, shown by the grey line in Fig, 7.12(b), features a primary component at 61 MHz and a secondary component at 100 MHz. By careful optimization of the circuitry and excitation pulse, we could preferentially excite the 100 MHz component so as to ensure maximal imaging resolution with optimal signal-to-noise. However, having a range of well-defined frequencies present provides versatility to vary the operating frequency as desired, for example for the examination of frequency resonances or attenuation spectra.

Signal properties
We insonify a flat surface of polytetrafluoroethylene (PTFE) placed in water at 30 °C. PTFE is employed because its reflection function is approximately flat over the range of ray angles generated by our transducer (Briggs and Kolosov, 2010) and will not generate surface acoustic waves (which would complicate our results). The signal at focus is shown by the black line in Fig. 7.12(a). The Fourier spectrum of the focal signal is shown by the black line in Fig. 7.12(b). The primary frequency component is now at 55 MHz and the secondary component is at 83 MHz. We can readily explain and estimate this frequency shift by considering a generic pulse with a Gaussian frequency distribution:

$$V(z = 0) = V_0 \exp\left[-\frac{(f - f_0)^2}{2\sigma^2}\right] \qquad [7.15]$$

where V_0 is the initial peak voltage of the pulse, f_0 is the initial centre frequency and σ is the width of the frequency distribution. If the transducer-surface distance z is non-zero, the returning voltage will be attenuated as per Equation 7.5 leading to a modified voltage:

$$V(z) = V_0 \exp\left[-\frac{(f - f_0)^2}{2\sigma^2} - 2\alpha' f^2 z\right] \qquad [7.16]$$

The peak frequency will occur when the exponent is minimal, that is, when,

$$f_0(z) = \frac{f_0}{1 + 4\alpha' z \sigma^2} \approx f_0(1 - 4\alpha' z \sigma^2) \qquad [7.17]$$

where we have used the Taylor series expansion to give an approximate form valid when $4\alpha' z \sigma^2 \ll 1$. Acoustic attenuation thus causes the peak frequency to

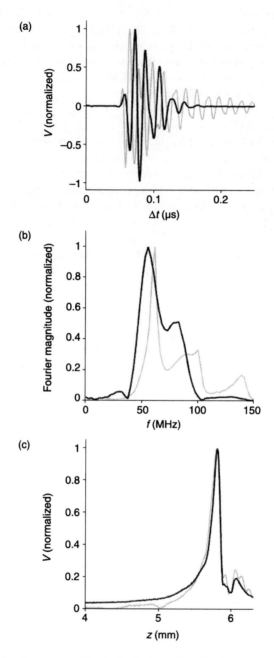

Fig. 7.12 (a) Signals corresponding to the first reverberation in the transducer unit (grey) and the reflection from a PTFE surface (black) located at the focal point in water at 30 °C. The amplitude of the focal pulse is 62 millivolt; (b) magnitude of the Fourier spectrum of the pulses in (a); (c) peak voltage of the raw pulse (black line) and the 55 MHz component (grey line) as the transducer–sample distance z is varied (Parker *et al.*, 2010).

shift to lower frequencies during propagation. Taking $\alpha' = 18 \times 10^{-15}\,\mathrm{Np\ m^{-1}\ Hz^{-2}}$, $\sigma = 20\,\mathrm{MHz}$ and $z = 5.8\,\mathrm{mm}$ (see below), then Equation 7.17 predicts that the 100 MHz peak will shift to 83 MHz and the 60 MHz peak will shift to 51 MHz. This is consistent with the frequency shifts observed in Fig. 7.12(b).

Note that we will henceforth consider the operating frequency of our Versatile Scanning Acoustic Platform (VSAP) to be 55 MHz. The 6 dB bandwidth is measured as 21 MHz.

7.4.2   Detection properties

Our signal is detected and undergoes an analogue-to-digital conversion by an oscilloscope (Lecroy Waverunner Xi-64). We typically average over 100 sweeps to reduce random noise. The dynamic range of the oscilloscope data is measured as 90.5 dB. However, spurious signals of the order of 100 μV limit the lower end of this range. Hence, for a typical voltage amplitude of 100 mV, we obtain a signal-to-noise ratio of 60 dB.

7.4.3   Focal position

The true focal position can be conveniently located by considering how the pulse amplitude varies with the transducer-surface separation, as demonstrated by Lee and Bond (1994). In Fig. 7.12(c), we show the so-called $V(z)$ curve from the PTFE surface. It is sharply peaked in the focal region, where the reflecting waves are in phase and give maximal signal. The voltage decreases away from focus due to the dephasing of the focused waves and, for surfaces beyond the focal point, due to the geometric fact that a proportion of the reflecting rays fall outside the lens (Lee and Bond, 1994). We can thus estimate the focal point as the point at which the peak voltage is a maximum, giving $z_F = 5.791\,\mathrm{mm}$. This definition is crude, since the focal point will vary with frequency due to chromatic lens aberrations and frequency-dependent attenuation in the medium. More precisely, the focal point should be measured at the operating frequency (Lee and Bond, 1994). The magnitude of the 55 MHz pulse component is shown by the grey dashed line in Fig. 7.12(c). It shows a small deviation from the earlier value, giving an improved focal position of $z_F = 5.795\,\mathrm{mm}$.

7.4.4   Resolution

The lateral resolution is estimated from Equation 7.5 to be $d = 23\,\mu\mathrm{m}$ (assuming water at 30 °C and a frequency of 55 MHz). The lateral resolution/beam spot size can be determined experimentally by considering the beam interaction with sharp objects such as needles, edges or line objects (Shiloh et al., 1991). Of these, the edge response is particularly convenient and accurate, and has been employed successfully elsewhere (Shiloh et al., 1991; Lee and Bond, 1993, 1994). Glass provides a sharp edge and here we employ a glass microscope slide (Fig. 7.13(a)). The upper surface of the slide is located in the focal plane and the beam is scanned

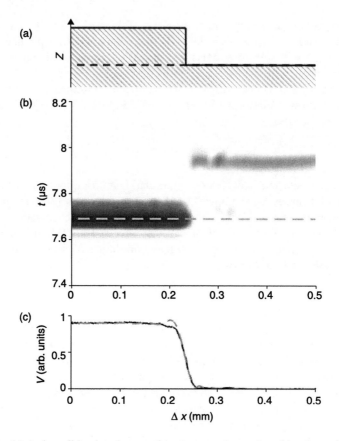

Fig. 7.13   (a) A glass slide placed on another creates a sharp edge; (b) voltage detected as a function of time and lateral position across the edge. The dashed line denotes the focal plane; (c) received voltage in the focal plane, and the theoretical prediction (dashed line) of Equation 7.12 (Parker *et al.*, 2010).

across the edge (in the *x*-direction). The voltage signal in *x-t* space (Fig. 7.13(b)) clearly shows a step in the echo time due to presence of the edge. The voltage profile in the focal plane, which is presented in Fig. 7.13(c), shows a smooth transition to zero due to finite beam width. An excellent description of the theoretical edge response for different ultrasonic imaging systems is presented in Lee and Bond (1993). For a confocal system, the received voltage at focus is given by

$$V(x, z = F) = A \left| \frac{J_1(\pi Dx / F\lambda)}{\pi Dx / F\lambda} \right|^2 * H(-x_s) \qquad [7.18]$$

Here $H(-x_s)$ is the Heaviside step function ($H=1$ for $x<x_s$ and $H=0$ for $x>x_s$), which models the step and A is a normalisation factor that incorporates the electro-mechanical response of the system. The Bessel function J_1 was defined earlier in

Equation 7.11 and is squared here due to the compounded effects of emission and receiving. The beam width is determined by fitting Equation 7.17 to the measured $V(x,z = F)$ data. For a beam width of 25 µm, we get an excellent agreement between the experimental data (solid line) and theoretical fit (dashed line). This width is also in good agreement with the estimated value of 23 µm (see above). Note that by differentiating the edge response (Drescher-Krasicka and Willis, 1996), we arrive at the line spread function and by Fourier transforming this result we can also derive the modulation transfer function (Shilo et al., 1991).

The axial resolution is set by the pulse length. The −6 dB width of the focal pulse in Fig. 7.7(a) is 25.6 ns, corresponding to a distance in 30 °C water of 39 µm. Meanwhile the depth of focus is given by the expression (Maev, 2008):

$$F_z = \frac{8F^2 v}{D^2 f + 2Fv}$$
[7.19]

For water at 30 °C, this gives $F_z = 200$ µm.

7.4.5   Point spread function

The spatial distribution of the beam can also be examined through the point spread function (PSF). This is the signal returned by a point object as a function of the lateral distance between the lens and object. This approach has been considered elsewhere to characterise an ultrasound beam (Shiloh et al., 1991; Jensen and Svendsen, 1992). To make a quantitative measure of beam resolution, the needle tip must be significantly smaller than the beam resolution. Since it is difficult to obtain needles this small for our interests, we are limited here to obtaining qualitative information about the beam profile and spreading. We image the tip of a wire, which is 80 µm in diameter. At a given sample distance, we scan laterally across the object. The position of the sample is expressed relative to the focal point by the parameter Z, with $Z < 0$ corresponding to the sample being in front of the focal point. The demodulated signal detected as a function of lateral displacement is presented in Fig. 7.14. At the focal point the PSF, and therefore the beam, is at its narrowest, and is single peaked. Note that the PSF at the focal point is limited by the size of the pin tip, and so does not reach the true minimal value. The lateral extent of the PSF/beam increases as we move away from the focal point. Lobes form due to diffraction effects and the PSF develops a curvature due to the lens curvature.

In addition, we have calculated the PSF theoretically using the Field II simulation (Jensen and Svendsen, 1992), with the results presented in the right-hand column of Fig. 7.9. This method assumes linear acoustics and derives the PSF by summing the spatial impulse response from small segments of the concave lens, for which an analytic form exists (Tuphome, 1969; Stepanishen, 1971). The PSF is then the convolution of the spatial impulse response with the excitation function, for which we assume some approximate form. Our experimental results are in good agreement with the theoretical predictions, including the beam width,

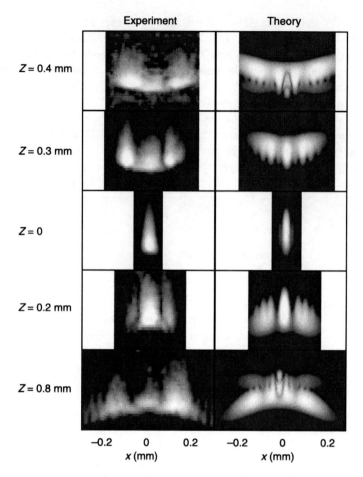

Fig. 7.14 Point spread function as measured experimentally (left column) using a pin point and determined theoretically (right column) using the Field II simulation. The *y*-axis of each image represents the time scale of the returning signal.

curvature and presence of lobes. Deviations arise from a number of sources, including irregularities in the pin tip and lens apodisation.

7.4.6   Angular reflection

Another important consideration in beam detection is that of angular reflection from a tilted surface. For a surface which is normal to the incident beam, the waves are reflected back to the transducer. However, a tilted surface results in angular reflection, which reduces the proportion of the reflected energy that falls back on the transducer lens. In this manner, the surface topography modulates the returning acoustic amplitude. From simple geometrical arguments, we expect that the received signal becomes negligible when the surface is tilted away from the horizontal by around 20 degrees, and this is consistent with that observed experimentally.

7.4.7   Temperature bath

It is essential for a uniform ambient temperature between the transducer and the sample to prevent the thermal image aberrations. It is this, above all else, which controls resolution; for this reason, we present some detail in order to illustrate the issues. We have mapped the temperature distribution in the sample unit using a four wire platinum resistance thermometer (Hart Scientific 5612 probe), accurate to 0.01 °C (Parker *et al.*, 2010), traceable to ITS-90. After setting the desired temperature on the water circulator and allowing sufficient time to equilibrate, we measured the actual temperature at five positions across the sample unit (see inset of Fig. 7.15). This was performed for set temperatures of 30, 40, 50 and 60 °C.

Note that for access, the transducer could not be put in place during these measurements. The deviation between the set temperature and the measured temperature ΔT is shown in Fig. 7.15. At point A, which is adjacent to the temperature probe regulating the bath, the measured temperature is in excellent agreement with the set temperature. Furthermore, the correspondence between the temperatures at points A and E show that the annular temperature bath is of approximately uniform temperature. The inner sample well is lower in temperature than the annular water bath, due to imperfect thermal contact. Furthermore, within the sample well there exists a temperature distribution due to thermal losses and currents. At a set temperature of 30 °C, this temperature variation is approximately 0.01 °C.

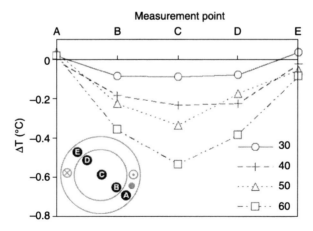

Fig. 7.15 Difference between the actual temperature and the set temperature at five positions A–E across the sample unit, as indicated in the inset. The set temperatures (°C) are specified in the key.

Taking into account the temperature dependence of the speed of sound (Fig. 7.6), this corresponds to a 0.003% variation in the speed of sound and therefore distance measurements that are negligible. The temperature variation grows at larger ambient temperatures, being 0.2 °C at 60 °C. However, the corresponding variation in the speed of sound is approximately 0.01%, which is still small enough to be negligible for most purposes. The presence of the foam insulation and sealing film is crucial in preventing evaporation and heat losses, giving an order of magnitude improvement in the thermal variations; in their absence the temperature variations in the inner well were approximately 0.2 °C at 30 °C and 2 °C at 60 °C. Hence, through careful control of temperature, we have achieved at least a ten-fold improvement in resolution. Further reduction in thermal variations could be made by improved thermal insulation and the inclusion of stirring to suppress thermal currents.

7.4.8   Positioning system

Each rotational increment on the stepper motor corresponds to a 50 nm translation of the axle. A large backlash occurs in the belt and its connection with the gears and is particularly evident during a change of direction. This is negated by performing saw tooth raster scanning where, after a full scan in one direction, the system is reversed back to its start point, ready for the next scan (rather than the conventional back-and-forth scanning). The start point can be conveniently defined by one of the motion limiters. Accuracy of motion is readily estimated in the z-direction by independently measuring the distance from the pulse time-of-flight.

7.5   Exemplars of acoustic microscopy and applications to food structure

7.5.1   Exemplars of acoustic microscopy

Onion skin
One of the first suggested applications of scanning acoustic microscopy was in the study of biological matter (Lemons and Quate, 1975), due to its non-invasiveness and beneficial imaging contrast. This area has now grown, with major applications being found in the study of bone and teeth, cells and soft tissue samples such as the eye and skin (Briggs and Kolosov, 2010; Kundu, 2004; Foster *et al.*, 2000). Figure 7.16(a) presents an acoustic image of onion skin. This is a C-scan image from the focal plane of the transducer, and was achieved by time-gating the signal over a 50 ns gate about the focal plane. The focal plane was 30 microns beneath the uppermost cell surface, and so corresponds approximately to the mid-plane of the cell. Individual onion cells are clearly visible, even at modest imaging resolution. Excellent contrast of the macroscopic cellular structure is observed, with the membrane/wall exhibiting the strongest reflection.

The image took approximately two hours to acquire. It is worth noting that the mechanical properties of cells and soft tissues can be extracted from acoustic microscopy using a variety of mathematical techniques (Kundu, 2004; Briggs and Kolosov, 2010).

Integrated circuits
One of the largest applications of scanning acoustic microscopy has been in non-destructive testing of electronic circuitry (Kundu, 2004). Figure 7.16(b,c) presents a comparison between the acoustic and optical image of an integrated circuit. The acoustic image plots the amplitude of the reflection from the circuit surface. It is

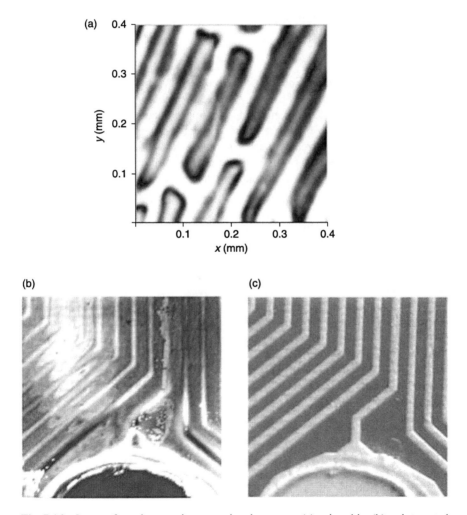

Fig. 7.16   Images from the scanning acoustic microscope: (a) onion skin; (b) an integrated circuit (22 mm) with the corresponding optical image presented in (c).

important to note that the image is taken through a plastic protective layer over the circuit. The oval fringes in the top left of Fig. 7.16(b) arise from the interference between successive echoes from the sample. It is clear that the acoustic image contains greater detail than the optical image, as noted elsewhere (Lemons and Quate, 1975), likely to arise from acoustic-sensitive aberrations in the coating or sub-surface details.

7.5.2   New information which has improved understanding of food structure

Acoustic microscopy is a relatively new field of study in foods and little has been published specifically in this area. Here we present so-far unpublished results obtained whilst developing acoustic microscopical applications for the food industry.

Fatty materials

In processed meats, the distribution of the fat is an important consideration, as is characterization of the origin of the fat. We have studied, together with colleagues at the University of Valencia (Antonio Mulet and José Benedito), ultrasound methods for the characterization of fat. In recent work we have shown that we can image the distribution of protein (strongly backscattering – dark image) and fat (weakly scattering – lighter image) in Chorizo (Fig. 7.17) and Iberian Ham (Fig. 7.18). Note the layered structure of the meat, arising from the processing method.

Emulsion stability

The volume fraction (and particle size) of fat/oil can be determined from measurements of the speed of sound, here plotted as a function of height in a sample creaming under the influence of gravity (Fig. 7.19). A version of the VSAP platform is available commercially, called the Acoustiscan, which has a number of advantages with respect to optical techniques such as the Turbiscan. In any optically opaque sample, measurements are confined to a thin layer close to the surface of the material in the case of optical methods, whilst the acoustic technique gives a bulk measurement that is directly related to the concentration of the dispersed phase. In any optically turbid sample manifesting multiple scattering, the backscattered light intensity is not simply related to the sample concentration, whilst the acoustic method can work even in the case of optically highly absorbing materials (black), such as crude oils, carbon particle suspensions and concentrated emulsions.

7.6   Conclusion and future trends

7.6.1   Acousto-optical microscopy

The coupling of optical and acoustic waves opens up further possibilities for gathering optical and functional information through photo acoustics (Tam, 1986)

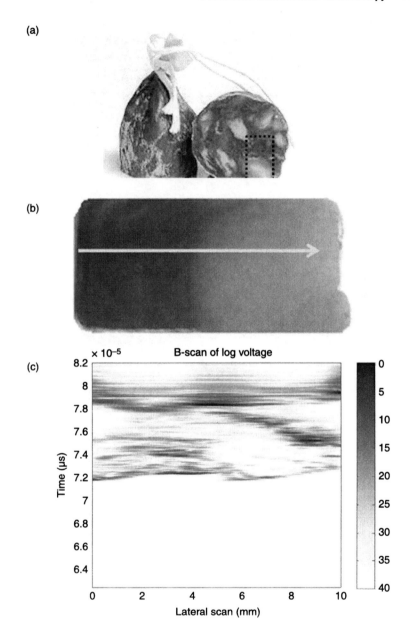

Fig. 7.17 Acoustic image of fat in chorizo sausage. Darker colours represent more intense back scattering from protein (acoustically brighter image). (a) Packaged chorizo with region of slice dotted; (b) optical image of slice; (c) acoustic image of the same slice.

(a)

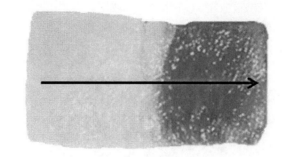

(b)

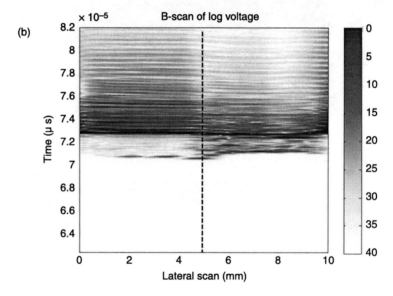

Fig. 7.18 Fat in a slice of Iberian ham. Darker colours represent stronger backscattering (acoustically brighter image). (a) Optical image of ham slice; (b) acoustic image of the same slice.

and ultrasound-modulated optical tomography (USMOT) (Marks *et al.*, 1993; Wang *et al.*, 1995; Leutz and Maret, 1995; Kempe *et al.*, 1997). In the latter case, a precision acoustic beam is employed to 'tag' light travelling through a turbid medium. The tagging occurs through the acoustic modulation of such optical properties as refractive index and the concentration of scatterers and absorbers. This technique enables the scattered light to be spatially localised within the sample to the depth and resolution of the acoustic beam. Importantly, this has enabled optical (including fluorescence) imaging in biological samples to depths that are well beyond the conventional scattering limit. While early work on USMOT employed low acoustic frequencies of around 1 MHz (Marks *et al.*,

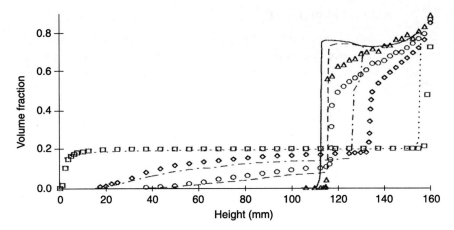

Fig. 7.19   Volume fraction determined from speed of sound in a freely creaming emulsion (Pinfield *et al.*, 1994).

1993; Wang *et al.*, 1995; Leutz and Maret, 1995; Kempe *et al.*, 1997), recent work is moving towards the higher frequencies found in acoustic microscopes. For example, working at 75 MHz, optical imaging in tissue phantoms at a resolution of around 30 microns and to a depth of over 2 mm has recently been achieved through this method (Kothapalli and Wang, 2008). This micro-scale resolution marks the advent of acousto-optical microscopy (Fig. 7.20).

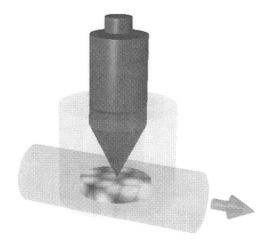

Fig. 7.20   Diagram of scanning acoustic microscope with laser diffuse illumination of the focal region.

7.7   Acknowledgements

We thank Edith Corona Jiménez for the images of Chorizo and Iberian Ham, measured whilst working in our laboratory. We thank the Biotechnology and Biological Science Research Council for financial support (BB/F004923/1 and BB/F0F/302).

7.8   References

ATALAR, A. (1978), An angular-spectrum approach to contrast in reflection acoustic microscopy, *J. Appl. Phys.*, **49**, 5130–9.

BILANIUK, N. and WONG, G. S. (1993), Speed of sound in pure water as a function of temperature, *J. Acoust. Soc. Am.*, **93**, 1609–12.

BILANIUK, N. and WONG, G. S. (1996), Erratum: Speed of sound in pure water as a function of temperature (*J. Acoust. Soc. Am.* **93**, 1609–12, 1993), *J. Acoust. Soc. Am.*, 3257.

BRIGGS, G. A. D. and KOLOSOV, O. (2010), *Acoustic Microscopy*, New York, Oxford Science Publications.

CHALLIS, R. E., POVEY, M. J. W., MATHER, M. L. and HOLMES, A. K. (2005), Ultrasound techniques for characterizing colloidal dispersions, *Rep. Prog. Phys.*, **68**, 1541–637.

CHARLIER, J. P. and D CROWET, F. (1986), Wave equations in linear viscoelastic materials, *J. Acoust. Soc. Am.*, **79**, 895–900.

CHEEKE, J. N. (2002), *Fundamentals and Applications of Ultrasonic Waves*, Boca Raton, FL, CRC Press.

DEL GROSSO, V. A. and MADER, C. W. (1972), Speed of sound in pure water, *J. Acoust. Soc. Am.*, **52**, 1442–6.

DRESCHER-KRASICKA, E. and WILLIS, J. R. (1996), Mapping stress with ultrasound, *Nature* 384, 52–55.

DUCK, F. A., *et al.* (1998), *Ultrasound in Medicine*, Bristol, Institute of Physics.

DUKHIN, A. S. and GOETZ, P. J. (2002), *Ultrasound for Characterizing Colloids: Particle Sizing, Zeta Potential, Rheology*, Oxford, Elsevier.

DUKHIN, A. S. and GOETZ, P. J. (2010), *Characterization of Liquids, Nano- and Microparticulates, and Porous Bodies using Ultrasound* (*Studies in Interface Science*), Oxford, Elsevier.

FOSTER, J. (1984), High resolution acoustic microscopy in superfluid helium, *Physica B + C*, **126**, 199–205.

FOSTER, F. S., PAVLIN, C. J., HARASIEWICZ, K. A, CHRISTOPHER, D. A. and TURNBULL, D. H. (2000), Advances in ultrasound biomicroscopy, *Ultrasound in Medicine and Biology*, **26**, 1–27.

HADIMIOGLU, B. and QUATE, C. F. (1983), Water acoustic microscopy at suboptical wavelengths, *Appl. Phys. Lett.*, **43**, 1006–7.

JENSEN, J. A. and SVENDSEN, N. B. (1992), Calculation of pressure fields from arbitrarily shaped, apodized, and excited ultrasound transducers, *IEEE Trans. Ultrason. Ferroelectr. Freq. Control*, **39**, 262–7.

JIPSON, V. and QUATE, C. F. (1978), Acoustic microscopy at optical wavelengths, *Appl. Phys. Lett.*, **32**, 789–91.

KAYE and LABE (1995), Available from: http://www.kayelaby.npl.co.uk/ (Accessed 3 June 2013).

KEMPE, M., LARIONOV, M., ZASLAVSKY, D. and GENACK, A. Z. (1997), Acousto-optic tomography with multiply scattered light, *J. Opt. Soc. Am. A*, **14**, 1151–8.

KINO, G. S. (1980) *Scanned Image Microscopy*, E. A. Ash (ed.), London, Academic Press, 1–21.

KOTHAPALLI, S-R. and WANG, L. V. (2008), Ultrasound-modulated optical microscopy, *J. Biomed. Optics*, **13**, 054046.

KRAUTKRÄMER, J. and KRAUTKRÄMER, H. (1990), *Ultrasonic Testing of Materials*, Berlin, Springer-Verlag.

KUNDU, T. (2004), *Ultrasonic Non-destructive Evaluation: Engineering and Biological Material Characterization*, London, CRC Press.

LANDAU, L. D. and LIFSCHITZ, E. M. (1970), *Theory of Elasticity*, 2nd English edition, Oxford, Pergamon.

LEE, U. W. and BOND, L. J. (1993), Characterization of ultrasonic imaging systems using transfer functions, *Ultrasonics*, **31**, 405–15.

LEE, U. W. and BOND, L. J. (1994), Characterising the performance of a confocal acoustic microscope, *IEEE Proc.-Sci. Meas. Technol.*, **141**, 48–56.

LEMONS, R. A. and QUATE, C. F. (1974a), Acoustic microscope – scanning version, *Appl. Phys. Lett.*, **24**, 163–5.

LEMONS, R. A. and QUATE, C. F. (1974b), Integrated circuits as viewed with an acoustic microscope, *Appl. Phys. Lett.*, **25**, 251–3.

LEMONS, R. A. and QUATE, C. F. (1975), Acoustic microscopy: biomedical applications, *Science*, **188**, 905–11.

LEUTZ, W. and MARET, G. (1995), Ultrasonic modulation of multiply scattered light, *Physica B: Condensed Matter*, **204**, 14–19.

LIANG, K. K., KINO, G. S. and KHURI-YAKUB, B. T. (1985), Material Characterization by the Inversion of V(z), *IEEE Trans. Sonics Ultrasonics*, **32**, 213–24.

LOVE, A. E. (1944), *Treatise on the Mathematical Theory of Elasticity*, New York, Dover Publications.

MAEV, R. G. (2008), *Acoustic Microscopy: Fundamentals and Applications*, Weinheim, Wiley-VCH.

MARKS, P. A. *et al.* (1993), in *Photon Migration and Imaging in Random Media and Tissues*, R. Alfano and B. Chance (eds), Proc. SPIE 1888 500-10.

MCCLEMENTS, D. J. and POVEY, M. J. W. (1989), Scattering of ultrasound by emulsions, *J. Phys. D: Applied Phys.*, **22**, 38–47.

MORSE, P. M. and INGARD, K. U. (1968), *Theoretical Acoustics*, New York, McGraw Hill.

MUHA, M. S., MOULTHROP, A. A., KOZLOWKSI, G. C. and HADIMIOGLU, B. (1990), Acoustic microscopy at 15.3 GHz in pressurized superfluid helium, *Appl. Phys. Lett.*, **56**, 1019–21.

PARKER, N. G., NELSON, P. V. and POVEY, M. J. (2010), A versatile scanning acoustic platform, *Measur. Sci. and Tech.*, **21**, 045901doi:10.1088/0957-0233/21/4/045901.

PIERCE, A. D. (1981), *Accountics: An Introduction to its Physical Principles and Applications*, New York, McGraw-Hill.

PINFIELD, V. J. and CHALLIS, R. E. (2011), Acoustic scattering by a spherical obstacle: Modification to the analytical long-wavelength solution for the zero-order coefficient, *J. Acoust. Soc. Am.*, **129**(4), 1851–6.

PINFIELD, V. J., DICKINSON, E. and POVEY, M. J. W. (1994), Modelling of concentration profiles and ultrasound velocity profiles in a creaming emulsion: importance of scattering effects, *J. Coll. Interface Sci.*, **166**, 363–74.

PINFIELD, V. J., HARLEN, O. G., POVEY, M. J. W. and SLEEMAN, B. D. (2006), Scattering of ultrasound by particles, *SIAM J. Appl. Math.*, **66**, 489–509.

POVEY, M. J. W. (1997), *Ultrasonic Techniques for Fluids Characterization*, San Diego, Academic Press.

POVEY, M. J. W. (2000), *Crystallization Processes in Fats and Lipid Systems*, N. Garti and K. Sato (eds), New York, Marcel Dekker Inc, 255–88.

POVEY, M. J., MOORE, J. D., BRAYBROOK, J., SIMONS, H., BELCHAMBER, R. *et al.* (2011), Investigation of bovine serum albumin denaturation using ultrasonic spectroscopy, *Food Hydrocolloids*, **25**(5), 1233–41.

RUGAR, D. (1984), Resolution beyond the diffraction limit in the acoustic microscope: A nonlinear effect, *J. Appl. Phys.*, **56**, 1338–46.

SHILOH, K., SOM A. K. and BOND, L. J. (1991), Characterisation of high frequency focused ultrasonic transducers using modulation transfer function: concept and experimental approach, *IEEE Proc.-A-Sci. Meas. Technol.*, **138**, 205–12.

STEPANISHEN, P. R. (1971), Transient radiation from pistons in an infinite planar baffle, *J. Acoust. Soc. Am.*, **49**, 1629–38.

STRUTT, BARON RAYLEIGH, J. W. (1872), Investigation of the disturbance produced by a spherical obstacle on the waves of sound, *Proceedings of the London Mathematical Society*, **4**, 253–83.

STRUTT, BARON RAYLEIGH, J. W. (1877), *Theory of Sound*, London, Macmillan.

STRUTT, BARON RAYLEIGH, J. W. (1896), *The Theory of Sound*, 2nd edition, London, Macmillan.

TAM, A. C. (1986) Applications of photoacoustic sensing techniques, *Rev. Mod. Phys.*, **58**, 381–431.

TIMOSHENKO, S. P. and GOODIER, J. N. (1970), *Theory of Elasticity*, 3rd edition, New York, McGraw-Hill International Editions.

TSCHOEGL, N. W. (1989), *The Phenomenological Theory of Linear Viscoelastic Behavior*, New York, Springer Verlag.

TUPHOLME, G. E. (1969), Generation of acoustic pulses by baffled plane pistons, *Mathematika*, **16**, 209–24.

UBERALL, H. E. (1992), *Acoustic Resonance Scattering*, Philadelphia, PA, Gordon and Breach Science.

WANG, L., JACQUES, S. L. and ZHAO, X. (1995), Continuous-wave ultrasonic modulation of scattered laser light to image objects in turbid media, *Opt. Lett.*, **20**, 629–31.

YU, Z. and BOSECK, S. (1995), Scanning acoustic microscopy and its applications to material characterization, *Rev. Mod. Phys.*, **67**, 863–91.

ZININ, P. V. and WEISE, W. (2004), Theory and applications of acoustic microscopy, in T. Kundu (ed.), *Ultrasonic Non Destructive Evaluation: Engineering and Biological Material Characterization*, Ch. 11, Boca Raton, FL, CRC Press.

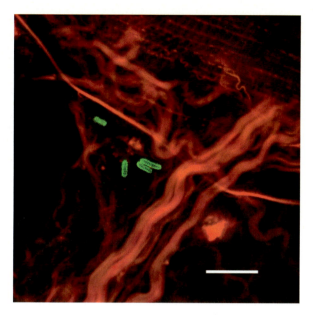

Plate I (Chapter 4) *Eschericia coli* OH157 immunolabelled with FITC (green) in connective tissue of beef muscle, collagen is labelled with Fast Green FCF (red); scale bar = 5 µm.

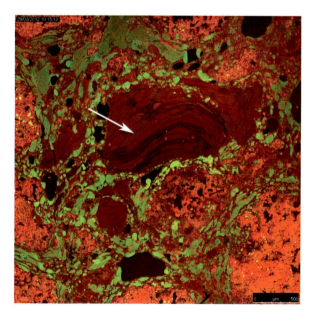

Plate II (Chapter 4) Raw pork sausage. Cryostat section labelled with Nile Blue to show fat (green), proteins (dark red) and starch (bright red). Note muscle fibres (arrow); scale bar = 500 µm. Auty and Doran, unpublished data.

Plate III (Chapter 4) Three-dimensional CSLM projection of raw dessert apple parenchyma, showing cell walls labelled with a 0.1% aqueous solution of Trypan Blue (red); scale bar = 50 μm.

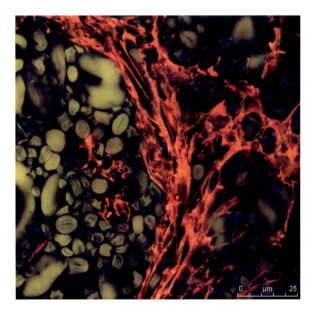

Plate IV (Chapter 4) CSLM image of white bread showing gluten network (red), partially gelatinised starch grains (grey) and lipid droplets (purple); scale bar = 25 μm.

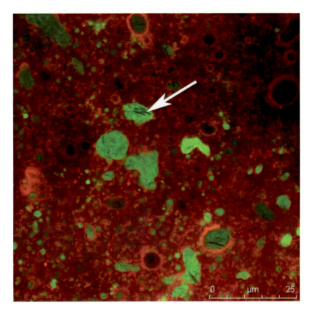

Plate V (Chapter 4) Confocal micrograph of confectionery product: caramel (toffee). Dual-labelled to show liquid fat (green) and protein (bright red). Arrow indicates fat crystals.

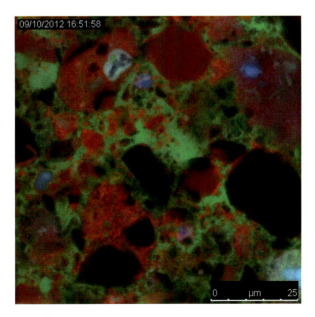

Plate VI (Chapter 4) Confocal micrograph of confectionery product: milk chocolate. Dual-labelled, 3 channel imaged; image shows fat (green), protein (red), cocoa solids (blue – autofluorescence), sugar crystals (black); scale bar = 25 μm.

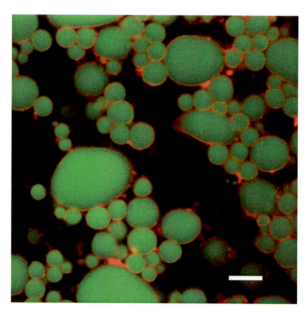

Plate VII (Chapter 4) Confocal micrograph of a dairy product, highlighting typical phase structures: mayonnaise, an oil-in-water emulsion; image shows fat droplets stabilised with a thin protein interface. Dual-labelled with Nile Red and Fast Green FCF to show fat (green) and protein (red); scale bar = 10 μm.

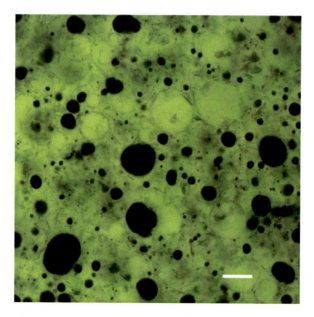

Plate VIII (Chapter 4) Confocal micrograph of a dairy product, highlighting typical phase structures: butter, a water-in-oil emulsion. Image shows water droplets as dark circles and liquid fat continuous phase. Fat crystals are visible as thin dark lines by negative contrast. Labelled with Nile Red to show liquid fat; scale bar = 10 μm.

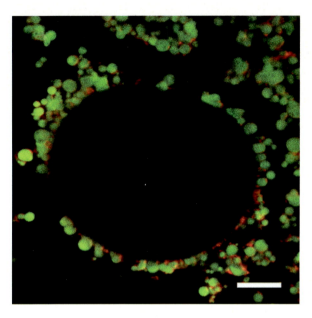

Plate IX (Chapter 4) Confocal micrograph of a dairy product, highlighting typical phase structures: whipped cream, a liquid foam. Image shows partially coalesced fat droplets and protein at the air–water interface. Dual-labelled with Nile Red and Fast Green FCF to show fat (green) and protein (red); scale bar = 10 μm.

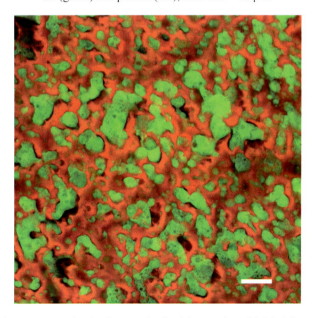

Plate X (Chapter 4) Confocal micrograph of a dairy product, highlighting typical phase structures: cheddar cheese, a fat-filled viscoelastic gel. Image shows a continuous protein phase with entrapped fat globules. Dual-labelled with Nile Red and Fast Green FCF to show fat (green) and protein (red); scale bar = 10 μm.

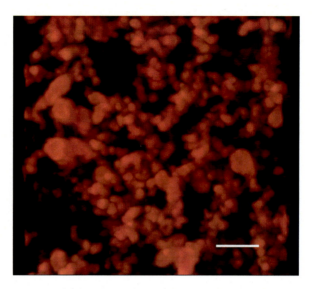

Plate XI (Chapter 4) Three-dimensional CSLM projection of whey protein microspheres produced by segregative phase separation with konjac glucomannan and labelled with Fast Green FCF; scale bar = 5 μm.

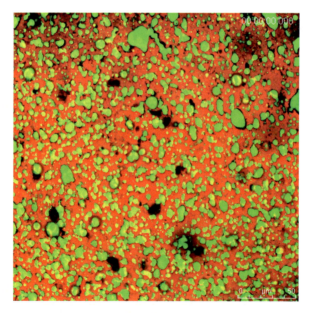

Plate XII (Chapter 4) Simple measurement of fat globule size (equivalent circular diameter) obtained from confocal micrograph of Cheddar cheese dual-labelled to show fat (green) and protein (red). This colour plate also appears in black and white in Chapter 4, Fig. 4.6(a).

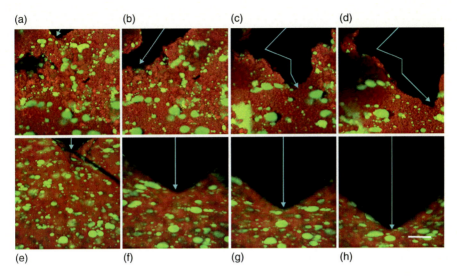

(a)   (b)   (c)   (d)

(e)   (f)   (g)   (h)

Plate XIII (Chapter 4) CSLM image showing fracture propagation through a fat-filled whey protein gel. Protein labelled with Fast Green FCF (red) and fat labelled with Nile Red (green); scale bar = 25 μm. For explanation of panels (a) to (h), see Section 4.8.3. Abhyankar and Auty, unpublished data.

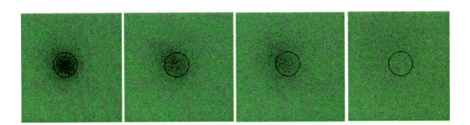

Plate XIV (Chapter 4) Fluorescence recovery after photobleaching (FRAP). Confocal image of bleached (Ar laser, 10 s, 100 mW) region of interest (ROI, dark circle), in a whey protein gel containing 4 kDa FITC-dextrans, showing gradual fluorescence recovery. This colour plate also appears in black and white in Chapter 4, Fig. 4.10(a).

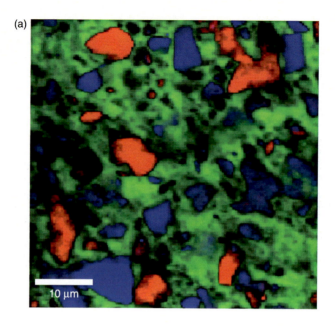

(a)

10 μm

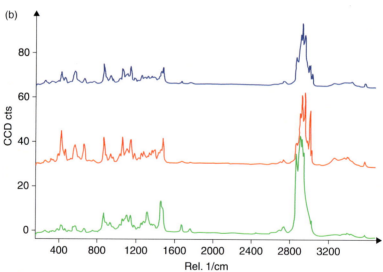

(b)

Plate XV (Chapter 4) Confocal Raman micrograph of chocolate (a) showing fat (green), sugar (blue) and milk solids (red) discriminated based on their Raman scattering spectra (b). Image reproduced with kind permission of Lot Oriel Ltd, (Witech Germany)

8

Using magnetic resonance to explore food microstructures*

P. S. Belton, University of East Anglia, UK

DOI: 10.1533/9780857098894.1.223

Abstract: This chapter reviews the ways in which magnetic resonance may be used to explore food microstructure. The basic concepts of nuclear magnetic resonance (NMR) and electron magnetic resonance are explained and the relevant theoretical background to the use of NMR relaxometry in one and two dimensions, self-diffusion measurements and imaging is given, as well as the background to dynamic electron spin resonance (ESR) measurements. The use of these techniques is described in characterising food microstructure.

Key words: nuclear magnetic resonance, electron spin resonance, relaxation, diffusion, imaging.

8.1   Introduction

At first sight, using radio frequency spectroscopy to study food microstructure seems unlikely. Typically in microscopy the resolution of an image is in the order of the wavelength of the observed radiation. So for magnetic resonance it would be expected that possible resolutions would vary from centimetres (electron spin resonance) to metres (nuclear magnetic resonance (NMR)). However, this is not so due to the special properties of these kinds of spectroscopies, which arise because of the low energy of the transitions involved. These transitions are caused by the quantum mechanical properties of nuclei and electrons, the so-called 'spin'. This is analogous but not equivalent to the macroscopic properties of rotating bodies. Spins in electrons and the nuclei are quantised in units of ½, and electrons

---

* It was originally intended that this chapter should be written by Brian Hills. Unfortunately, Brian became very ill and was unable to start work on it. Brian died on the 29th October 2012. He was an outstanding scientist and good friend with whom I was privileged to work for many years. This chapter is dedicated to him.

and the hydrogen nucleus have a spin of just ½. Other nuclei, such as deuterium, have greater spin values: deuterium has spin 1; sodium has spin ¾; etc. However, as most of this review will concentrate on hydrogen nuclei, the effects of higher spin will not be considered in detail.

Spinning electric charges generate a magnetic moment. Thus when a particle with spin ½ is exposed to a magnetic field, the degeneracy of the magnetic quantum levels is lifted and two levels of differing energy are formed. These correspond to spins aligned parallel and anti-parallel to the field. The parallel alignment is the lower energy state and therefore is in excess. The energy difference between the two levels is given by the relationship:

$$\Delta E = h\frac{\omega}{2\pi} \tag{8.1}$$

and

$$\omega = \gamma B_0 \tag{8.2}$$

where ω is the Larmor or resonance frequency in radians s^{-1}, h is Plank's constant, B_0 is the applied magnetic field flux density in Tesla and γ is the magnetogyric ratio of the nucleus. This value is expressed slightly differently for electrons, but the same fundamental relationship applies.

ω may be considered as the frequency with which the magnetic moments rotate around the applied magnetic field. Since ω is in the radio frequency range, the energy difference ΔE is small. This has a number of important consequences (Belton, 1995, 2010; Hills, 1998). First, the population difference between the two levels is very small, hence the signal strength is small. This is often said to be the most important effect of low transition energy. In fact, it is the least important since the number of naturally occurring quanta of radio frequency at ambient temperatures is very small, hence background noise to signals is also very small. The low energy also means spontaneous transitions between energy levels is slow so relaxation processes are also slow, typically of the order of microseconds to seconds. This means that they may be easily measured and are on the same timescale as chemical exchange and diffusion in pores. This has important consequences for the investigation of microstructures.

Finally, the low energy of the quanta means that the huge numbers of quanta may be generated easily. Under these circumstances, the uncertainty principle may be expressed as

$$\Delta n \Delta \theta \cong 1 \tag{8.3}$$

where Δn is the uncertainty in the number of quanta and $\Delta \theta$ is the uncertainty in the phase of the spin. Since n is large, a large absolute uncertainty in n is a small relative uncertainty, hence the phase is tightly controlled. This means that magnetic resonance is coherence spectroscopy in which the state of the spin system may be manipulated by the spectroscopist. Hence an almost infinite number of experiments are possible in which the spin system may be manipulated in arbitrary ways to explore different facets of the system of interest.

The very prolixity of possible NMR experiments makes comprehensive reviewing difficult. In this chapter an attempt is made to explain the basic principles of the experiments and to illustrate them with selected examples. Pointers are given to the wider literature, but no attempt to give comprehensive coverage is made.

8.2   The magnetic resonance experiment

8.2.1   Nuclear magnetic resonance (NMR)

All NMR experiments require the sample to be exposed to a magnetic field. This results in a net spin magnetisation along the direction of the applied field. This is very small and hard to measure, so the usual procedure is to excite the magnetisation with a pulse of radio frequency energy so that all the individual magnetisations rotate in phase. This generates a net transverse magnetisation, which may be detected in a coil system tuned to resonate at the frequency of rotation. Since the magnetisation has effectively been tipped from a direction along the applied field to one transverse to it, the pulse is known as a '90-degree' or '$\pi/2$' pulse. Similarly, the magnetisation can be completely inverted by application of a '180-degree' or 'π' pulse. Following these pulses, the system has been perturbed from equilibrium and the resulting relaxation process may be measured. The process of relaxation from a perturbation of the populations of the energy levels is described by a time constant T_1, which is usually referred to as the spin lattice relaxation time. For spin ½ systems, this is an exponential process.

Transverse relaxation is more complicated. Immediately following the 90-degree pulse, all the spins precess in phase so in effect they all have the same frequency. If the sample is in a magnetic field that is not homogeneous, then each spin will precess at a frequency determined by the local field.

The dephasing of spins in magnetic field gradients is central to the use of NMR for imaging and the measurement of diffusion. In order to understand more about this process, it is necessary to consider the rotating frame of reference. In this, the z-axis is aligned along the direction of the main magnetic field and the x- and y-axes are rotated at the Larmor frequency. This means that the net magnetisation at equilibrium can be represented as a magnetisation vector pointing along the z-direction. Application of a radio frequency field at the Larmor frequency transversely to the z-direction generates a magnetic field rotating at the Larmor frequency in the x,y plane. Since the plane is rotating at the Larmor frequency, this field is static in the rotating frame. The response of the magnetisation vector to the radio frequency field is to rotate about it. The rotation rate is given by

$$\omega_1 = \gamma B_1 \qquad\qquad [8.4a]$$

If the field is held for time, t, the angle, θ, swept out is given by

$$\theta = \gamma B_1 t \qquad\qquad [8.4b]$$

Suitable adjustments of t and B_1 can rotate the magnetisation through 90 or 180 degrees. Figure 8.1 shows the effect of these pulses on the equilibrium magnetisation. After the 90-degree pulse, the spins may dephase due to magnetic field inhomogeneities or intrinsic relaxation processes. The signal following the 90-degree pulse is called the free induction decay. The effects of magnetic field inhomogeneities are shown in Fig. 8.2. In the x,y plane, the effect is to make the spins rotate at frequencies more or less than the Larmor frequency. After some time, τ, the spins have dephased, a 180-degree pulse applied along y reverses the direction of travel of the spins and at time, 2τ, they become aligned again. This is the so-called spin echo. The evolution of the magnetisation is shown in Fig. 8.2. A train of 180-degree pulses called a Carr-Purcell-Gill-Meiboom sequence (CPMG) removes the effect of field inhomogeneities and extracts the intrinsic relaxation time (Farrar and Becker, 1971; Callaghan, 1995). Figure 8.3 shows the train of spin echoes resulting from a CPMG sequence.

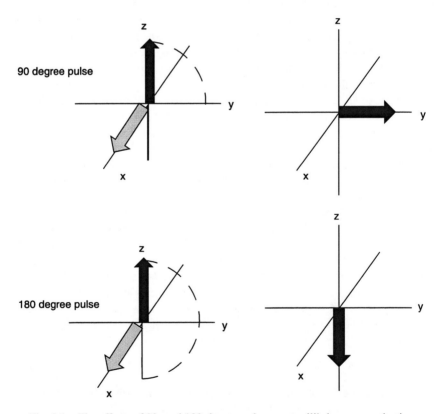

Fig. 8.1   The effects of 90- and 180-degree pulses on equilibrium magnetisation.

(a)

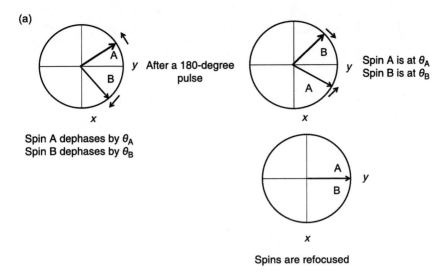

(b)

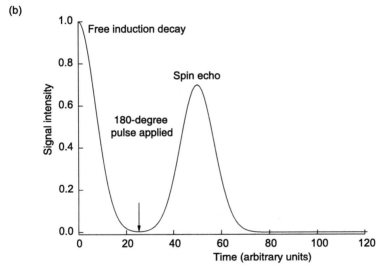

Fig. 8.2 Effects of magnetic field inhomogeneities: (a) effects of dephasing due to magnetic field gradients on the spins A and B which are in different parts of the gradient and the refocusing of the magnetisation by a 180-degree pulse in the rotating frame; (b) observed change in magnetisation during a 90- to 180-degree pulse sequence showing the formation of a spin echo.

8.2.2   Electron spin resonance (ESR)

In electron spin resonance (also known as electron paramagnetic resonance), the magnetic species of interest is the unpaired electron. Although the basic response of the unpaired electron to a magnetic field is the same as that of the nucleus, for historical reasons the terminology used is slightly different to that used in NMR.

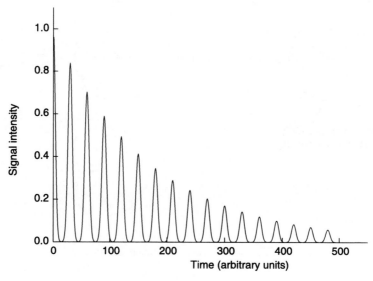

Fig. 8.3   The refocusing of magnetisation by a series of 180-degree pulses in a CPMG sequence.

The basic interaction is written in terms of the electronic factor, *g*, and the Bohr magneton β_e. Equation 8.4a is thus rewritten as

$$\omega = \frac{g\beta_e B_0}{\hbar} \qquad \qquad [8.4c]$$

The value of the term $g\beta_e/\hbar$ is about 1000 times greater than the value of γ for the proton, so the resonance frequency is equivalently higher. Typical relaxation rates and line widths are much greater in ESR experiments than NMR experiments. Consequently, pulsed ESR is more difficult experimentally than pulsed NMR and a continuous wave detection method is more widely used. However, it should be noted that pulsed ESR can be useful as a technique in biochemical and biological applications (Reginsson and Schiemann, 2011; Tsvetkov and Grishin, 2009; van Gastel, 2009).

In the continuous wave application, the main magnetic field is slowly scanned at constant frequency and the resonance of the spin system is observed. Typically the line observed is the first derivative of the in-phase resonance line shape and often contains more than one line. The reason for this is the formation of a hyperfine structure, which is due to the interactions of the electron spins with the local nuclear spins. Unpaired electrons on organic compounds typically lead to chemical instability, so although weak ESR signals can be detected from food, unless specifically investigating these, it is usual to add a spin label in the form of a stable free radical such as 2,2-diphenyl-1-picrylhydrazyl (usually referred to as DPPH). The spectra of DPPH in a number of media are shown in Fig. 8.4.

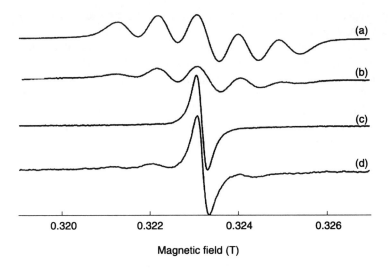

Fig. 8.4   ESR spectra of a spin label in: (a) chloroform/methanol; (b) a lipid bilayer; (c) water; (d) liposomes. The spin label is insoluble in water and the signal is featureless solid signal, in the liposomes the spin label is soluble in the lipid bilayer but insoluble in the water; hence a mixed solid and weak mobile signal is observed.

8.3   Theoretical background

8.3.1   Relaxation in NMR

This discussion is limited to proton relaxation, as this is the principle nucleus used in the study of food microstructure. We consider two main types of relaxation– spin lattice relaxation and transverse, sometimes called spin–spin relaxation. Spin lattice relaxation is the response to the perturbation of the equilibrium distribution of spins between the two energy levels. The mechanism for this is fluctuation of local magnetic fields due to motion of the molecules bearing the nuclei. Spin lattice relaxation is an exponential process and can be described by an equation that relates the magnetisation at time t to that at equilibrium.

Where M_∞ is the equilibrium magnetisation, M_0 is the magnetisation immediately after perturbation, M_t is the magnetisation at time t, and T_1 is the spin lattice relaxation time, then

$$(M_\infty - M_t) = (M_\infty - M_0)\exp\left(-\frac{t}{T_1}\right) \hspace{2cm} [8.5]$$

T_1 is typically of the order of hundreds of milliseconds to seconds. To be effective, the fluctuations in the local magnetic fields must have components of frequency at the Larmor frequency or twice the Larmor frequency, such that the relaxation rate $(R_1 = 1/T_1)$ is given by

$$R_1 \propto [J(\omega_0) + J(2\omega_0)] \hspace{2cm} [8.6]$$

where J represents the spectral density function. This represents the intensity of motions at a frequency, ω.

Transverse relaxation is more strongly dependent on motion than spin lattice relaxation. When motion is slow (local magnetic field fluctuation rates $< {\sim}10^5$ s^{-1}), spin interactions are strong and direct quantum mechanical exchange occurs via the so-called flip-flop interaction. This regime is referred to as the rigid lattice regime. Under these circumstances, the relaxation is non-exponential and is of the order of tens of microseconds or less. In contrast, when motions are significantly greater than rates of 10^5 s^{-1}, the relaxation is exponential and typically of the order of hundreds of microseconds to hundreds or thousands of milliseconds. This regime is the motionally narrowed regime. In this case, the relaxation is exponential and is given by

$$M_t = M_0 \exp\left(-\frac{t}{T_2}\right) \qquad [8.7]$$

The relevant spectral densities are such that the rate R_2 $(= 1/T_2)$ is given by

$$R_2 \propto [J(0) + J(\omega)_0 + J(2\omega_0)] \qquad [8.8]$$

$J(0)$ is the spectral density at zero frequency and implies that T_2, in contrast to T_1, is very sensitive to slow motions and is always less than or equal to T_1.

Both T_1 and T_2 are sensitive to the motional state of the molecules in which the nuclei are embedded, so there is considerable sensitivity to factors such as viscosity in liquid media or temperature in almost any phase. When two materials with different motional characteristics are in close proximity to one another and exchange of nuclei is possible, then dramatic effects, particularly on T_2, can be seen. Figure 8.5 illustrates the effects of the addition of protein to water transverse relaxation. The reason that the relaxation rate increases so dramatically is that the water protons are in fast exchange with the exchangeable protons on the protein (Belton, 2011; Hills et al., 1989a,b). In the fast exchange condition, the relaxation rate is given by

$$1/T_{2obs} = P_{water}/T_{2\ water} + P_{protein}/T_{2\ protein} \qquad [8.9]$$

where P_i is the fractional population of the ith component.

Since the T_2 of water is about 3 s and that of protein is about 15 µs, the effects of a very small pool of exchangeable protein protons is considerable.

This sensitivity indicates how relaxation time measurements may be of use in the examination of microstructural features. If different regions exist, where different relaxation times for water occur because of local exchange effects, then they become distinguishable by their relaxation times. The rate of the exchange can depend both on chemical exchange rates and the rates of diffusion of the molecules of interest to the relaxation sites. The simplest case given above is where the rates of exchange are fast. However, even in this case of fast exchange, it is possible to see the effects of exchange rate by examination of the dependence of the relaxation rate on the pulse spacing

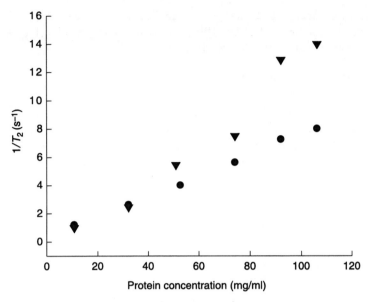

Fig. 8.5 A plot of the transverse relaxation rate vs. protein concentration for water protons in a protein solution. Circles are native protein and triangles are denatured protein.

in a CPMG sequence (Hills *et al.*, 1989a,b; Hills, 1998; Hills and Nott, 1999).

In general, for the chemical exchange to take place, a molecule must diffuse from one site to another. In some cases, where the molecule diffuses to the paramagnetic site, relaxation may be enhanced merely by proximity and no exchange step needs to happen. The situation for a two-site system is illustrated in Fig. 8.6. Site A will have different relaxation times, chemical shifts and diffusion coefficients to site B. Molecules close to the boundary between A and B can diffuse across it quickly. However, molecules distant from the boundary must

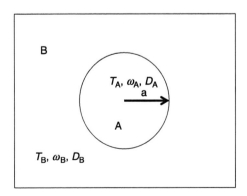

Fig. 8.6 The relevant parameters for the description of a two-site exchange process.

traverse the distance, a, to exchange. A detailed analysis of the effects of diffusion on exchange (Belton and Hills, 1987) showed that the relationship between diffusion and relaxation would take the form:

$$M_t = \sum_{i=1}^{\infty} ([A_i \exp(-S_i)], t) \qquad [8.10]$$

M_t is the observed magnetisation at time t, and A_i and S_i are the intensities and rates of an infinite series of relaxations. Both of these terms are complicated functions of the relaxation rates, diffusion coefficients, distances and relative populations in the sites. In practice, only two or three relaxation processes are observed, but it is possible to observe three processes where only two sites exist (Hills *et al.*, 1989c). If the condition:

$$2\sqrt{2}\left(\frac{a^2}{\pi^2 D}\right)\|R\| \geq 1 \qquad [8.11]$$

is met, then only one relaxation time would be observed. This is the fast exchange limit for diffusive exchange. Where a is the dimension of the region in which diffusion is taking place, D is the diffusion coefficient and $|R|$ is the modulus of the difference in relaxation rates of the two sites. Implicit in Equations 8.10 and 8.11 is the possibility of the measurement of distance by relaxation time measurement. In the general case given by Equation 8.10, modelling is required to extract information about the dimensions of the sites (Hills *et al.*, 1989c). However, observation of the condition in Equation 8.11 does allow an estimate of the upper limit of the size of the sites, which can be useful (Gao *et al.*, 2006). Depending on the values of R and D, the length scales explored can range from nanometres to microns.

8.3.2   The NMR spectrum and magnetic field gradients

In Section 8.2, the effects of local fields was discussed. Local fields can arise from local magnetic field inhomogeneities or from the effects of the electrons surrounding the nuclei. The effect of electrons gives rise to the phenomenon of chemical shift. Since the local electronic behaviour is strongly affected by the local chemical bonding environment, the chemical shift is an excellent indicator of local chemistry. Its measurement has proved to be of enormous power in analysis and structure determination (Richards and Hollerton, 2011; Spyros and Dais, 2012). Since chemical shifts arise from local fields, they are manifested in the free induction decay by sets of spins precessing at different frequencies. The free induction decay signal is related to the spectrum by a Fourier transform relationship:

$$S(\omega) = \frac{1}{2\pi} \int F(t)e^{-i\omega t} dt \qquad [8.12]$$

where $S(\omega)$ is the spectrum and $F(t)$ is the free induction decay.

In the example above, the range of frequencies is due to a range of chemical shifts. However, if a field gradient is applied to the system, then a range of frequencies will be due to the field gradient and the position of the spins in the gradient. In the most straightforward case, a linear 1D gradient is applied. In this case, the frequency (ω_x) of a spin along some coordinate, x, will be given by

$$\omega_x = \gamma B_0 + \gamma Gx \qquad [8.13]$$

where G is the field gradient in Tesla m^{-1}. If the gradient is applied for a time, t, the phase angle, ϕ, through which the spin is shifted is given by

$$\phi = \gamma Gtx \qquad [8.14]$$

The term, γGt, is the wave vector and usually represented by the letter k. It has units of reciprocal length and is analogous to the wave vector in X-ray diffraction. By analogy with Equation 8.12, the spin density across the dimension x, $S(k)$, is related to the signal generated by the gradient:

$$S(k) = \frac{1}{2\pi} \int F(k)e^{-i2\pi kx} dk \qquad [8.15]$$

In essence, by labelling the spins by phase in a field gradient, the resulting spectrum gives a plot of spin density against position, which is an image. By generalising this to three dimensions, a complete image may be generated. Equation 8.15 is the basic equation that is often referred to as k-space imaging.

There are many possible imaging experiments that weight the image by relaxation time, diffusion coefficient, chemical shift, velocity, etc. The actual process of labelling also has an enormous number of variations. More information is given by Hills (1998) and Callaghan (1995, 2011). The factors limiting resolution in k-space imaging have been discussed in detail by Callaghan (1995). So far the best results obtained are of the order of a few microns resolution (Flint et al., 2012).

k-space imaging creates an map analogous to a visual image but weighted by appropriate NMR parameters. This can be immensely useful but in the case, for example, of a porous material, the information required may be the pore size distribution. Thus counting and measuring the pores in the image would be tedious and, in the case of small pores, would not be possible because of the lack of resolution of the image. A similar consideration applies to emulsions and questions of the distribution of droplet size. A possible route to exploring these parameters lies in the measurement of restricted diffusion. For 3D diffusion, the root mean squared distance, x, moved in time, t, by a molecule with a diffusion coefficient, D, is given by

$$x^2 = 6Dt \qquad [8.16]$$

For water, $D = 2.5 \times 10^{-9} \text{m}^2\text{s}^{-1}$, hence in 1 s a diffusing water molecule can explore a space of about 120 microns. If a pulse of field gradient is applied (Equation 8.14), the spins are shifted in phase according to their position in space. If no diffusion takes place, then an equal and opposite pulse, at a later time Δ, will result in an exact reversal of the phase shifts and the original signal will be recovered as a spin echo. This assumes that no relaxation has taken place, but this can be taken into account of by carrying out the experiment with no pulsed field gradient applied and taking the ratio of the two intensities. However, if diffusion takes place, then the second pulse of field gradient will not fully refocus the spins, as they are no longer in the position they were in when they were first labelled. Writing $q = \gamma Gt$, where t is the duration of the pulse of field gradient, the signal attenuation, $E(R,\Delta)$, can be written as

$$P(R,\Delta) = \int E(R,\Delta)e^{-i.q.r}dq \qquad [8.17]$$

$P(R,\Delta)$ is the probability of a spin moving a distance R in time Δ.

The similarity of Equation 8.17 to Equation 8.15 is obvious and using this approach to explore space is often referred to as q-space imaging. However, the parameter measured is the change in signal attenuation with changes in q and/or Δ. The problem is to find the change in R that corresponds to the change in E and hence find $P(R,\Delta)$. In principle, however, $P(R,\Delta)$ will contain information about the geometry of the environment if sufficient time is allowed for a diffusing molecule to explore the confining space.

In the simple case of unrestricted diffusion, $E(R, \Delta)$ is given by

$$E(R,\Delta) = e^{-4\pi^2q^2D\Delta} \qquad [8.18]$$

Where diffusion is restricted, the problem becomes more difficult. Analytical solutions do exist. For a number of cases with well-defined geometries, see Callaghan (1995, 2011), Hills *et al.* (1990) and Hill (1998), and more general solutions often need numerical modelling. The general effect of restriction on diffusion is to reduce the apparent diffusion coefficient with time.

8.3.3   Multi-dimensional methods in NMR relaxometry

In general, k-space imaging is a 3D, or more, method. Diffusion too is in principle a tensor value and anisotropic materials will have different values in different directions. However, a relatively novel application of 2D methods uses the Laplace transform to generate two or more dimensional maps (Callaghan, 2011; Godefroy and Callaghan, 2003; Godefroy *et al.*, 2003; Hills *et al.*, 2004; Hills, 2010). Conceptually, the idea is similar to a multi-dimensional Fourier transformation in that it allows the correlation of two or more separate parameters. We might expect that a food material containing a number of different environments would exhibit a number of relaxation times or ranges of apparent diffusion coefficients. Typically these would be apparent as a multicomponent relaxation decay. Such decays are notoriously difficult to analyse and the most used method of analysis is the 1D Laplace transform using the CONTIN algorithm (Provencher, 1982). Such an analysis results in a relaxation time spectrum, which is a plot of the relaxation times expressed as a continuous distribution, in which the peaks may be interpreted as being indicative of discrete relaxation processes (Fig. 8.7). Whilst the use of CONTIN is helpful in resolving individual processes, the

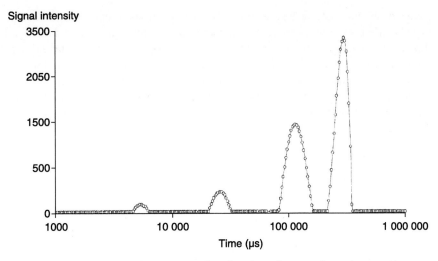

Fig. 8.7   The distribution of transverse relaxation times for water in apple parenchyma as calculated using the CONTIN 1D Laplace transform (courtesy of B. P. Hills).

combination of two or more processes increases the discrimination considerably. A good example of this is the combined T_1–T_2 experiment, in which the magnetisation is inverted and allowed to recover for a variable time, t_1. At time t_1, the magnetisation is recovered and a CPMG sequence applied for a time t_2. The magnetisation, M_{t_1, t_2} is given by

$$M_{t_1, t_2} = M_{\infty} \iint [P(T_1, T_2)] \left[1 - 2\exp\left(-\frac{t_1}{T_1} \right) \right] \exp\left(-\frac{t_2}{T_2} \right) dt_1 dt_2 \qquad [8.19]$$

assuming that $M_0 = -M_{\infty}$.

The term $P(T_1, T_2)$ represents the correlation map of T_2 versus T_1. The transform may be generalised to any coupling of processes that may be described by exponentials. Thus coupling diffusion and transverse relaxation uses the term q^2 from Equation 8.18 as the exponential rate constant.

Unfortunately the extraction of the correlation function by performing the inverse Laplace transform of Equation 8.19 or its generalised form is by no means simple. Numerically the inverse transform is ill conditioned and very sensitive to experimental noise. However, an algorithm has been developed (Song et al., 2002), which does provide a computable and robust approach. However, artefacts are common in the plots so considerable care is required in their interpretation (Marigheto et al., 2007; Callaghan, 2011).

8.3.4   ESR spectrum

The general shape of the ESR spectrum can be related to the dynamics of the spin probe. In most practical situations, the spin probes are very dilute, so that probe–probe interactions can be ignored and translational motion has little effect on relaxation. Two factors have to be taken into account: the modulation of the dipolar interactions by rotation and the modulation of the hyperfine interaction by rotation. Both interactions are tensor quantities and their principle components must be known before correlation times can be calculated. For a three-line ESR spectrum, provided the rotational correlation time is greater than about 1 nanosecond, two correlation times can be calculated, which are designated $\tau_c(B)$ and $\tau_c(C)$, and defined by

$$\tau_c(B) = \frac{15B}{4b\Delta B_0} \quad \text{and} \quad \tau_c(C) = \frac{8C}{28020000b^2} \qquad [8.20]$$

where

$$\Delta = 2\pi\beta_e \, [g_{zz} - (g_{xx}-g_{yy})/2]/h; \qquad b = 4\pi \, [A_{zz} - (A_{xx}-A_{yy})/2]/3$$

and

$$B = 0.866W_{+1}[1 - (h_{+1}/h_{-1})^{1/2}], \qquad C = 0.866W_0[(h_0/h_{+1})^{1/2}+(h_0/h_{-1})^{1/2}-2] \qquad [8.21]$$

The parameters $h_{0,+1,-1}$ refer to the line heights of the three transitions and $W_{0,+1}$ are the line widths with 0 representing the central transition. B_0 is the magnetic field strength. $\tau_c(B)$ corresponds to the averaging of the anisotropic hyperfine tensor (A) and the g tensor, whereas $\tau_c(C)$ corresponds to the averaging of the hyperfine interaction only. $\tau_c(B)$ and $\tau_c(C)$ are not necessarily equal and may have slightly different values (Fairhurst et al., 1983). If this is the case, then usually the slopes of the plots of correlation time versus reciprocal temperature are parallel and the activation energy may be calculated. Generally, $\tau_c(B)$ is regarded as the more reliable value (Belton et al., 1999; Fairhurst et al., 1983). More sophisticated line shape analysis is needed for the slow motion regime (Budil et al., 1996).

8.4   Practical applications of magnetic resonance systems

The potential applications of magnetic resonance to systems are many and varied. The following sections do not attempt to be exhaustive in their account but rather to illustrate the ways in which the approaches outlined in the theoretical background section can be applied. More comprehensive accounts may be found in the articles and books listed in the sources of further information section of this chapter.

8.4.1   Relaxometry

Early work on muscle tissue (Belton et al., 1972; Derbyshire and Parsons, 1972) demonstrated the existence of multiple exponential transverse relaxation decays of water protons. Attempts were made to relate these fractions to compartments within the muscle. More modern approaches have proved much more sophisticated, with a more detailed understanding of the relationship between fibre diameter and relaxation (Christensen et al., 2011) and the role of rigor (Aursand et al., 2009).

Plant tissues and starch have proved particularly susceptible to study by relaxometry and the group led by Hills has been most active in developing this area. In the case of starch, early work using a combination of relaxation measurements showed that in a sample containing 10% water, the water was very mobile with a water lifetime at a site being of the order of 10^{-7} s (Tanner *et al.*, 1991). In more complex systems, such as potato starch suspensions and potato cell walls and potato tissue (Hills and Lefloch, 1994), multiple relaxation processes for water were observed and a 1D Laplace transform was used to separate them. It was possible to assign water to extra- and inter-granular sites in starch. However, the intra-granular water itself occurs in three separate sites (Tang *et al.*, 2000), as water in the amorphous growth rings, the semi-crystalline lamellae, and as 'channel water' in the hexagonal channels of B type crystalline regions. This work has been extended to examine starch gelation and linterisation. It was possible to distinguish amorphous and crystalline regions in the granules and to observe the changes in these regions with starch treatment (Tang *et al.*, 2001). Finally, solid state ^{13}C NMR was used to distinguish two sets of ordered structures in types A, B and C starches (Tang and Hills, 2003).

Figure 8.7 is an example of a 1D Laplace transform of transverse relaxation data from apple parenchyma. Four separate relaxation process are identified and are assigned to vacuole, cytoplasm and extracellular compartments, and mobile components of cell wall material (Hills and Remigereau, 1997). The assignment is based on previous work by Snaar and van As (1992) and a numerical model developed by Hills and Remigereau (1997). The assignments were then used to observe the effects of drying and freezing on ice formation and the integrity of the various cellular components.

Whilst 1D data sets are obviously valuable, their resolution is limited. Two-dimensional methods should be able to generate much better discrimination of various components. The advantages and disadvantages of 2D Laplace transforms are well illustrated in a detailed study of sucrose solutions (Marigheto *et al.*, 2007). Using T_1–T_2 correlation peaks arising from different sugars, CH protons were resolved. Because of the exchange of protons between OH groups on the sugar and water, there was a strong dependency of the correlation plot, both on pulse spacing in the CPMG sequence and spectrometer frequency. In addition, the exchange process itself gave rise to additional peaks. A theoretical analysis showed that two-site exchange with an intermediate exchange rate should give rise to four peaks at the corners of a square. However, both numerical simulation and experiment showed that although four peaks could be seen, they did not appear at the corners of a square. It is not clear why precisely this occurs, but it may be due to the numerical instability of the inverse Laplace transform. Another effect of exchange is to generate apparently non-physical peaks where T_1 is smaller than T_2. The appearance of these peaks is indicative of exchange processes, but they might also arise from artefacts due to the transform. It has been suggested that the problem of artefacts may be improved by the use of the PARAFAC algorithm (Tonning *et al.*, 2007), which extracts systematic variation in the data whilst leaving non-systematic data in the residuals.

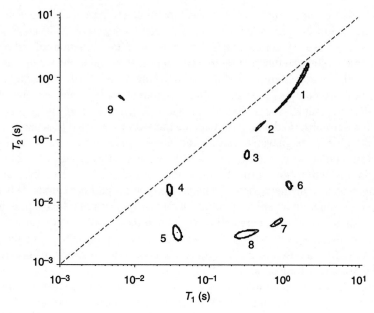

Fig. 8.8   A 2D T_1–T_2 Laplace transform plot of water in apple parenchyma (courtesy of B. P. Hills).

An example of the increased information content of the 2D approach may be visualised by comparing Fig. 8.8, which is a 2D T_1–T_2 plot of apple parenchyma, with Fig. 8.7. Note especially the non-physical peak 9 in Fig. 8.8, which is the result of exchange effects. The set of 2D data can be expanded (Marigheto *et al.*, 2008) to include diffusion correlated relaxation, which resulted in the observation of two peaks due to water and two due to sucrose. Since water is a major component of the signal, it can obscure less intense signals. Suppression of the water signal can be achieved by applying a field gradient; since water has the largest diffusion coefficient, it dephases most rapidly so that its signal is lost more quickly than other components.

A number of other studies have been reported using 2D methods (Godefroy *et al.*, 2003; Hills *et al.*, 2005; Marigheto *et al.*, 2005) and many useful applications have been summarised in reviews (Hills, 2006, 2010; van Duynhoven *et al.*, 2010).

8.4.2   *q*- and *k*-space imaging

The translational motion of a molecule explores physical space. In the case of a randomly moving molecule confined to a pore, given sufficient time, it will visit every possible location in the pore. Its probability distribution will not then be a function of its diffusion coefficient but of the pore geometry. In principle therefore, in the long time limit (i.e. when Δ is sufficiently large and signal attenuation due to relaxation is not too great) the geometry of the pore may be determined. A plot of ln (E) versus q gives a diffraction-like curve analogous to an X-ray diffraction powder pattern intensity plot (Hills, 1998; Callaghan, 2011).

A good example of the use of pulse field gradient methods is given by its application to emulsions (Bernewitz *et al.*, 2011; Hills, 1998; Voda and van Duynhoven, 2009). In the limit of diffusion being completely restricted within the droplet, an estimate of the distribution of droplet sizes can be obtained. However, often the situation is more complicated, because if the droplets are small, then droplet self-diffusion may contribute to the apparent internal diffusion of the molecules. Also there may be inter-droplet transport and finally the emulsion may not be a simple two-phase one. Usually it is possible to suppress intra-droplet diffusion and droplet self-diffusion by lowering the temperature and increasing the viscosity of the continuous phase by adding a thickening agent (Voda and van Duynhoven, 2009). Under these circumstances, characterisation of the droplet size on a length of submicron to 100 microns is in principle possible.

In the case of water in oil emulsions, diffusion between droplets can be used to study the characteristics of water transport across the droplet wall and in the continuous phase (Bernewitz *et al.*, 2011; Voda and van Duynhoven, 2009). This approach can also be applied to double emulsions (Guan *et al.*, 2010) and interpreted in terms of a model involving the diffusion coefficients and residence times of water molecules in the specific compartments and the dispersal phase.

The foregoing discussion assumes that the molecules have sufficient time to explore the geometry of their environment. Where the time is less, it would be expected that the only measurement possible would be the free diffusion coefficient within the confines of the pore or droplet. However, Latour *et al.* (1993) have shown that this is not the case when the diffusion time (Δ) is very short, since then only the molecules close to the confining edge sense its presence and its restriction on diffusion. In this case, the measured diffusion coefficient ($D\Delta$) is given by (Hills, 1998):

$$\frac{D(\Delta)}{D_0} = 1 - \left(\frac{4}{9\pi^{\frac{1}{2}}}\right)(D\Delta)^{\frac{1}{2}}\left(\frac{S}{V}\right)_{+\,\text{higher order terms}} \qquad [8.22]$$

where D_0 is the unrestricted diffusion coefficient, S is the surface area and V is the volume of the pore or droplet. Thus, direct information may be obtained about the surface to volume ratios of confining geometries.

The usefulness of k-space imaging in measuring microstructure directly depends on the exact definition of microstructure. As pointed out above, the practical limit for conventional MRI is of the order of 5 microns. However, such resolution requires long acquisition times, so that imaging droplets in an emulsion is not realistic. In practice, microstructural information on the micron scale is inferred by modelling rather than by direct measurement. Nevertheless, on a larger length scale, microstructurally related parameters, such as relaxation times or diffusion coefficients, may be mapped within the sample.

An interesting example of the use of velocity mapping by MRI is given by Klemm *et al.* (2001), where the velocity profiles on the scale of 300 to 400 microns were mapped in systems where percolation was possible and the fractal dimensions of the systems were calculated.

On the larger length scale, MRI has been used in the study of ripening in fruits such as tomatoes (Ciampa *et al.*, 2010; Zhang and McCarthy, 2012), browning in pears (Hernandez-Sanchez *et al.*, 2007) and in cheeses (Altan *et al.*, 2011; Mariette *et al.*, 1999). Imaging has also proved useful in understanding processes in meat (Bertram and Andersen, 2004; Bertram *et al.*, 2004; Bonny *et al.*, 2007) and fish (Aursand *et al.*, 2009; Erikson *et al.*, 2012).

A particular area of activity has been the study of process relevant treatments, such as the cooking and drying of grains, which has required detailed modelling and led to a better understanding of the subtlety of the process (Ghosh *et al.*, 2008a,b, 2009; Horigane *et al.*, 2006; Stapley *et al.*, 1999). Applications have also been made to the drying of pasta and freezing in potato (Hills *et al.*, 1997a,b).

8.4.3   ESR methods

Unlike NMR, ESR requires the addition of a spin probe if it is to be used in foods for the exploration of microstructure. This has the disadvantage that it is invasive, but has the advantage that the probe may be synthesised with specific properties that enable its localisation in one phase of the foodstuff. Typically, probes can be synthesised so that they are soluble in the lipid or the water phase of the food (Belton *et al.*, 1999). This can be exploited to determine local microviscosities in the lipid and water phases (Rozner *et al.*, 2010; Yucel *et al.*, 2012). More subtle information is available by using the spin probe 4-phenyl-2,2,5,5-tetramethyl-3-imidazoline-1-oxyl (PTMIO), which partitions 70 to 80% in oil and 20 to 30% in water. Yucel *et al.* (2012) were able to show that crystallisation of the lipid in an emulsion-based delivery system excluded the spin probe onto the surface of the emulsion droplets. As the spin probe was a model for solutes in the droplets, this illustrated the way in which hydrophobic solutes could be made more available in drug and nutrient delivery systems.

In single-phase systems, spin probes can still yield useful information. Using a water soluble probe, Gillies *et al.* (1996) explored both the rotational and translational diffusion of a spin probe in a water-gelatine system. When the water content was below 35%, translational diffusion was undetectable. Rotation was detectable up to 0% water, but below 26%, water rotation was slow with a sharp increase in rate above 26%. It was calculated that at this point free rotation occurred and that the volume of the restrictions to rotation must be greater than the molecular volume of gyration of the spin probe ($\sim 10^{-28}$ m^3).

8.5   Nano-scale magnetic resonance

The combination of atomic force microscopy (AFM) and nuclear magnetic resonance (termed magnetic resonance force microscopy, MRFM) promises the possibility of nanometre resolution of food and biological samples (Degen *et al.*, 2009). The process is based on methods very different from conventional NMR. Essentially, the system consists of an ADM placed inside a magnet. The sample is attached to the cantilever tip of the AFM. Placed below the tip is a magnetic cone of about 200 nm diameter that creates a vertical magnetic field gradient. This ensures that in the sample a slice of only a few nanometres thick is in resonance. This determines the vertical resolution. Horizontal resolution is determined by the lateral motion of the AFM tip. The sample is irradiated with a radio frequency field oscillating in and out of resonance frequency. When the section of the sample comes into resonance, the AFM tip moves due to the induced magnetic moment and is detected optically. Using this technique on tobacco mosaic virus, samples images were created with a resolution of 10 nm (Degen *et al.*, 2009).

8.6   Conclusion and future trends

One-dimensional relaxometry is now a reasonably mature area and future developments seem likely to be in the realm of better numerical methods to interpret the data. Two-dimensional methods still have some way to go. The Laplace inversion is numerically unstable and difficult to implement. There are still artefacts that appear in the plots and the problem of the difference between theoretical and actual exchange peak patterns is troubling. More work is clearly needed in this area. Resolution in k-space imaging is probably nearing its practical limits using conventional methods. The use of AFM methods may enable much greater resolution in some samples. However, it has to be recognised that simple reduction in available pixel size does not necessarily mean more useful data. As resolution increases, image blurring processes due to diffusion and thermal motion become more important, so that very high image resolution may only be possible in systems that have been immobilised in some way by lowering the temperature or using a fixative. In some systems, the use of these measures may destroy the structures of interest.

8.7   Sources of further information and advice

I have relied heavily on the books by Hills (1998) and Callaghan (1995, 2011), which cover many of the fundamentals of the subject with a plethora of examples. A snapshot of current trends in magnetic resonance in food is given by the proceedings of the biennial conference 'Applications of Magnetic Resonance in Food Science'. The proceedings of the 10th conference are available (Renou *et al.*, 2011). Reviews by Hills (2006, 2010), van Duynhoven *et al.* (2010) and Voda and van Duynhoven (2009) are extremely valuable. A general review of the applications of magnetic resonance in food is given by a recent special edition of the journal '*Magnetic Resonance in Chemistry*' (2011, 49, Supplement 1, S1–S134). Applications of NMR to food analysis, including a useful general introduction to NMR, are covered in Spyros and Dias (2012).

8.8   Acknowledgement

The author would like to thank the late Brian Hills for permission to use his original drawings for Figs 8.7 and 8.8.

8.9   References

ALTAN, A., OZTOP, M. H., MCCARTHY, K. L. and MCCARTHY, M. J. (2011), Monitoring changes in feta cheese during brining by magnetic resonance imaging and NMR relaxometry, *J. Food. Eng.*, **107**, 200–7.

AURSAND, I. G., VELIYULIN, E., BOCKER, U., OFSTAD, R., RUSTAD, T. and ERIKSON, U. (2009), Water and salt distribution in atlantic salmon (salmo salar) studied by low-field ^{1}H NMR, ^{1}H and ^{23}Na MRI and light microscopy: effects of raw material quality and brine salting, *J. Agric. Food Chem.*, **57**, 46–54.

BELTON, P. S. (1995), NMR in context, *Ann. Rep. on NMR Spectrosc.*, **31**, 18.

BELTON, P. S. (2010), Magnetic resonance in food science – twenty years forward and twenty years back, in J-P. RENOU, P. S. BELTON and G. WEBB (eds), *Magnetic Resonance in Food Science*, Cambridge, UK, Royal Society of Chemistry, 1–10.

BELTON, P. S. (2011), Spectroscopic approaches to the understanding of water in foods, *Food Revs. Int.*, **27**, 170–191.

BELTON, P. S. and HILLS, B. P. (1987), The effects of diffusive exchange in heterogeneous systems on nmr line-shapes and relaxation processes, *Molec. Phys.*, **61**, 999–1018.

BELTON, P. S., JACKSON, R. R. and PACKER, K. J. (1972), Pulsed NMR studies of water in striated-muscle, Part I: Transverse nuclear spin relaxation-times and freezing effects, *Biochim. Biophys. Acta*, **286**, 16–22.

BELTON, P. S., GRANT, A., SUTCLIFFE, L. H., GILLIES, D. G. and WU, X. (1999), Selected spin probes for the electron spin resonance study of the dynamics of water and lipids in doughs, *J. Agric. Food Chem.*, **47**, 4520–4.

BERNEWITZ, R., GUTHAUSEN, G. and SCHUCHMANN, H. P. (2011), NMR on emulsions: characterisation of liquid dispersed systems, *Magn. Reson. Chem.*, **49**, S93–104.

BERTRAM, H. C. and ANDERSEN, H. J. (2004), Applications of NMR in meat science, in G. A. WEBB (ed.), *Ann. Rep. NMR Spectrosc*, **53**, 157–202

BERTRAM, H. C., WHITTAKER, A. K., SHORTHOSE, W. R., ANDERSEN, H. J. and KARLSSON, A. H. (2004), Water characteristics in cooked beef as influenced by ageing and high-pressure treatment – an NMR micro-imaging study, *Meat Sci.*, **66**, 301–6.

BONNY, J. M., FOUCAT, L., MOUADDAB, M., SIFRE-MAUNIER, L., LISTRAT, A. and RENOU, J. P. (2007), Advances in the magnetic resonance imaging of extracellular matrix of meat, in I. A. FARHAT, P. S. BELTON and G. A. WEBB, G. A. (eds), *Magnetic Resonance in Food Science: From Molecules to Man*, Cambridge, UK, Royal Society of Chemistry, 84–8.

BUDIL, D. E., LEE, S., SAXENA, S. and FREED, J. H. (1996), Nonlinear-least-squares analysis of slow-motion EPR spectra in one and two dimensions using a modified Levenberg-Marquardt algorithm, *J. of Magn. Reson. Ser. A*, **120**, 155–89.

CALLAGHAN, P. T. (1995), *Principles of Nuclear Magnetic Resonance Microscopy*, Oxford, Oxford University Press.

CALLAGHAN, P. T. (2011), *Translational Dynamics and Magnetic Resonance*, Oxford, Oxford University Press.

CHRISTENSEN, L., BERTRAM, H. C., AASLYNG, M. D. and CHRISTENSEN, M. (2011), Protein denaturation and water-protein interactions as affected by low temperature long time treatment of porcine *Longissimus dorsi*, *Meat Sci.*, **88**, 718–22.

CIAMPA, A., DELL'ABATE, M. T., MASETTI, O., VALENTINI, M. and SEQUI, P. (2010), Seasonal chemical-physical changes of PGI Pachino cherry tomatoes detected by magnetic resonance imaging (MRI), *Food Chem.*, **122**, 1253–60.

DEGEN, C. L., POGGIO, M., MAMIN, H. J., RETTNER, C. T. and RUGAR, D. (2009), Nanoscale magnetic resonance imaging, *Proc. Nat. Acad. Sci.*, **106**, 1313–17.

DERBYSHIRE, W. and PARSONS, J. L. (1972), NMR investigations of frozen muscle systems, *J. Magn. Reson.*, **6**, 344.

ERIKSON, U., STANDAL, I. B., AURSAND, I. G., VELIYULIN, E. and AURSAND, M. (2012), Use of NMR in fish processing optimization: a review of recent progress, *Magn. Reson. Chem.*, **50**, 471–80.

FAIRHURST, S. A., PILKINGTON, R. S. and SUTCLIFFE, L. H. (1983), Rotational correlation times and radii of dithiazol-2-yl and dithiazolidin-2-yl free-radicals, *J. Chem. Soc.-Farad. Trans. I*, **79**, 439–52.

FARRAR, T. C. and BECKER E. D. (1971), *Pulse and Fourier Transform NMR*, San Diego, Academic Press.

FLINT, J. J., HANSEN, B., PORTNOY, S., LEE, C. H., KING, M. A. *et al.* (2012), Magnetic resonance microscopy of human and porcine neurons and cellular processes, *Neuroimage*, **60**, 1404–11.

GAO, C. L., STADING, M., WELLNER, N., PA, MILLS, E. N. C. and BELTON, P. S. (2006), Plasticization of a protein-based film by glycerol: A spectroscopic, mechanical, and thermal study, *J. Agric. Food Chem.*, **54**, 4611–16.

GHOSH, P. K., JAYAS, D. S., SMITH, E. A., GRUWEL, M. L. H. and WHITE, N. D. G. (2008a), Mathematical modelling of wheat kernel drying with input from moisture movement studies using magnetic resonance imaging (MRI), Part II: Model comparison with published studies, *Biosystems Engineering*, **100**, 547–54.

GHOSH, P. K., JAYAS, D. S., SMITH, E. A., GRUWEL, M. L. H., WHITE, N. D. G. and ZHILKIN, P. A. (2008b), Mathematical modelling of wheat kernel drying with input from moisture movement studies using magnetic resonance imaging (MRI), Part I: Model development and comparison with MRI observations, *Biosystems Engineering*, **100**, 389–400.

GHOSH, P. K., JAYAS, D. S. and GRUWEL, M. L. H. (2009), Measurement of water diffusivities in barley components using diffusion weighted imaging and validation with a drying model, *Drying Technol.*, **27**, 382–92.

GILLIES, D. G., SUTCLIFFE, L. H., WU, X. and BELTON, P. S. (1996), Molecular motion of a water-soluble nitroxyl radical in gelatin gels, *Food Chem.*, **55**, 349–52.

GODEFROY, S. and CALLAGHAN, P. T. (2003), 2D relaxation/diffusion correlations in porous media, *Magn. Reson. Imag.*, **21**, 381–3.

GODEFROY, S., CREAMER, L. K., WATKINSON, P. J. and CALLAGHAN, P. T. (2003), The use of 2D Laplace inversion in food material, in P. S. BELTON, A. M. GIL, G. A. WENN and D. RUTLEDGE (eds), *Magnetic Resonance in Food Science: Latest Developments*, Cambridge, UK, Royal Society of Chemistry, 85–92

GUAN, X. Z., HAILU, K., GUTHAUSEN, G., WOLF, F., BERNEWITZ, R. and SCHUCHMANN, H. P. (2010), PFG-NMR on W-1/O/W-2-emulsions: Evidence for molecular exchange between water phases, *European J. Lipid Sci. Tech.*, **112**, 828–37.

HERNANDEZ-SANCHEZ, N., HILLS, B. P., BARREIRO, P. and MARIGHETO, N. (2007), An NMR study on internal browning in pears, *Postharvest Bio. Tech.*, **44**, 260–70.

HILLS, B., BENAMIRA, S., MARIGHETO, N. and WRIGHT, K. (2004), T_1-T_2 correlation analysis of complex foods, *Appl. Magn. Reson.*, **26**, 543–60.

HILLS, B., COSTA, A., MARIGHETO, N. and WRIGHT, K. (2005), T_1-T_2 NMR correlation studies of high-pressure-processed starch and potato tissue, *Appl. Magn. Reson.*, **28**, 13–27.

HILLS, B. P. (1998), *Magnetic Resonance in Food Science*, New York, Wiley.

HILLS, B. P. (2006), Applications of Low-Field NMR to Food Science, in G. A. WEBB (ed.), *Ann. Rep. NMR Spectrosc.*, **58**, 177–230

HILLS, B. P. (2010), *Relaxometry*, in G. A. MORRIS and J. W. EMSLEY (eds), *Multi-Dimensional Methods for the Solution State*, Chichester, UK, John Wiley and Sons, 533–42

HILLS, B. P. and LEFLOCH, G. (1994), NMR studies of non-freezing water in cellular plant-tissue, *Food Chem.*, **51**, 331–6.

HILLS, B. P. and REMIGEREAU, B. (1997), NMR studies of changes in subcellular water compartmentation in parenchyma apple tissue during drying and freezing, *Int. J. Food Sci. Tech.*, **32**, 51–61.

HILLS, B. P. and NOTT, K. P. (1999), NMR studies of water compartmentation in carrot parenchyma tissue during drying and freezing, *Appl. Magn. Reson.*, **17**, 521–35.

HILLS, B. P., TAKACS, S. F. and BELTON, P. S. (1989a), The effects of proteins on the proton nmr transverse relaxation-time of water. Part II: Protein aggregation, *Molec. Phys.*, **67**, 919–37.

HILLS, B. P., TAKACS, S. F. and BELTON, P. S. (1989b), The effects of proteins on the proton nmr transverse relaxation-times of water. Part I: Native bovine serum-albumin, *Molec. Phys.*, **67**, 903–18.

HILLS, B. P., WRIGHT, K. M. and BELTON, P. S. (1989c), NMR studies of water proton relaxation in sephadex bead suspensions, *Molec. Phys.*, **67**, 193–208.

HILLS, B. P., WRIGHT, K. M. and BELTON, P. S. (1990), The effects of restricted diffusion in nuclear-magnetic-resonance microscopy, *Magn. Reson. Imag.*, **8**, 755–65.

HILLS, B. P., GODWARD, J. and WRIGHT, K. M. (1997a), Fast radial NMR microimaging studies of pasta drying, *J. Food Eng.* **33**, 321–35.

HILLS, B. P., GONCALVES, O., HARRISON, M. and GODWARD, J. (1997b), Real time investigation of the freezing of raw potato by NMR microimaging, *Magn. Reson. Chem.*, **35**, S29–S36.

HORIGANE, A. K., TAKAHASHI, H., MARUYAMA, S., OHTSUBO, K. and YOSHIDA, M. (2006), Water penetration into rice grains during soaking observed by gradient echo magnetic resonance imaging, *J. Cereal Sci.*, **44**, 307–16.

KLEMM, A., KIMMICH, R. and WEBER, M. (2001), Flow through percolation clusters: NMR velocity mapping and numerical simulation study, *Phys. Rev. E*, **63**, 041514.

LATOUR, L. L., MITRA, P. P., KLEINBERG, R. L. and SOTAK, C. H. (1993), Time-dependent diffusion-coefficient of fluids in porous-media as a probe of surface-to-volume ratio, *J. Magn. Reson. Ser. A*, **101**, 342–6.

MARIETTE, F., COLLEWET, G., MARCHAL, P. and FRANCONI, J. M. (1999), Internal structure characterization of soft cheeses by MRI, in P. S. BELTON, B. P. HILLS and G. A. WEBB (eds), *Advances in Magnetic Resonance in Food Science*, Cambridge, UK, Royal Society of Chemistry, 24–34

MARIGHETO, N., DUARTE, S. and HILLS, B. P. (2005), NMR relaxation study of avocado quality, *Appl. Magn. Reson.*, **29**, 687–701.

MARIGHETO, N., VENTURI, L., HIBBERD, D., WRIGHT, K. M., FERRANTE, G. and HILLS, B. P. (2007), Methods for peak assignment in low-resolution multi-dimensional NMR cross-correlation relaxometry, *J. Magn. Reson.*, **187**, 327–42.

MARIGHETO, N., VENTURI, L. and HILLS, B. (2008), Two-dimensional NMR relaxation studies of apple quality, *Postharvest Biol. Tech.*, **48**, 331–40.

PROVENCHER, S. W. (1982), CONTIN – a general-purpose constrained regularization program for inverting noisy linear algebraic and integral-equations, *Comp. Phys. Commun.*, **27**, 229–42.

REGINSSON, G. W. and SCHIEMANN, O. (2011), Pulsed electron-electron double resonance: beyond nanometre distance measurements on biomacromolecules, *Biochem. J.*, **434**, 353–63.

RENOU, J.-P., BELTON, P. and WEBB, G. (eds) (2011), *Magnetic Resonance in Food Science – An Exciting Future*, Cambridge, UK, Royal Society of Chemistry.

RICHARDS, S. A. and HOLLERTON, J. C. (2011), *Essential Practical NMR for Organic Chemistry*, Chichester, UK, Wiley.

ROZNER, S., SHALEV, D. E., SHAMES, A. I., OTTAVIANI, M. F., ASERIN, A. and GARTI, N. (2010), Do food microemulsions and dietary mixed micelles interact? *Colloid Surf. B-Biointerfaces*, **77**, 22–30.

SNAAR, J. E. M. and VANAS, H. (1992), Probing water compartments and membrane-permeability in plant-cells by h–1-NMR relaxation measurements, *Biophys. J.*, **63**, 1654–8.

SONG, Y. Q., VENKATARAMANAN, L., HURLIMANN, M. D., FLAUM, M., FRULLA, P. and STRALEY, C. (2002), T$_1$-T$_2$ correlation spectra obtained using a fast two-dimensional Laplace inversion, *J. Magn. Reson.*, **154**, 261–8.

SPYROS, A. and DAIS, P. (2012), *NMR Spectroscopy in Food Analysis*, Cambridge, UK, Royal Society of Chemistry.

STAPLEY, A. G. F., LANDMAN, K. A., PLEASE, C. P. and FRYER, P. J. (1999), Modelling the steaming of whole wheat grains, *Chem. Engin. Sci.*, **54**, 965–75.

TANG, H. R. and HILLS, B. P. (2003), Use of C–13 MAS NMR to study domain structure and dynamics of polysaccharides in the native starch granules. *Biomacromolecules*, **4**, 1269–76.

TANG, H. R., GODWARD, J. and HILLS, B. (2000), The distribution of water in native starch granules – a multinuclear NMR study, *Carbohydr. Polymers*, **43**, 375–87.

TANG, H. R., BRUN, A. and HILLS, B. (2001), A proton NMR relaxation study of the gelatinisation and acid hydrolysis of native potato starch, *Carbohydr. Polymers*, **46**, 7–18.

TANNER, S. F., HILLS, B. P. and PARKER, R. (1991), Interactions of sorbed water with starch studied using proton nuclear-magnetic-resonance spectroscopy, *J. Chem. Soc.-Faraday Trans.*, **87**, 2613–21.

TONNING, E., POLDERS, D., CALLAGHAN, P. T. and ENGELSEN, S. B. (2007), A novel improved method for analysis of 2D diffusion-relaxation data–2D PARAFAC-Laplace decomposition, *J. Magn. Reson.*, **188**, 10–23.

TSVETKOV, Y. D. and GRISHIN, Y. A. (2009), Techniques for EPR Spectroscopy of Pulsed Electron Double Resonance (PELDOR): A review, *Instruments and Experimental Techniques*, **52**, 615–36.

VAN DUYNHOVEN, J., VODA, A., WITEK, M. and VAN AS, H. (2010), Time-domain NMR applied to food products, in G. A. WEBB (ed.), *Annual Reports on NMR Spectroscopy*, vol 69, San Diego: Elsevier Academic Press Inc.

VAN GASTEL, M. (2009), Pulsed EPR spectroscopy, *Photosynthesis Research*, **102**, 367–73.

VODA, M. A. and VAN DUYNHOVEN, J. (2009), Characterization of food emulsions by PFG NMR, *Trends Food Sci. Tech.*, **20**, 533–43.

YUCEL, U., ELIAS, R. J. and COUPLAND, J. N. (2012), Solute distribution and stability in emulsion-based delivery systems: An EPR study, *J. Colloid Interface Sci.*, **377**, 105–13.

ZHANG, L. and MCCARTHY, M. J. (2012), Measurement and evaluation of tomato maturity using magnetic resonance imaging, *Postharvest Biol. Tech.*, **67**, 37–43.

9

X-ray micro-computed tomography for resolving food microstructures

M. Barigou and M. Douaire, University of Birmingham, UK

DOI: 10.1533/9780857098894.1.246

Abstract: X-ray computed tomography is a 3D imaging technique for the non-invasive, non-destructive visualisation and measurement of the internal microstructure of materials. Using X-rays, a series of radiographs of a sample are recorded from different angles, and then used to reconstruct the internal 3D microstructure by means of a suitable reconstruction algorithm. The method has a high penetrating power and probing efficiency, and is unlimited by morphological complexity. This chapter describes the technique, its advantages and limitations, the latest technological advances, and reviews of applications in food systems.

Key words: absorption, attenuation, computed tomography, microstructure, X-ray, X-ray scan.

9.1 Introduction

The term 'tomography' originates from the Greek words *tomos*, meaning 'slices', and *graphos*, meaning 'imaging'. X-ray tomographic imaging was introduced in the early 1970s, with its theory being first applied for clinical purposes by Godfrey Hounsfield and Allan Cormack, for which they shared a Nobel Prize. X-ray tomography is a non-invasive technique that allows the visualisation of the interior of a specimen via the generation of cross-sectional data. Several non-invasive techniques measuring different physical excitations have been developed, such as magnetic resonance imaging and nuclear magnetic imaging (using the magnetic field), positron emission tomography (using gamma rays) and X-ray tomography. In the case of X-ray tomography, an X-ray beam is focused on the studied sample and a shadow image reflecting X-ray attenuation along the beam path is recorded. The rotation of the sample generates successive images that are stored and subsequently analysed by computer assisted tomography or CAT scanning.

The development of food microstructure engineering raised the need for imaging techniques that could accurately describe product microstructure. Following the development and standardisation of benchtop tomography equipment from the 1990s, X-ray tomography is gradually becoming an important tool in this area, complementing other available techniques such as light microscopy and electron microscopy. The technique has a high penetrating power and probing efficiency, and is unlimited by the complexity of internal and external surfaces. It is non-invasive and non-destructive, which is a key advantage over more traditional techniques, as it provides images free of sample preparation artefacts. However, benchtop equipment generates polychromatic X-rays that tend to complicate quantitative analysis and produce some image artefacts, but some of these problems can be overcome with the use of synchrotron facilities.

Absorption mode tomography is most beneficial when characterising high contrast objects such as extruded products and dry food and foams, although a survey of the literature presented in this chapter shows that the range of applications is much wider as high moisture products such as fruits or meat can also be described.

9.2   Description of X-ray techniques

9.2.1   X-ray principles

X-rays are a form of electromagnetic radiation, with a much shorter wavelength than visible light. The shorter wavelength means that they carry more energy than visible light, ranging from approximately 0.5 keV to several MeV. The distance and direction travelled by an X-ray beam through a sample depends on the energy content of the beam and the composition of the sample. Interactions of X-rays with matter can result in their transmission, absorption or scattering (Fig. 9.1):

(i)   **Transmission**: the X-rays pass through the sample unaffected;
(ii) **Absorption**: the energy carried by the X-rays is transferred to the sample (the photon disappears completely);
(iii) **Scattering (Compton scatter)**: the photons are diverted in a different direction. This can be combined with a loss of energy, whereby the diverted stream is referred to as secondary radiation.

Absorption and scattering are stochastic processes; it is impossible to predict which individual photon will be scattered or transmitted. X-ray radiation is also ionizing, meaning that the X-rays carry enough energy to ionise an atom when absorbed. Following absorption and scattering, the emerging beam contains fewer photons as it is attenuated by these two processes. In both X-ray microscopy and X-ray tomography the image is formed by the transmitted photons, and the grey levels reflect local X-ray attenuation. The coefficient of attenuation

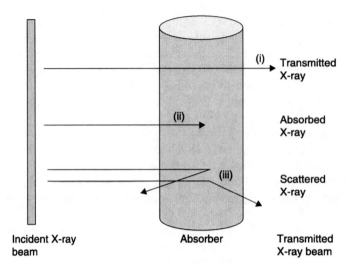

Fig. 9.1   Interaction of X-rays with matter.

(μ) depends on the beam energy (E), the material density (ρ) and atomic number (Z), thus:

$$\frac{\mu}{\rho} = K\frac{Z^4}{E^3}$$ [9.1]

where K is a constant (Maire et al., 2001). At energy levels above 200 keV, the attenuation is strongly dependent on both absorption and Compton scatter, as shown by the following correlation:

$$\mu = \rho\left(a + b\frac{Z^{3.8}}{E^{3.2}}\right)$$ [9.2]

where a and b are energy dependent coefficients (Dyson, 1990). For low density materials (small atomic number), the attenuation is dominated by scattering. This is the basis of phase-contrast tomography, which will be discussed in more detail later in this chapter.

For a monoenergetic beam (consisting of X-rays carrying the same energy), the half-value layer (HVL) is the thickness of a given material that results in an emerging beam of half its original energy. The linear attenuation coefficient is then given by (Allisy-Roberts and Williams, 2008):

$$\mu = \frac{0.693}{HVL}$$ [9.3]

The attenuation process is exponential and the X-ray beam can never be absorbed completely, regardless of the material thickness. The Beer–Lambert law describes the relationship between the number of transmitted photons and the number of incident photons carrying the same amount of energy, without predicting the secondary radiations. The intensity of the transmitted X-ray beam, I, is described as:

$$I = I_0 e^{-\mu d} \qquad\qquad [9.4]$$

where I_0 is the intensity of the incident beam and d is the thickness of the sample (Allisy-Roberts and Williams, 2008). As the attenuation is not uniform throughout the sample however, the transmission path along x must be considered, so that:

$$I = I_0 e^{-\int \mu(x)dx} \qquad\qquad [9.5]$$

The attenuation coefficient increases with material density and atomic number, and decreases as the radiation energy increases. Because of their higher density, materials such as clay, iron and organic matter exhibit a much higher absorption than water, thus leading to a contrast in the eventual image. Biological materials, for which the attenuation coefficient is small, require weaker X-rays (0.2–5 nm wavelength). Such beams are also referred to as soft X-rays, as opposed to hard X-rays (hundreds of keV) used for metallurgical specimens.

However, in practice, the beams produced (e.g. by X-ray tubes) are often wide and polyenergetic, unless a synchrotron facility is used. Therefore, their behaviour differs slightly from the theory described above, and the Beer–Lambert law has to be integrated over the whole energy spectrum. As the beam travels through the sample, the low energy photons are attenuated, leading to an increased concentration of high energy photons, known as beam hardening. The increased width of the beam also leads to the capture of more scattered radiation by the detector. The scattered photons are less energetic, and the softening effect is greatest for high-energy X-rays.

9.2.2   Overview of X-ray techniques

In X-ray microscopy, the sample image obtained from the transmitted X-ray beam is projected onto a 2D plane. Each pixel of the image reflects the sum of the attenuation coefficients along the X-ray path, ideally free of noise. X-ray tomography and X-ray absorption micro-tomography, otherwise called X-ray micro-CT which is an extension of X-ray tomography adapted for small samples ($< \sim$3–4 cm), provide 3D information through multiple projections of the object taken from many different angles. This can be achieved by rotating either the sample or the source-detector pair (e.g. medical tomography). Provided the radiographs of the object are taken at sufficiently small angular increments, it is possible to predict an attenuation value for every point (often called voxel) of the 3D matrix created to represent the sample. A filtered back projection algorithm is then used to reconstruct the 3D information from the 2D slices. X-ray micro-tomography uses high-energy X-rays to obtain a finer resolution reaching μm or even nm (Maire *et al.*, 2001; Wang *et al.*, 2011). X-ray tomography can be based on either absorbance or the phase shift of incident X-rays, extending its use to a wider range of applications, and can therefore be performed in three different modes (Salvo *et al.*, 2003):

1. **Absorption**, based on the Beer–Lambert law (Equation 9.4), is probably the most commonly used mode. Performed in a synchrotron facility or with a previously calibrated system, this mode allows quantitative measurement of the absorption coefficient on the μm scale (Elliott *et al.*, 1990).
2. **Phase contrast tomography** is required for samples with low absorption (Bronnikov, 2002; Groso *et al.*, 2006). In this case, variations in refractive index as well as interference effects are used to recreate an image. Phase contrast tomography takes advantage of the refractive index of X-rays for all materials being slightly less than unity (alteration of the path of the X-ray beam). As two parallel X-rays pass on each side of an interface, the modulation of the beam induces a difference in their optical phase based on the difference in refractive index between the two materials. This mode is generally more sensitive than the absorption mode and is used when the attenuation contrast is too small (e.g. low-density materials with poor absorption), or when two distinct materials exhibit the same attenuation.
3. **The holotomography mode** is derived from the phase contrast mode and requires at least two complete scans at various distances so that quantitative analysis can be performed. This mode can be useful when the absorption mode fails to provide a decent image, for example, when the absorption contrast is too low. The technique has been described by Cloetens *et al.* (1999), but hitherto does not seem to have been applied to food materials.

9.2.3   Resolving microstructure: advantages of X-ray techniques

The highest achievable resolution for any microscopy technique is proportional to the wavelength of the light source. X-rays have a wavelength of a few nanometres and are highly energetic, which gives X-ray microscopy two advantages over visible light microscopy: a better achievable resolution (nm scale) and the ability to visualize samples that are opaque to visible light.

It is worth noting that Transmission Electron Microscopy (TEM) and Scanning Electron Microscopy (SEM), which use accelerated electron beams, offer similar or better resolution than X-ray microscopy. Table 9.1 summarises the main advantages and drawbacks of electron microscopy and X-ray tomography. SEM can be used to look at surface details with a very high resolution (a few nm) but it requires:

- adequate contrast between the feature and the surrounding area. Often, the contrast is insufficient and coatings must be added to achieve an acceptable contrast level; and
- working in vacuum conditions. This limits the use of conventional electron microscopy to vacuum-proof samples with coated surfaces.

Comparison between SEM and X-ray imaging for foam microstructure characterisation can be found in the work of Trater *et al.* (2005).

TEM can accurately describe 3D structure with nm resolution (Koster *et al.*, 2000), but the thickness of the sample is limited due to the inelastic scattering of electrons. However, X-ray samples can be twice as thick.

Table 9.1 Comparison of X-ray micro-CT and electron microscopy

Technique	Advantages	Drawbacks
XCT	3D microstructure sample in its natural state, high and low moisture	Need for stability acquisition time (up to hours for XCT performed with beams from X-ray tubes, reduced to minutes at synchrotron facilities)
SEM and TEM	High magnification straightforward picture, easy to read	Artefacts induced by sample preparation vacuum High moisture samples have to be dried beforehand

X-ray microscopy, being almost free of sample preparation, avoids artefacts caused by drying, fixating, freezing or staining of the sample. These preparation techniques are commonly required for other high resolution imaging techniques including confocal laser scanning microscopy (CLSM), TEM and SEM.

Interpreting images resulting from the projection of a 3D sample on a 2D plane can be difficult, as the projection results from a sum of overlapping features. One way to overcome this problem is to assemble images of 2D slices. For example, successive 2D pictures could be obtained by polishing/removing a few microns of sample between each shot. However, this is invasive, tedious and not very accurate, as preparation artefacts are likely.

Confocal laser scanning microscopy
Confocal laser scanning microscopy makes 3D reconstruction possible by providing 2D images of thin slices of the sample, which can then be assembled in order to create the structure. The resolution is on the order of microns, but for many samples the laser cannot penetrate deeply inside the sample, so the depth of measurement is limited. Specific staining can be applied in order to label components such as proteins, DNA and fat for an easily readable picture. However, the staining procedure can produce its own specific artefacts and induce structure denaturation.

Hassan *et al.* (1995) visualised yogurt microstructure (casein micelles and live microorganisms) by means of CLSM, and Kanit *et al.* (2006) successfully used confocal data to model food structure. The development of confocal Raman Spectroscopy has been used to provide a more precise description of the chemical composition of a sample. Such a level of qualitative description can also be achieved with X-ray tomography after calibration. Ghorbani *et al.* (2011), for instance, linked the grey levels in the tomographic image of a mineral sample to the individual components of the structure.

Ultrasound and acoustic technology
Ultrasound and acoustic technology can be operated in two different modes. Amplitude modulation in one dimension is used for the measurement of tissue or

fluid depth, and brightness modulation can discriminate between tissues of different densities. One of the main food applications is in the meat industry, where ultrasonic techniques are used for the determination of fat thickness, yield evaluation and general meat quality assessment (Chen and Sun, 1991).

Magnetic resonance imaging
Magnetic resonance imaging (MRI) is based on the absorption and emission of energy by a sample subjected to a magnetic field. Carefully chosen settings enable the acquisition of images with good contrast between specific areas. Seidell *et al.* (1990) showed that measurement of adipose tissue distribution via MRI and X-ray tomography produces similar results. Baete *et al.* (2008) measured the distribution of bubble size and dispersion in porous hydrogel foams by R2 Nuclear Magnetic Resonance and also showed that the results matched X-ray tomography data. Of the two techniques, MRI can provide more information on composition; however, it does not provide a detailed image of the internal structure.

Electrical tomography
Electrical tomography is based on impedance, capacitance or electromagnetic inductance measurements. The technique has been used to measure moisture content in grains (Abdullah *et al.*, 2004) and for on-line process monitoring (Boonkhao *et al.*, 2011), but only global data can be obtained as the spatial resolution is poor (Du and Sun, 2004).

Mercury intrusion porosimetry
Mercury intrusion porosimetry is a traditional method for measuring the porosity and pore connectivity of samples (Portsmouth and Gladden, 1991). It is widely used in industry to measure the porosity of catalysts. However, this technique relies on the permeability of the sample and is therefore less sensitive than CT imaging (Rigby *et al.*, 2006; Ghorbani *et al.*, 2011).

9.3   Theory of X-ray tomography

9.3.1   Tomographic scan acquisition
X-ray tomography is now a fairly mature technique and various configurations are currently available. All set-ups comprise an X-ray source, a sample holder and an X-ray detector. By adjusting the distances between the sample, source and detector, a magnification can be selected. A standard set-up is schematically represented in Fig. 9.2.

Source
The energy spectrum provided by the source, usually described in terms of peak energy, determines the type of material that can be analysed. High-energy beams penetrate matter more easily but are less sensitive to changes in material density. X-ray tubes are used as a source in most laboratory scanners. In these devices, the

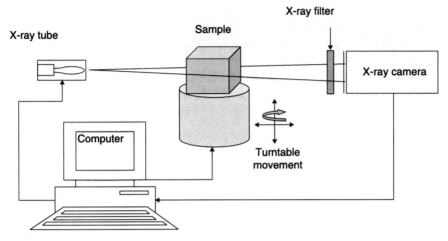

Fig. 9.2 Diagram of an X-ray micro-CT system.

loss of kinetic energy of an electron beam hitting a thin metal layer is partly translated into X-rays, generating a conical X-ray beam that is then collimated to a fixed width. For this beam geometry, the magnification depends on the position of the object relative to the beam source: the closer it is to the source, the higher the magnification. The focal spot size limits the maximum achievable resolution. A beam forming a large spot will result in a blurred image, whereas smaller spots result in sharper images. The generated X-ray beam is polychromatic, causing beam hardening, as discussed below, and complicating quantitative analysis. To limit artefacts such as beam hardening, a filter can be placed between the source and the sample.

In contrast, the X-ray radiation issued by a synchrotron, in which the acceleration of the electrons is realised through a high-energy magnetic field, is composed of almost parallel rays. This allows a better resolution (hundreds of nm) due to the homogeneous signal obtained. Furthermore, the use of a monochromator allows the selection of a specific X-ray energy, which creates the potential for further qualitative as well as quantitative analysis. A synchrotron source therefore provides the best quality images in terms of signal/noise ratio. The beam size is smaller, usually 1 mm × 30 mm, thus restricting its use to the study of small samples (dimensions equivalent to beam size). However, developments in synchrotron technology have enabled more precise mapping, made possible by the higher resolution and improved qualitative and quantitative analysis achieved due to the monochromatic nature of the X-ray beam. Falcone *et al.* (2005) described bread crumb microstructure with X-ray tomography using a synchrotron as a light source. However, accessing a synchrotron is not as easy as using a benchtop unit, due to the limited number of synchrotron facilities available around the world.

Sample holder

Several generations of X-ray scanners have recently been developed. The latest versions have the sample placed on a rotating round metal plate while the source and detector are fixed. This configuration makes the use of filters easier as they can be positioned behind the source. Scan time can be significantly reduced if multiple X-ray source-sensor pairs are used. This configuration also allows the analysis of non-static structures (Williams and Jia, 2003).

Detector

Detection of the attenuated beam is realised by X-ray sensors placed on the opposite side, which must have a stable free noise response and a wide dynamic range. The earliest detectors were ionization chambers containing high-pressure Xenon gas. They have since been replaced by scintillation detectors; solid-state devices with greater detection efficiency. A fluorescent screen (e.g. containing Caesium Iodide) converts the transmitted X-ray photons into visible light, which is then transferred by optical lenses to a cooled CCD camera that creates a digitized image. To avoid reconstruction artefacts, it is important to stay within the linear zone of the detector response, by selecting either an appropriate beam energy or sample thickness.

X-ray radiographs from the sample are taken from different viewing angles, in small angular increments over 180- or 360-degree rotation. A shadow image is produced at each angle and the images are then numerically processed to reconstruct a 3D density map (Fig. 9.3). The scanning time depends on parameters such as sample size, desired resolution, averaged number of frames, and size of the angular increments. Esveld *et al.* (2012) used a Skyscan 1072 desktop CT system (Skyscan, Belgium) to characterize the internal structure of crackers. A

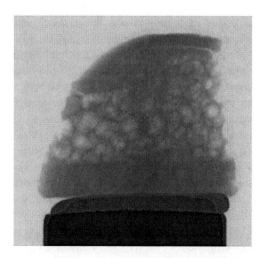

Fig. 9.3   Shadow image of an aerated chocolate bar.

6 mm × 6 mm sample and 180-degree rotation, in increments of 0.45 degrees, required about 40 minutes for acquisition and read-out, which is equivalent to 2.8 seconds per projection.

9.3.2   Principle of tomographic reconstruction

Tomographic reconstruction refers to the mathematical procedure that integrates a series of projections taken from different viewing angles into a 3D image. Thus, tomographic images are recreated from a stack of conventional 2D images obtained from projections of a 3D object. Each 2D image is the result of the summed attenuation through the sample, and is composed of pixels each having a value related to the intensity of the signal received. The number of projections needed and the increment value depend on the desired number of pixels in the reconstructed image as well as on the scanner used. When considering the reconstructed 3D image, each pixel is often named a voxel, as it refers to a volume element with a depth equal to the thickness of the layer considered.

Several reconstruction methods are available, which are derived from the Fourier theorem. For an infinite number of projections, the Fourier transform of the object is deduced and the inverse Fourier transform produces the attenuation map of the sample. However, this method presents errors, as the number of projections is finite and the Fourier space has to be re-sampled from polar to Cartesian coordinates. The filtered back-projection method (Ramachandran and Lakshminarayanan, 1971), which is derived from the Fourier theorem, has proved to be numerically more accurate, can be easily computed and is consequently one of the most common methods. It compensates for the error due to the finite number of projections by filtering each projection – several filters exist, for example, Hanning filter, Weiner filter – before performing the Fourier transformation.

It is also possible to reconstruct the 3D map of attenuation using algebraic methods (Natterer, 2008). One example of algebraic reconstruction uses an analytical definition of a Radon inverse transform. It defines a linear system:

$$\vec{p} = M\vec{x} \tag{9.6}$$

where $\vec{p}$ is the projection, $\vec{x}$ the image pixels value and M the contribution matrix of the projection points to $\vec{x}$.

In practice, back-projection software maps estimates of attenuation coefficients (the level of accuracy is higher for small increments) into a series of 2D slices. Each point is convolved with its neighbours through a seed algorithm using a Fourier transformation. The stack of 2D slices is then simply reconstructed into a 3D image that can be presented as a whole, or as virtual slices along a user-defined cutting plane. The minimum distance between slices is limited to the size of a pixel. Once the shadow images are reconstructed into a 3D image, it can be numerically sliced and analysed. Figure 9.4 shows a slice taken from the 3D scan of an aerated chocolate bar at a depth indicated by the horizontal line on the left of the picture.

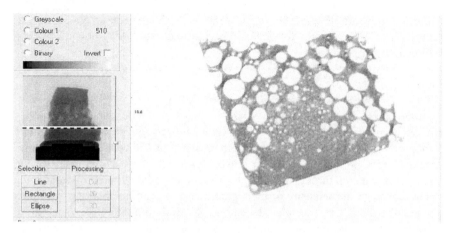

Fig. 9.4 Screenshot showing slicing of reconstructed 3D image of an aerated chocolate bar.

9.3.3 Image post-processing

Once the recorded radiographs have been reconstructed into a 3D image, several post-processing steps are usually implemented, as follows.

Background subtraction

A background subtraction step is often used to reduce background noise. Raw images can also be corrected against the background to take account of the non-uniform response of the CCD detector and the fact that the X-ray beam can be polychromatic. This involves subtracting the average flat field acquired at the beginning or end of the experiment.

Windowing

Pixels in an image obtained by X-ray tomography are displayed according to a greyscale, which varies between −1000 and +3000, far too many different levels for the human eye to distinguish. It is, therefore, common to set a window containing the attenuation coefficient of the researched feature, and a black or white value on the outside. The grey levels inside that window are then rescaled so that the feature appears more clearly.

Thresholding

Thresholding procedures are used to segment the feature or section of interest from the background. For a chosen area of interest, a user defined binary threshold can be used to convert greyscale images into binary images (black and white) by assigning the value 0 to all pixels whose intensity is above the chosen value, and 1 to every other. Image segmentation is often used to separate phase materials, for example, solid-liquid boundaries, or cell walls in foams. However, segmentation using a single fixed threshold can be sensitive to background noise, and small contrast differences in the studied features can make direct threshold segmentation difficult. It should be noted that segmentation is a subjective method as the threshold is set by the user. Parameters such as average porosity are therefore highly dependent on the set threshold value. To overcome this issue, Sahoo *et al.* (1997) developed an automated threshold selection method using the entropy of the histogram.

Segmentation Kriging or indicator Kriging, is derived from segmentation procedures (Oh and Lindquist, 1999). Two thresholds are fixed and a 1 value is assigned for voxels with intensities above the higher threshold, while voxels with intensities below the lower thresholds are assigned the value 0. For voxels whose value lies in between the two thresholds, a most probable value is computed from the value of the surrounding voxel. This technique presents the advantage of being less sensitive to background noise than segmentation. However, it requires more computing time and, more importantly, can miss features that are of voxel size as several voxels are required to define a particular trend. Debaste *et al.* (2010) compared the segmentation and Kriging methods and showed that unless the noise is kept very low, the Kriging method led to an error reduction of up to 25%.

In the case of phase-contrast tomography, phase retrieval methods have to be applied prior to reconstruction, since both absorption and phase contrast contribute to the image formed (Bronnikov, 2002). Whilst the reconstructed 3D image can be used to qualitatively describe the microstructure, further data processing can be applied to calculate a number of quantitative parameters. Lim and Barigou (2004) demonstrated how such a quantitative analysis could be conducted on cellular materials by using image analysis combined with stereology, but the capability of commercial data processing software has been enhanced such that most of these extensive calculations can now be done at the touch of a button, including:

- **Percentage object volume**: proportion of volume occupied by solid objects;
- **Fragmentation index**: reflects the structural connectivity of an object. Developed by Hahn *et al.* (1992), it can be used to describe the shape of the surface (concave or convex);
- **Structure model index**: estimates the characteristic form of the features within the sample; e.g. it can describe forms within a cellular foam, i.e. spherical, octagonal, rod-like, plate-like, etc.;
- **Structure thickness**;
- **Object surface/volume ratio**: basic parameter allowing the characterization of structure complexity, i.e. the size and distribution of pores. It is also a basis for the model-dependent estimation of structure thickness, i.e. size and distribution of structure composing elements;
- **Structure separation index**: reflects the size of the space between pores;
- **Degree of anisotropy**: a measure of the preferential alignment of pores;
- **Medial axis analysis:** traces the fundamental characteristics of the void pathways and reflects the skeleton of the voidage. It refers to a one-voxel thick structure and characterises the paths present in the sample network. This is

achieved using algorithms that iteratively modify the segmented image (Thovert *et al.*, 1993; Lee *et al.*, 1994; Lindquist *et al.*, 1996; Mendoza *et al.*, 2007).

9.3.4   Equipment and settings

Several companies provide computed tomography equipment, including Skyscan (Belgium), Scanco (Brütisellen, Switzerland), Toshiba (medical application), Pheonix X-ray (Nanotom; Rahmanian *et al.*, 2009), PreXion (Japan, specialised in dentistry application), Nikon (electronics) and Siemens Somatom Emotion CT. Most of the available literature concerning food applications, which employed benchtop equipment, used Skyscan or Scanco instruments.

Since the early 1970s and the invention of computerized tomography, four generations of scanners have been developed. Up to the third-generation scanners, it was more common for the sample to rotate whilst the source and detector were fixed. This configuration requires significant care in the positioning of the sample: if the rotation centre is misaligned the resulting image will be blurred and noisy, leading to errors. However, the fourth-generation scanner solved this problem as the ensemble source-detector system is made to rotate around the object.

Every tomographic set-up has to be accurately calibrated and sample characteristics such as size and rotation axis have to be known, to ensure that the X-rays are appropriate for the sample and to reduce geometric uncertainties. The energy content of the X-rays is set to optimise contrast, with the optimum beam conditions and exposure time depending on the sample composition. The beam characteristics (voltage and intensity) have to be set before image acquisition. Other important settings and configurations for benchtop equipment include:

- **Resolution**: determined by the size of a pixel, or volume of a voxel. Typical resolution lies between 100 and 200 µm, whereas high resolution equipment goes down to ~100 nm. The maximum achievable spatial resolution is governed by the available photons and by the configuration itself;
- **Rotation increment**: (usually a fraction of a degree;
- **Number of frames**: to be averaged for a shadow image. The use of multiple images improves the signal-to-noise ratio;
- **Magnification**: adjusted by the distance between the X-ray source and sample;
- **Noise reduction**: ring artefacts and beam hardening steps. Noise reduction is achieved by correcting the image against a flat field. Beam hardening reduction involves introducing a filter between the source and the sample. This reduces polychromatism and diminishes the beam-hardening artefact. Aluminium filters (0.5 or 1 mm) absorb low energy radiations, which are transmitted by aluminium-copper filters.

The resulting scan time is a function of the desired resolution, the sample size and the set exposure time. Obviously, a 360-degree scan will be twice as long as a 180-degree scan with the same rotation increment, but the resulting image is usually sharper and contains more information. As a rule of thumb, a 360-degree scan will be necessary for density studies, whereas a 180-degree scan is generally sufficient for porosity studies. Typical scanning time for benchtop X-ray tomographs is about one hour. The scanning time reduces to approximately one minute or less with the use of synchrotron facilities.

9.4   Contrast, resolution and sample preparation techniques

As previously noted, X-ray tomography is an accurate technique for imaging the inside of a material sample. The image quality is determined by a combination of contrast (how well adjacent areas can be distinguished) and spatial resolution (necessary to capture fine details). This section contains a description of the most common artefacts, as well as providing additional details regarding data acquisition.

9.4.1   Artefacts

X-ray tomography, like other visualisation methods, is not exempt from artefacts, the most common being described below (Vidal *et al.*, 2005).

Beam hardening

Due to X-ray beams being polyenergetic, that is, containing photons with different energies, a common artefact is due to beam hardening. Beam hardening occurs when a heterogeneous beam passes through a material. Low-energy photons are absorbed more quickly, resulting in an attenuated emerging beam with a higher mean energy. Therefore, the longer the matter is exposed to the beam, the harder the X-rays become. The centre of a sample is exposed to the radiation for much longer and can appear less absorbent in reconstruction. This is called the cupping effect. One way of dealing with this kind of artefact is to place a filter between the source and the sample so that the beam can be pre-hardened.

Hot points

Hot points are another common artefact. They appear when X-ray photons pass through the scintillator and hit the CCD camera, resulting after reconstruction in streaks across the volume. Such artefacts can be removed by the use of conditional median filtering.

Ring artefacts

Ring artefacts are named after the circular arc they create around the rotation axis, are caused by the detector response not being linearly proportional to the received light intensity, for example due to dead pixels in the CCD. The software provided by the system manufacturer would usually deal with this artefact.

Partial volume effects, whereby a high contrast object smaller than the voxel in which it is contained can appear to be the same size as the voxel, can also affect the CT image and hence its size can be overestimated. This is caused by the scanner being unable to differentiate between two distinct zones when within the same voxel; for example, when a small amount of high density material smaller than a voxel is included in a low-density matrix. In the same way, partial volume effects can reduce the visibility of low contrast details.

9.4.2   Contrast

The contrast in the CT image results from the different levels of X-ray attenuation throughout the sample. Contrast resolution refers to the ability to distinguish sharply between regions of the image. The image is usually calculated on a pixel matrix (typically ~2000 × 1000). The CT number (CT_n, sometimes referred to as the Hounsfield unit scale) is a value corresponding to each voxel. It represents the average attenuation coefficient of the beam through the matter contained within the voxel and is calculated as:

$$CT_n = 1000 \times \frac{(\mu - \mu_w)}{\mu_w}$$

[9.7]

where μ_w is the linear attenuation coefficient of water. Thus, CT_n is 0 for water and −1000 for air ($\mu \sim 0$). The CT image is represented by pixels having $-1000 < CT_n < 3000$, resulting in 4000 grey levels, which is why windowing is often applied. This technique, fairly common in digital imaging, helps accentuate certain details by setting a reduced interval of CT numbers to be displayed.

Noise tends to limit the quality of a CT scan. It originates either from random variation in the number of detected photons (quantum noise), from electronic noise (produced in the measuring system) or from the reconstruction algorithm itself (structural noise). One way of reducing quantum noise is to increase the number of detected photons by increasing the scan time or the beam energy. Note that increasing the slice width effectively reduces the noise but also decreases the spatial resolution and so a compromise has to be made in the selection of parameters. Alternatively, reducing the field of view means that there is a smaller detector area for each pixel. Increasing the beam intensity can also result in a lower contrast and be detrimental to the quality of the image. Thus, a fine balance has to be found to achieve an acceptable signal/noise ratio.

Contrast is also affected by scattered photons. Whilst the primary radiation carries the information to be imaged, scattered photons act like a veil and reduce contrast. The noise resulting from photon scattering is homogeneous due to the stochastic nature of the process. Reducing the volume of the exposed sample decreases the amount of scattered radiation. A contrast medium can also be introduced to improve contrast; so far however, this has been mostly used in medical applications.

9.4.3   Spatial resolution

Spatial resolution determines the ability to resolve fine microstructural detail. It is mostly controlled by the size and number of detector elements, the size of the X-ray focal spot and the relative distances between the source, object and detector. Fine detail is of course most clearly seen when the contrast is high, contrast and resolution being inter-related. Spatial resolution is affected by the source – as a beam of almost parallel rays will lead to a better spatial resolution – and the sample area imaged in one pixel. For X-ray tubes, the maximum resolution is linked to the size of the microfocus, which has to be as small as possible in order to obtain the best resolution. An acceptable compromise has to be found here, as the microfocus size is proportional to the delivered power. High resolution can, therefore, be achieved at the expense of a long scanning time.

The maximum achievable resolution is also a function of the size of the imaged sample. As a rule of thumb, common instruments will achieve approximately 100 to 200 μm resolution for samples up to 10 cm, down to 10 μm for smaller objects below 2 cm. X-ray tomographs acquired using a benchtop system have a resolution of several μm for small samples on the order of about 1 cm, for example, 8 μm for Maire *et al.* (2001), 7.8 μm for Falcone *et al.* (2005) and Esveld *et al.* (2012), whilst Bellido *et al.* (2006) achieved a resolution of 10 μm^3 per voxel for a sample size of 7.32 × 2.00 × 1.20 mm. For micro-tomography performed in synchrotron facilities, using dedicated beamlines, the resolution is less than 1 μm (Salvo *et al.*, 2003); Mebatsion *et al.* (2009) achieved a resolution of 700 nm per pixel. More recent benchtop systems (nano-CT systems) have a similar nanometre range resolution but their capabilities are more limited, including longer scan times, lower penetration energy and the inability to capture dynamic phenomena.

9.4.4   Sample preparation techniques

A major advantage of CT is that the technique requires minimal preparation, thus avoiding artefacts due, for example, to sample fixation. Indeed, there is no need for chemical fixation or sample preparation, apart from cutting the sample to give it an appropriate shape and size so that it fits inside the field of view. Care should be exercised when cutting or shaping the sample, as this can alter its microstructure, for example, through the creation of cracks. For instance, Bellido *et al.* (2006) reported bubble deformation artefacts caused by squashing a sample of dough between two hard layers to adjust its height.

In X-ray micro-CT it is imperative that the sample is stable for the time required to acquire the scan, for example, a foam with rising bubbles will result in a blurred image. For dynamic processes, synchrotron tomography may be more appropriate as the scan acquisition time is much shorter (on the order of 30 s). A good example is the work of Babin *et al.* (2006), where the rising of bread in an oven was successfully monitored.

For moist samples, such as fruits, appropriate measures have to be taken to avoid dehydration through sample heating: these may involve covering the sample in polymer foil (Verboven *et al.*, 2008) or enclosing it in a plastic tube (Mendoza *et al.*, 2007).

9.5   Applications to food

Modern food processing is increasingly concerned with the production of products with complex microstructures that determine their mechanical and aesthetic properties. There is an established need to characterise food microstructure using non-destructive and non-invasive methods, either to improve quality control methodologies, or as a tool to investigate the complex relationship between microstructure, material (physical) as well as sensoric properties, and processing. The success of X-ray tomography techniques in medical, geological, biological and other material sciences has led to its application in food science and technology. Lim and Barigou (2004) were amongst the first to use X-ray micro-CT to investigate a number of cellular food materials (aerated chocolate, strawberry mousse, honeycomb chocolate bar, chocolate muffin and marshmallow) and to present a rigorous quantitative analysis of their measurements. By combining image analysis with a stereological technique, quantitative information was obtained on a number of parameters, including spatial cell size distribution, cell wall-thickness distribution, connectivity and voidage.

Compared to other commonly used techniques such as bright-field imaging, light microscopy, CLSM, TEM, SEM, or MRI, X-ray micro-CT provides a much more detailed and more reliable description of 3D structure and is, thus, gradually becoming a much used tool in the study of food materials. Most of the studies reported to date are reviewed below to illustrate the power of the technique and its wide range of applications.

9.5.1   Meat and related products

X-ray tomography has been successfully used to evaluate the overall fat content of meat products. Cardinal *et al.* (2001) showed that the X-ray data correlate with a chemical analysis of the fat content of raw salmon. Rørå *et al.* (2005) estimated the fat deposits in salmon using the same technique, in order to investigate the effect of the farmed salmon diet. In the same manner, Furnols *et al.* (2009) determined the percentage of lean meat in pig carcasses using X-ray micro-CT.

X-ray micro-CT has also been applied to study the myofilament structure and fat distribution in meat products. Because of the different absorption coefficients for bone, lean meat and fat, quantitative measurements are possible via the measurement of energy attenuation (Damez and Clerjon, 2008). Frisullo *et al.* were able to obtain an accurate quantification of the fat content and its spatial distribution in Italian salami (Frisullo *et al.*, 2009) and in beef muscle (Frisullo *et al.*, 2010b). The salt distribution in dried cured ham has also been quantified through computed tomography analysis, as X-ray attenuation increases with salt concentration in the tissue (Vestergaard *et al.*, 2005).

Jensen *et al.* (2011) investigated the distribution of porcine fat and rind with phase-contrast CT. Phase-contrast CT gave better results than absorption-based CT, because it is based on the use of the refractive index for the X-ray rather than the absorption coefficient, thus providing a better contrast in such a low density material as well as reducing exposure time. The technique also allowed the investigation of the fatty acid composition of the fat fraction, and the density variation in the meat fraction.

9.5.2   Fruits and vegetables

The description of the internal structure of fruits is a key issue for those who wish to understand the relationship between material properties, fruit quality and post-harvesting transformations. X-ray tomography proved accurate in determining a realistic representation of the porous structure and volume of the intercellular space in apple tissue (Mendoza *et al.*, 2007, 2010). Synchrotron tomography images were used by Verboven *et al.* (2008) and Mebatsion *et al.* (2009) to model fruit tissue, including fruit cortex, cell wall, pore network and cells. Visualisation of the cell walls was made possible by the high resolution of the synchrotron facility (0.7 μm per pixel). However, it should be noted that X-ray tomography offers a relatively small field of view compared to the MRI technique. The latter enables an overview of the entire fruit (Musse *et al.*, 2010), which can be beneficial in some applications such as the measurement of the spatial distribution of core breakdown in pears (Lammertyn *et al.*, 2003). Description of wood and plant microstructure at the micron scale has also been achieved through phase-contrast tomography (Mayo *et al.*, 2010).

X-ray micro-CT has been used to probe fruits and vegetables, either in their natural moist state or dried. Brown *et al.* (2008, 2010) studied the shrinkage and collapse of carrots using different drying methods by analysing the microstructure of initial and dried samples. Similar studies have been conducted by Léonard *et al.* (2008) to investigate the effect of far-infrared radiation-assisted drying on the microstructure and porosity of banana tissue.

9.5.3   Low moisture food: porous materials

X-ray micro-CT is particularly suited to the characterisation of the internal morphology of dry foods such as cereals, bread, crackers and biscuits that exhibit a cellular structure (Van Dalen *et al.*, 2007). It provides an accurate description of the internal microstructure, which can be characterized and quantified by means of 3D image analysis techniques and compared with computerized models (Esveld *et al.*, 2012). Such models can then be used, for example, for the characterization of water sorption (Debaste *et al.*, 2010).

Other applications include the study of the role of fat in air trapping and the effect of sugar on cookie structure and surface cracking (Pareyt *et al.*, 2009), and frying times for bread-coated chicken nuggets (Adedeji and Ngadi, 2009).

9.5.4   Granular products

X-ray micro-CT is also a powerful tool for studying the porosity and morphology of granules (Farber *et al.*, 2003). Several food studies have been reported. For example, pore networks of pea and wheat bulks have been successfully investigated by Neethirajan *et al.* (2008). Mohorič *et al.* (2009) used X-ray micro-CT to establish the relationship between rice kernel microstructure and cooking behaviour, whereas Zhu *et al.* (2012) used a combination of X-ray micro-tomography and Environmental Scanning Electron Microscopy (ESEM) to compare the kernel structure of a wild type of rice and a genetically modified version lacking two starch branching enzymes. Thus, significant differences in the kernel structure were shown; the wild-type kernel was dense with polygonal starch granules, whereas the modified kernel producing high amylase, exhibited more elongated granules with larger air volumes between them.

X-ray micro-CT has also been used to investigate the development of the porous structure in coffee beans and quantify alterations resulting from different roasting processes (Frisullo *et al.*, 2012). Neethirajan *et al.* (2006) resolved the total air volume and air path network in several grains to explain the difference in airflow resistance between them: whilst Debaste *et al.* (2010) used tomographic imaging data to produce models of the porous structure of dried yeast grains.

9.5.5   Bread and bread crumbs
Bread crumb quality can be rapidly assessed via 2D microscopic imaging (Lassoued *et al.*, 2007). However, resolving the microstructure is more challenging and has received significant attention in the X-ray micro-CT literature (Babin *et al.*, 2007; Guessasma *et al.*, 2008). Falcone *et al.* (2005) studied the microstructure of bread crumb and correlated it to compression test results and, thus demonstrated its large-scale deformation properties. The technique proved accurate for describing bread as a whole, including properties such as the maze of voids and air cells (Bellido *et al.*, 2006), the bubble size distribution in wheat flour dough (Wang *et al.*, 2011), the pore structure (Falcone *et al.*, 2004) and dough rise during baking (Babin *et al.*, 2006; van Dalen *et al.*, 2009). The latter study provided data for numerical modelling of bubble growth, as well as a better understanding of the bubble formation process. However, this type of dynamic study could only be achieved with the help of a synchrotron, as the scanning rate of a benchtop apparatus would have been much too slow to capture the important changes that occur in the product during scanning.

9.5.6   Extruded products: foams
The description of 3D foam microstructure has been made possible by developments in X-ray tomography to the extent that density and density contrast can accurately be determined. Fast X-ray micro-CT performed at synchrotron facilities enabled the monitoring of foam formation and foam microstructure during a bread-making process (Babin *et al.*, 2006). Foam microstructure can be quantified in terms of properties such as cell size distribution, wall thickness, volume fraction, number of interconnected cells and void fraction.

Tomography is a powerful tool for the description of extruded foam products. A number of studies have been reported, which have correlated foam microstructure to its mechanical properties (Trater *et al.*, 2005; Agbisit *et al.*, 2007; Cho and Rizvi, 2009; de Mesa *et al.*, 2009; Robin *et al.*, 2010; Karkle *et al.*, 2011). Some studies have also related microstructure to the extrusion process parameters (Chaunier *et al.*, 2007) and extrudate composition (Parada *et al.*, 2011). Babin *et al.* (2007) related the final microstructure of extruded starch foams to the extrusion parameters, biochemical content of the foam, as well as its physicochemical properties. The effects of wheat bran concentration on the mechanical properties of extruded foam have also been characterized. The mechanical strength was correlated to the fine microstructure observed with X-ray micro-CT (Robin *et al.*, 2011a,b).

Licciardello *et al.* (2012) used the technique to determine the effect of sugar and egg white on meringue microstructure. As previously cited, the structure of aerated chocolate and other food foams has also been explored by Lim and Barigou (2004). Haedelt *et al.* (2007) linked the geometric characteristics of aerated chocolate to its sensory properties, whereas Frisullo *et al.* (2010a) attempted a qualitative description of the microstructure of manufactured chocolate. They showed that air bubbles can clearly be discriminated from sugar crystals and cocoa mass (which has greater density, therefore a higher attenuation coefficient) by comparing tomographic data with chemical analyses. The method provided a good estimation of the sugar concentration and the shape of solid materials, and results showed no relationship between sugar crystal content and pore structure.

9.5.7   Other miscellaneous foods

Laverse *et al.* (2012a) correlated mayonnaise microstructure with its rheological properties and quantified the fat content and its distribution in mayonnaise. However, Mousavi *et al.* (2005) showed that X-ray micro-CT, preceded by a freeze drying step to remove solid water, provides an accurate means of describing the shape and size of ice crystals in frozen food. X-ray micro-CT has also been utilised for the characterisation of the microstructure of yogurt and cream cheese-type products, as well as for the quantitative study of the fat present therein (Laverse *et al.*, 2011a,b).

9.5.8   *In vivo* applications

To date, there have been no *in vivo* applications reported in the area of food microstructure. The closest applications mainly concern the evaluation of body composition in livestock used for research purposes (Hollo *et al.*, 2010; Bunger *et al.*, 2011). Despite the fact that live animals cannot be kept perfectly still, X-ray micro-CT seems to provide reasonable accuracy (Nade *et al.*, 2005). However, ultrasound measurement, even though less accurate, is often preferred as it is much more suitable for field applications (Szabo *et al.*, 1999).

9.6   Conclusion and future trends

A number of challenges still face X-ray tomography in food science and technology. These include, for example, dense food with a high water content and little difference between the attenuation coefficients of different phases; dehydration of moist materials during scanning; inability to capture dynamic processes; and the large number of radiographs needed to obtain a high level of accuracy, which makes the technique presently too slow for online 'real-time' imaging.

However, ongoing developments, including more powerful computers and improved algorithms, will increase the capability of the technique and improve its probing efficiency, making it more accurate and affordable both in terms of processing time and capital cost. Further improvements in noise reduction, computing time, detector sensitivity, reaction time, as well as novel tomographic reconstruction and analysis methods for improved quantification of food microstructure parameters, are to be expected. Innovations in instrument development will, in turn, incite the development of enhanced mathematical models, improved understanding of process-microstructure-property relationships and, thus more rational product and process design. The recent introduction of fast X-ray scanners in the medical field will also probably lead to the development of affordable on-line food inspection instruments in food processing plants.

The use of synchrotron radiation sources instead of X-ray tubes provides better resolution for more accurate and detailed quantitative studies, and the possibility of real-time acquisition to enable application to dynamic processes. The use of phase-contrast synchrotron tomography would broaden the scope of X-ray tomography applications to enable the differentiation between objects of similar X-ray absorption properties but different microstructure (Betz *et al.*, 2007). Potential applications would include detection of fatty acid composition, differentiation between cartilage and soft bones, connective tissue in meat, and protein-lipid-carbohydrate interactions in bread-making (Jensen *et al.*, 2011). Stereoscopic dual energy X-ray is employed to target material identification and allows the extraction of atomic number in layers of different materials (Wang and Evans, 2003).

Coupling X-ray micro-CT with other analytical techniques broadens the application range and enchances the ability to obtain a more complete picture of food microstructure. Development of X-ray adsorption and emission spectroscopy enables the differentiation of different chemical species; for example, X-ray micro-CT has been successfully used in combination with mercury porosimetry (Rigby *et al.*, 2006) and with Raman spectroscopy (Crean *et al.*, 2010). Soft X-ray cryo-microscopy/tomography has recently been developed to study the internal structure of biological cells (Wang *et al.*, 2000, 2001; Steven and Aebi, 2003). Several studies have also coupled X-ray micro-CT with ESEM (Pauwels *et al.*, 2010), and such systems are commercially available. Coupling these techniques simplifies identification of the observed features.

9.7   References

ABDULLAH, M. Z., GUAN, L. C., LIM, K. C. and KARIM, A. A. (2004), The applications of computer vision system and tomographic radar imaging for assessing physical properties of food, *Journal of Food Engineering*, **61**, 125–35.

ADEDEJI, A. A. and NGADI, M. O. (2009), 3D imaging of deep-fat fried chicken nuggets breading coating using X-Ray micro-CT, *International Journal of Food Engineering*, 5. DOI 1110.2202/1556-3758.1452.

AGBISIT, R., ALAVI, S., CHENG, E., HERALD, T. and TRATER, A. (2007), Relationships between microstructure nd mechanical properties of cellular cornstarch extrudates, *Journal of Texture Studies*, **38**, 199–219.

ALLISY-ROBERTS, P. J. and WILLIAMS, J. R. (2008), *Farr's Physics for Medical Imaging*, Edinburgh, Saunders.

BABIN, P., DELLA VALLE, G., CHIRON, H., CLOETENS, P., HOSZOWSKA, J. *et al.* (2006), Fast X-ray tomography analysis of bubble growth and foam setting during breadmaking, *Journal of Cereal Science*, **43**, 393–7.

BABIN, P., DELLA VALLE, G., DENDIEVEL, R., LOURDIN, D. and SALVO, L. (2007), X-ray tomography study of the cellular structure of extruded starches and its relations with expansion phenomenon and foam mechanical properties, *Carbohydrate Polymers*, **68**, 329–40.

BAETE, S. H., DE DEENE, Y., MASSCHAELE, B. and DE NEVE, W. (2008), Microstructural analysis of foam by use of NMR R2 dispersion, *Journal of Magnetic Resonance*, **193**, 286–96.

BELLIDO, G. G., SCANLON, M. G., PAGE, J. H. and HALLGRIMSSON, B. (2006), The bubble size distribution in wheat flour dough, *Food Research International*, **39**, 1058–66.

BETZ, O., WEGST, U., WEIDE, D., HEETHOFF, M., HELFEN, L. *et al.*, (2007), Imaging applications of synchrotron X-ray phase-contrast microtomography in biological morphology and biomaterials science. Part I: General aspects of the technique and its advantages in the analysis of millimetre-sized arthropod structure, *Journal of Microscopy*, **227**, 51–71.

BOONKHAO, B., LI, R. F., WANG, X. Z., TWEEDIE, R. J. and PRIMROSE, K. (2011), Making use of process tomography data for multivariate statistical process control, *AIChE Journal*, **57**, 2360–8.

BRONNIKOV, A. V. (2002), Theory of quantitative phase-contrast computed tomography, *J. Opt. Soc. Am. A.*, **19**, 472–80.

BROWN, Z. K., FRYER, P. J., NORTON, I. T., BAKALIS, S. and BRIDSON, R. H. (2008), Drying of foods using supercritical carbon dioxide – Investigations with carrot, *Innovative Food Science and Emerging Technologies*, **9**, 280–9.

BROWN, Z. K., FRYER, P. J., NORTON, I. T. and BRIDSON, R. H. (2010), Drying of agar gels using supercritical carbon dioxide, *Journal of Supercritical Fluids*, **54**, 89–95.

BUNGER, L., MACFARLANE, J. M., LAMBE, N. R., CONINGTON, J., MCLEAN, K. A. *et al.* (2011). Use of X-ray computed tomography (CT) in UK sheep production and breeding, in, S. Karuppasamy (ed.), *CT Scanning – Techniques and Applications*, INTECH Open access Publisher.

CARDINAL, M., KNOCKAERT, C., TORRISSEN, O., SIGURGISLADOTTIR, S., MØRKØRE, T. *et al.* (2001), Relation of smoking parameters to the yield, colour and sensory quality of smoked Atlantic salmon (*Salmo salar*), *Food Research International*, **34**, 537–50.

CHAUNIER, L., DELLA VALLE, G. and LOURDIN, D. (2007), Relationships between texture, mechanical properties and structure of cornflakes, *Food Research International*, **40**, 493–503.

CHEN, P. and SUN, Z. (1991), A review of non-destructive methods for quality evaluation and sorting of agricultural products, *Journal of Agricultural Engineering Research*, **49**, 85–98.

CHO, K. Y. and RIZVI, S. S. H. (2009), 3D microstructure of supercritical fluid extrudates. Part II: Cell anisotropy and the mechanical properties, *Food Research International*, **42**, 603–11.

CLOETENS, P., LUDWIG, W., BARUCHEL, J., VAN DYCK, D., VAN LANDUYT, J. *et al.* (1999), Holotomography: quantitative phase tomography with micrometer resolution using hard synchrotron radiation X-rays, *Applied Physics Letters*, **75**, 2912–14.

CREAN, B., PARKER, A., ROUX, D. L., PERKINS, M., LUK, S. Y. *et al.* (2010), Elucidation of the internal physical and chemical microstructure of pharmaceutical granules using X-ray micro-computed tomography, Raman microscopy and infrared spectroscopy, *European Journal of Pharmaceutics and Biopharmaceutics*, **76**, 498–506.

DAMEZ, J.-L. and CLERJON, S. (2008), Meat quality assessment using biophysical methods related to meat structure, *Meat Science*, **80**, 132–49.

DEBASTE, F., LÉONARD, A., HALLOIN, V. and HAUT, B. (2010), Microtomographic investigation of a yeast grain porous structure, *Journal of Food Engineering*, **97**, 526–32.

DE MESA, N. J. E., ALAVI, S., SINGH, N., SHI, Y.-C., DOGAN, H. and SANG, Y. (2009), Soy protein-fortified expanded extrudates: baseline study using normal corn starch, *Journal of Food Engineering*, **90**, 262–70.

DU, C.-J. and SUN, D.-W. (2004), Recent developments in the applications of image processing techniques for food quality evaluation, *Trends in Food Science andamp; Technology*, **15**, 230–49.

DYSON, N. A. (1990), *X-rays in Atomic and Nuclear Physics*, Cambridge, UK, Cambridge University Press.

ELLIOTT, J. C., ANDERSON, P., DAVIS, G., DOVER, S. D., STOCK, S. R. *et al.*. (1990), Application of X-ray microtomography in materials science illustrated by a study of a continuous fiber metal matrix composite, *Journal of X-Ray Science and Technology*, **2**, 249–58.

ESVELD, D. C., VAN DER SMAN, R. G. M., VAN DALEN, G., VAN DUYNHOVEN, J. P. M. and MEINDERS, M. B. J. (2012), Effect of morphology on water sorption in cellular solid foods. Part I: Pore scale network model, *Journal of Food Engineering*, **109**, 301–10.

FALCONE, P. M., BAIANO, A., ZANINI, F., MANCINI, L., TROMBA, G. *et al.* (2004), A novel approach to the study of bread porous structure: phase-contrast X-ray microtomography (eeprint), *Journal of Food Science*, **69**.

FALCONE, P. M., BAIANO, A., ZANINI, F., MANCINI, L., TROMBA, G. *et al.*, (2005), Three-dimensional quantitative analysis of bread crumb by X-ray microtomography, *Journal of Food Science*, **70**, E265–72.

FARBER, L., TARDOS, G. and MICHAELS, J. N. (2003), Use of X-ray tomography to study the porosity and morphology of granules, *Powder Technology*, **132**, 57–63.

FRISULLO, P., LAVERSE, J., MARINO, R. and NOBILE, M. A. D. (2009), X-ray computed tomography to study processed meat microstructure, *Journal of Food Engineering*, **94**, 283–9.

FRISULLO, P., LICCIARDELLO, F., MURATORE, G. and DEL NOBILE, M. A. (2010a), Microstructural characterization of multiphase chocolate using X-Ray microtomography, *Journal of Food Science*, **75**, E469–76.

FRISULLO, P., MARINO, R., LAVERSE, J., ALBENZIO, M. and DEL NOBILE, M. A. (2010b), Assessment of intramuscular fat level and distribution in beef muscles using X-ray microcomputed tomography, *Meat Science*, **85**, 250–5.

FRISULLO, P., BARNABÀ, M., NAVARINI, L. and DEL NOBILE, M. A. (2012), Coffea arabica beans microstructural changes induced by roasting: An X-ray microtomographic investigation, *Journal of Food Engineering*, **108**, 232–7.

FURNOLS, M., TERAN, M. F. and GISPERT, M. (2009), Estimation of lean meat content in pig carcasses using X-ray Computed Tomography and PLS regression, *Chemometrics and Intelligent Laboratory Systems*, **98**, 31–7.

GHORBANI, Y., BECKER, M., PETERSEN, J., MORAR, S. H., MAINZA, A. and FRANZIDIS, J. P. (2011), Use of X-ray computed tomography to investigate crack distribution and mineral dissemination in sphalerite ore particles, *Minerals Engineering*, **24**, 1249–57.

GROSO, A., ABELA, R. and STAMPANONI, M. (2006), Implementation of a fast method for high resolutionphase contrast tomography, *Opt. Express*, **14**, 8103–10.

GUESSASMA, S., BABIN, P., VALLE, G. D. and DENDIEVEL, R. (2008), Relating cellular structure of open solid food foams to their Young's modulus: Finite element calculation, *International Journal of Solids and Structures*, **45**, 2881–96.

HAEDELT, J., BECKETT, S. T. and NIRANJAN, K. (2007), Bubble-included chocolate: relating structure with sensory response, *Journal of Food Science*, **72**, E138–42.

HAHN, M., VOGEL, M., POMPESIUS-KEMPA, M. and DELLING, G. (1992), Trabecular bone pattern factor – a new parameter for simple quantification of bone microarchitecture, *Bone*, **13**, 327–30.

HASSAN, A. N., FRANK, J. F., FARMER, M. A., SCHMIDT, K. A. and SHALABI, S. I. (1995), Formation of yogurt microstructure and three-dimensional visualization as determined by confocal scanning laser microscopy, *Journal of Dairy Science*, **78**, 2629–36.

HOLLO, G., SOMOGYI, T., HOLLO, I., ANTON, I. and REPA, I. (2010), Application of X-ray Computer Tomography (CT) in cattle production, in, *Farm Animal Imaging Congress*, 2010 Rennes, France.

JENSEN, T. H., BÖTTIGER, A., BECH, M., ZANETTE, I., WEITKAMP, T. *et al.* F. (2011), X-ray phase-contrast tomography of porcine fat and rind, *Meat Science*, **88**, 379–83.

KANIT, T., N'GUYEN, F., FOREST, S., JEULIN, D., REED, M. and SINGLETON, S. (2006), Apparent and effective physical properties of heterogeneous materials: representativity of samples of two materials from food industry, *Computer Methods in Applied Mechanics and Engineering*, **195**, 3960–82.

KARKLE, E. L., ALAVI, S. and DOGAN, H. (2011), Cellular architecture and its relationship with mechanical properties in expanded extrudates containing apple pomace, *Food Research International*. DOI 10.1016/j.foodres.2011.11.003.

KOSTER, A. J., ZIESE, U., VERKLEIJ, A. J., JANSSEN, A. H. and DE JONG, K. P. (2000), Three-dimensional transmission electron microscopy: a novel imaging and characterization technique with nanometer scale resolution for materials science, *The Journal of Physical Chemistry B*, **104**, 9368–70.

LAMMERTYN, J., DRESSELAERS, T., VAN HECKE, P., JANCSÓK, P., WEVERS, M. and NICOLAÏ, B. M. (2003), MRI and X-ray CT study of spatial distribution of core breakdown in 'Conference' pears, *Magnetic Resonance Imaging*, **21**, 805–15.

LASSOUED, N., BABIN, P., DELLA VALLE, G., DEVAUX, M.-F. and REGUERRE, A.-L. (2007), Granulometry of bread crumb grain: Contributions of 2D and 3D image analysis at different scale, *Food Research International*, **40**, 1087–97.

LAVERSE, J., MASTROMATTEO, M., FRISULLO, P., ALBENZIO, M., GAMMARIELLO, D. and DEL NOBILE, M.A. (2011a), Fat microstructure of yogurt as assessed by X-ray microtomography, *Journal of Dairy Science*, **94**, 668–75.

LAVERSE, J., MASTROMATTEO, M., FRISULLO, P. and DEL NOBILE, M. A. (2011b), X-ray microtomography to study the microstructure of cream cheese-type products, *Journal of Dairy Science*, **94**, 43–50.

LAVERSE, J., MASTROMATTEO, M., FRISULLO, P. and DEL NOBILE, M. A. (2012), X-ray microtomography to study the microstructure of mayonnaise, *Journal of Food Engineering*, **108**, 225–31.

LEE, T. C., KASHYAP, R. L. and CHU, C. N. (1994), Building skeleton models via 3D medial surface axis thinning algorithms, *CVGIP: Graphical Models and Image Processing*, **56**, 462–78.

LÉONARD, A., BLACHER, S., NIMMOL, C. and DEVAHASTIN, S. (2008), Effect of far-infrared radiation assisted drying on microstructure of banana slices: An illustrative use of X-ray microtomography in microstructural evaluation of a food product, *Journal of Food Engineering*, **85**, 154–62.

LICCIARDELLO, F., FRISULLO, P., LAVERSE, J., MURATORE, G. and DEL NOBILE, M.A. (2012), Effect of sugar, citric acid and egg white type on the microstructural and mechanical properties of meringues, *Journal of Food Engineering*, **108**, 453–62.

LIM, K. S. and BARIGOU, M. (2004), X-ray micro-computed tomography of cellular food products, *Food Research International*, **37**, 1001–12.

LINDQUIST, W. B., LEE, S.-M., COKER, D. A., JONES, K. W. and SPANNE, P. (1996), Medial axis analysis of void structure in three-dimensional tomographic images of porous media, *J. Geophys. Res.*, **101**, 8297–310.

MAIRE, E., BUFFIÈRE, J. Y., SALVO, L., BLANDIN, J. J., LUDWIG, W. and LÉTANG, J. M. (2001), On the application of X-ray microtomography in the field of materials science, *Advanced Engineering Materials*, **3**, 539–46.

MAYO, S. C., CHEN, F. and EVANS, R. (2010), Micron-scale 3D imaging of wood and plant microstructure using high-resolution X-ray phase-contrast microtomography, *Journal of Structural Biology*, **171**, 182–8.

MEBATSION, H. K., VERBOVEN, P., ENDALEW, A. M., BILLEN, J., HO, Q. T. and NICOLAI, B. M. (2009), A novel method for 3D microstructure modeling of pome fruit tissue using synchrotron radiation tomography images, *Journal of Food Engineering*, **93**, 141–8.

MENDOZA, F., VERBOVEN, P., MEBATSION, H. K., KERCKHOFS, G., WEVERS, M. and NICOLAI, B. (2007), Three-dimensional pore space quantification of apple tissue using X-ray computed microtomography, *Planta*, **226**, 559–70.

MENDOZA, F., VERBOVEN, P., HO, Q. T., KERCKHOFS, G., WEVERS, M. and NICOLAÏ, B. (2010), Multifractal properties of pore-size distribution in apple tissue using X-ray imaging, *Journal of Food Engineering*, **99**, 206–15.

MOHORIČ, A., VERGELDT, F., GERKEMA, E., DALEN, G. V., DOEL, L. R. V. *et al.* (2009), The effect of rice kernel microstructure on cooking behaviour: a combined μ-CT and MRI study, *Food Chemistry*, **115**, 1491–9.

MOUSAVI, R., MIRI, T., COX, P. W. and FRYER, P. J. (2005), A novel technique for ice crystal visualization in frozen solids using X-Ray micro-computed tomography, *Journal of Food Science*, **70**, e437–42.

MUSSE, M., DE GUIO, F., QUELLEC, S., CAMBERT, M., CHALLOIS, S. and DAVENEL, A. (2010), Quantification of microporosity in fruit by MRI at various magnetic fields: comparison with X-ray microtomography, *Magnetic Resonance Imaging*, **28**, 1525–34.

NADE, T., FUJITA, K., FUJII, M., YOSHIDA, M., HARYU, T. *et al.* (2005), Development of X-ray computed tomography for live standing cattle, *Animal Science Journal*, **76**, 513–17.

NATTERER, F. (2008), X-ray tomography, *Inverse Problems and Imaging*, **1943**, 17–34.

NEETHIRAJAN, S., KARUNAKARAN, C., JAYAS, D. S. and WHITE, N. D. G. (2006), X-ray computed tomography image analysis to explain the airflow resistance differences in grain bulks, *Biosystems Engineering*, **94**, 545–55.

NEETHIRAJAN, S., JAYAS, D. S., WHITE, N. D. G. and ZHANG, H. (2008), Investigation of 3D geometry of bulk wheat and pea pores using X-ray computed tomography images, *Computers and Electronics in Agriculture*, **63**, 104–11.

OH, W. and LINDQUIST, W. B. (1999), Image thresholding by indicator kriging, *IEEE Transactions on Pattern Analysis and Machine Intelligence*, **21**, 590–602.

PARADA, J., AGUILERA, J. M. and BRENNAN, C. (2011), Effect of guar gum content on some physical and nutritional properties of extruded products, *Journal of Food Engineering*, **103**, 324–32.

PAREYT, B., TALHAOUI, F., KERCKHOFS, G., BRIJS, K., GOESAERT, H. *et al.* (2009), The role of sugar and fat in sugar-snap cookies: Structural and textural properties, *Journal of Food Engineering*, **90**, 400–8.

PAUWELS, B., LIU, X. and SASOV, A. (2010), X-ray nanotomography in a SEM, in, S. R. Stock (ed.), *Developments in X-Ray Tomography*, vol. VII.

PORTSMOUTH, R. L. and GLADDEN, L. F. (1991), Determination of pore connectivity by mercury porosimetry, *Chemical Engineering Science*, **46**, 3023–36.

RAHMANIAN, N., GHADIRI, M., JIA, X. and STEPANEK, F. (2009), Characterisation of granule structure and strength made in a high shear granulator, *Powder Technology*, **192**, 184–94.

RAMACHANDRAN, G. N. and LAKSHMINARAYANAN, A. V. (1971), Three-dimensional reconstruction from radiographs and electron micrographs: application of convolutions instead of Fourier Transforms, *Proceedings of the National Academy of Sciences*, **68**, 2236–40.

RIGBY, S. P., WATT-SMITH, M. J., CHIGADA, P., CHUDEK, J. A., FLETCHER, *et al.* (2006), Studies of the entrapment of non-wetting fluid within nanoporous media using a synergistic combination of MRI and micro-computed X-ray tomography, *Chemical Engineering Science*, **61**, 7579–92.

ROBIN, F., ENGMANN, J., PINEAU, N., CHANVRIER, H., BOVET, N. and VALLE, G. D. (2010), Extrusion, structure and mechanical properties of complex starchy foams, *Journal of Food Engineering*, **98**, 19–27.

ROBIN, F., DUBOIS, C., CURTI, D., SCHUCHMANN, H. P. and PALZER, S. (2011a), Effect of wheat bran on the mechanical properties of extruded starchy foams, *Food Research International*, **44**, 2880–8.

ROBIN, F., DUBOIS, C., PINEAU, N., SCHUCHMANN, H. P. and PALZER, S. (2011b), Expansion mechanism of extruded foams supplemented with wheat bran, *Journal of Food Engineering*, **107**, 80–9.

RØRÅ, A. M. B., BIRKELAND, S., HULTMANN, L., RUSTAD, T., SKÅRA, T. and BJERKENG, B. (2005), Quality characteristics of farmed Atlantic salmon (Salmo salar) fed diets high in soybean or fish oil as affected by cold-smoking temperature, *LWT – Food Science and Technology*, **38**, 201–11.

SAHOO, P. K., SLAAF, D. W. and ALBERT, T. A. (1997), Threshold selection using a minimal histogram entropy difference, *Optical Engineering*, **36**, 1976–81.

SALVO, L., CLOETENS, P., MAIRE, E., ZABLER, S., BLANDIN, J. J. *et al.* (2003), X-ray micro-tomography an attractive characterisation technique in materials science, *Nuclear Instruments and Methods in Physics Research Section B: Beam Interactions with Materials and Atoms*, **200**, 273–86.

SEIDELL, J. C., BAKKER, C. J. G. and VANDERKOOY, K. (1990), Imaging techniques for measuring adipose tissue distribution – a comparison between computed tomography and 1.5-T magnetic resonance, *American Journal of Clinical Nutrition*, **51**, 953–57.

STEVEN, A. C. and AEBI, U. (2003), The next ice age: Cryo-electron tomography of intact cells, *Trends in Cell Biology*, **13**, 107–10.

SZABO, C., BABINSZKY, L., VERSTEGEN, M. W. A., VANGEN, O., JANSMAN, A. J. M. and KANIS, E. (1999), The application of digital imaging techniques in the in vivo estimation of the body composition of pigs: a review, *Livestock Production Science*, **60**, 1–11.

THOVERT, J. F., SALLES, J. and ADLER, P. M. (1993), Computerized characterization of the geometry of real porous media: Their discretization, analysis and interpretation, *Journal of Microscopy*, **170**, 65–79.

TRATER, A. M., ALAVI, S. and RIZVI, S. S. H. (2005), Use of non-invasive X-ray microtomography for characterizing microstructure of extruded biopolymer foams, *Food Research International*, **38**, 709–19.

VAN DALEN, G., NOOTENBOOM, P., VAN VLIET, L. J., VOORTMAN, L. and ESVELD, E. (2007), 3D imaging, analysis and modelling of porous cereal products using X-ray microtomography, *Image Analysis Stereology*, **26**, 169–77.

VAN DALEN, G., NOOTENBOOM, P., DON, A., DEN ADEL, R. and ROIJERS, E. (2009), 3D Imaging of the solid phase of porous bakery products using synchrotron X-ray microtomography, in, V. Capasso (ed.), *The 10th European Congress of Stereology and Image Analysis*, 22–26 June 2009 Milan, Italy. Societa Editrice Esculapio-Progetto Leonardo, 335–41.

VERBOVEN, P., MEBATSION, H., MENDOZA, F., TEMST, K., WEVERS, M. *et al.* (2008), Comparison of different X-ray computed tomography techniques for the quantitative characterization of the 3D microstructure of pear fruit tissue, in G. Busse (ed.), *Emerging Technologies in Non-Destructive Testing*, 2–4 April 2008, Stuggart, Taylor & Francis Ltd, 331–6.

VESTERGAARD, C., ERBOU, S. G., THAULAND, T., ADLER-NISSEN, J. and BERG, P. (2005), Salt distribution in dry-cured ham measured by computed tomography and image analysis, *Meat Science*, **69**, 9–15.

VIDAL, F. P., LÉTANG, J. M., PEIX, G. and CLOETENS, P. (2005), Investigation of artefact sources in synchrotron microtomography via virtual X-ray imaging, *Nuclear Instruments and*

Methods in Physics Research Section B: Beam Interactions with Materials and Atoms, **234**, 333–48.

WANG, S., AUSTIN, P. and BELL, S. (2011), It's a maze: the pore structure of bread crumbs, *Journal of Cereal Science,* **54**, 203–10.

WANG, T. W. and EVANS, J. P. O. (2003), Stereoscopic dual-energy X-ray imaging for target materials identification, *Vision, Image and Signal Processing, IEE Proceedings,* **150**, 122–30.

WANG, Y., JACOBSEN, C., MASER, J. and OSANNA, A. (2000), Soft X-ray microscopy with a cryo scanning transmission X-ray microscope. Part II: Tomography, *Journal of Microscopy-Oxford,* **197**, 80–93.

WANG, Y. X., DE CARLO, F., MANCINI, D. C., MCNULTY, I., TIEMAN, B. *et al.* (2001), A high-throughput X-ray microtomography system at the Advanced Photon Source, *Review of Scientific Instruments,* **72**, 2062–28.

WILLIAMS, R. A. and JIA, X. (2003), Tomographic imaging of particulate systems, *Advanced Powder Technology,* **14**, 1–16.

ZHU, L.-J., DOGAN, H., GAJULA, H., GU, M.-H., LIU, Q.-Q. and SHI, Y.-C. (2012), Study of kernel structure of high-amylose and wild-type rice by X-ray microtomography and SEM, *Journal of Cereal Science,* **55**, 1–5.

Part II

Measurement, analysis and modelling of food microstructures

10

Food microstructure and rheology

M. A. Rao, Cornell University, USA

DOI: 10.1533/9780857098894.2.275

Abstract: Well-defined geometries, in which the flow fields can be analyzed for shear rates and shear stresses, can be used to determine the viscous and viscoelastic behavior of foods and dispersions of food polymers. A large-gap vane-in-cup geometry with at least four vanes can be used to determine values of yield stress. These tests require millilitres of a sample. Microrheology is based on using embedded micron-sized probes. The mean square displacement (MSD) of probe particles suspended in a sample is measured by dynamic light scattering as a function of time. From the generalized Stokes–Einstein relationship, the frequency dependent viscoelastic moduli are obtained from the MSD.

Key words: rheological properties measurement, plate-cone, concentric cylinder, direct yield stress, microrheology, mean square displacement.

10.1   Introduction

Traditionally, rheology deals with the application of stress, strain and strain/shear rate to test samples and their units. Data needed to characterize solid and viscoelastic foods is stress-strain data, and for fluid foods shear rate versus shear stress, and small-amplitude oscillatory data.

In this chapter, traditional rheological methods that have been used on foods over a wide range of concentrations are reviewed first, followed by microrheological techniques, based on light scattering, which have been introduced recently.

10.2   Traditional rheological methods and food structure

Typically, bulk rheological data on fluid foods are obtained using well-defined geometries, in which the flow fields can be analyzed for shear rates and shear stresses. The concept of steady simple shear or uniform deformation means that each fluid element undergoes exactly the same deformation and stresses and also the tangential stresses are independent of position in space. A viscosmetric flow may be viewed as a flow which, from the point of view of a fluid element, is indistinguishable from steady simple shear. A theoretical discussion of viscometric flow can be found in Truesdell (1974).

The measuring geometries commonly used to achieve viscometric flows are concentric cylinder (Couette), cone-plate and parallel plate, and capillary/tube (Figs 10.1 to 10.3). It is noted that these systems, especially the capillary/tube geometry, require sample volumes of several millilitres. Thus, they preclude the study of rare or precious materials, including many biological samples that are difficult to obtain in such quantities. However, they have been used to characterize numerous commercially-important foods and dispersions of food polymers. The important assumptions for the fluid conditions in these geometries are no-slip at the fluid–solid interfaces, steady, laminar flow, and constant and uniform temperature (Rao, 2007).

In these geometries, the shear stress is calculated from the total torque sensed by the fluid. The pertinent equations for the calculation of shear rate for each type of geometry can be derived from the equations of continuity and motion (Rao, 2007). It is noted that for the cone-plate geometry, with small cone angles (~2–4 degrees), the shear rate in the test fluid is uniform; in contrast, for the other geometries, the shear rate depends on the dimensions of the test geometry. The equations for shear stress and shear rate for the four geometries are:

- **Concentric cylinder geometry**
  In concentric cylinder geometry (Fig. 10.1), the shear stress can be determined from the total torque (M):

$$\sigma = \frac{M}{2\pi r_i^2 h} \qquad [10.1]$$

  where r_i is the radius of the rotating bob.

- **Newtonian shear rate**
  The Newtonian shear rate in a concentric cylinder geometry, $\dot{\gamma}_N$, can be calculated exactly from the expression:

$$\dot{\gamma}_N = \frac{2\Omega}{\left[1 - (r_i / r_o)^2\right]} \qquad [10.2]$$

  where Ω is the angular velocity of the rotating bob, r_i is the radius of the bob and r_o is the radius of the cup. The shear rate of a non-Newtonian food cannot be determined from a simple expression involving the angular velocity and for large-gap geometries we must use a suitable relationship between rotational speed and shear stress to correct for non-Newtonian behavior (Rao, 2007).

- **Cone-plate geometry**
  In a cone-plate geometry (Fig. 10.2), the shear stress is obtained from

$$\sigma_{\theta\phi} = \frac{3M}{2\pi r_o^3}.$$

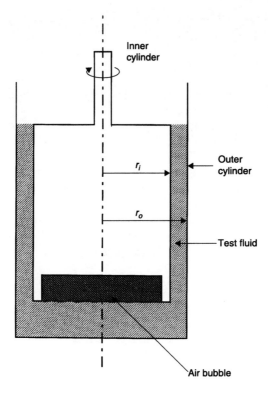

Fig. 10.1 Concentric cylinder measuring system. A narrow gap between the cylinders is preferred but this precludes studies on large particles. The particles may also settle (Rao, 2007).

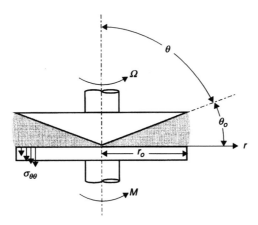

Fig. 10.2 The cone-plate geometry is popular because at low cone angle, θ_o, the shear rate is uniform in the gap. This geometry may not be suitable for dispersions with large particles (Rao, 2007).

Defining T_{cn} as the torque per unit area, the above equation can be written in terms of the cone diameter D:

$$\sigma_{\theta\phi} = \frac{3T_{cn}}{D}$$

[10.3]

For low values of the cone angle θ_o, the angular velocity and viscosity are related by

$$\Omega = \frac{3T}{4\eta r_o}(\theta_o + \theta_o) = \frac{3T\theta_o}{\eta D}$$

[10.4]

Because, for small cone angles, the shear rate is uniform in the flow field, the cone-plate geometry is used extensively in continuous shear and oscillatory shear experiments; it is emphasized that the latter are to be conducted within the linear viscoelastic range. One drawback of the cone-plate geometry is that because the truncation gap is relatively small, about 60 µm, it should not be used for studying foods with larger particles, such as gelatinized starch granules.

- **Parallel disk geometry**
  The parallel disk geometry (Fig. 10.3), also called the parallel plate geometry, consists of two disks with radius, r_o, separated by the gap, h. Assuming steady, laminar and isothermal flow, the expression for shear rate is

$$\dot{\gamma} = \frac{\Omega r_o}{h}$$

[10.5]

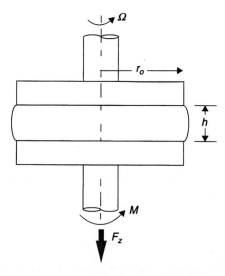

Fig. 10.3 The parallel plate geometry can be used with dispersions containing large particles by adjusting the gap (Rao, 2007).

The shear stress can be determined from the measured torque, M:

$$\sigma = \frac{3M}{2\pi r_o^3}\left[1+\frac{1}{3}\frac{d\ln M}{d\ln\gamma}\right]$$ [10.6]

where Ω is the angular velocity, r_o is the disk radius and h is the gap between the disks. Although the shear field is not homogeneous, the parallel disk geometry is useful in handling dispersions that contain relatively large size particles, such as gelatinized starch dispersions.

10.2.1   Capillary or tube viscometer

In capillary flow (Fig. 10.4), the pressure drop, Δp (Pa), over a known length of tube, L (m) and diameter (D), and the corresponding volumetric flow rate, Q ($m^3\ s^{-1}$), of the food in laminar flow are measured. The equation for the shear stress, σ_w, at the wall is

$$\sigma_w = \frac{D\Delta p}{4L}$$ [10.7]

The general equation for the shear rate in tube flow is

$$\left(\frac{dv_z}{dr}\right)=\left(\frac{3}{4}\right)\frac{4Q}{\pi r_o^3}+\frac{\sigma_w}{4}\frac{d\left(4Q/\pi r_o^3\right)}{d\sigma_w}$$ [10.8]

where v_z is the axial velocity of the fluid and r_o is the tube radius. Glass capillary viscometers, with small radii (3–5 mm) are used extensively for determining the viscosity of Newtonian fluids and for the determination of intrinsic viscosity.

10.2.2   Measuring systems for foods exhibiting slip

For fluid foods without large particles, concern is whether slip occurs at the fluid food–solid surface of the measurement geometry. With several foods, such as

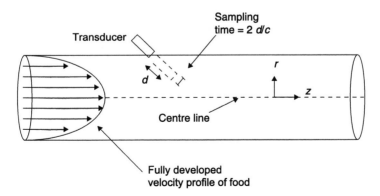

Fig. 10.4   Flow in a capillary/tube viscometer. Small-diameter glass units are used for Newtonian fluids (Rao, 2007).

apple sauce and mustard, often a thin layer of aqueous fluid accumulates at the solid surfaces of the measuring geometry; with melted cheese, the fluid is an oil layer. This phenomenon results in a finite slip velocity of the fluid, as opposed to the zero fluid velocity assumed in the no-slip boundary condition.

The effects of wall slip may be overcome in two measuring systems. System 1 is the lubricated biaxial strain rheometer (Chatraei et al., 1981). This geometry, consisting of parallel disks (not shown here) is well-suited for foods for which the no-slip boundary conditions are of concern and which exhibit yield stress; the test material is squeezed. However, the strain rates obtained are low, and only the biaxial strain viscosities are obtained. System 2 is the vane-in-cup geometry, which imitates flow in a concentric-cylinder system, and is well-suited for foods for which slip effects cannot be minimized, and for foods containing large particles that may tend to settle (Barnes and Carnali, 1990). An additional desirable feature is that when the vane is inserted into the test fluid, minimum disturbance is caused by the thin blades. Barnes and Carnali (1990) determined flow properties using the measuring geometries, concentric cylinder (bob-in-cup) and a four-bladed vane-in-cup. Both types of sensor systems were made from stainless steel to identical diameters and heights, the vane blades were 1 mm thick and the hub had a radius of 3 mm. Both systems had an annular gap of 1 mm and a sample depth equal to twice the height of the bob or vane. The most significant result of this study was the identification of the vane-in-cup as a viable measuring geometry for the study of very shear thinning, power-law fluids, with small power-law exponent, which often exhibit a yield stress.

Another concern with foods containing particles is that the particles settle in the measurement geometry, which leads to a change in the concentration of particles in the sample, especially in a concentric cylinder system with significant height, during the waiting period prior to experiment and during the conduct of experiment. Other concerns are whether the particle shape or size is altered during shearing when the particles are large; for parallel plates, the gap between the plates is recommended to be more than ten times the particle diameter.

10.2.3   Vane yield stress

A large-gap vane-in-cup geometry with a minimum of four vanes can be used to determine values of yield stress directly (Dzuy and Boger, 1985) and the role of structure on yield stress (Genovese and Rao, 2005). In the measurements on yield stress, systems with ratios of cup-to-vane radii of about 1.8 have been used. Figure 10.5 is a diagram of a six-blade vane that has been used in studies on yield stress in the author's laboratory (Qiu and Rao, 1989; Genovese and Rao, 2005).

The vane yield stress, σ_0, is calculated from the equation:

$$T_m = \frac{\pi D_v^3}{2}\left(\frac{H}{D_v} + \frac{1}{3}\right)\sigma_0 \qquad [10.9]$$

where T_m is the maximum torque reading (N m), D_v (m) is the diameter and H (m) the height of the paddle, respectively. The method is simple to use and consistent

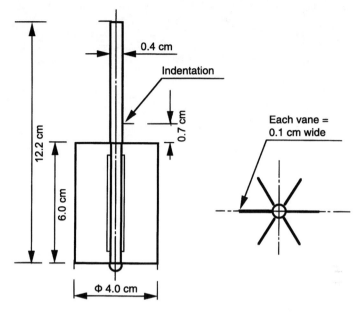

0.4 cm

Indentation

0.7 cm

12.2 cm

6.0 cm

Φ 4.0 cm

Each vane =
0.1 cm wide

Fig. 10.5   A six-blade vane. A narrow-gap vane-in-cup configuration can be used to obtain viscosity data on foods that exhibit slip at the wall. A large-gap vane-in-cup configuration can be used to obtain yield stress data on high-solids dispersions (Rao, 2007).

results were obtained for a wide range of food products (Rao, 2007). The structure of commercial food dispersions is influenced strongly by specific unit operations during manufacture: homogenization for mayonnaise, ketchup and mustard; finishing for apple sauce, and finishing and evaporation for tomato puree. Furthermore, the static (σ_{0s}) and dynamic (σ_{0d}) yield stress values of a food product can be determined before (static) and after breaking down (dynamic) its structure under continuous shear, respectively (Fig. 10.6). When the yield stress values are plotted against the apparent shear rate of the vane, the resulting texture map reflects the influence of the structure-inducing operations (Fig. 10.7; Genovese and Rao, 2005).

10.2.4   Use of rheology to monitor structural changes during food processing and storage

Generally, small-amplitude oscillatory techniques have been used to monitor structural changes during food processing. Using this technique in the linear viscoelastic range, we can obtain the two viscoelastic parameters, the storage modulus, G' (Pa) and the loss modulus, G'' (Pa); in addition, the ratio of the two moduli = G''/G' is also a useful parameter (Rao, 2007).

Typical examples of this approach are the use of time-sweep and temperature-sweep measurements at a fixed oscillatory frequency to follow gelation and

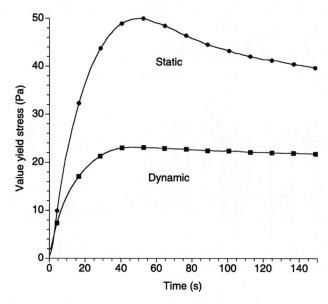

Fig. 10.6   Illustration of static and dynamic yield stresses of a commercial tomato ketchup sample (data from Genovese and Rao, 2005).

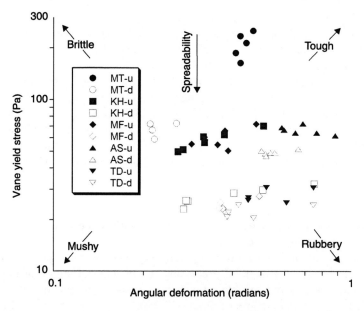

Fig. 10.7   Texture map of commercial dispersions using a large-gap vane-in-cup configuration. Abbreviations used: MT, mayonnaise; KH, ketchup; MF, mustard; AS, apple sauce; TD, tomato puree. The letters u and d indicate static and dynamic data, respectively (Genovese and Rao, 2005).

softening. In addition, the time for gelation and minimum polymer concentration for gelation can be determined (Kavanagh *et al.*, 2000; Rao, 2007).

Gelation is a critical phenomenon where the transition variable is the connectivity of the physical or chemical bonds linking the basic structural units of the material. Therefore, rheological properties are sensitive indicators of the critical gel point. The concept of gel point has received much attention in synthetic and biopolymer gels. Here, the symbols t_c and T_{gel} will be used for the gel time and gel temperature, respectively. Much focus has been placed on the use of the small amplitude oscillatory shear technique to measure the dynamic moduli during the gelation process, in order to identify the gel point. By this method, the continuous evolution of the viscoelastic properties throughout the gelation process can be followed. Specific criteria for identifying the gel point are listed and discussed elsewhere (Rao, 2007): these include G' and G'' crossover, power law behavior of the shear modulus and threshold G' value.

As one example, gel cure, G' versus time, data on fibrils of β-lactoglobulin were extrapolated using the relationship (Kavanagh *et al.*, 2000):

$$G' \approx G'_{inf} \exp(-B/t) \tag{10.10}$$

where t is the time in seconds, B is an empirical parameter and G'_{inf} is the value of G' at infinite time. As can be seen in Fig. 10.8, Equation 10.10

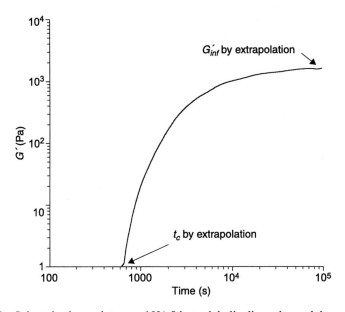

Fig. 10.8   Selected gel cure data on a 10% β-lactoglobulin dispersion and the curve fitted to obtain the gel time, t_c, and the storage modulus at infinite time, G'_{inf}. This figure has been redrawn and is not an identical reproduction from the original reference (with kind permission from Kavanagh *et al.*, 2000).

predicted satisfactorily both the asymptotic limit of G' as $t \to \infty$ and the logarithmic singularity as $t \to t_c$, the gelation time of a 10% β-lactoglobulin dispersion. Both of these limits were used in subsequent testing of the data against physical models, such as a percolation-based kinetic gelation model (Gosal et al., 2004).

10.2.5   Structural characteristics of protein fibrils

Techniques, such as the use of fractal dimensions, have been developed to characterize compact particles in food dispersions (Rao, 2007). Filament-shaped protein particles have been characterized by two length scales: the persistence length, l_p, and the contour length, L_c. The persistence length, defined as $l_p = \kappa/(k_B T)$, is the typical length at which thermal fluctuations begin to bend the polymer in different directions; it is used to characterize the flexibility or rigidity of a filament (MacKintosh, 1998). Increase in filament rigidity results in a decrease in the shear modulus of a gel, because of reduced entanglement. The contour length, L_c, of a filament is its length at maximum extension. A filament is considered flexible when $l_p \ll L_c$, and rigid in the opposite situation, $l_p \gg L_c$; many biological filaments are in a third intermediate category, semiflexible filaments with l_p and L_c of comparable magnitude (Storm et al., 2005).

The persistence length has been obtained using various experimental techniques, including dynamic light scattering, microscopic observation of thermal fluctuations and transmission electron microscopy (TEM) data (Loveday et al., 2012). Loveday et al. (2011) studied the effect of added calcium on the morphology and functionality of whey protein isolate (WPI) nanometre-scale fibrils. They found that the gel times decreased with the amount of calcium added, from 476 s at 0 mM to 82 s at 120 mM. G'_{inf} increased to a maximum at 80 mM $CaCl_2$ and then decreased with more added $CaCl_2$. WPI fibrils formed without $CaCl_2$ were long and semi-flexible and associated in large entangled networks more than 10 µm across on the TEM grid. Fibrils formed in the presence of 40 mM $CaCl_2$ were shorter, and were bent and twisted.

10.3   Microrheology

Traditionally, viscoelastic properties of a fluid food have been obtained from small-amplitude oscillatory shear measurements in a mechanical rheometer. Most commercial rheometers require several millilitres of the material to probe frequencies up to tens of Hz. The upper range is limited by the onset of inertial effects, when the oscillatory shear wave decays appreciably before propagating throughout the entire sample (Gardel et al., 2005).

It is difficult to obtain many biological materials in millilitre quantities. Microrheology can probe a material's response at micrometre length scales with microlitre sample volumes in a non-invasive manner (Mason and Weitz, 1995; Mason *et al.* 1997). The techniques use embedded micron-sized probes to locally deform a sample. There are two broad classes of microrheology techniques:

1. those involving the active manipulation of probes by the local application of stress; and
2. those measuring the passive motions of particles due to Brownian fluctuations.

In both cases, when the embedded particles are much larger than any structural size of the material, particle motions measure the macroscopic stress relaxation. Smaller particles measure the local mechanical response and also probe the effect of steric hindrances caused by local microstructure. The use of small colloidal particles theoretically extends the accessible frequency range by shifting the onset of inertial effects to the MHz regime (Gardel *et al.*, 2005).

10.3.1   Single particle microrheology

One-particle microrheology assumes that the local environment surrounding the bead reflects that of the bulk. In general, it is advisable to test for probe surface chemistry effects in a new system. It is noted that the bulk response is measured only if the probe size is larger than the length scale of heterogeneity in the sample. These length scales are often unknown prior to a microrheology experiment. Agarose is an example of a material containing many smaller voids, or pores, through which smaller probe particles may move.

The essential physics behind this technique was summarized by Dasgupta *et al.* (2001): the mean square displacement (MSD) reflects the response of a fluid to the stress applied to it by the thermal motion of the bead. The MSD, $(\Delta r^2(\tau))$, of probe particles suspended in a sample as a function of time has been measured by dynamic light scattering. In a purely viscous fluid, the probe particles diffuse through it and the MSD increases linearly with time $(\Delta r^2(\tau)) = 6D\tau$. After determining the diffusion coefficient D, we can calculate the viscosity of the material $\eta = (k_B T/(6\pi Da))$, where a is the radius of the beads.

In a viscoelastic fluid, the motion of the probe particles is constrained and the MSD reaches an average plateau value (Δr_p^2) that is set by the elastic modulus of the material. By equating the thermal energy $k_B T$ of each bead with its elastic energy $\frac{1}{2}\kappa(\Delta r_p^2)$, where κ is the effective spring constant that characterizes the elasticity of the surrounding medium, an expression for the spring constant $\kappa \sim k_B T/(\Delta r_p^2)$ can be obtained. The elastic modulus $G'(\omega)$ is related to the spring constant by a factor of length, which is the bead radius, a, and we can obtain a relationship between the elastic modulus and the MSD, $G'(\omega) \sim k_B T/((\Delta r_p^2)a)$. (Dasgupta *et al.*, 2001).

In general, the full frequency dependence of the viscoelastic moduli is obtained from the MSD by using the generalized Stokes–Einstein relationship (GSER) (Gardel et al., 2005):

$$\tilde{G}(s) = \frac{k_B T}{\pi a s (\Delta r^2(s))}$$
[10.11]

where $(\Delta r^2(s))$ is the Laplace transform of $(\Delta r^2(\tau))$ and $\tilde{G}(s)$ is the viscoelastic spectrum as a function of the Laplace frequency, s.

For comparison with mechanical spectroscopy data, the Fourier space representation of the GSER is used:

$$G^*(\omega) = \frac{k_B T}{\pi a i \omega \mathcal{F}\{(\Delta r^2(\tau))\}}$$
[10.12]

where $\mathcal{F}\{(\Delta r^2(\tau))\}$ is the Fourier transform of the MSD. A local power law expansion for $(\Delta r^2(\tau))$ leads to expressions for the elastic $G'(\omega)$ and the loss moduli $G''(\omega)$ (Mason, 2000).

Experimentally, in a dynamic light scattering experiment, a laser beam impinges on a sample and is scattered by the particles into a detector placed at an angle, θ, with respect to the incoming beam (Gardel et al., 2005). The intensity of light that reaches the detector fluctuates in time as the particles diffuse and rearrange in the sample. Dasgupta et al. (2001) used three different bead sizes, ranging between 0.46 μm and 0.97 μm at 0.0025 wt% concentration. The time averaged intensity correlation functions were collected for 2 to 3 hours at room temperature:

$$g_1(\tau) = \exp\left[-\frac{q^2(\Delta r^2(\tau))}{6}\right]$$
[10.13]

where $(\Delta r^2(\tau))$ is the ensemble averaged 3D MSD, and q is the scattering wave vector given by

$$q = \frac{4\pi n}{\lambda} \sin\left(\frac{\theta}{2}\right)$$
[10.14]

where n is the refractive index of the sample, λ the wavelength of the laser in vacuum and θ the scattering angle.

Once the MSD is obtained, as described above, the GSER can be applied to extract the frequency dependent viscoelastic moduli. Single light-scattering techniques are typically sensitive to frequencies in the range of 0.01 to 10 Hz, similar to the frequency range available with a conventional mechanical rheometer. However, these techniques are practical only when a significant fraction of the scattered light that is collected has been scattered by single probe particles and there is little multiple scattering.

10.3.2   Diffusive wave spectroscopy (DWS)

In DWS, a laser beam impinges on an opaque sample and the light is scattered multiple times before exiting the sample (Gardel *et al.*, 2005). It is important to note that because DWS measures the properties of multiply scattered light in suspensions where all photons have been multiply scattered, all the scattering-vector information is lost. As a consequence, only two experimental modes, transmission and backscattering, are used (Weitz and Pine, 1992). DWS treats the photon path through the sample as a diffusive or random walk phenomenon. Also, it extends the conventional DLS to systems that are dominated by multiple scattering, and hence to systems that are essentially opaque. Therefore, DWS requires a turbid or concentrated sample and the technique can be used to study practical phenomena, such as sol–gel transitions and glassy states (Alexander and Dalgleish, 2006). In DWS experiments, the volume fraction of the embedded colloidal spheres is large, approximately 10^{-2}, in order to insure that the transport mean path, l^*, is a small fraction of the length of the sample chamber, l. The ratio of l/l^* is typically greater than 5 (Gardel *et al.*, 2005).

The MSD is obtained from the field autocorrelation function, $g_1(\tau)$ at a delay time τ is (Gardel *et al.*, 2005):

$$g_1(\tau) = \int_0^\infty P(s)\,exp\left[-k_o^2\langle\Delta r^2(\tau)\rangle\frac{s}{3l^*}\right]ds \qquad [10.15]$$

where $P(s)$ is the probability of light travelling a path of length s, $k_o = 2\pi/\lambda$ is the wave vector of the incident light, and λ is the wavelength of light in the medium. The transport mean free path of the light, l^*, is a characteristic of the sample itself and reflects the amount of scattering; it is the length light must travel before its direction is randomized. It is emphasized that the key to the solution of the above equation is the determination of $P(s)$ for the experimental geometry with the correct boundary conditions (Pine *et al.*, 1988). The transport mean free paths for the samples used in these experiments are roughly 4 to 10 times smaller than the cell thickness, depending on the bead size used, ensuring strong multiple scattering. Because in DWS the number of scattering events is large, there is no appreciable angular dependence of the light emerging from the sample, and measurement at a precisely defined angle is not necessary.

Much of the research work on food systems with DWS has been on milk proteins. DWS is capable of following the kinetics of destabilization of concentrated food colloids and detects differences at the microstructural level (Corredig and Alexander, 2008). The onset of aggregation always appears earlier in DWS measurements than in oscillatory-shear measurement, because DWS is a microscopic measurement and is more sensitive to particle changes than rheology (Alexander and Dalgleish, 2006). Dhabi *et al.* (2010) determined that the jamming concentration of casein was 18.9 w/w%. In addition, changes in l^* can be taken as indications of changing organization within the suspension (Alexander and Dalgleish, 2006).

Based on microrheological experiments, Corrigan and Donald (2009) estimated the critical concentration for gelation of β-lactoglobulin as below 3%, lower than the 5.2% estimated by fitting the G'_{inf} versus concentration data to the cascade model (Gosal *et al.*, 2004). They pointed out that oscillatory shear bulk rheological tests may disrupt either the fibril or gel formation. Moschakis *et al.* (2010) also noted that the particle tracking method has higher sensitivity and can detect changes in the structuring of the system before these are registered by the bulk rheological measurement.

10.3.3   Multiparticle video microscopy

In a homogeneous, isotropic material, it is sufficient to examine the MSD of a single particle. However, in heterogeneous materials, it would be useful to obtain the MSD of two or more particles and this can be facilitated by the increasing use of video microscopy. The main advantage in using video microscopy for microrheology is the potential for following the motions of as many as 100 colloidal particles simultaneously, and the ability to obtain the ensemble averaged MSD while still retaining each of the individual particle trajectories (Gardel *et al.*, 2005). Algorithms have been developed to automate the process of particle-tracking, finding particle centres and accurately finding particles (Gardel *et al.*, 2005; Apgar *et al.*, 2000; Crocker and Weeks, 2000). Two-dimensional trajectories obtained using video microscopy have been used by Xu *et al.* (2007) and Moschakis *et al.* (2010).

Examples of studies using such direct particle tracking to determine the microheterogeneity include work on dispersions of wheat gliadin (Xu *et al.*, 2002) and β-glucan (Xu *et al.*, 2007), and on a phase separating emulsion (Moschakis *et al.*, 2006). Xu *et al.* (2002) evaluated the contributions of the 10, 25 and 50% highest MSD values to the ensemble-averaged MSD in 0.25, 0.50, 1.0 and 2% wheat gliadin dispersions. For β-glucan solutions of $\leq$ 1%, those contributions were about 11, 26 and 52%, respectively, close to those of glycerol, a homogeneous liquid. In contrast, for a 2% β-glucan solution, those contributions were 20, 39 and 65%, respectively, indicating that the 2% β-glucan solution exhibited a greater degree of heterogeneity than a homogeneous liquid. Examples of studies in which viscoelastic moduli were determined using direct particle tracking include dispersions of wheat gliadin (Xu *et al.* 2002) and β-glucan (Xu *et al.*, 2007), and the gel points and kinetics of acid-induced sodium caseinate gelation (Moschakis *et al.*, 2010). In Fig. 10.9, the storage, G' (dynes cm^{-2}) and loss G'' (dynes cm^{-2}) moduli of a 400 mg mL^{-1} gliadin dispersion (Xu *et al.*, 2002) are shown. The necessary particle trajectories and their MSDs were obtained using a well-dispersed suspension of 0.97 µm diameter fluorescent polystyrene microspheres (0.1 vol.%); the values of the viscoelastic moduli were obtained from the MSDs, as described earlier.

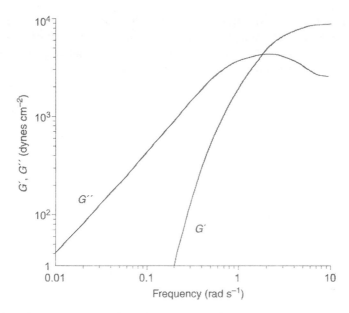

Fig. 10.9 Viscoelastic properties of a wheat gliadin dispersion 400 mg mL^{-1} using 0.97 μm diameter, fluorescent polystyrene microspheres, and video microscopy (with kind permission from Xu *et al.*, 2002). The figure was redrawn and is not an identical reproduction from the original reference.

10.4   Conclusion

Mechanical rheological tests require millilitre-volume samples. However, they can provide flow rheological parameters (apparent viscosity and yield stress) and oscillatory rheological parameters (G', G'') over a wide range of temperatures on samples with a wide range of concentrations; they are well-suited for obtaining data on high-solids materials, such as mayonnaise. They can also be used to follow phase transitions, particularly gelation.

Single- and multi-particle light-scattering techniques and DWS probe particle movements non-invasively at the molecular level. Together with the single-particle technique, the frequencies covered by DWS are about 10^{-1} to 10^{5} radian s^{-1} (Dasgupta *et al.*, 2001). DWS allows for the collection of low-noise MSD spectra, from which rheological data can be generated over a wide frequency range. However, in contrast to video microscopy, much information is lost, because DWS measurements are based on ensemble-averages of the Brownian motion of thousands of particles and cannot be used to measure the MSD distribution (Apgar *et al.*, 2000; Xu *et al.*, 2002). Recently, Moschakis *et al.* (2010) obtained G'-time data on a 10% sodium caseinate dispersion. However, DWS and other light-scattering experiments seem to have been limited to dispersions and polymer solutions having concentrations of approximately 4%.

Invariably, they are restricted to fixed ambient temperatures, so that temperature sweep data (e.g. G'-temperature) have not been obtained. In addition, they do not seem to provide data on flow rheology so that two important properties, apparent viscosity and, when applicable, yield stress, cannot be obtained.

10.5   References

ALEXANDER, M. and DALGLEISH, D. G. (2006), Dynamic light scattering techniques and their applications in food science, *Food Biophysics*, **1**, 2–33.

APGAR, J., TSENG, Y., FEDOROV, E., HERWIG, M. B., ALMO, S. C. and WIRTZ, D. (2000), Multiple-particle tracking measurements of heterogeneities in solutions of actin filaments and actin bundles, *Biophysical Journal*, **79**, 1095–106.

BARNES, H. A. and CARNALI, J. O. (1990), The vane-in-cup as a novel rheometer geometry for shear thinning and thixotropic materials, *J. Rheol.*, **34**, 841–65.

CHATRAEI, S. H., MACOSKO, C. W. and WINTER, H. H. (1981), A new biaxial extensional rheometer, *J. Rheol.*, **25**, 433–43.

CORREDIG, M. and ALEXANDER, M. (2008), Food emulsions studied by DWS: recent advances, *Trends in Food Science & Technology*, **19**, 67–75.

CORRIGAN, A. M. and DONALD, A. M. (2009). Particle tracking microrheology of gel-forming amyloid fibril networks, *European Physical Journal E*, **28**(4), 457–62.

CROCKER, J. C. and WEEKS, E. R. (2006), Software package for particle tracking. Available from: *http://www.physics.emory.edu/~weeks/idl/tracking.html* (cited in Moschakis *et al.*, 2006).

CROCKER, J. C., VALENTINE, M. T., WEEKS, E. R., GISLER, T., KAPLAN, P. D. *et al.* (2000), Two-point microrheology of inhomogeneous soft materials, *Physical Review Letters*, **85**, 888–91.

DASGUPTA, B. R., TEE, S-Y., CROCKER, J. C., FRISKEN, B. J. and WEITZ, D. A. (2001), Microrheology of polyethylene oxide using diffusing wave spectroscopy and single scattering, *Physical Review E*, **65**, 051505.

DHABI, L., ALEXANDER, M., TRAPPE, V., DHONT, J. K. G. and SCHURTENBERGER, P. (2010), Rheology and structural arrest of casein suspensions, *Journal of Colloid and Interface Science*, **342**, 564–70.

DZUY, N. Q. and BOGER, D. V. (1985), Direct yield stress measurement with the vane method, *J. Rheol.*, **29**, 335–47.

GARDEL, M. L., VALENTINE, M. T. and WEITZ, D. A. (2005), Microrheology, in *Microscale Diagnostic Techniques*, K. Breuer (ed.), New York, Springer-Verlag, 1–54.

GENOVESE, D. B. and RAO, M. A. (2005), Components of vane yield stress of structured food dispersions, *J. Food Sci.*, **70**(8), E498–504.

GOSAL, W. J., CLARK, A. H. and ROSS-MURPHY, S. B. (2004), Fibrillar β-lactoglobulin gels. Part II: Dynamic mechanical characterization of heat-set systems, *Biomacromolecules*, **5**(6), 2420–9.

KAVANAGH, G. M., CLARK, A. H. and ROSS-MURPHY, S. B. (2000), Heat-induced gelation of globular proteins. Part IV: Gelation kinetics of low pH β-Lactoglobulin gels, *Langmuir*, **16**, 9584–94.

LOVEDAY, S. M., SU, J., RAO, M. A., ANEMA, S. and SINGH, H. (2011), Effect of calcium on the morphology and functionality of whey protein nanofibrils, *Biomacromolecules*, **12**, 3780–8.

LOVEDAY, S. M., RAO, M. A. and SINGH, H. (2012), Nanoscale food protein particles: their dispersions and gels, in, B. BHANDARI and Y. ROOS (eds), *Food Materials Science and Engineering*, Oxford, Blackwell Publishing.

MACKINTOSH, F. C. (1998), Theoretical models of viscoelasticity of actin solutions and the actin cortex, *Biol. Bull.*, **19**(40), 351–3.

MASON, T. G. (2000), Estimating the viscoelastic moduli of complex fluids using the generalized Stokes-Einstein equation, *Rheol. Acta*, **39**, 371–8.

MASON, T. G. and WEITZ, D. A. (1995), Optical measurements of frequency-dependent linear viscoelastic moduli of complex fluids, *Physical Review Letters*, **74**(7), 1250–3.

MASON, T. G., GANESAN, K., VAN ZANTEN, J. H., WIRTZ, D. and KUO, S. C. (1997), Particle tracking microrheology of complex fluids, *Physical Review Letters*, **79**(17), 3282–5.

MOSCHAKIS, T., MURRAY B. S. and DICKINSON, E. (2006), Particle tracking using confocal microscopy to probe the microrheology in a phase-separating emulsion containing non-adsorbing polysaccharide, *Langmuir*, **22**, 4710–19.

MOSCHAKIS, T., MURRAY B. S. and DICKINSON, E. (2010), On the kinetics of acid sodium caseinate gelation using particle tracking to probe the microrheology, *J. Colloid and Interface Science*, **345**(2), 278–85.

PINE, D. J., WEITZ, D. A., CHAIKIN, P. M. and HERBOLZHEIMER, E. (1988), Diffusing-wave spectroscopy, *Physical Review Letters*, **60**(2), 1134–7.

QIU, C-G. and RAO, M. A. (1989). Effect of dispersed phase on the slip coefficient of apple sauce in a concentric cylinder viscometer, *J. Texture Studies*, **20**, 57–70.

RAO, M. A. (2007), *Rheology of Fluid and Semisolid Foods: Principles and Applications*, 2nd edition, New York, Springer, 483.

STORM, C., PASTORE, J. J., MACKINTOSH, F. C., LUBENSKY, T. C. and JANMEY, P. A. (2005), Nonlinear elasticity in biological gels, *Nature*, **435**, 191–4.

TRUESDELL, C. (1974), The meaning of viscometry in fluid dynamics, *Annual Review of Fluid Mechanics*, **6**, 111–46.

WEITZ, D. A. and PINE, D. J. (1992), *Dynamic Light Scattering*, W. Brown (ed.), Oxford, Oxford University Press.

XU, J., TSENG, Y., CARRIERE, C. J. and WIRTZ, D. (2002), Microheterogeneity and micro-rheology of wheat gliadin suspensions studied by multiple particle tracking, *Biomacromolecules*, **3**(1), 92–9.

XU, J., CHANG, T., INGLETT, G. E., KIM, S., TSENG, Y. and WIRTZ, D. (2007), Micro-heterogeneity and micro-rheological properties of high-viscosity oat β-glucan solutions, *Food Chemistry*, **103**, 1192–8. $G'(\omega) \sim k_B T / ((\Delta r_p^2) a) \eta = (k_B T / (6\pi D a))$

11

Tribology measurement and analysis: applications to food microstructures

T. B. Mills and I. T. Norton, University of Birmingham, UK

DOI: 10.1533/9780857098894.2.292

Abstract: This chapter details the development and use of tribology equipment as a method to explore food microstructures under thin film conditions. First, it discusses the emergence and importance of tribological methods, and their use alongside more established techniques. Current findings on various studies on food structure are then outlined. This is followed by an explanation of microstructural influences on tribological behaviour. The chapter then reviews current approaches and findings using the developed methods, before concluding with future trends and applications.

Key words: soft tribology, food tribology, food friction.

11.1   Introduction

There are a variety of techniques available to probe the microstructure of food products. Each of these techniques can tell us something about how food structures are made up, or how they will perform when consumed. This in turn can allow better understanding of mechanisms involved in microstructure formation and performance, leading to improved formulations in the future. Tribology, the study of the lubrication behaviour of food structures is discussed here. Whilst the field of tribology is well established and vast, here, consideration is only given to its use within the realm of food microstructures. The study of the lubrication and friction behaviour of food products is a relatively new pursuit, but has been shown to differentiate between food formulations under assumed simulated oral conditions (Malone and Appelqvist *et al.*, 2003). This field, studying phenomena akin to the mouth coating and lubrication of the oral surfaces by food structures, provides information not sufficiently covered by other complementary techniques such as rheology and microscopy.

This chapter will cover the basis of tribology, the importance of this field in food microstructure research, the techniques available and the findings of various studies on food structures. Further to this, the applications and future trends are speculated on, based on the direction of current research.

11.2   Background tribology

Tribology research is a wide field, which has been studied for many years. The primary application for work in this area is mechanical engineering and similar fields such as the study of combustion engine components and wheel traction (Barwell, 1974; Nakada, 1994). These applications are mainly concerned with lubrication and friction effects on surfaces in machinery susceptible to wear. Understanding and minimising friction effects can allow an increase in efficiency and lifetime of equipment. Several reviews have dealt with developments of tribology and novel lubricants, from bearings and large-scale industry to modern zero gravity and small-scale components such as magnetic discs (Archbutt and Deeley, 1900; Bartz, 1978; Roberts, 1986; Nakada, 1994; Spikes, 2001).

Friction is a force that is present in most everyday situations, whenever there is movement of one surface against another. Tribology deals with friction, lubrication and wear effects in these systems. Friction force is defined as

$$F = \mu W \hspace{4cm} [11.1]$$

where F is friction force (N), μ is the friction coefficient (or traction coefficient) and W is the normal force applied perpendicular to the direction of friction (N). Friction coefficient is often used to represent the characteristic friction of a system for comparison.

Friction for any system is dependent on a number of factors, which could include temperature, surface roughness, relative speeds, normal forces and presence of lubricant. To categorise lubricants, the friction coefficient across a range of relative speeds is often assessed, since the extent of lubrication present alters the experienced friction. The resulting data is presented as a Stribeck curve, which consists of the traction coefficient at low speeds, moving to high average speeds between surfaces at a constant normal force (Stribeck, 1902; Czichos, 1978). A sample schematic is presented in Fig. 11.1.

A lubricant's effect will differ depending on the extent of entrainment (the volume present) between the two surfaces. This gives rise to three main regimes in the Stribeck curve across the speed range studied. Initially at low speeds the lubricant is not entrained, or only in very small volumes, and as such does not have enough force to move the surfaces apart. As such boundary lubrication is present, where both surfaces contact causing high friction. With increasing speed, fluid is forced between the surfaces beginning to separate them and causing partial contact from larger asperities, but reducing friction overall; this is termed the mixed regime. This behaviour continues with increasing speed, until surfaces become completely separated, entering the hydrodynamic regime (elasto-hydrodynamic regime when soft surfaces are investigated). At this point, friction between the surfaces is determined by the drag caused by the fluid, which is influenced by its viscosity and structure.

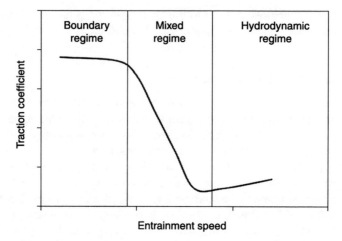

Fig. 11.1   Characteristic Stribeck curve displaying the different lubrication regimes with increasing speed; from left to right: the boundary, mixed and hydrodynamic regime.

11.2.1   Importance of friction and lubrication in food science technology

The performance and perception of food products in the mouth is a complex mix of factors, coupling the physical phenomena occurring with our interpretation of the information provided by the senses. The ability to predict performance would allow better tailored formulation of future products, reducing the need for sensory trials. As such, a number of techniques are used to gather information about food microstructure, which are the subject of other chapters. Tribology specifically provides information on thin film friction behaviour. This area of processing is associated with spreading and squeeze flow of material between the tongue and the oral surfaces during assessment and processing. The information that can be obtained differs from the related rheology of food microstructures in that individual components can greatly affect behaviour, rather than the overall bulk properties dictating performance. Previous studies have had some success in relating sensory attributes of liquid products such as thickness to rheology, although the shear rate used varies greatly between studies from $10\,s^{-1}$ in the work by Cutler *et al.* (1983) to up to $1000\,s^{-1}$ in the studies of Shama and Sherman (1973). This variation, as well as limited range of sensory attributes, highlights the need for more complementery techniques to be applied.

11.3   Techniques for measuring tribological parameters

There are a relatively small number of techniques currently being used in the field of food research to study tribological properties of structures. These range from simple configurations sliding one surface over another at a fixed speed, through rheometer geometries to custom tribology equipment, allowing a full range of parameters such as surfaces, speeds, and geometries to be controlled.

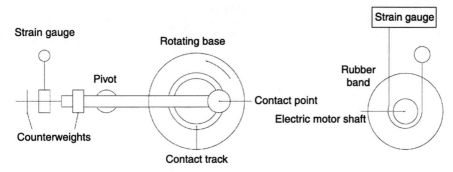

Fig. 11.2   Friction apparatus from Fort (1962) and Prinz and Lucas (2000).

The first class of equipment is seen in some of the earliest food-based work by Kokini *et al.* (1977), who in turn used methodology from earlier work concerning adsorption and boundary friction on polymer surfaces (Fort, 1962). The method consists of a weighted platform attached to a strain gauge for recording friction forces; this platform rests on a rotating turntable, which provides the moving surface for frictional studies (Fig. 11.2). Prinz and Lucas (2000) also used a simple friction system developed from earlier work by Halling (1975), The friction equipment consists of a rubber band held against the rotating cylinder of an electric motor; friction is recorded by a load cell attached to one end of the rubber band (Fig. 11.2). These configurations allowed significant correlations to be carried out for their respective systems and are discussed later; however, the range of behaviour that can be explored is relatively limited, given a single moving surface in both cases.

Dresselhuis *et al.* (2008a,b,c) developed a tribology rig (Fig. 11.3), termed the optical tribological configuration (OTC). Here the configuration consisted of a lower plate glass surface that oscillates over a 16 mm length against a stationary upper probe to which a second surface is attached, in this case a pig's tongue section, or rubber surface. The use of a glass lower plate allows for the use of a confocal laser scanning microscope (CLSM) to observe the sample while undergoing testing, although this configuration does limit the range of contact conditions that can be studied to sliding behaviour. The main focus of this work was to highlight the fact that the selection of surfaces that best represent oral surfaces is important in order to draw the best correlations, although the authors do point out that the use of actual oral tissues does present a number of problems in terms of availability and variability. Clear differences in emulsion behaviour under shear were observed between use of a pig's tongue and PDMS (polydimethylsiloxane), a common soft surface used in food tribology applications, suggesting further development of the PDMS surfaces is needed to produce a representative model oral surface.

For some time, rheology of food systems has been used to characterise and predict material behaviour. With the relatively recent interest in tribological

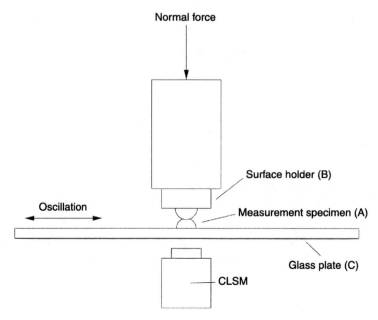

Fig. 11.3 Optical tribological configuration (OTC): A, sample; B, upper contact; C, lower glass plate (Dresselhuis *et al.*, 2008).

properties, there have been attempts to extend rheology equipment to measure friction. Anton Paar has produced the first system of interest: the equipment consists of a ball loaded onto three flat plates arranged in the form of an inverted pyramid. There is very little published work currently available for this configuration and thus it is only possible to comment on the original development data. The system has been used to look at the lubrication of lubrication greases, motor oils and dairy fat-based food samples. The main advantage claimed for this configuration is the precision in the control of the speed and force associated with the rheometer configuration. The principle of operation allows only for the study of sliding friction and limits the available materials that can be used for the upper surface to metal spheres. An issue highlighted with the configuration is that, due to the small volume, the entire cell content is subject to a range of shear, which could influence the recorded forces (Heyer and Läuger, 2008, 2009). Thermo Scientific have produced similar equipment, although in this case the bottom plate remains fixed, whilst the ball shaft is free to move. This design limits the materials that can be used for the lower plate, since they must be machined to a specific design, and potentially suffers once again from the same issues as the Anton Paar system (Fig. 11.4). No specific publications have been found for food-related studies using this system.

In addition to these commercial systems, two in-house designs have been developed and used. Kavehpour and McKinley (2004) have described the first of

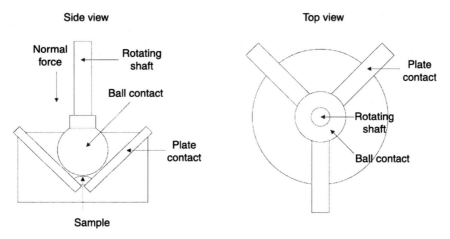

Fig. 11.4   Diagram of the Anton Paar tribology equipment (Heyer and Läuger, 2009).

these designs. Relatively less modification of the traditional rheology geometry has been applied in this case; the lower plate remains unchanged, allowing for a range of materials to be placed on it, whilst the upper plate consists of a steel disk. An illustration of the effect of different surface roughness materials is presented, along with data for commercial lubricants. Once again no complex or food structures are considered in this study, although they are suggested as a future direction.

Goh *et al.* (2010) have developed the last and most recent rheology attachment; once again the lower plate is relatively unchanged, allowing for attachment of any flat material. The upper contact probe consists of two balanced hemispheres equidistant from the central rheometer shaft, giving two contact points for measurement. This study specifically considers the tribological behaviour of food systems and presents preliminary data for corn syrup that is compared with previously published work obtained using full-scale tribology equipment (Fig. 11.5). Differentiation between samples was very good and comparable with similar work obtained on more sophisticated equipment (De Vicente *et al.* (2005a)). The work did not attempt to include sensory information and, whilst similar behaviour was seen as in previous work, data was only presented for simple model food systems.

The final class of equipment is, inevitably, the most expensive, but it does offer exploration of a greater range of variables. Several authors have reported use of a similar design, the Mini Traction Machine (PCS Instruments) being most common (Fig. 11.6). This equipment consists of independently driven ball and disk specimens, with the ball contacting the disk at an angle of 45 degrees. A load beam controls the normal force acting through the ball, perpendicular to the disk at the point of contact. A force transducer located on the ball arm records the friction force. This configuration allows nearly any surface to be substituted for

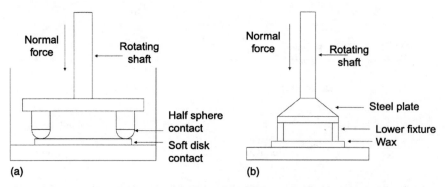

Fig. 11.5 Diagrams of: (a) rheology equipment developed by Goh *et al.* (2010); and (b) rheology equipment developed by Kavehpour and McKinley (2004).

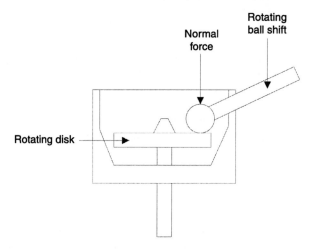

Fig. 11.6 Tribology equipment developed by PCS Intruments (London): the Mini Traction Machine (MTM).

the ball or the disk, as well as allowing a whole range of sliding and rolling friction behaviour to be explored. A number of studies have been reported using this equipment, which are described in more detail in the following section.

11.4   Microstructural influences on tribological behaviour

In the studies carried out on the equipment described previously, a number of different materials have been investigated ranging from single-phase systems through to mixed and particulate systems. As tribology focuses on thin film effects, microstructure not only plays a part through bulk material properties, but also via individual components of the microstructure (e.g. particulates or droplets). The push to understand the physical mechanism for material lubrication is something that is becoming more widespread.

In initial studies by Kokini *et al.* (1977), the authors tried to predict three key attributes for model food products by correlating sensory evaluation with physical measurements of viscosity and of friction. Sixteen liquids were tested in this study, eleven of which were formulations of cereal hydrolysate (Maltrin 05, Grain Processing Co.), guar gum (Stein Hall Co.) and carboxymethylcellulose (CMC7HOXF, Hercules, Inc.), with the remaining five being commercial syrups and honey. Samples were assessed for a number of sensory attributes with 'thick', 'smooth' and 'slippery' chosen for more detailed analysis, since these allowed prediction of the remaining attributes. The mouth was assumed to be represented by two parallel plates with a variable gap and relative movement between the surfaces, and with this arrangement the attributes 'smooth' and 'slippery' were assumed to be proportional with boundary friction force and with the sum of the viscous force and friction force, respectively. With these assumptions, viscosity and friction measurements were carried out and correlated with sensory scores. 'Smooth' ratings gave a correlation coefficient of −0.94 and 'slippery' a correlation coefficient of −0.88 (Kokini *et al.*, 1977). While these correlations are strong for these materials, they are relatively simple and homogeneous systems.

This work was extended further to include correlations with actual food products, predominantly dairy-based, in an effort to address more complex structures. Once again the same methodology was used to assess and compare the three key sensory attributes to friction data. A friction attachment on an Instron texture analyser, which consists of a pulley and sledge system to measure friction, was used to assess tribological parameters. Similar success was reported in that correlations with smoothness and friction existed; however, for the majority of the samples tested, the ranges of smoothness were relatively small, producing a cluster rather than a clear linear relationship: thus the correlation coefficient quoted at 0.82 relies heavily on two outlying samples to form a linear progression (Kokini and Cussler, 1983).

The first instance of the use of the MTM equipment for food applications reported by Malone *et al.* (2003) compared sensory attributes for a small range of guar gum and emulsion samples with tribology measurements. Initially, for different guar concentrations, a rating of oral slipperiness was correlated with the friction coefficient across a range of relative surface speeds; a significant correlation in the mixed lubrication regime was present, suggesting this is the most relevant set of conditions to relate to behaviour in-mouth. Following this, a correlation with Stribeck curves for emulsions at increasing oil contents and perceived fattiness was carried out; here a more complex situation existed, but those samples with similar friction behaviour were rated similarly in the mouth.

De Vicente *et al.* (2005a) also used the MTM to study lubrication for food systems. Initially they used the system with one soft surface, an elastomer disk, to study the full range of lubrication behaviour of a range of samples, including

Newtonian corn syrup, simple hydrocolloids, emulsions and Carbopol suspensions. The authors normalised the data for all systems by multiplying speed with viscosity to produce a Newtonian master curve, highlighting that the two hydrocolloid systems tested behave as Newtonian materials in tribology when normalised at high shear viscosities. More details of the systems studied by this technique are detailed in subsequent articles by the authors (De Vicente *et al.*, 2005b,c,d).

Building on previous work, De Wijk and Prinz (2005) made basic friction measurements on a range of oil-containing samples (including custards and mayonnaises) to compare with sensory evaluation. The key result from this work showed that lower fat samples were rated rougher and less creamy than the higher fat content samples, although only concentrations up to 20% oil are presented. In addition, they reported the effect of droplet size in constant fat content samples, as well as the effect of the inclusion of different sized silica particles. Once again a pattern was observed in that with increasing droplet and particle size an increase in measured friction, and feelings of roughness were seen. However, the range of droplet sizes was relatively narrow, lying between 2 and 6 microns. The authors later attempted to increase the range of fat contents studied (0–72%), as well as to test the validity of the findings for a wider range of food samples including sauces and desserts. Fat content was found to relate well to perceived fattiness, and correlations with friction measurements for fattiness and creaminess were reasonable, with creaminess giving the weaker correlation of the two attributes (De Wijk and Prinz, 2007).

Bongaerts *et al.* (2007a) looked at changing the contact surfaces and modifying surface roughness and hydrophobicity, and its effects on lubrication. This study builds on earlier work by De Vincente (2005b), using the same samples for comparison with the present study. The use of a soft PDMS ball and a range of defined roughness disks were compared. The main finding was that increasing surface roughness increased the transition point between the mixed and hydrodynamic regime. The influence of surface hydrophobicity was also noted, with aqueous samples entraining more readily on hydrophilic surfaces, leading to much lower friction coefficients in the boundary and mixed regime. Ranc *et al.* (2006) further explored the effect of surface characteristics by creating silicone surfaces with similar topography to the tongue surface in the sub-millimetre range. In lubricated environments, surfaces with designed topographies performed very differently to smooth surfaces, with high pillar densities corresponding to large friction coefficients. However, this study showed that the effects are not simply a function of surface roughness, but due to the contact angle and spreading behaviour that is altered by the topography.

Prinz and Lucas (2000) used simple tribology equipment to investigate how sensations of astringency in the mouth are affected as a result of a reduction in saliva's ability to lubricate when reacting with tannins in foods. Friction measurements of saliva and saliva tannin mixes were taken at a fixed speed, and the addition of tannin was found to greatly increase friction as well as reduce viscosity. Bongaerts *et al.* (2007b) carried out a further study on saliva using the

MTM system. In general, in tribology studies on food systems, the use of saliva in the system is intermittent, with most authors agreeing that saliva itself can play an important role in lubrication; however, its complexity and variability make it difficult to include in all studies.

The study by Bongaerts *et al.* (2007b) specifically focused on the lubricating properties of human saliva. The initial findings showed that fresh saliva is an excellent boundary lubricant and that it forms a film on the contact surfaces reducing friction and that this film tends to be consistent, even for aged and centrifuged saliva, although greatly affecting bulk properties. The authors also highlighted the fact that whilst reliable results were gathered using the two subjects in the study, variability in the saliva behaviour was present between subjects and a much larger sample group would be needed. Vardhanabhuti *et al.* (2011) used the MTM system to look at saliva lubrication and the links with astringency following on from work by Prinz and Lucas (2000), which looked at saliva and tannin interactions on astringency with simpler equipment. The lubrication of saliva films between PDMS surfaces was recorded and, when β-lactoglobulin at pH 3.5 was added to the contact region, a rapid increase in friction was observed in line with the previously reported astringency. However, at pH 7, where no astringency is associated, the change in friction was smaller and changed more slowly, indicating that astringency is strongly linked to friction.

In previous studies by Rossetti *et al.* (2009), the authors tried to directly measure the effect of astringent compounds on lubrication, by applying them to a pre-absorbed layer of saliva on the measurement surfaces. A large increase in friction was observed when these compounds were introduced, indicating that the saliva layer was being compromised. The study also included sensory data for comparison, although it concluded that the lubrication measurements are not sufficient to predict astringency, and that a more complex mix of lubrication, mechanosensation and chemosensation may be responsible for the actual perception of astringency.

Increasingly studies are seeking to explain the mechanisms underlying the observed differences in lubrication behaviour. Garrec and Norton (2011) looked at simple hydrocolloid samples in an effort to link lubrication properties with hydrocolloid chain structure and length. These authors showed that the conformation of the sample in solution affected its ability to lubricate in the boundary and mixed regimes, with random coil solutions hindering friction far more than rigid-rod and extended coil systems. At this scale, it is hypothesised that the random coil samples are excluded from the measurement gap, since they cannot orientate to fit into the gap, and thus differences, even at the molecular scale, can affect lubrication behaviour. This led to an additional study of the mechanism of boundary lubrication by sodium salts, with different anions in water and guar gum systems (Garrec and Norton, 2012a). For simple salt water systems, boundary lubrication was affected by the type of anion with chaotropic anions having relatively little effect, whereas kosmotropic anions led to a reduction in friction. In this case, the lubrication is considered to be the result of hydrated ions binding to the surfaces, widening the distance between them and reducing friction. Similar behaviour was observed in guar gum tested systems. However, in this case an alternative mechanism is proposed; here the salts are considered to induce varying degrees of guar gum deposition, which provide efficient lubrication.

Stokes *et al.* (2011) also studied hydrocolloid systems. A number of polysaccharides were tested for their tribological behaviour in soft contacts, taking into consideration fluid rheology and absorbed film properties. Of the polysaccharides studied, differences in hydrodynamic lubrication were explained purely by viscosity effects, whereas for mixed and boundary lubrication, the absorbed film dictated behaviour. For smooth surfaces, a dependency on total absorbed mass as well as the absorbed film storage modulus was observed for some speeds. For the rough surfaces studied, the total amount of polymer absorbed was found to be the dominating factor. Pectin was noted to be the most effective lubrication, due to its maintenance of a high viscosity at high shear, allowing full hydrodynamic lubrication at lower speeds, as well as providing a large absorbed layer lowering friction in boundary and mixed environments.

Zinoviadou *et al.* (2008) carried out further studies on polysaccharides by investigating the tribological behaviour of cross-linked starch and locust bean gum with similar rheological profiles. Despite this similarity, starch was shown to lubricate more efficiently. In addition, the effect of saliva was considered, and starch was shown to maintain its lubrication properties, despite the saliva reducing its bulk viscosity. It is hypothesised that the starch granules are responsible for effective lubrication and that the addition of saliva, while affecting bulk properties, still leaves sufficient starch particles to form a lubricating layer.

Gabriele *et al.* (2010) studied the lubrication properties of fluid gels. These systems are highly concentrated particulate gel systems, which have potential (and current) uses in food and consumer products, where control of rheological and tribological properties is desirable without the need for excessive additives. The article by Garrec and Norton (2012b) provides the most recent explanation of the formation and subsequent behaviour of such particulate gels, through exploring their rheological properties as a function of production technique. The authors showed that a fluid gel system is created by using shear in order to limit the ordering process to relatively small discrete particles, in line with previous studies (Norton *et al.*, 1999).

Further exploration of the formation process, by controlling exit temperature and therefore the ratio of ordering under shear against quiescent ordering, revealed that the rheological properties of the samples could be manipulated using this method. The lubrication behaviour of a fluid gel system across the different regimes has been shown to be complex, when compared to a single-phase system; at slow entrainment speeds, since particle exclusion occurs, the continuous medium of the sample dominates control of lubrication (Fig. 11.7). With increasing speed, the particles begin to be entrained, creating a monolayer of particles that increases friction. At a critical speed, particle entrainment is unrestricted, allowing full lubrication of the contact region by the bulk material. This critical value is suggested to be dependent on the elasticity of the gel particles and on the normal force applied in the study. Further study of gel particles of overall smaller

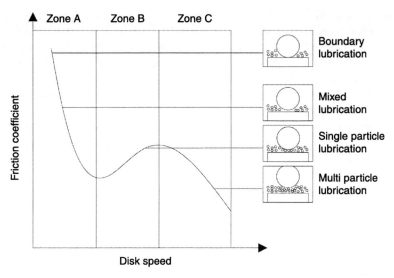

Fig. 11.7   Diagram of the proposed fluid gel lubrication mechanism.

dimensions was also subsequently shown to further reduce friction (Gabriele *et al.*, 2010).

Current work by the authors of this chapter is exploring the lubrication behaviour of fluid gel systems as a function of solvent quality, including the effects of both salt and alcohol content. Initial results for salt water systems agree with previous work by Garrec and Norton (2012a), whereby friction is decreased in the boundary regime by the addition of kosmotropic anions, but increasing concentrations of these anions has little effect. For systems containing alcohols, a trend of decreasing friction with increasing alcohol chain length was observed, primarily in the boundary regime. With the addition of agar to form fluid gel systems, the lubrication behaviour was affected greatly by both changes in solvent. For ethanol systems, the lubrication pattern for fluid gels previously established is present; however, a shift in the second maximum friction is observed, together with an increase in friction across the boundary and mixed regimes with increasing ethanol concentration. This is thought to be primarily a result of the change in particle sizes produced in the presence of ethanol during ordering, creating wider less well-defined particles with increasing ethanol concentration. The same effect is observed for systems containing salts with much greater sized particles created with increasing concentrations of salts due to the increased speed of ordering. This greatly delayed the onset of particle entrainment, with large differences observed between samples in the boundary regime.

Work has been carried out on oil in water (O/W) emulsions, which aims to investigate links between emulsion instability and coalescence behaviour and lubrication and sensory aspects associated with fats and oils. Relatively similar emulsions, with the exception of their expected stability to coalescence (controlled primarily by emulsifier), were shown to differ in their sensory and tribological behaviour. A tendency towards increased coalescence increased the perception of fattiness and creaminess in conjunction with improved lubrication in tribology measurements. The hypothesis proposed to explain the phenomenon is that either a layer of coalesced fat forms in the less stable emulsions, creating a film that lowers friction, or that larger droplets are formed, increasing hydrostatic pressure which separates the oral surfaces decreasing friction.

Further work by the authors explored the tendency of the emulsions to adhere and spread on the oral surfaces *in-vivo* and *ex-vivo*, by imaging the droplets with confocal laser scanning microscopy. The authors concluded that droplets first adhere, then spread on the oral surfaces forming a film of oil lowering friction, and that unstable droplets do this more readily than stable ones (Dresselhuis *et al.*, 2008a,b,c). Chojnicka *et al.* (2008) have studied the lubrication behaviour of dispersions of protein aggregate, whey protein isolate and fibrillar aggregates of ovalbumin from egg white, using the MTM system. The lubrication behaviour was then compared with properties such as the concentration, aggregate size and shape. Aggregate size and shape were shown to affect friction, with higher concentrations and larger sizes providing better lubrication. These differences could not be explained simply by bulk rheology alone and thus the aggregates themselves could be determining lubrication.

The group extended this work to look at the lubrication behaviour of emulsions, gels (broken up through shearing in a syringe) and emulsion filled gels (Chojnicka *et al.*, 2009). For the emulsion systems, similar results to previous studies were observed; the systems had lower friction coefficients than their continuous phases, with emulsions containing 40% oil overlapping the behaviour for pure oil. It is suggested that lubrication is determined by the ratio of the viscosities of the two phases: if the oil is more than four times the viscosity of the continuous phase, the oil will determine the lubrication properties. For the gel systems studied, whey protein isolate (WPI) and gelatine had similar patterns of behaviour, but with gelatine generating much lower friction coefficients.

Boundary regime behaviour was observed until approximately a speed of 25 mm/s was achieved, and beyond this a gradual decrease in friction was observed until the maximum speed was reached. For carrageenan, mixed lubrication was present over the whole speed range studied, giving decreasing friction coefficients with increasing speed. The difference in behaviour is attributed to the structure of the gel: for example, WPI releases water when broken, which would be expected to impact on the friction recorded over the other gel systems. In addition, both WPI and gelatine were observed to create inhomogeneous gel particles when broken, and this indicates that it is uncertain if all or only some particles affect its lubrication properties.

The final stage of testing introduced emulsions into the gel formulations and a decrease in friction was observed when these systems were compared to the pure gel systems, although any affects of varying oil content were masked by the gel lubrication behaviour. This was shown in part to be an effect of changing the gel properties and subsequently the creation of broken gel samples. Throughout the study, Stribeck curves were used to compare samples, and good repeatability was shown along with the results of some CLSM images before and after processing in the tribometer, as evidence that no structural change was caused by processing. Because of the high shear environment in the measurement contact region and the relatively low volumes that would pass through it, it is likely that further breakdown of the gel would occur and so the lubrication behaviour could change during the measurement, and its behaviour is likely to be a result of a small volume of sheared gel rather than the bulk gel present.

11.5   Conclusion and future trends

The main application for the information obtained from the discussed methods is the ability to predict performance of food microstructures and to build new microstructures to perform a desired function.

Increasingly it is being noted that physical measurements can be related to actual in-mouth processes, but that it is not a simple relationship. In the case of tribology, it is generally agreed that lubrication and friction are key phenomena in oral processing, whereas in the past only rheology was considered. In reality a combination of physical properties are needed to understand material behaviour at the various stages of oral processing, and this needs to be coupled with physiological aspects in order to predict actual sensory perception. Further studies will need to include more interdisciplinary efforts, coupling lubrication measurements as close to the oral environment as is feasible, sensory testing and other physical characterisation, such as rheology and volatile release. An approach that tries to understand how microstructure and individual components affect lubrication will become more common for future work, since it has been shown in a number of previous studies that differences at this level can affect behaviour.

Early studies using tribology focused on gathering absolute values of friction for well-defined samples. These studies did manage to draw correlations with sensory attributes but the method was not perfect. Further development of tribology equipment to look at more time-dependant measurements, where materials can change as they are measured, needs more focus, as this behaviour is similar to mixing, saliva interactions and structure breakdown in the mouth. This approach has been recognised in some studies (Dresselhuis et al., 2008a,b,c), where emulsion coalescence is studied as a factor for perception. Although friction measurements are carried out at fixed speeds over time, for samples which are expected to coalesce, friction values are still presented and compared as single values rather than changes with time; this is partly due to only small differences being present, and partly that the study focused on perception and emulsion droplet sizes in vivo.

Current work by the authors of this chapter is attempting to extend the measurement capability of the commonly-used MTM system. Time-dependent tribology could be used to study material structuring and breakdown, such as the melting of gel suspensions or fats and its association with creamy and mouth coating experiences. This approach is also considered to differentiate material lubrication behaviour based on mixing ability. Since lubrication is a measure of thin film behaviour, in the mouth when an inhomogeneous mixture of material is assessed, it is not always a good representation of this environment to compare it to a pure mixed sample in measurement. It is possible under these conditions that one section of material has a preference to be entrained and assessed, which can bias any assessments made both in the mouth and by tribology, leading to artificially high/low frictions.

As the systems under study increase in complexity, so do their lubrication mechanisms: this is something that has been highlighted in this chapter. As such it is increasingly important to understand the influence of testing parameters on the behaviour of materials. It is often assumed that normal load has little influence on the pattern of entrainment and lubrication; for the studies in the scope of this chapter, consideration to normal force is mostly given with a view to gaining a desired contact pressure that is applicable to the mouth or actual *in-vivo* conditions and therefore is often fixed. Recently, some studies have included the dependency of friction coefficient on the normal load, and for the model food systems often studied, a dependency can be observed (Gabriele *et al.*, 2010).

These authors studied the effect of normal force on the complex entrainment behaviour of fluid gel systems (defined previously). With increasing normal force, the speed required to start entraining the particles also increased, although the magnitude of the friction coefficient was largely unaffected (Fig. 11.8). Garrec and Norton (2011) also considered the effect of the normal force for a series of hydrocolloid systems: a non-linear dependency on normal load was present from values greater than 0.7 N, which varied for different hydrocolloids due to their individual deposition behaviour. It is necessary for the effect of normal force on entrainment to be studied for future work in this area, since it can influence lubrication behaviour and not just the overall magnitude of the friction force.

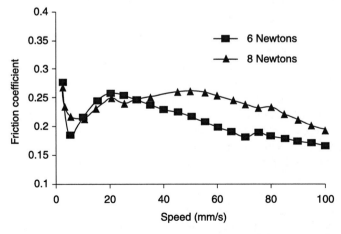

Fig. 11.8   Agarose fluid gel lubrication properties at different normal forces.

Currently, a number of groups work in the field of food tribology and their work provides a source of further information in this area. First, the authors of this chapter and the associated group at the University of Birmingham, headed by Professor Ian Norton, research this topic and a publication list can be obtained from Professor Norton's profile page on the University's website. A wider view of oral processing and *in-vitro* measurements, including tribology, was recently presented in Spyropoulos *et al.* (2011). Second, Professor Jason Stokes, currently at the University of Queensland, Australia, has published a series of articles regarding food tribology, and is currently active in this area. A recent short review with Dr Jianshe Chen, University of Leeds, highlights some tribology work and their perceived role of tribology within the oral processing framework (Chen and Stokes, 2011). Dr Chen is active across the field of food performance and processing, and more details can be obtained for both academics at their respective institution websites. A number of publications and current research has also come from the Top Institute Food and Nutrition, formerly Wageningen Centre for Food Sciences (WCFS), and the latest paper in the area examines lubrication and deposition behavior of polysaccharide solutions (Stokes *et al.*, 2011).

11.6   References

ARCHBUTT, L. and R. M. DEELEY (1900), *Lubrication and lubricants, a treatise and practice of lubrication*, London, C. Griffin & Co., Ltd.

BARTZ, W. J. (1978), Tribology, lubricants and lubrication engineering – a review, *Wear*, 49(1), 1–18.

BARWELL, F. T. (1974), The tribology of wheel on rail, *Tribology*, 7(4), 146–50.

BONGAERTS, J. H. H., FOURTOUNI, K. and STOKES, J. R. (2007a), Soft-tribology: Lubrication in a compliant pdms-pdms contact, *Tribology International*, 40(10–12), 1531–12.

BONGAERTS, J. H. H., ROSSETTI, D. and STOKES, J. R. (2007b), The lubricating properties of human whole saliva, *Tribology Letters*, 27(3), 277–87.

CHEN, J. and STOKES, J. R. (2012), Rheology and tribology: Two distinctive regimes of food texture sensation, *Trends in Food Science & Technology*, 25(1), 4–12.

CHOJNICKA, A., DE JONG, S., DE KRUIF, C. G. and VISSCHERS, R. W. (2008), Lubrication properties of protein aggregate dispersions in a soft contact, *Journal of Agricultural and Food Chemistry*, 56(4), 1274–82.

CHOJNICKA, A., SALA, G., DE KRUIF, C. G. and VAN DE VELDE, F. (2009), The interactions between oil droplets and gel matrix affect the lubrication properties of sheared emulsion-filled gels, *Food Hydrocolloids*, 23(3), 1038–46.

CUTLER, A. N., MORRIS, E. R. and TAYLOR, L. J. (1983), Oral perception of viscosity in fluid foods and model systems, *Journal of Texture Studies*, 14(4), 377–95.

CZICHOS, H. (1978), *Tribology: A systems approach to the science and technology of friction, lubrication, and wear*, Amsterdam, Elsevier.

DE VICENTE, J., STOKES, J. R. and SPIKES, H. A. (2005a), Behaviour of complex fluids between highly deformable surfaces: Isoviscous elastohydrodynamic lubrication, *Marie Curie Fellowship Association Annals*, 4.

DE VICENTE, J., STOKES, J. R. and SPIKES, H. A. (2005b), The frictional properties of newtonian fluids in rolling–sliding soft-ehl contact, *Tribology Letters*, 20(3), 273–86.

DE VICENTE, J., STOKES, J. R. and SPIKES, H. A. (2005c), Lubrication properties of non-adsorbing polymer solutions in soft elastohydrodynamic (EHD) contacts, *Tribology International*, 38(5), 515–26.

DE VICENTE, J., STOKES, J. R. and SPIKES, H. A. (2005d), Soft lubrication of model hydrocolloids, *Food Hydrocolloids*, **20**(4), 483–91.

DE WIJK, R. A. and PRINZ, J. F. (2005), The role of friction in perceived oral texture, *Food Quality and Preference*, **16**(2), 121–9.

DE WIJK, R. A. and PRINZ, J. F. (2007), Fatty versus creamy sensations for custard desserts, white sauces, and mayonnaises, *Food Quality and Preference*, **18**(4), 641–50.

DRESSELHUIS, D. M., DE HOOG, E. H. A., COHEN STUART, M. A. and VAN AKEN, G. A. (2008a), Application of oral tissue in tribological measurements in an emulsion perception context, *Food Hydrocolloids*, **22**(2), 323–35.

DRESSELHUIS, D. M., DE HOOG, E. H. A., COHEN STUART, M. A., VINGERHOEDS, M. H. and VAN AKEN, G. A. (2008b), The occurrence of in-mouth coalescence of emulsion droplets in relation to perception of fat, *Food Hydrocolloid*, **22**(6), 1170–83.

DRESSELHUIS, D. M., STUART, M. A. C., VAN AKEN, G. A., SCHIPPER, R. G. and DE HOOG, E. H. A. (2008c), Fat retention at the tongue and the role of saliva: Adhesion and spreading of 'protein-poor' versus 'protein-rich' emulsions, *Journal of Colloid and Interface Science*, **321**(1), 21–9.

FORT, T. (1962), Adsorption and boundary friction on polymer surfaces, *The Journal of Physical Chemistry*, **66**(6), 1136–43.

GABRIELE, A., SPYROPOULOS, F. and NORTON, I. T. (2010), A conceptual model for fluid gel lubrication, *Soft Matter*, **6**(17), 4205–13.

GARREC, D. A. and NORTON, I. T. (2011), The influence of hydrocolloid hydrodynamics on lubrication, *Food Hydrocolloids*, **26**(2), 389–97.

GARREC, D. A. and NORTON, I. T. (2012a), Boundary lubrication by sodium salts: A Hofmeister series effect, *Journal of Colloid and Interface Science*, **379**(1), 33–40.

GARREC, D. A. and NORTON, I. T. (2012b), Understanding fluid gel formulation and properties, *Journal of Food Engineering*, **112**(3), 175–82.

GOH, S. M., VERSLUIS, P., APPELQVIST, I. A. M. and BIALEK, L. (2010), Tribological measurements of foods using a rheometer, *Food Research International*, **43**(1), 183–6.

HALLING, J. (1975), *Introduction to Tribology*, New York, Springer-Verlag.

HEYER, P. and LÄUGER, J. (2008), A flexible platform for tribological measurements on a rheometer, *AIP Conference Proceedings*, **1027**(1), 1168–70.

HEYER, P. and LÄUGER, J. (2009), Correlation between friction and flow of lubricating greases in a new tribometer device, *Lubrication Science*, **21**(7), 253–68.

KAVEHPOUR, H. P. and MCKINLEY, G. H. (2004), Tribo-rheometry: From gap-dependent rheology to tribology, *Tribology Letters*, **17**(2), 327–35.

KOKINI, J. L., KADANE, J. B. and CUSSLER, E. L. (1977), Liquid texture perceived in the mouth, *Journal of Texture Studies*, **8**(2), 195–218.

KOKINI, J. L. and CUSSLER, E. L. (1983), Predicting the texture of liquid and melting semi-solid foods, *Journal of Food Science*, **48**(4), 1221–5.

MALONE, M. E., APPELQVIST, I. A. M. and NORTON, L. T. (2003), Oral behaviour of food hydrocolloids and emulsions. Part I: Lubrication and deposition considerations, *Food Hydrocolloids*, **17**(6), 763–73.

NAKADA, M. (1994), Trends in engine technology and tribology, *Tribology International*, **27**(1), 3–8.

NORTON, I. T., JARVIS, D. A. and FOSTER, T. J. (1999), A molecular model for the formation and properties of fluid gels, *International Journal of Biological Macromolecules*, **26**(4), 255–61.

PRINZ, J. F. and LUCAS, P. W. (2000), Saliva tannin interactions, *Journal of Oral Rehabilitation*, **27**(11), 991–4.

RANC, H., SERVAIS, C., CHAUVY, P. F., DEBAUD, S. and MISCHLER, S. (2006), Effect of surface structure on frictional behaviour of a tongue/palate tribological system, *Tribology International*, **39**(12), 1518–26.

ROBERTS, W. H. (1986), Some current trends in tribology in the UK and Europe, *Tribology International*, **19**(6), 295–311.

ROSSETTI, D., BONGAERTS, J. H. H., WANTLING, E., STOKES, J. R. and WILLIAMSON, A. M. (2009), Astringency of tea catechins: More than an oral lubrication tactile percept., *Food Hydrocolloids*, **23**(7), 1984–92.

SHAMA, F. and SHERMAN, P. (1973), Identification of stimuli controlling the sensory evaluation of viscosity. Part II: Oral methods, *Journal of Texture Studies*, **4**(1), 111–18.

SPIKES, H. (2001), Tribology research in the twenty-first century, *Tribology International*, **34**(12), 789–99.

SPYROPOULOS, F., HEUER, E. A. K., MILLS, T. B. and BAKALIS. S. (2011), Protein-stabilised emulsions and rheological aspects of structure and mouthfeel, in I. T Norton, F. Spyropoulos and P. Cox (eds), *Practical Food Rheology*, Oxford, UK, Wiley-Blackwell, 193–218.

STOKES, J. R., MACAKOVA, L., CHOJNICKA-PASZUN, A., DE KRUIF, C. G. and DE JONGH, H. H. J. (2011), Lubrication, adsorption, and rheology of aqueous polysaccharide solutions, *Langmuir*, **27**(7), 3474–84.

STRIBECK, R. (1902), Die wesentlichen eigenschaften der gleit- und rollenlager. *Zeitschrift des Vereins Deutscher Ingenieure*, **36**, 1341–8 1432–8, 1463–70.

VARDHANABHUTI, B., COX, P. W., NORTON, I. T. and FOEGEDING, E. A. (2011), Lubricating properties of human whole saliva as affected by α-lactoglobulin, *Food Hydrocolloids*, **25**(6), 1499–506.

ZINOVIADOU, K., JANSSEN, A. and DE JONGH, H. (2008), Tribological properties of neutral polysaccharide solutions under simulated oral conditions, *Journal of Food Science*, **73**(2), E88–94.

12

Methods for modelling food cellular structures and the relationship between microstructure and mechanical and rheological properties

S. J. Cox, Aberystwyth University, UK

DOI: 10.1533/9780857098894.2.310

Abstract: Foams are found in many foodstuffs, for example to make a product larger, to insulate it, or to change its taste and feel without any increase in weight. Understanding the local structure of a foam and the ways in which it evolves in time offers possibilities to control these functions. This chapter describes the mesoscopic structure of foam at the level of films, Plateau borders and bubbles, and relates it to the many useful properties that a foam exhibits. Relevant dynamical properties include changes in bubble size due to inter-bubble gas diffusion (coarsening), film rupture, the re-distribution of liquid through the foam under gravity, and the effect of disorder and liquid content on a foam's solid and liquid-like rheology.

Key words: foam, diffusion-driven coarsening, gravity-driven drainage, film rupture, foam rheology.

12.1 Introduction

Liquid foams are familiar materials: they are used in the home in a variety of cleaning applications, and have many uses in ore-separation and enhanced oil recovery (Bikerman, 1953; Weaire and Hutzler, 1999; Cantat *et al.*, 2010). They are also widely used as food products, such as forming the top of a cappuccino or a pint of beer, or a chocolate mousse. Moreover, many of the solid foams that are found in foods, such as bread, cakes and meringues, are formed from a liquid foam and then solidified, perhaps by freezing or baking; so the structure of a liquid foam is an important 'ingredient' of the solidified foam structure. Finally, concentrated emulsions such as mayonnaise, and even rather more complicated foods such as ice cream, have a structure and/or dynamical properties very similar to foams; thus, although in the following we will concentrate on liquid foams, the science of foams has much wider applicability.

Foams are typically two-phase materials, usually a gas (the discrete phase) included within a liquid matrix (the continuous phase). Air is included in foods for a number of reasons, in particular because it alters mouth-feel and appearance at very low cost, but also because it can keep foods warm (thermal insulation, as in a cappuccino). It is possible to foam many foodstuffs, often after pureeing them, but the danger is that the foam will collapse before it reaches the consumer.

Our goal here is to describe the structure and dynamics of a foam, to quantify how parameters such as bubble volume and liquid content affect the feel, stability and lifetime of these widely-used materials.

12.2   Foam structure

A liquid foam can be produced by first creating a solution which contains some stabilising molecules, or surfactants. The latter could be a detergent, such as sodium dodecyl sulphate (SDS), in washing applications, or a protein, such as egg-white (albumin) or wheat (as in beer), in food applications. Then gas, usually air, must be added in any of a number of ways: beating, injecting, etc. (Weaire and Hutzler, 1999).

12.2.1   Liquid fraction

Perhaps the most significant control parameter for a foam is its liquid content, or liquid fraction, ϕ_l, defined as the ratio of liquid volume to foam volume. Due to the effects of gravity, this generally decreases over time as liquid drains through the foam and collects beneath it. This drainage process is described in more detail below. A foam in which very little liquid remains (ϕ_l up to ~0.01) is termed 'dry', otherwise it is 'wet'. The wet limit is the liquid fraction at which the bubbles become spherical and lose contact with each other (close to $\phi_l = 0.36$); then the foam becomes a bubbly liquid, and a different set of rules come into play.

12.2.2   Geometry

The liquid structure is, to a good approximation, governed by surface tension: each film tries to minimise its surface energy, which is equal to the product of its surface tension γ and its area. Over short time-scales, in which the surfactants do not move significantly, the surface tension is constant, and thus a foam is a network of thin films that minimises its surface area, subject to the constraint that the volume of the gas enclosed in each cell, or bubble, is constant. Again, the latter constraint is only an approximation, valid at short times, as will be explained below. Note that each film actually consists of two gas–liquid interfaces, so that the film tension γ is equal to twice the interface tension, the quantity that is more usually measured (e.g. with the pendant drop method).

The Belgian scientist Plateau (1873) realised that the films in a foam do not meet at random. Instead, there are particular rules for how films join together, and it was later shown by mathematicians that these rules are direct consequences of area minimisation (Taylor, 1976; Morgan, 2008). Plateau's rules are only applicable when a foam has reached mechanical equilibrium and when it is dry. They state that:

1. films only meet in threes, at equal angles of 120 degrees, along lines known as Plateau borders; and
2. these lines only meet in fours, at the tetrahedral angle, at vertices (Fig. 12.1).

The local geometry of a foam is therefore well defined, and in practice it is often apparent that these rules are obeyed for liquid fractions of up to a few percent, although there are exceptions (Brakke, 2005; Barrett *et al.*, 2008).

In the dry limit, the Plateau borders are lines, but in a wet foam they are liquid-carrying channels, carrying a high proportion of the liquid. They have a curved triangular cross-section (Fig. 12.2a), and so the swollen vertices in a wet foam have a highly non-trivial shape. Predicting the flow of liquid in the vertices remains a challenge, and a significant one, since the flow through the network of

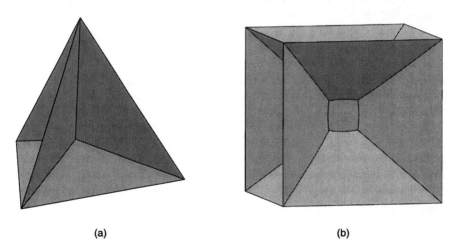

(a)   (b)

Fig. 12.1 Illustrating Plateau's laws with soap films in wire frames. These simulated images can be reproduced with a bent coat-hanger and a bowl of soapy water (Isenberg, 1992): (a) in a tetrahedral frame, there are 6 films, 4 Plateau borders where the films meet at 120 degrees, and 1 central vertex where 4 Plateau borders meet at the tetrahedral angle; (b) in a cubic frame, there are 13 films: the cubic symmetry is broken because the minimum film area is obtained when Plateau borders meet 4-fold; that is, instead of one 8-fold vertex in the centre of the frame, there are four 4-fold vertices surrounding a rectangular face, which can be found in three different orientations.

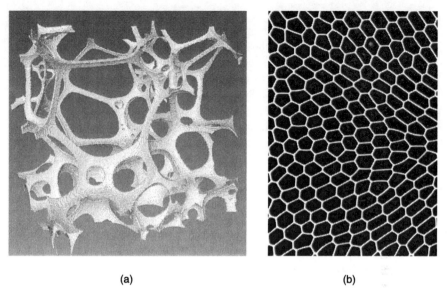

(a)   (b)

Fig. 12.2   (a) Reconstructed image from tomography of a 3D foam (Lambert *et al.*, 2005). The films are not shown. Under gravity, the size of the Plateau borders is larger at the bottom of the foam; (b) photograph of a dry 2D foam between parallel glass plates 1.5 mm apart (Ran *et al.*, 2011). The bubble areas are around 0.1 cm^2.

Plateau borders plays an important role in the lifetime of a foam, as we shall see below.

To Plateau's laws we must add the Laplace–Young law, which relates the curvature of a film to the difference in pressure, p, between the bubbles on either side of it. To be more precise, the pressure difference, Δp, across a film is equal to the product of surface tension and the mean curvature, κ, of the film:

$$\gamma\kappa = \Delta p \qquad\qquad\qquad\qquad\qquad [12.1]$$

Viewed another way, the law tells us that because pressure is constant in each bubble, the curvature of each film must be constant. Thus a single bubble of volume $4/3\pi$ has unit radius and its mean curvature (defined as the sum of the reciprocals of the two principal radii of curvature) is $\kappa = 2$, giving a pressure 2γ greater than atmospheric pressure. If the bubble is larger, then κ decreases and the pressure goes down.

These geometric rules can more easily be illustrated with a 2D foam, such as can be made by trapping a foam between two parallel glass plates (Fig. 12.2b). Provided the separation is small enough, each bubble in the foam touches both plates and therefore appears as a polygonal prism with curved sides. The Laplace–Young law tells us that each film is an arc of a circle, and Plateau tells us that they meet in threes at 120 degrees. Two-dimensional foams are now widely used in the physics community, to demonstrate and to understand the laws governing both the statics and dynamics of foams.

12.2.3   Disorder

A great deal of progress has been made by studying ordered foams, most notably by Princen (2000), who studied the 2D dry foam consisting of hexagons, and then the wet version of this, with Plateau borders at each vertex. In 3D, microfluidic devices now allow the creation of ordered (crystalline) wet foams (van der Net *et al.*, 2006; Höhler *et al.*, 2008). However, foams are disordered in most applications, and it is important not only to describe the disorder, but also to be aware of the effects that it has on foam properties.

There are two standard measures of disorder: the topological disorder is the variation in the number of faces, *F*, of a bubble, while the geometric disorder is the variation in the volumes, V_b, of the bubbles. These can be expressed as standard deviations or, more usually, second moments. Thus:

$$\text{Topological disorder: } \mu_2(F) = <(F - <F>)^2> = <F^2> - <F>^2 \qquad [12.2]$$

$$\text{Geometrical disorder: } \mu_2(V_b) = <(V_b - <V_b>)^2> = <V_b^2> - <V_b>^2 \qquad [12.3]$$

where $<\cdot>$ denotes an average over all bubbles in the foam. In addition, both of these measures have a natural 2D interpretation, in terms of number of sides and areas, respectively. It has recently been shown that, under certain conditions, these two measures are correlated in 2D (Durand *et al.*, 2011).

12.2.4   Simulating static foam structure

A dry foam shares its topology with the well-known Voronoi tesselation of space: so-called seed points, which will represent the centres of the bubbles, are scattered in space and then the Voronoi cell surrounding that seed-point consists of all of the space that is closer to this seed-point than any other. In this way, close-packed polyhedra are created. To turn this into a foam requires that the faces and edges of the cells are allowed to curve, so as to satisfy Plateau's laws and the Young–Laplace law.

To perform the latter step, it is possible, at least in 2D, to iteratively solve for the curvatures of the edges and the positions of the vertices where they meet. This was accomplished in the late 1980s (Kermode and Weaire, 1990) and later extended to wet 2D foams (Bolton and Weaire, 1992) by calculating, in addition, the position and curvature of the liquid–gas interfaces with tension ½γ that form the sides of the Plateau borders (Fig. 12.3).

However, these codes have largely been replaced by the Surface Evolver (Brakke, 1992), which works on the basis of minimising the energy of a triangulated (meshed) surface. This is free software, that requires as input the topology of the foam (e.g. from a Voronoi construction); it works in both 2- and 3D and can be used to determine properties of the static structure of both dry and wet disordered foams. The user supplies target bubble volumes and boundary conditions (e.g. fixed walls or periodicity), decides on a level of mesh refinement that balances the desired accuracy with available computational time, and iterates ('evolves') to a minimum of surface area. It is also possible to consider forces other than those due to surface tension, such as gravity.

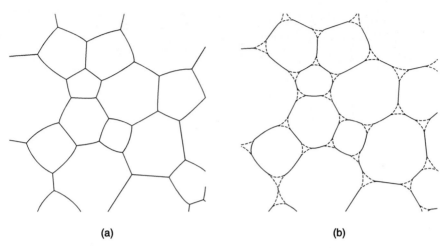

(a)   (b)

Fig. 12.3   Disordered 2D foam, simulated in the Surface Evolver: (a) dry, with all interfaces having tension γ; (b) wet, with a liquid fraction of 0.03. Here the interfaces (dashed lines) which surround the Plateau borders have tension $\frac{1}{2}\gamma$. Note that the whole foam has dilated due to the addition of liquid.

12.3   Dynamic properties of foams

12.3.1   Coarsening and Ostwald ripening

In the absence of film rupture (breakage/bursting), the volume of the bubbles in a foam changes only very slowly. This change is due to the diffusion of gas through the liquid, from one bubble to another, and becomes significant over time-scales of minutes to hours (e.g. in beer foams). The gas moves from high pressure bubbles to those with low pressure, and the net effect is that some disappear while others grow, and so the average bubble size increases (Fig. 12.4). Hence the process is known as coarsening.

In dry foams, the gas molecules move through the films. In wet foams, the liquid acts as a reservoir of gas, and molecules diffuse in and out of the liquid, in a process known as Ostwald ripening. Clearly in real foams, which are likely to have a gradient of liquid fraction, both processes are at work, although the time-scales of each may be different. In general terms, ripening slows down as liquid fraction increases, so this effect can be reduced by keeping the liquid fraction high (see the section on drainage below).

A remarkable result for coarsening in dry 2D foams is due to von Neumann (1952): the rate of change of area of a bubble does not depend upon the bubble's area, A_b, but is proportional to the bubble's number of sides, n:

$$\frac{\mathrm{d}A_b}{\mathrm{d}t} = \gamma\zeta\frac{\pi}{3}(n-6) \qquad\qquad [12.4]$$

Thus bubble pressure is determined by n. Here ζ is the permeability of the gas in the liquid; since this varies for different gases, it allows the rate of coarsening to

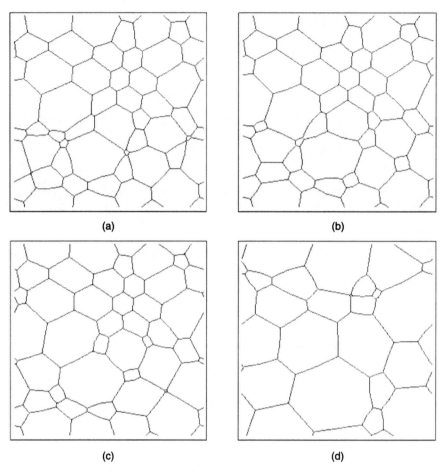

Fig. 12.4 The coarsening of a 2D foam, simulated with the Surface Evolver. Images are shown at time-steps, arbitary units: (a) 400; (b) 450; (c) 500; and (d) 1000. Bubbles with fewer than six sides shrink and bubbles with more than six sides grow. Note the region of mostly hexagonal bubbles in the upper left quadrant, which only start to evolve once nearby bubbles disappear and disrupt the hexagonal order.

be controlled. Clearly the effect of disorder on coarsening is significant, and the coarsening process can also be slowed by reducing the disorder of the foam.

In 3D there is no formula relating volume change to number of faces. A scaling argument shows that the average bubble diameter increases with the square-root of time, $d_b \sim t^{\frac{1}{2}}$, but to investigate the fate of individual bubbles it is necessary to return to the fact that, because the pressure difference is proportional to the mean curvature, the rate of change of bubble volume depends on the integral of κ over the surface of the bubble. Formulae have been derived to express this in different forms (MacPherson and Srolovitz, 2007), but not in a way that is any more tractable.

A simulation of coarsening would proceed from a static initial configuration, calculating the area of each face and its curvature (or the pressure difference across it), and then adjusting the target volume for each bubble accordingly. For such dynamic simulations, the accuracy of Surface Evolver is not always required; indeed, the computational time required to achieve high accuracy for a large number of bubbles is usually prohibitive, and other methods such as the Potts model, which represents bubbles on a lattice and uses Monte Carlo sampling to minimise surface area, are available (Jiang *et al.*, 1999).

Ostwald ripening is well-described by the Lifshitz–Slyozov–Wagner mean-field theory, which shows that the average bubble diameter increases with the cube-root of time, $d_b \sim t^{1/3}$ The intermediate case, of moderately wet foams, is only just starting to yield to theory, supported by Potts model simulations (Fortuna *et al.*, 2012).

12.3.2   Film rupture
When foam films thin, for example due to liquid drainage or the presence of dirt, they fail. The effect is to increase the average bubble size, just as for coarsening, but in a rather more dramatic way. Understanding the process of rupture not only means that it could be reduced, but that it is also necessary on occasion to remove an unwanted foam, and inducing rupture is by far the fastest way to do this, for example through the use of antifoams (Garrett, 1993).

Stabilisation
To generate a foam from a mixture of gas and liquid requires the addition of some material to stabilise the interfaces once formed. Most familiar in the bathroom are surfactant-stabilised products (shampoo), but the same rules apply to food foams. A sufficient quantity of surfactant, protein, or other stabilising agent is required, often measured through the critical micelle concentration (or cmc, not to be confused with constant mean curvature, which describes the geometry of the films!) which, when exceeded, means that all surfaces are covered. Its function is to lower the surface tension.

The foamability of a solution, as might be expected, is a measure of how much foam can be made with a given concentration of a given product. Tumbling tubes, or even just vigorous controlled shaking, are often used to test foam height as a function of chemical constitution. In addition to surface coverage, important factors are the extent to which the foam coarsens during the test and how quickly liquid drains out of the films and Plateau borders under gravity.

Rupture mechanisms
In the absence of a sufficient quantity of the stabiliser, the surface tension in the film remains too high and it tears itself apart. This is the reason why pure water cannot be foamed, and why a mountain stream that carries foam on its surface must be contaminated in some way. As a foam evolves, the films change their size and shape, causing the stabilising agent to move around. During this motion it is possible that the local concentration will change, giving rise to changes in surface

tension. Over time, surfactant diffusion will often reduce these differences, but if the motion is sufficiently fast, the surface tension may rise high enough to cause the film to fail.

Another mechanism by which a film may fail is the introduction of other material. The most obvious one is dirt: a small dust particle can enter the film and, depending on the contact angle at which it meets the liquid–gas interface, can cause the two sides of the film to meet, and the film fails. The same process may occur, in foods such as ice-cream, where fat or ice particles are present.

Finally, both gravity and capillary suction draw liquid from the films into the Plateau borders. If they thin sufficiently, then the disjoining pressure in the films is insufficient to keep the two sides of the film apart, and the result is rupture.

12.3.3   Liquid drainage

Clearly the force of gravity has a significant effect on a foam. Bubbles immersed in a liquid are buoyant and, conversely, the liquid descends between the bubbles. Most of the liquid in a foam resides in the network of Plateau borders and the vertices where they meet, although when the bubbles are very small there may be a significant proportion of liquid in the films (Carrier *et al.*, 2002); gravity causes the liquid to move through the foam, and this motion is resisted by the liquid viscosity and smoothed by capillary forces. The liquid fraction decreases and the films also thin, and hence become more prone to rupture, dictating foam lifetime. Theories of foam drainage have been developed to predict the rate at which liquid moves through and then leaves a foam, with the aim of predicting foam lifetime.

Drainage models

The model of foam drainage that we describe here (Weaire and Hutzler, 1999) attributes the viscous dissipation to the motion of liquid through long, narrow, Plateau borders, appropriate to a fairly dry foam with immobile interfaces. The latter condition means that the molecules stabilising the interfaces are sufficiently rigid that they do not move with the bulk liquid flow, and manifest a no-slip boundary condition; protein-stabilised foams are good examples of this (Hutzler *et al.*, 2000). The other limit, of mobile interfaces, is better described with a model that attributes dissipation to the flow in the junctions of the Plateau borders (Koehler *et al.*, 1999). The resulting difference is often explained by recourse to an experiment in which liquid is added to the top of a standing foam at a flow-rate Q_{in}; the velocity, v, at which the liquid descends through the foam is expressed as a power law, $v \sim Q_{in}^{\alpha}$, with $\alpha = \frac{1}{2}$ for immobile and $\alpha = \frac{1}{3}$ for mobile interfaces. Nonetheless, the two models are similar, and the reader is referred to Koehler *et al.* (2000) for further details.

The model assumes that the liquid flows through a static foam (no bubble motion and no deformation of the foam by the liquid motion) except that, unlike in porous media, the foam can swell to accommodate more liquid. The significant variable is then taken to be the cross-sectional area of a Plateau border $A(x,t)$ at position x and time t. For simplicity we will consider the 1D case, appropriate for a foam confined within a narrow vertical cylinder. A is a local measure of the foam's liquid fraction: $\phi_1 = cA/V_b^{2/3}$, with the constant $c \approx 5.35$ depending weakly on the foam structure.

The analysis proceeds by considering a force balance on a Plateau border, relating the weight of the liquid, ρg, to the viscous dissipation and the pressure drop:

$$\rho g = \frac{\eta f <v>}{A} + \frac{\partial}{\partial x}\left(\frac{\gamma}{r}\right)$$

[12.5]

Here η is the viscosity of the liquid, $f \approx 49$ is a constant and r is the radius of curvature of the border, which is related to the pressure through the Laplace–Young law. A further geometric calculation shows that $A = C^2 r^2$, with $C^2 = 0.161$. Combining Equation 12.5 with an equation expressing conservation of liquid leads to

$$3\eta f \frac{\partial A}{\partial t} + \frac{\partial}{\partial z}\left(\rho g A^2 - C\gamma \frac{\sqrt{A}}{2}\frac{\partial A}{\partial z}\right) = 0$$

[12.6]

where the factor of 3 arises from a consideration of the possible orientations of the Plateau borders.

This non-linear partial differential equation, often referred to as the channel-dominated drainage equation, is certainly not easy to solve. Although it is usually solved numerically, there is a small number of analytic and asymptotic solutions (Kraynik, 1983; Weaire and Hutzler, 1999; Koehler *et al.*, 2000) applicable to steady flows and to 'forced' or 'free' drainage, in which the input flow-rate is constant or zero respectively. It performs remarkably well in predicting the results of experiments (Fig. 12.5).

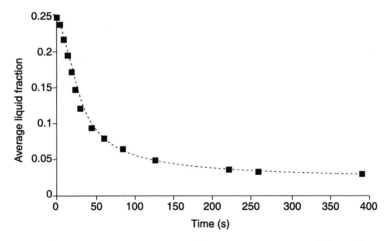

Fig. 12.5 The channel-dominated foam drainage equation successfully predicts the rate at which liquid drains under gravity. Equation 12.6 is solved numerically and compared with an experiment which tracks the amount of liquid in a detergent foam confined within a vertical cylinder (see Cox *et al.* (2000) for further details). Note in particular that the liquid content of the foam does not asymptote to zero: a small amount of liquid remains at steady state, distributed in a way predicted by an analytic solution of Equation 12.6.

12.4   Rheology

Rheology is the branch of fluid mechanics concerned with the deformation and flow of non-Newtonian, or complex, fluids. The high interfacial area of a foam, which consists only of gas, liquid and a small quantity of some stabilising molecules, confers on it remarkable rheological properties (Höhler and Cohen-Addad, 2005). Under small applied strains, a liquid foam responds as an elastic solid, with a well-defined shear modulus. However, at high strain-rate, a foam yields and then flows like a liquid. A rheologist would say that a foam is a shear-thinning, yield stress fluid. The shear thinning reflects the way in which the foam viscosity, which is very different to the viscosity of the solution from which it is made, decreases with increasing shear rate. An alternative is to call a foam a visco-elasto-plastic (VEP) fluid, recognising that all three processes play a role in its behaviour.

Before discussing ways in which foam rheology can be simulated, and the dependence of shear modulus and yield stress on material parameters, we outline the dissipative mechanisms at work in a flowing foam.

12.4.1   Dissipation

For a foam to flow, the bubbles must pass each other. They do this in what are known as T1 topological changes, where the applied deformation causes a film to shrink and disappear. This leaves a situation which is energetically unfavourable (e.g. a vertex connecting more than four Plateau borders in 2D), and so a new film is formed (Fig. 12.1b). The process is illustrated in 2D in Fig. 12.6; note that it is associated with a drop in energy (and stress) and is therefore an elementary form of dissipation.

To a first approximation, we can view a foam as being always close to equilibrium: it hops from one state to the next in a quasi-static fashion, appropriate to slow flows, with T1s being the only source of dissipation. More realistically, we should take into account the viscous flow of liquid through the Plateau border network and during the generation of new films (Khan and Armstrong, 1986; Kraynik and Hansen, 1987; Schwartz and Princen, 1987). Finally, as laid out by Buzza et al. (1995), there are contributions from the way in which the stabilising molecules move around, both within the films and between them, in response to film stretching and compression, generation of new films and gradients in concentration.

12.4.2   Simulation methods

The quasi-static rheology of dry foams can be simulated with the Surface Evolver (Brakke, 1992; Wyn et al., 2008), by assuming that the foam is always in equilibrium (a minimum of surface energy). T1s are accomplished by deleting small faces and splitting the resulting vertices. In the other limit of liquid fraction, wet foams are simulated with variants of Durian's bubble model (Durian, 1995), or soft-disk models (Langlois et al., 2008; Rognon and Gay, 2009), more familiar in the field of granular materials. In its most basic incarnation, the bubble model

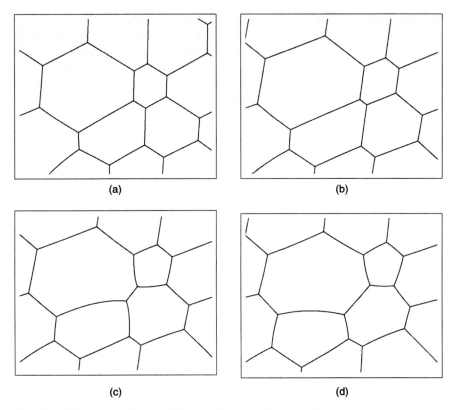

(a)   (b)

(c)   (d)

Fig. 12.6   Topological changes (T1s) in a disordered dry foam. Time increases from panels
(a) to (d) in these snapshots from a quasi-static simulation of simple shear. The small film
in the centre disappears, to be replaced with a new film almost perpendicular to it.

incorporates a repulsive force between (spherical) bubbles and a friction force
proportional to their relative motion.

The middle ground – foams of intermediate liquid fraction and/or those with
dissipative flow in the films – is less well described, mainly because of the
difficulty in resolving the different length-scales (films on one hand and the
Plateau border network on the other). A notable recent addition to the literature is
the work of Cantat (2010), which augments a dry 2D foam simulation using the
vertex model (in which films are assumed straight) with a surface tension that
varies with film stretching rate. Another is the Lattice–Boltzmann method (Benzi
et al., 2009), which may offer the ability to investigate the coupling between the
flow in the films and in the Plateau border network.

12.4.3   Shear modulus and yield stress

The shear modulus G measures the stiffness of a material, measured from a rest
(zero stress) state by determining, in the case of a foam, how the surface area

increases with deformation. Princen's prediction (Princen, 2000) for the dry 2D hexagonal honeycomb is $G=0.93\gamma/\sqrt{A_b}$, which shows that increasing the surface tension and/or reducing the bubble size will increase the stiffness. The other major contributor to the value of G is the liquid fraction, since G must tend to zero as ϕ_l approaches the wet limit, where the bubbles come apart; experiments suggest that in 3D this dependence is (Princen, 2000):

$$G \approx \frac{\gamma}{R_{32}} (1-\phi_l)\frac{1}{3} (0.36-\phi_l) \tag{12.7}$$

where R_{32} is the Sauter mean radius of the bubbles.

Disorder also has an effect, as 2D calculations (Cox and Whittick, 2006) (Fig. 12.7) indicate that G can change by 20% as the bubble-size distribution becomes wider.

The yield stress, τ_y, is the applied stress at which a foam goes from solid to liquid-like behaviour as the strain or strain-rate is increased, and is therefore a measure of the dynamic response of a foam. It too depends strongly on the liquid fraction, since in the wet limit it must reduce to zero, as well as bubble size and surface tension. Phenomenological expressions of the form:

$$\tau_y \approx \frac{\gamma}{R_{32}} f(\phi_l) \tag{12.8}$$

are often used (Princen, 2000), where f is a function that goes to zero at $\phi_l \approx 0.36$. It is also true that disorder has an effect on τ_y, but this has yet to be clearly understood.

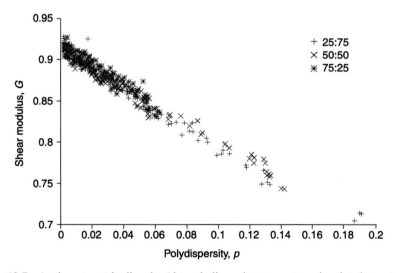

Fig. 12.7 As the geometric disorder (the polydispersity parameter p is related to $\mu_2(A_b)$) increases, simulations of bidisperse, dry, 2D, foams show that the shear modulus decreases from the honeycomb value. The legend denotes different proportions of large to small bubbles.

12.5   Conclusion

Although this survey of food foams is necessarily brief (the material can, and does, fill a 300-page book by Cantat *et al.* (2010), now translated into English), we hope that it gives a flavour of the ways in which foam structure and properties interact. Combining knowledge of mathematics, physics and chemistry allows foam properties to be controlled and, in many cases, tailored to applications.

Here, we have taken the view that it is the mesoscopic structure – films, Plateau borders, bubbles – that gives foams their special and useful properties. A chemist would probably take a more microscopic view but, particularly in the dry limit, the ability to predict the dynamics of drainage and coarsening provides a practical understanding of foam lifetime and clues as to how to control it.

Not that there is room for complacency: little consensus exists on how to predict the properties of foams in between the wet and dry limits, and in the mesoscopic picture described here, there is still no clear understanding of rupture. There are also few simulations that combine processes such as drainage and rheology, so that the interactions between them can be understood.

Although they are widely used, there is still much to understand about foams.

12.6   References

BARRETT, D. T., KELLY, S., DOLAN, M. J., DRENCKHAN, W., WEAIRE, D. and HUTZLER, S. (2008), Taking Plateau into microgravity, *Microgravity Sci. Tech.*, **20**, 17–22.

BENZI, R., CHIBBARO, S. and SUCCI, S. (2009), Mesoscopic Lattice–Boltzmann modeling of flowing soft systems, *Phys. Rev. Lett.*, **102**, 026002.

BIKERMAN, J. J. (1953), *Foams: Theory and Industrial Applications*, New York, Reinhold.

BOLTON, F. and WEAIRE, D. (1992), The effects of Plateau borders in the two-dimensional soap froth. Part II: General simulation and analysis of rigidity loss transition, *Phil. Mag. B*, **65**, 473–87.

BRAKKE, K. (1992), The Surface Evolver, *Exp. Math.*, **1**, 141–65.

BRAKKE, K. (2005), Instability of the wet cube cone soap film, *Coll. Surf. A*, **263**, 4–10.

BUZZA, D. M. A., LU, C.-Y. and CATES, M. E. (1995), Linear shear rheology of incompressible foams, *J. Phys. II France*, **5**, 37–52.

CANTAT, I. (2010), Gibbs elasticity effect in foam shear flows: a non quasi-static 2D numerical simulation, *Soft Matter*, **7**, 448–55.

CANTAT, I., COHEN-ADDAD, S., ELIAS, F., GRANER, F., HÖHLER, R. et al. (2010), *Les mousses – structure et dynamique*, Paris, Belin.

CARRIER, V., DESTOUESSE, S. and COLIN, A. (2002), Foam drainage: a film contribution, *Phys. Rev. E*, **65**, 061404.

COX, S. J. and WHITTICK, E. L. (2006), Shear modulus of two-dimensional foams: The effect of area dispersity and disorder, *Eur. Phys. J. E*, **21**, 49–56.

COX, S. J., WEAIRE, D., HUTZLER, S., MURPHY, J., PHELAN, R. and VERBIST, G. (2000), Applications and generalizations of the Foam Drainage Equation, *Proc. R. Soc. Lond. A*, **456**, 2441–64.

DURAND, M., KÄFER, J., QUILLIET, C., COX, S., TALEBI, S. A. and GRANER, F. (2011), Statistical mechanics of two-dimensional shuffled foams: Prediction of the correlation between geometry and topology, *Phys. Rev. Lett.*, **107**, 168304.

DURIAN, D. J. (1995), Foam mechanics at the bubble scale, *Phys. Rev. Lett.*, **75**, 4780–3.

FORTUNA, I., THOMAS, G. L., DE-ALMEIDA, R. M. C. and GRANER, F. (2012), Growth laws and self-similar growth regimes of coarsening two-dimensional foams: Transition from dry to wet limits, *Phys. Rev. Lett.*, **108**, 248301.

GARRETT, P. R. (1993), *Defoaming: Theory and Industrial Applications, Surfactant Science*, New York, Marcel Dekker.

HÖHLER, R. and COHEN-ADDAD, S. (2005), Rheology of liquid foam, *J. Phys.: Condens. Matter*, **17**, R1041–69.

HÖHLER, R., SANG, Y. Y. C., LORENCEAU, E. and COHEN-ADDAD, S. (2008), Osmotic pressure and structures of monodisperse ordered foam, *Langmuir*, **24**, 418–25.

HUTZLER, S., COX, S. J., WEAIRE, D. and WILDE, P. J. (2000), New developments in foam drainage, in P. Zitha, J. Banhart and G. Verbist (eds), *Foams, Emulsions and their Applications*, Bremen, MIT-Verlag, 5–12.

ISENBERG, C. (1992), *The Science of Soap Films and Soap Bubbles*, New York, Dover.

JIANG, Y., SWART, P. J., SAXENA, A., ASIPAUSKAS, M. and GLAZIER, J. A. (1999), Hysteresis and avalanches in two-dimensional foam rheology simulations, *Phys. Rev. E*, **59**, 5819–32.

KERMODE, J. P. and WEAIRE, D. (1990), 2D-FROTH: a program for the investigation of 2-dimensional froths, *Comp. Phys. Commun.*, **60**, 75.

KHAN, S. A. and ARMSTRONG, R. C. (1986), Rheology of foams. Part I: Theory for dry foams, *J. Non-Newt. Fl. Mech.*, **22**, 1–22.

KOEHLER, S. A., HILGENFELDT, S. and STONE, H. A. (1999), Liquid flow through aqueous foams: The node-dominated foam drainage equation, *Phys. Rev. Lett.*, **82**, 4232–5.

KOEHLER, S. A., HILGENFELDT, S. and STONE, H. A. (2000), A generalized view of foam drainage: Experiment and theory, *Langmuir*, **16**, 6327–41.

KRAYNIK, A. M. (1983), *Foam Drainage*, Sandia Report (SAND83-0844).

KRAYNIK, A. M. and HANSEN, M. G. (1987), Foam rheology: a model of viscous phenomena, *J. Rheol.*, 31, 175–205.

LAMBERT, J., CANTAT, I., DELANNAY, R., RENAULT, A., GRANER, F. *et al.* (2005), Extraction of relevant physical parameters from 3D images of foams obtained by X-ray tomography, *Coll. Surf. A*, **263**, 295–302.

LANGLOIS, V. J., HUTZLER, S. and WEAIRE, D. (2008), Rheological properties of the soft-disk model of two-dimensional foams, *Phys. Rev. E*, **78**, 021401.

MACPHERSON, R. D. and SROLOVITZ, D. J. (2007), The von Neumann relation generalized to coarsening of three-dimensional microstructures, *Nature*, **446**, 1053–5.

MORGAN, F. (2008), *Geometric Measure Theory: A Beginner's Guide*, 4th edition, San Diego, Academic Press.

PLATEAU, J. A. F. (1873), Statique Expérimentale et Théorique des Liquides Soumis aux Seules Forces Moléculaires, Paris, Gauthier-Villars.

PRINCEN, H. M. (2000), The structure, mechanics and rheology of concentrated emulsions and fluid foams, in J. Sjöblom (ed.), *Encyclopedic Handbook of Emulsion Technology*, New York, Marcel Dekker, 243–78.

RAN, L., JONES, S. A., EMBLEY, B., TONG, M., GARRETT, P. *et al.* (2011), Characterisation, modification & mathematical modelling of sudsing, *Coll. Surf. A*, **382**, 50–7.

ROGNON, P. and GAY, C. (2009), Soft dynamics simulation. Part 2: Elastic spheres undergoing a T1 process in a viscous fluid, *Eur. Phys. J. E*, **30**, 291–301.

SCHWARTZ, L. W. and PRINCEN, H. M. (1987), A theory of extensional viscosity for flowing foams and concentrated emulsions, *J. Coll. Interf. Sci.*, **118**, 201–11.

TAYLOR, J. E. (1976), The structure of singularities in soap-bubble-like and soap-film-like minimal surfaces, *Ann. Math.*, **103**, 489–539.

VAN DER NET, A., DRENCKHAN, W., WEAIRE, D. and HUTZLER, S. (2006), The crystal structure of bubbles in the wet foam limit, *Soft Matter*, **2**, 129–34.

VON NEUMANN, J. (1952), *Discussion*, Cleveland, American Society for Metals, 108–10.

WEAIRE, D. and HUTZLER, S. (1999), *The Physics of Foams*, Oxford, Clarendon Press.

WYN, A., DAVIES, I. T. and COX, S. J. (2008), Simulations of two-dimensional foam rheology: localization in linear Couette flow and the interaction of settling discs, *Euro. Phys. J. E*, **26**, 81–9.

13

Granular and jammed food materials

G. C. Barker, Institute of Food Research, UK

DOI: 10.1533/9780857098894.2.325

Abstract: Food, in powdered or granular solids form, is very common, relatively easy to handle, easy to store and transport, shelf stable, hygienic, versatile and convenient. However, in many respects the behaviour of granular solids, during handling and processing, is difficult to understand and has features that can resemble both liquids and solids. Investigations into granular microstructures, such as those that occur in packing and jamming, have begun to reveal some underlying principles, which promise to make the mechanical and flow properties of powders and grains more predictable.

Key words: food, powder, microstructure, packing, jamming.

13.1 Introduction

Granular foods are ubiquitous; coffee and cereals are widely traded on global markets, powdered starch and crystalline salt are used in large volumes for food manufacture, small quantities of finely divided spices and ingredients are regularly added to complex food mixtures to enhance quality or performance, and millions of vending machines store coffee, milk and sugar as ready-made solids mixtures for immediate hydration. Food, in powdered or granular solids form, is relatively easy to handle, easy to store and transport, shelf stable, hygienic, versatile and convenient, so that detailed knowledge of the bulk properties has a significant commercial and technological value (Ortega-Rivas, 2009). In many respects, the ability to turn food into powder, without significant loss of nutritional value or functionality, is an element of food security; in times of crisis, powdered milk and starch, stored for extended periods in central locations, are often the dominant elements in emergency food relief operations.

Granular materials and powders have equally prominent roles in other industries, including pharmaceuticals, agrochemicals, engineering and construction but until recently many important material properties, such as mechanics and transport, have remained poorly understood. In many cases predictive models for granular flows, mixing, wetting, failures, etc., which all have significant industrial and economic impact, have been based solely on strong empiricism. The absence of theoretical understanding of the behaviour of granular materials stems from high levels of complexity, large variability and the absence of well-defined model systems; increasingly there are advanced visualisations, innovative measurements and, in particular, computer modelling that have begun to make some progress (Jaeger and Nagel, 1992; Kadanoff, 1999).

The variety of granular food materials is huge; sugar is used commercially with particle sizes that vary over three orders of magnitude (from fine icing sugar to coarse granulated sugar), and cornflakes, frozen vegetables and infant formula all have highly irregular particle shapes. These differences in particle properties are in addition to the obvious variations in chemistry, and surface chemistry, between milk, potato, vanilla or sugar in powdered forms. When the individual particles are very small, typically less than 100 microns, divided solids are usually called 'powder' as opposed to granular material, but this distinction is arbitrary (Brown and Richards, 1966). More significantly, for very small particles, van der Waals forces are strong so that solid surfaces stick together and the powder is 'cohesive' rather than free flowing. A similar effect arises for grains surrounded by small amounts of water; the liquid forms small 'bridges' between neighbouring particles and surface tension manifests as an attractive inter-particle force (Scheel *et al.*, 2008). In the absence of these mechanisms only hard, contact, inter-particle forces are usually relevant.

However, it is the ability of granular material to switch between a free flowing, liquid-like state and a static load bearing, solid-like state without apparent changes in the particle properties, which makes powders and grains so distinctive (Liu and Nagel, 2010). Predicting this transition for a particular material and a particular configuration, for example in response to shear stress or as a result of ageing, has proved elusive. Classical physicists such as Coulomb, Faraday, Janssen and Reynolds identified a few fundamental aspects of powder mechanics including dissipation, static friction and dilatancy, but these principles have not been applied systematically for solids handling applications and are obscured by the variety of materials (soil science is one possible exception).

Dissipation is a fundamental aspect of granular dynamics. Every time that two grains of powder collide (grain inertia regime), or move while they are in contact (quasi-static regime), some of the kinetic energy is lost to internal degrees of freedom. This means that, as a collection, grains slow down rapidly unless there is a constant source of kinetic energy (driving). Dissipation emphasises the role of 'packing'; when powder is poured from one container into another, the grains quickly come to rest and form a static bed or a pile. The individual powder grains are trapped in a fixed configuration or packing.

Static friction is most easily visualised as the force that resists the motion of an object down a sloping surface, but for close packed granular materials, where there are thousands of point contacts, the situation is far more complex. When solid grains are poured to make a pile, the maximum slope of the free surface, resisting the gravitational stress, originates from solid-solid friction at grain-grain and grain-floor contacts. The angle made by the free surface to the horizontal is called the angle of repose. From Coulomb onwards, scientists have failed to find a simple relationship that links the slope for which objects begin to slide with the angle of repose that forms during deposition of granular material and this remains an area of active research. The relationship between friction and repose is crucial for rational design of hoppers, chutes and other process equipment. Currently, appropriate hopper angles and profiles for discharge chutes, for particular solids materials, are found by incorporating the results from (expensive) laboratory tests into continuum mechanics models that describe stresses in particular geometries.

Friction between particles, and between particles and walls, is central to most of the observed properties of granular materials and crucial in understanding solids handling (Dziugys and Navakas, 2009). Friction between the grains and the container walls is responsible for highly non-linear variations in internal pressure as well as fluctuating wall stress. There are numerous engineering standards to ensure that containers have sufficient strength to cope with a broad range of conditions but occasionally unpredicted, eccentric or dynamic loads arise and cause catastrophic failure. Many current estimates of the wall loads are based on assumptions originally proposed by Janssen more than a century ago. Some modern analyses provide a critique of this model, but there is still insufficient understanding of the friction forces in particulate materials to provide improved, alternative inputs into most process designs.

Reynolds observed that static granular material must expand, at least momentarily, before it can flow. This phenomenon, called dilatancy, has simple geometrical origins and can be observed for a wide variety of disordered materials that are composed from close packed, non-deformable particles. Counter intuitively dilatancy causes water to flow towards stressed regions in granular material, rather than being squeezed out. Dilatancy is particularly important for sheared materials because the expansion is often localised in a set of bands, a few particles in depth, which run parallel to the shear plane. The inhomogeneous nature of the dilatant expansion can often be observed in flowing granular materials as 'shear bands' (Khidas and Jia, 2012). Powders that flow along chutes, through valves and out of bins, are subject to shear forces so that an improved appreciation of dilatancy has a significant impact on many areas of powder handling.

Dissipation, friction and dilatancy impinge on every aspect of granular materials behaviour but, because of their apparent dependence on particular materials properties and on process history, corresponding models have not been used to develop rationalised approaches for materials handling. One reason for recent improved understanding of powders and grains comes from an appreciation, driven by physics, that there are some important commonalities, as well as differences, among granular systems.

Possibly most significant in understanding the range of granular material response is the absence of thermal fluctuations. Granular materials of all kinds are athermal in the sense that changes in temperature are not significant with respect to the particulate motions – in the absence of a mechanical driving force, such as shaking or stirring, the 'granular temperature' is effectively zero. As a consequence of zero temperature, particle configurations are frozen or 'jammed' which, in turn, leads to observed variability of material properties like density or stress-strain relations. For each packing of coffee granules into a jar or for each sprinkling of spice into a complex mixture, the resulting behaviour, bulk density or mixing will depend on the precise details of the process and will not be averaged, rapidly, by microscopic fluctuations. Jammed configurations have very specialised statistical properties, driven by geometry and process history, and recent progress in understanding universal aspects of granular materials behaviour is associated with a corresponding statistical mechanical approach for understanding of jammed configurations.

13.2   Packing of granular food material

For granular material, the bulk density or the volume fraction is used to quantify the dispersed nature of the solid component. Bulk density is defined as the total mass of solids per unit volume of the divided material, whereas a volume fraction refers to the volume of solid particles per unit volume of material. Apart from their units, these may also differ when the particulates are porous or aggregated. However, even for a particular material, these descriptors are not unique or stationary. For one material, the range of particle sizes can easily affect the packing fraction; often materials with a wide range of particle sizes, polydispersity, have higher density. Mechanical excitations, such as gentle tapping, cause packed material to consolidate and so density also depends strongly on process history (Richard et al., 2005). Variable density has a significant impact on food manufacture; clearly gravimetric methods rather than volumetric methods are necessary to ensure strong process control during volume partition but, in addition, uncontrolled changes of material volume with time are detrimental to consumer appreciation; no one likes their jar to appear partially empty at purchase! In practical situations, changes in volume fraction are rarely reversible, so that properties like dispersion or flowability can be compromised by extended periods of storage or vibrations (Mohammadi and Harnby, 1997).

Prediction of packing density, based on material descriptions and process history, is difficult and it is often practical to consider only limiting values. In general, high energy processes, such as pouring or violent shaking, lead to open, low density, static configurations and hence a minimum volume fraction. In contrast, slow, low energy driving, such as tapping a container, causes the density to increase (consolidation) and after an extended period the volume fraction is considered to obtain a maximum value. Historically these limits are called the random loose packing fraction and the random close packing fraction, and for hard, monodispersed, spherical particles, the fractions are about 55% and 64%. These values can be compared with volume fractions of 74% that correspond with close packed ordered systems of spheres (Aste and Weaire, 2000). The limiting volume fractions are also affected by the inter-particle forces. Cohesive powders develop very open structures when poured or stirred and do not consolidate so easily in response to tapping. In industrial applications, the ratio (The Hausner ratio) between the maximum packing fraction, achieved after a prescribed set of taps, and the minimum fraction is sometimes used to quantify the cohesiveness of a powdered materials.

Slow densification caused by tapping is made up of many small changes in the particle positions, biased by downward movements, which become possible during the small, local volume expansions that are driven by a slow input of kinetic energy. Reorganisation of the packing is complex and includes both the independent motion of individual particles as well as collective motion of groups or clusters. The non-equilibrium statistical mechanics of this relaxation has been the subject of intense investigations. Nowak *et al.* (1998) studied the density of a packing of monodispersed spherical particles in a tall cylinder that was subjected to a series of discrete, vertical shakes or 'taps'. In response to tapping there were density fluctuations throughout the tube, but the overall trend was a slow densification of the packing with a logarithmic dependence of density on time. The slow increase of the volume fraction does not depend fundamentally on the initial value of the volume fraction, but the final steady state fraction decreases monotonically as the tapping intensity is increased. The kinetics also depend on the tapping intensity and, in particular, the rate of approach to the steady state volume fraction slows as the tapping becomes weaker. For particularly weak tapping, below a threshold strength, the packing becomes trapped in metastable configurations and progress towards a steady state is interrupted.

The threshold behaviour, which appears during the controlled shaking of granular material, leads to a novel view of the compaction process. In the presence of the threshold, the only way to reach a steady state for low intensity excitation involves a careful pre-treatment of the sample with stronger vibrations – an annealing process. Nowak *et al.* (1998) used a ramped variation of the tapping intensity with time to examine the low intensity shaking steady state. Starting with low-density material, tapping is applied with slowly increasing intensity. The material initially consolidates but, when the shaking strength exceeds the threshold strength, the steady packing fraction decreases monotonically with increasing intensity. When the ramp on the shaking intensity is reversed, the density of the material rises as the intensity decreases, but now it continues to rise when the shaking strength dips below the threshold strength. In this regime, consolidation is reversible along a continuous, monotonic curve relating the packing density and the excitation strength. The initial evolution, including an apparent compaction with increasing intensity, exists as a separate, irreversible branch of the density-shaking relationship.

The slow dynamics associated with granular compaction are in many ways reminiscent of the glassy dynamics that describe dense thermal systems. This comparison has stimulated many interpretations and analogies, involving annealing, historical effects and non-equilibrium thermodynamics for granular materials, as well as detailed investigations of granular density fluctuations and their statistical connection with density relaxation, such as fluctuation-dissipation. However, a simple relationship that predicts density changes from a few parameters remains to be found, even without the added complications that result from elongated particle shapes, inter-particle forces or interstitial fluid.

Nothing exemplifies the athermal properties of granular materials and the metastable nature of particle packing better than particle segregation. In a variety of geometries and configurations, the application of an apparently random mechanical driving force to a mixture of different grains can lead to strong separation, such as spontaneous ordering. Whereas thermodynamics predict an inevitable progression to molecular chaos, the special conditions of driven granular systems often lead to decreasing entropy and large-scale pattern formation. Most visibly segregation relates to grain size, large particles rising to the top as a result of shaking, but external forces can equally cause separation by density, by shape or by a combination of factors. In the course of processing operations, such as feeding or filling, it is impossible to shield mechanical vibrations completely so that maintaining the precise composition of a granular mixture is difficult; this inadvertent segregation phenomenon is the cause of many industrial failures and problems.

Whilst the percolation of very small particles through a stationary bed of larger particles causing a bed of 'fines' is easily understood, the driven segregation phenomena that causes rapid demixing in many processing conditions, operating for very small particle size difference and for very small driving intensity, is difficult to explain. Using a very simple Monte Carlo computer simulation model, Rosato et al. (1987) identified a statistical mechanism for vibration driven particle size segregation. In the simulations, isolated particles larger than their neighbours rose to the top of a particle bed in response to simple vertical shaking. Similarly shaken mixtures, with two distinct particle sizes, developed a clear separation, with a layer of larger particles resting on a bed of smaller ones. The simulations, although abstract, provided a clear picture of the separation phenomenon and gave detailed information about particle positions during the shaking process. The downwards motion of smaller particles was favoured by geometry and was identified as the underlying driver of 'the Brazil nut' effect. This mechanism drives separation for any shaken mixture as long as there is a disparity of particle sizes. The rate of separation increases with the size difference.

Rosato's simple simulations stimulated a large variety of experiments, including advanced imaging, and several other mechanisms leading to size separation for vibrated powders were discovered. These included internal avalanches, arching and mass convection. Although granular convection had been observed many times, particularly in fluidised systems, new non-invasive magnetic resonance imaging experiments by Knight et al. (1993) revealed the dimensions for convective flows and indicated a threshold particle size for which segregation effects arise. It became clear that there were many mechanisms that drove granular separation (Schröter et al., 2006) and that in practical situations these could coincide and produce counterintuitive results – such as the reverse Brazil nut effect!

Segregation phenomena are not restricted to vertical shaking regimes; particulate mixtures poured to form a pile on an open surface often develop stratifications with alternate layers of large and small particles. A drum-like vessel, rotating about a horizontal axis, is one of the most common configurations used for mixing and blending particulates. The particles move in complex trajectories, sometimes stationary with respect to the drum and sometimes in avalanches or ballistically down the free surface. A variety of segregation patterns can be observed in drum mixers, both axial and radial banding, and extended residence time or increased speed cannot guarantee superior mixing (Shinbrot and Muzzio, 2000).

Non-invasive observations, such as NMR imaging, show that distinct, microscopic particulate transport regimes, such as percolation, convection, avalanching or diffusion, are always in competition in driven granular flows. Banding occurs when rapid mixing, for example resulting from a convective mechanism, is restricted by static cores or non-interacting segments of material that are set up by the flow. The diffusion mechanism in driven particulate materials is very inefficient at breaking the barriers to mixing that are imposed by strong flows. In industrial configurations baffles, additional agitators or complex 'double-cone' geometries are usually added to restrict the onset of organised flows, but this process is largely empirical and may result in other drawbacks or unforeseen consequences. The rational design of powder mixing processes, and in particular the ability to scale up observations made in a benchtop blender to industrial sizes, is still a major objective for the food and other process industries. Non-invasive observations and computer simulations of driven granular materials (Laurent and Cleary, 2012) show that intermittent trapped particle configurations or packings are central to the control of mixing and to prediction based on fundamental understanding.

13.3   Jamming in granular materials

For free flowing granular material, one of the most familiar and irritating phenomenon is a sudden, unforeseen cessation of the flow. For no apparent reason, a previously smooth continuous stream of granular material stops and becomes blocked, often at a flow constriction or outlet. The flow is said to be 'jammed' and, in practice, is often returned to a free flowing state by a single extraneous force or 'poke'. Although jamming is widely regarded as a major nuisance, recent developments indicate that the jams may be fundamental and may provide a mechanism for systematic understanding of granular dynamics.

13.3.1   Cohesive arches

There are at least two types of jamming phenomena, also called bridging or arching, that occur in granular materials. First, 'cohesive' arches can occur when dense particulates, with attractive interactions, enter a convergent flow pattern such as that at the approach to the outlet of a conical or wedge-shaped hopper. The analysis of cohesive arches combines a continuum model of granular material with extensive flow property measurements and gives an established methodology, developed by Jenike, for the design of a hopper (Fitzpatrick *et al.*, 2004). The minimum opening size that prevents arching is established from the yield strength of the material and from an expression for the variation of the vertical stress with height in the particular hopper geometry – cones and wedges can be very different. While the vertical stress exceeds the material strength, cohesive arches do not interrupt the discharge. Although the Jenike methodology is an accepted industry standard, it inevitably involves time-consuming and expensive shear cell testing. Currently it is impossible to develop theoretical expressions that allow flow properties to be predicted *a priori*, from simple physical properties of granular materials such as particle size, temperature, moisture content and time of storage at rest.

13.3.2   Interlocking arches

The second kind of jamming involves 'interlocking' arches or bridges, which are formed when a group of hard particles become trapped, that is, their relative motion is impossible because of particle-particle contacts. These configurations do not rely on the details of the inter-particle interactions or other material properties and so may have some very general properties. The simplest visualisation of an interlocking arch is a chain of particles, bowed slightly upwards, that bridge over the outlet of a hopper. In this configuration, the particles in the bridge are jammed together tightly and none of them can move downwards; this structure can clearly hinder hopper discharge. In two dimensions, jamming bridges have been examined by To *et al.* (2001). These experiments showed that the statistics of the bridges, formed by disks at the outlet of a 2D hopper, has a relatively simple form (constrained, self avoiding, random walks) and can be used to predict the probability of hopper jamming events. For example, the random walk model showed that 2D hoppers, with openings that are between three and four particles wide, are most often jammed by five particle bridges, although six, seven and eight particle bridges jam the outlet with smaller frequencies.

Interlocking bridges are not restricted to hopper discharge sites and can be observed throughout packed structures. Barker and Mehta (1991) have examined the internal bridge structures in granular media, using hard particle computer simulations and Pugnaloni *et al.* (2001) have identified scaling distributions for bridge sizes; similar distributions quantify the structures in other systems including dense colloidal suspensions. Interlocking bridges reflect cooperative phenomena, particularly multi- particle stabilisations, which occur during the relaxation of close packed configurations (Fig. 13.1). In each bridge there are at least two

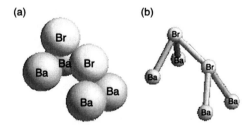

Fig. 13.1   (a) One of the simplest non-sequential structures in a packing involves two bridge particles (Br) supported by four base particles (Ba). Assuming the base particles are already in stable locations, the bridge particles can only achieve stability simultaneously. Panel (b) shows the 'contact network', the set of contacts for the bridge particles.

particles that could not have reached their stable position in a process of sequential deposition. Pouring, stirring, shaking and any other driving force for granular media cause cooperative motion of particles and so generate a granular structure full of interlocking bridges.

13.3.3   Influence on material properties

The bridge structures in granular media influence many of the material properties, including the mechanical and transport properties, but most importantly these structures highlight a weakness in the accepted understanding of close packing (Torquato and Stillinger, 2010). Most granular packings, and particularly those that define the loose and close packing limits, are assumed to be disordered or 'random' on the particle scale, but there is no formal definition of local ordering. Clearly there is always a strong interplay between disorder and packing for granular structures, small elements of order consistently increasing the overall volume fraction, so that without a corresponding measure of the amount of order the packing fraction is difficult to interpret. However, an alternative picture of close packed structures, in terms of jamming configurations, can be much clearer. Using a geometrical picture, a random close packing corresponds with a configuration that has a minimum value of an order parameter subject to a condition that the structure is jammed. Although there are many possible order parameters for granular configurations, some scalar metrics based on the orientations of the links in the contact network are possibly the simplest.

A bond orientation order parameter, Q_6, averaged over all the links in the contact network of the packing and normalised by the value for a face centred cubic lattice, has become an accepted global measure of order in granular structures. The geometrical picture of close packing also requires a more precise definition of jamming; several distinct jamming categories have been identified including locally collectively and strictly jammed structures. However, the interpretation of jamming is relatively intuitive and, given precise details of the configuration, the nature of the jamming can be evaluated algorithmically.

The geometrical picture of close packing, based on jamming and ordering, is a major step forwards, because it means that, for the first time, the relationship between two different random close packings and the theoretical limits for granular structures can be appreciated. The full range of packings, at least for frictionless convex particles, can be represented by continuous regions within a volume fraction and order parameter phase space and different process histories have corresponding trajectories. This space of configurations can be compared with the zero temperature plane of a jamming phase diagram, which is spanned by the density and the load (stress). Together these views indicate a connection between structural variations driven by packing constraints and complex rheology in athermal, amorphous systems that can drive rapid improvements in the development and control of granular materials processing.

13.4   Research and developments in the study of granular systems

In parallel with improved understanding and modelling, there has been a significant increase in modern imaging and visualisation of dense granular systems. Granular structures have been measured in 3D, using confocal light microscopy (Jenkins *et al.*, 2011), nuclear magnetic resonance, positron emission tomography and X-ray micro-tomography, etc. (Fu *et al.*, 2006). Crucially each of these techniques can provide information, such as particle coordinates, with sufficient accuracy to facilitate a reconstruction of the contact network (or fabric tensor) of the structure and can identify individual non-sequential particle stabilisations. Increasingly, these information capture methods can be extended to real granular flow regimes to provide visualisation of transitions between distinct packings.

Over a relatively short period, the behaviour of granular materials, and particularly the statistical properties of trapped hard particle systems, has become a focus for many diverse research activities from the study of complexity to glass formation; with applications that stretch from geophysics to crowd control. It is clear that several of the challenges identified for improved handling of granular materials, including advanced and hygienic food processing, predictive strategies for mixing, prevention of the ageing caused by consolidation, and controlled discharge from bulk storage, are within the reach of fundamental science. These advances will consolidate the ongoing development of new granular food materials such as colours, flavours and functional ingredients that are driven by consumer choice.

13.5   Conclusion

This short review has pursued a limited range of science that relates to food powders; powder fabrication methods such as spraying, drying and freezing fracture and functionality such as dispersion, rehydration, coating or encapsulation and release have not been examined. Similarly, we have not explored aspects of safety such as dust formation or allergenic effects, and we have not examined the chemistry of finely divided solid food materials. However, this review does indicate that science of microstructures has a role to play in the developing use of food powders. Increasingly, a detailed understanding of particle technology is leading the drive to replace empiricism by rational design in powder processing.

13.6   References

ASTE, T. and WEAIRE, D. (2000), *The Pursuit of Perfect Packing*, Bristol, UK, Institute of Physics.

BARKER, G. C. and MEHTA, A. (1991), Vibrated powders: a microscopic approach, *Phys. Rev. Lett*, **67**, 394–7.

BROWN, R. L. and RICHARDS, J. C. (1966), *Principles of Powder Mechanics*, Oxford, Pergamon Press.

DZIUGYS, A. and NAVAKAS, R. (2009), The role of friction in mixing and segregation of granular material, *Granular Matter*, **11**, 403–16.

FITZPATRICK, J. J., BARRINGER, S. A. and IQBAL, T. (2004), Flow property measurement of food powders and sensitivity of Jenike's hopper design methodology to measured values, *J. Food Eng.*, **61**, 399–405.

FU, X., DUTT, M., BENTHAM, A. C., HANCOCK, B. C., CAMERON, R. E. and ELLIOTT, J. A. (2006), Investigation of particle packing in model pharmaceutical powders using X-ray microtomography and discrete element method, *Powder Technology*, **167**, 134–40.

JAEGER, H. M. and NAGEL, S. R. (1992), Physics of the granular state, *Science*, **255**, 1523–31.

JENKINS, M. C., HAW, M. D., BARKER, G. C., POON, W. C. K. and EGELHAAF, S. U. (2011), Finding bridges in packings of colloidal spheres, *Soft Matter*, **7**, 684–90.

KADANOFF, L. P. (1999), Built upon sand: theoretical ideas inspired by granular flows, *Rev. Mod. Phys.*, **71**, 435–44.

KHIDAS, Y. and JIA, X. (2012), Probing the shear-band formation in granular media with sound waves, *Phys. Rev. E*, **85**, 051302.

KNIGHT, J. B., JAEGER, H. M. and NAGEL, S. R. (1993), Vibration-induced size separation in granular media: the convection connection, *Phys. Rev. Lett.*, **70**, 3728–31.

LAURENT, B. F. C. and CLEARY, P. W. (2012), Comparative study by PEPT and DEM for flow and mixing in a ploughshare mixer, *Powder Tech.*, **228**, 171–86.

LIU, A. J. and NAGEL, S. R. (2010), Granular and jammed materials, *Soft Matter*, **6**, 2869–70.

MOHAMMADI, M. S. and HARNBY, N. (1997), Bulk density modelling as a means of typifying the microstructure and flow characteristics of cohesive powders, *Powder Tech.*, **92**, 1–8.

NOWAK, E. R., KNIGHT, J. B., BEN-NAIM, E., JAEGER, H. M. and NAGEL, S. R. (1998), Density fluctuations in vibrated granular materials, *Phys. Rev. E.*, **57**, 1971–82.

ORTEGA-RIVAS, E. (2009), Bulk properties of food particulate materials: an appraisal of their characterisation and relevance in processing, *Food and Bioprocess Technology*, **2**, 28–44.

PUGNALONI, L. A., BARKER, G. C. and MEHTA, A. (2001), Multi-particle structures in non-sequentially reorganised hard sphere deposits, *Adv. Complex Systems*, **4**, 289–97.

RICHARD, P., NICODEMI, M., DELANNAY, R., RIBIERE, P. and BIDEAU, D. (2005), Slow relaxation and compaction of granular systems, *Nature Materials*, **4**, 121–8.

ROSATO, A., STRANDBURG, K. J., PRINZ, F. and SWENDSEN, R. H. (1987), Why the Brazil nuts are on top: size segregation of particulate matter by shaking, *Phys. Rev. Lett.*, **58**, 1038–40.

SCHEEL, M., SEEMANN, R., BRINKMANN, M., DI MICHIEL, M. and SHEPPARD, A. *et al.* (2008), Morphological clues to wet granular pile stability, *Nature Materials*, **7**, 189–93.

SCHRÖTER, M., ULRICH, S., KREFT, J., SWIFT, J. B. and SWINNEY, H. L. (2006), Mechanisms in the size segregation of a binary granular mixture, *Phys. Rev. E*, **74**, 011307.

SHINBROT, T. and MUZZIO, F. J. (2000), Non-equilibrium patterns in granular mixing and segregation, *Physics Today*, **March**, 25–30.

TO, K., LAI, P-K. and PAK, H. K. (2001), Jamming of granular flow in a two-dimensional hopper, *Phys. Rev. Lett.*, **86**, 71–4.

TORQUATO, S. and STILLINGER, F. H. (2010), Jammed hard-particle packings: from Kepler to Bernal and beyond, *Rev. Mod. Phys.*, **82**, 2633–72.

14

Modelling and computer simulation of food structures

S. R. Euston, Heriot-Watt University, UK

DOI: 10.1533/9780857098894.2.336

Abstract: Computer simulation is a technique that has been taken up relatively slowly by the food science community, partly due to the complex nature of food systems. However, progress has been made. It is the aim of this review to highlight some of the contributions that simulation has made to the study of food systems over the past few years, and to highlight simulation studies that could easily be adapted to investigate food systems. The review concentrates mainly on protein-containing systems, since this reflects the author's own interests, although many studies have been made on carbohydrates and polysaccharides, and to a lesser degree on triglycerides, and these are also discussed.

Key words: molecular simulation, protein structure, protein adsorption, protein gels, carbohydrate structure, polysaccharide structure, triglyceride structure.

14.1 Introduction

Computer simulation as a technique has been used almost since the first computers were constructed, although constraints on early computing power limited their applications. The Monte Carlo (MC) method as a way of carrying out molecular simulation was first suggested in the 1950s (Metropolis *et al.*, 1953), although Metropolis and Ulam had developed the general methodology during the Manhattan Project in the 1940s (Metropolis and Ulam, 1949). The development of molecular dynamics (MD) simulation for molecular systems followed soon after, with Alder and Wainright's seminal calculations on the phase diagram of hard spheres in the 1950s (Alder and Wainwright, 1957, 1959), although the application of the methodology to proteins had to wait another 20 years (McCammon *et al.*, 1977).

Advances in computing over the last 30 years have greatly accelerated the use of molecular simulation techniques, and with the wide availability of pre-programmed software packages, the simulation of molecular systems is now commonplace.

In this chapter, the common methods used in molecular simulation (MC, MD, Brownian dynamics (BD) and dissipative particle dynamics (DPD)) will be outlined, along with the Lattice–Boltzmann method (LBM), which is expected to increase in popularity for simulating food systems over the next few years. This will be followed by a review of the application of these methods to food relevant systems. The emphasis will be on protein-containing food systems, as this reflects the author's own interests. However, mention will be made of some of the more important studies of carbohydrate and triglyceride systems. Finally, comment will be made on the future prospects for simulation of food systems.

14.2   Molecular simulation methodology

Molecular simulation is generally used to describe the use of computers to mimic the motion of atoms, particles or molecules from which physicochemical properties (thermodynamic and structural) can be derived. However, the term modelling is usually reserved for when mathematical equations are used to describe a system and solutions of these provide physicochemical and structural data on that system. This review will mainly be confined to molecular simulation. The application of finite element methods to, for example, heat transfer problems in food systems (Wang and Su, 2003) will not be covered, nor will the methods from computational fluid dynamics (CFD), which have also been used extensively for various problems in food engineering (Norton and Sun, 2007).

Molecular simulation can be divided into two general methodologies, MC simulation and dynamic simulations methods. MC simulations are probabilistic in nature, that is, they rely on generating a sequence of conformations according to the probability with which they are likely to occur. Time is not an explicit variable in these simulations. However, dynamics simulation methods are deterministic. The methods involve solving one or other forms of Newton's second law of motion (F = ma). This allows calculation of the forces on each atom in a system from which a prediction of the displacement of the atoms during a given time interval can be made and a trajectory of the system followed over time. These methods will now be discussed in more detail.

14.2.1   Monte Carlo methods

Monte Carlo (MC) simulation uses random numbers to generate and evolve conformations of a molecular system. In mathematical terms, MC simulations are an algorithmic representation of a Markov chain (Robert and Casella, 2004). These are a sequence of 'events' where the chain moves from one state to the next based on a transition probability. The mathematics of Markov chains are complex and draw heavily on ideas from statistics. It is not necessary to understand the detailed mathematics of Markov chains to be able to use MC methods. However, an important point that comes from the study of these processes is that it is possible to show that any Markov chain will approach an equilibrium condition, which is independent of the starting conformation if allowed to evolve for long enough. The implication of this for molecular simulations is that MC methods can be used to generate a molecular system at equilibrium, and that any properties of this system (e.g. thermodynamic properties) will be the equilibrium properties for that system.

To carry out an MC simulation is simple in principle. Taking as an example a system of spheres, the initial positions of the particles in the box are defined either at random, using random numbers to define the Cartesian coordinates, or in an ordered array. Movement of the particles is effected by choosing one at random and then displacing its centre of mass by a small random displacement in the x, y and z coordinate directions. This sequence of random movements is repeated thousands or millions of times to generate a set of conformations that describe the system. In theory, the properties of the system (thermodynamic or structural) could be deduced by averaging the property over a large set of randomly generated conformations and applying a weighting factor to the average that takes account of the fact that some conformations are less probable than others (i.e. some have a particularly high internal energy due to a large number of less favourable contacts between particles).

This is an inefficient method for calculating average properties, and a much better method is to apply a weighting factor to the conformation itself, so that high energy conformations are only generated with a low probability in the first place. You can do this by using a special sampling technique within the general MC method and this allows conformations in the sequence to be generated only with the probability with which they are likely to occur. The most common sampling technique used is called 'importance sampling' (Robert and Casella, 2004). When importance sampling is used with the canonical ensemble (constant NVT), the acceptance or rejection of a new conformation is based on the change in energy (sum of the pair interactions between particles) that occurs when a particle is moved.

To code this into a moving-on algorithm for particles means that the transition from one conformation to another is only accepted with a probability equal to $\exp(-\Delta E/kT)$, where ΔE is the difference in energy between the new and old conformations. If this is repeated over a large number of attempted MC steps, then when the system reaches equilibrium a Boltzmann distribution of energy states is generated, which is the correct description for the canonical ensemble. Further details of the mathematics or coding of algorithms for MC simulations can be found in Frenkel and Smit (2002). Average properties of the system can be calculated once the simulation has reached equilibrium. Attainment of an equilibrium state is usually assessed by following the internal energy of the system until this reaches a constant value.

14.2.2   Molecular dynamics method

Molecular dynamics (MD) uses a different principle to MC simulation. This involves a direct evaluation of the forces on an individual particle or atom. The force can be calculated from the slope of the pair interaction potential between particles. Since the force is a vector quantity and has both a direction and magnitude, the net force on a particle can be used to predict the direction and distance a particle moves. Starting from Newton's second law of motion:

$$F = ma = m\frac{dv}{dt} = m\frac{d^2r}{dt^2} \qquad [14.1]$$

a relationship between forces and displacement r can be derived. This equation cannot be solved analytically and requires a numerical integration approach to calculate the displacement of the particle. Various algorithms have been derived to solve Equation 14.1, all of which involve expansion as a Taylor series. The simplest method, although not the most accurate, is the Verlet algorithm (Verlet, 1967).

In MD simulations, displacement of particles or atoms is carried out over small time steps, typically 1 to 2 fs. The size of the time step is extremely small because it is necessary to ensure that the magnitude of the force on the moving particle does not change significantly when the particle moves. Therefore a moving-on routine can be devised by considering a situation where the displacement changes by a small increment δr over a small time step δt. This can then be expanded as a Taylor series to derive the equation:

$$r(t+\delta t) = r(t) + \frac{dr(t)}{dt}\delta t + \frac{1}{2}\frac{d^2r(t)}{dt^2}\delta t^2 + \frac{1}{6}\frac{d^3r(t)}{dt^3}\delta t^3 + \cdots \qquad [14.2]$$

which can be simplified by substituting for v and a:

$$r(t+\delta t) = r(t) + v(t)\delta t + \frac{1}{2}a(t)\delta t^2 + \frac{1}{6}\frac{d^3r(t)}{dt^3}\delta t^3 + \cdots \qquad [14.3]$$

This equation cannot be used to predict displacement because the third-order (and higher) terms cannot be calculated. However, the motion of the particles is time reversible, and so a second equation can be defined that considers the displacement at a time, $t - \delta t$:

$$r(t-\delta t) = r(t) - v(t)\delta t + \frac{1}{2}a(t)\delta t^2 - \frac{1}{6}\frac{d^3r(t)}{dt^3}\delta t^3 + \cdots \qquad [14.4]$$

Equations 14.5 and 14.6 can be added together and rearranged to give a moving-on routine where the third-order terms cancel out:

$$r(t+\delta t) = 2r(t) - r(t-\delta t) + a(t)\delta t^2 \qquad [14.5]$$

This is the basis of the Verlet algorithm and to predict the new position of a particle only requires knowledge of the position at the previous time step $(t - \delta t)$, at the current time step (t) and the acceleration at the current time step (a_t). When using MD algorithms based on equations like Equation 14.4, the total energy of the system has to be calculated to ensure it is conserved. A limitation of the Verlet algorithm is that the velocity required for the calculation of the kinetic energy component of total energy has to be estimated from the displacement $t-\delta t$ and $t+\delta t$, which introduces an error that means that energy conservation is not exact. Other algorithms have been introduced, such as the leap-frog (Hockney, 1970) and the velocity-Verlet (Swope *et al.*, 1982) methods, which include the velocity as part of the moving-on routine and have better energy conservation in the simulation. Further mathematical details of the various integrators for MD simulations can be found in Frenkel and Smit (2002).

A recent addition to the MD methodology is the use of coarse-grained (CG) representations of molecules. In this approach, rather than including all atoms in a molecule in the simulation, molecules are divided into groups that are represented as single pseudo-atoms that have properties and interactions that are a combination of those of the constituent atoms. A common technique is to map four atoms onto a single particle, as is used in the popular Martini CG force field (Marrink *et al.*, 2007; Monticelli *et al.*, 2008; Lopez *et al.*, 2009). Such methods also typically use soft-interaction potentials. These decays more slowly than more realistic representations of non-bonded interactions, and therefore it is possible to use a longer time step, often a few tens of fs. The longer time steps and fewer atoms in the system (water is also coarse-grained) mean that simulations can be run for much longer time scales, albeit at the expense of atomic detail. Other recent advances in MD methodology have concentrated on speeding up simulations using so-called accelerated methods. These include essential dynamics (ED) (Amadei *et al.*, 1993), replica exchange molecular dynamics (REMD) (Sugita and Okamoto, 1999), hyperdynamics (Perez *et al.*, 2009), metadynamics (Laio and Gervasio, 2008) and temperature accelerated dynamics (Perez *et al.*, 2009). Some of these methods are more difficult to implement than classical MD, and to date only ED, REMD and metadynamics have been implemented on protein-sized molecules with varying degrees of success.

14.2.3   Brownian dynamics

The only forces that are important in atomic and molecular systems are the interactions forces (van der Waals, electrostatic and bonded forces). For particulate systems such as colloidal systems, other forces become important and Brownian dynamics (BD) simulation is a more appropriate methodology. In BD simulation, the interaction forces between each particle are still important and contribute to the motion of the particles. In addition, two other forces must be included that arise from the interactions between particles and the suspending medium, usually water. These extra forces are a hydrodynamic force due to friction between the particles and the solvent, and a stochastic force due to random collisions between solvent molecules and the particles. The first implementation of a BD simulation was made by Ermak and McCammon (1978). They used a form of Newton's second law called the Langevin equation to derive a moving-on routine for interacting particles undergoing Brownian motion as

$$r_i(t + \Delta t) = r_i(t) + \sum_j \nabla \underline{\underline{D}}_{ij}(t) + \frac{1}{kT} \sum_j \underline{\underline{D}}_{ij}(t) F_j^{nh}(t).\Delta t + R_i(\underline{\underline{D}}_{ij}(t), \Delta t) \qquad [14.6]$$

Here $\underline{D}_{ij}$ is the diffusion tensor and describes hydrodynamics interactions between all particles. The interaction forces (non-hydrodynamic, F^{nh}) are accounted for in the second term, and these are also dependent on the hydrodynamic interactions between the particles. The stochastic term R_i is also a function of the diffusion tensor, and the time step. Often hydrodynamic interactions are considered to be negligible, and are not included in the moving-on routine. If they are included, then suitable mathematical approximations have been proposed by Oseen (1924) and Rotne and Prager (1969).

14.2.4   Dissipative particle dynamics

Dissipative particle dynamics (DPD) (Hoogerbrugge and Koelman, 1992; Koelman and Hoogerbrugge, 1993; Español and Warren, 1995) is a method that has been devised to allow the simulation of the dynamics of mesoscopic particles, such as colloidal particles and molecules that would require extremely long simulations and very large systems to study with MD. In DPD, the constituent particles represent molecules (a CG model of the molecules) or a group of solvent molecules. The DPD particles interact through their centre of mass. The forces between the particles are summed and displacement is deduced using integrators that are the same as or similar to those used for MD. The major difference between MD and DPD, apart from the CG nature of the molecules, is the nature of the forces between them.

In DPD, the total force on a particle is the sum of three forces, a conservative force (F_C), which is the pair interaction between particles; a dissipative force (F_D), similar to the hydrodynamic force in BD; and a stochastic (random) force (F_R). F_D is a friction term that acts to push particles apart if they are approaching each other, and to pull them back together if they are moving apart. Although the F_D has a similar effect to the hydrodynamic term in BD, it is represented as a pair potential between the particles that conserves both angular and linear momentum. This frictional term leads to a gradual loss of kinetic energy in the system, which is compensated for by F_R to ensure conservation of energy. The F_R term gives the particles an extra kick of kinetic energy. The mathematical details of the method, including equations to predict the F_D and F_R terms, can be found in Groot and Warren (1997). Since the system is comprised of particles and not atoms, the form of the potentials for all three forces can be taken as soft potentials and it allows for large time steps to be used, thus allowing the dynamics of mesoscopic systems to be followed over relevant time scales as well as length scales.

14.2.5   Lattice–Boltzmann method (LBM)

There are three different ways (and length scales) in which the flow of fluids can be described. Continuum mechanics considers fluids in terms of quantities such as pressure, density and velocity, quantities that all vary continuously with time, and which vary over length scales much larger than those of the constituent particles of the fluid. This means that individual particle properties can be averaged over the whole system and the discrete nature of individual atoms/molecules is unimportant. CFD can be used to calculate continuum properties of fluids by solution of the Navier–Stokes equation. However, at the atomic level, fluid flow can be described by the movement of individual atoms or molecules. At these length scales, the properties vary abruptly due to the discrete size and velocities of the atoms.

At intermediate (mesoscopic) length scales, a third approach can be used to describe fluid properties, and it utilises the Boltzmann equation. This equation can be used to describe the statistical distribution of a particle in a system of other particles, such as the probability of finding a particle in a position r at a time t with a velocity v. In the Boltzmann equation, the probability is expressed as two terms, one of which describes the streaming (movement) of particles and the second takes account of the changes to the probability distribution function that occur when particles collide. The Lattice–Boltzmann equation can be thought of as a reduced form of the Boltzmann equation, where the motion of the particles is discretised so that they are confined to a lattice and are only able to move in a limited number of possible directions. The LBM describes the fluid as being composed of particles that move according to consecutive streaming and collision steps, as in the Boltzmann equation, but the number of degrees of freedom is restricted due to the lattice. Since the particles are represented as probability distribution functions, the individual lattice points should not be thought of as being populated by individual particles, but by distributions of particles. The general form of the Lattice–Boltzmann equation is

$$f(x+v\Delta t, t+\Delta t)=f(x,t)+\Omega \qquad [14.7]$$

where the new probability distribution (f) of particles is calculated over a small time step Δt, by taking account of the velocities v of the particles and the collisions through a collision operator Ω. Methods for defining the collision operator are outside the scope of this review and details can be found in Ladd and Verberg (2001).

The LB method is in effect a way of solving the Navier–Stokes equation in a simplified fashion, and it has a number of advantages over CFD. It is much easier to code the LBM, and algorithms are usually relatively few lines of code compared to CFD codes. Since the LBM considers particles (or populations of particles), it deals more easily with multi-component systems than CFD, and thus it is suitable for phase separation studies and systems such as oil-water-amphiphile systems where interfacial effects are critical. It is in this particular area of multiphase flow and structural properties that the LBM is likely to be of particular use in the modelling of food systems.

14.3   Food biomolecular structure and function: proteins

Food protein related studies are not that common, but where simulation has been used, it have helped elucidate some key issues relating to the function of protein in foods. MD and MC simulation have both been used to study food protein structure and properties, and a short review of these follows. In addition, mention will be made of studies carried out on non-food proteins, where these are particularly relevant or where the method could usefully be applied to food systems.

The first MD simulations on a protein were carried out in the 1970s (McCammon *et al.*, 1977), and since then the method has grown in popularity and ease of use. Nowadays, simulation is accepted as a complementary tool used to test experimental findings and to help elucidate the molecular mechanisms of structural changes in biological systems. For food proteins, MD simulation is more popular since it is more suited to investigating the dynamics of structural change. MC simulation is used less often, but is particularly useful where a CG representation of the protein is acceptable and large numbers of protein molecules are required for a realistic model of the system. As an example of this, we will discuss the application of MC simulation to investigate casein micelle structure.

The main use of MD simulation has been to study food protein denaturation. Denaturation is, of course, an important functional property of food proteins, where heated protein solutions are used to thicken foods or to form soft solid gels that give structure to liquid systems. Denaturation has also been studied widely using MD methods for non-food proteins, since the unfolding process can sometimes tell us something about protein folding mechanisms. The thermal unfolding of β-lactoglobulin (β-lac) has been studied by MD simulation (Euston *et al.*, 2007; Euston, 2012), mainly because there is a large catalogue of experimental results with which to compare unfolded trajectories.

β-lac is a globular protein with a molecular weight of about 18 kDa comprising 162 amino acids (Sawyer and Kontopidis, 2000). The central core of the molecule contains 8 β-strands (labelled A-H) that fold to form an anti-parallel β-sheet that makes up a β-barrel structure. Fatty acids or other hydrophobic ligands can bind in this central calyx. A ninth β-strand (I) does not contribute to the β-barrel and has been linked to dimer formation in β-lac solutions.

Heating of β-lac above a temperature of about 75 to 80 °C leads to denaturation. In this process the native structure goes through conformational changes, which include exposure of the interior hydrophobic residues (Hoffmann and van Mil, 1997) and changes in the secondary structure (Qi *et al.*, 1997). Several authors have hypothesised that the denatured conformation of β-lac should be thought of as a molten globule state (Qi *et al.*, 1995, 1997; Iametti *et al.*, 1996; Carrotta *et al.*, 2001).

MD simulation has been used to investigate the mechanisms of the unfolding transition. One of the problems with MD is that the time scale normally achievable is of the order of tens or maybe hundreds of ns. Since denaturation takes place over second time scales at temperatures just above the denaturation temperature of β-lac, it is unlikely that unfolding will be observed in a conventional MD simulation at realistic temperatures. This is not a new problem, and simulators who investigate thermal unfolding usually do so at elevated temperatures, typically as high as 500 K to accelerate the unfolding transitions that lead to denaturation (Day *et al.*, 2002). It has been shown for a small protein (the 64 amino acid chymotrypsin inhibitor 2) that the unfolding pathway at high temperatures is the same as at lower temperatures, and this justifies the use of elevated temperature simulations.

Recently, this method has been tested on the larger β-lac molecule along with other methods for accelerating simulations (ED and REMD) (Euston *et al.*, 2007; Euston, 2012). The simulations have highlighted possible mechanisms for the heat-induced unfolding pathway that are in general agreement with experimental observations. On heating, the simulated β-lac unfolds rapidly to an intermediate that has some characteristics of the molten globule state (Fig. 14.1). The secondary structure of the simulated β-lac is highly stable to heating, even at high temperature, and apart from the helical structure and the I-strand, much of the secondary structure is conserved.

The tertiary fold of the protein becomes distorted with heating for longer at high temperature, and this leads to distortion of the β-barrel. Analysis of the molecular motion of the heated β-lac using principle component analysis (Euston, 2012) reveals that during heating, random coil regions of the molecule, particularly the N- and C-terminal ends of the chains, and the loops that connect different β-strands in the β-barrel, are excited and lead to conformational change. As heating proceeds, the excitation starts to move into the B strand regions and slowly some of the β-sheet structure starts to break up. The least stable β-strand, other than the I-strand, has been found to be the B-strand, and this is in agreement with experimental observations of heated β-lac. Analysis of the unfolding pathway for the simulated β-lac at different temperatures (Euston, 2012) suggests that the same pathway is followed for a wide range of temperatures from 350 to 500 K, thus confirming the validity of the high temperature simulations.

The usefulness of other methods of accelerated MD methods (ED and REMD) was shown to be limited (Euston, 2012). In the case of ED simulations, this was due to the formation of elongated conformations due to stretching of parts of the tertiary fold. For REMD, the idea is to speed up simulations at realistic denaturation temperatures by generating high-temperature conformations and exploring the low-temperature energy landscape around these. Although the REMD simulation at 350 K showed more unfolding than a normal MD simulation at 350 K, the unfolding was still slow and the REMD simulation would have taken a long time to achieve denaturation (Euston, 2012).

A number of other studies have looked at various aspects of the denaturation of proteins that have relevance to food systems. Pressure-induced changes in the structure of proteins have been simulated in the case of hen egg white lysozyme (HEWL) (McCarthy and Grigera, 2006). High hydrostatic pressure can be used in the preservation of foods due to its lethal effect on pathogenic and spoilage micro-organisms (Heinz and Buckow, 2010). However, it is also known to lead to denaturation of proteins (Sasahara *et al.*, 2001) and so a molecular level understanding of this is important when attempting to develop high-pressure processes that preserve foods, but have minimal effect on the structure.

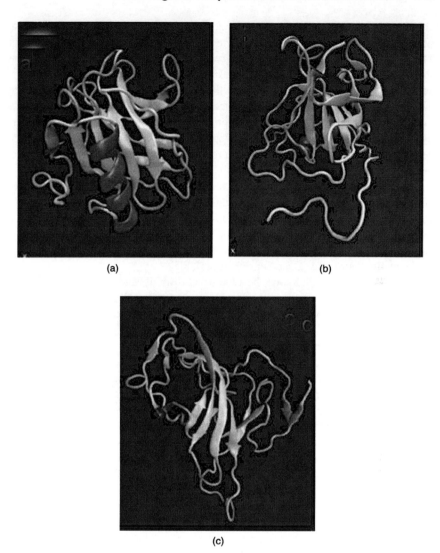

(a)   (b)

(c)

Fig. 14.1   Snapshot conformations from a 110 ns MD simulation of the thermal unfolding of β-lac A heated at 500 K. The native state (a) shows an 8-strand β-barrel, much of which is conserved throughout the simulation. Heating for 10 ns leads to a partially unfolded state (b) that has a loosened tertiary fold, but much of the secondary structure conserved. This may represent a molten globule state. Heating for 110 ns (c) leads to more extensive unfolding of the tertiary structure and loss of secondary structure.

McCarthy and Grigera (2006) have studied the effects of high pressure on the mobility of lysozyme. Their initial studies of HEWL were not able to reach the time scales required for denaturation of the protein to occur, and they were able only to probe the elastic response of the protein to pressures up to 3 kbar. They found that at 3 kbar, the mobility of most residues in the protein was reduced, although a few regions showed anomalous (more mobile) behaviour, which was believed to occur due to changes in the hydration shell around these residues. This is in agreement with experimental results (Sasahara *et al.*, 2001). Secondary structure was also stabilised at high pressure. They hypothesise that high pressure acts as a conformational selector, in that the number of conformations accessed by the protein at high pressure is reduced. Further analysis of the high and low conformations showed that at high pressure the ratio of hydrophobic to hydrophilic surface increases, which indicates that hydrophobic interactions become weaker. Presumably, this is the mechanism that drives unfolding of proteins under high pressure.

Not all food proteins are globular, and therefore susceptible to heat induced denaturation. Caseins are important functional molecules in formulated foods where they are good emulsifiers and foaming agents, and in their supramolecular form the casein micelle are able to form food gel structures such as are found in cheese and yoghurt. The milk caseins are examples of the group of proteins termed intrinsically disordered (Dyson and Wright, 2005).

Determining a structure for the caseins is difficult and still open to debate. There are two views on how the structure of caseins should be defined. The rheomorphic model views casein structure as a dynamic state that adopts a different set of possible conformations, dependent on external factors such as pH, ionic strength and temperature (Holt and Sawyer, 1993), and thus lacking a definable tertiary structure. However, Farrell *et al.* (2002) see the structure of the caseins as being closer to the molten globule state, an intermediate in the unfolding/ folding pathway observed for some globular proteins, where a tertiary fold exists but is more flexible and mobile than in the native state. To support their model for casein structure, Kumosinski *et al.* (1991, 1993a,b) have carried out a series of molecular modelling studies on various caseins combined with Fourier transform infra-red determination of secondary structure, and sequence prediction. They have found that the caseins differ markedly in the level of secondary structure, and that the amino acid side-chains tend to cluster together in the molecule in hydrophobic and hydrophilic blocks. Based on the modelling studies, they have proposed unique energy minimised structures for the various caseins.

The caseins are found in solution in a supramolecular form, the casein micelle where thousands of individual molecules join to form a nanoparticle of protein that can be up to 600 nm in diameter. To model the formation and structure of such a large aggregate is not feasible using MD simulation and other approaches have to be used. Two approaches have been taken to model the self-association of caseins as simple models for micelle structure: a BD model proposed by Dickinson and Krishna (2001) and an MC model put forward by Euston and Horne (2005) and Euston and Nicolosi (2007). Dickinson and Krishna (2001) used a mesoscopic BD model to probe β-casein solutions. In this simulation, the casein chain was a set of beads, where each bead represented several amino acids and took on an average 'character' of those beads with respect to hydrophobic/hydrophilic character and interactions. They did not observe self-association of the caseins with this model, a situation they attributed to using too high a casein concentration.

Euston and Horne (2005) and Euston and Nicolosai (2007) have simulated casein molecules as short block copolymers to investigate their tendency to form micelles in solution and to test some of the assumptions in the dual-binding model (Horne, 1998) for casein micelle formation. The dual-binding model makes the assumption that the caseins can be thought of as block copolymers, where the hydrophobic and hydrophilic amino acid side-chains are mostly found in blocks within the primary sequence. To form the casein micelle, they then self-associate through interaction between hydrophobic blocks, or through calcium bridging between phosphoserine residues in the hydrophilic blocks. Only κ-casein lacks phosphoserine and cannot associate via the latter mechanism, and this has been proposed as the reason why it is found at the surface of the micelle since it acts as a terminator of micelle growth; it only has one hydrophobic block and cannot propagate a self-association structure (Horne, 1998).

In the MC simulation, α_{s1}-casein was modelled as a triblock copolymer (with a central hydrophilic block and two terminal hydrophobic blocks), α_{s2}-casein as a tetrablock copolymer (alternating hydrophilic and hydrophobic blocks) and β-casein as a diblock copolymer (one hydrophilic and one hydrophobic). All polymer chains were comprised of 24 segments. Simulations of the individual caseins on a 2D square lattice (Fig. 14.2) show that α_{s1} and β-casein form large micelles, whilst α_{s2}-casein forms very small micelles (Fig. 14.2) through association of the hydrophobic blocks. The structure of the β-casein micelles was consistent with structures proposed from experiments, that is, in our 2D representation they were ellipsoid in shape with a hydrophobic core and a steric stabilising hydrophilic block outer layer. The α_{s1}-casein micelles were distinctive, with more than one hydrophobic core joined by a connecting hydrophilic block, such that some molecules contribute to more than one block.

The behaviour of the α_{s2}-casein can be explained by its block structure, where one of the hydrophobic blocks is sandwiched between two hydrophilic blocks and thus has difficulty in forming association structures with other hydrophobic blocks. Two and three component mixtures of the micelles were also simulated (Euston and Nicolosi, 2007). In two-component mixtures, only $\alpha_{s1}+\beta$ (1:1 ratio) mixtures formed true mixed micelles where the hydrophobic blocks appeared to mix freely in the hydrophobic core of the micelle. Whenever α_{s2}-casein is present in a mixture, it does not fully contribute to the hydrophobic core of the micelle, but sits close to the surface of a micelle composed of the other component(s). Since the second block of α_{s2}-casein has a low ability to self-associate, when the molecule sits at the surface of a larger micelle, it tends to prevent further growth of the micelle.

Of course, this is a very simple model, and the possibility of calcium bridging via phosphoserines, which would facilitate the formation of self-association structures by α_{s2}-casein, is not included in the model. Thus α_{s2}-casein is acting as a terminator of the micelle in a similar way to that proposed for κ-casein. An obvious extension of this model would introduce all four caseins, and the possibility of calcium bridging between hydrophilic blocks. Then we may be able to probe more accurately the early stages of self-association that leads to casein micelle formation.

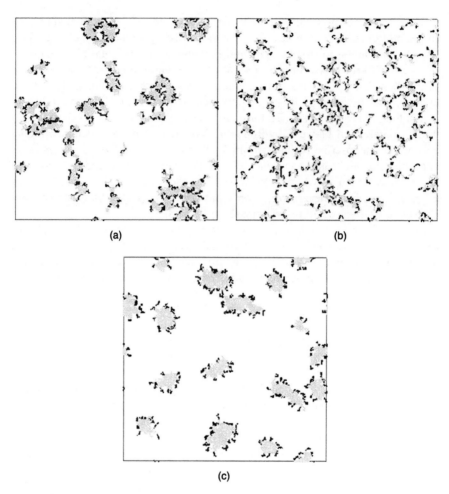

Fig. 14.2 Snapshot conformation for (a) model α_{s1}-casein; (b) model α_{s2}-casein; (c) model β-casein (grey = hydrophobic segments, black = hydrophilic segments). Reprinted from Euston and Nicolosi (2007) (http://www.dairy-journal.org/) with the permission of EDP Sciences.

A further area of interest concerning food proteins is their interaction with other molecules or ligands. For example, the protein β-lac is known to bind hydrophobic ligands in a binding site defined by the β-strands that define the β-barrel. When a bound ligand is present, this has been demonstrated to alter the heat stability and surface properties of the β-lac molecule. Gu and Brady (1992) used MD simulation to investigate the binding of retinol in the hydrophobic binding site of β-lac. When retinol is bound in the hydrophobic pocket, the EF loop undergoes a conformational change that closes the entrance to the binding pocket. *In vivo*, this may serve as a mechanism that can be used to identify ligand bound β-lac molecules. The food scientist is interested in this conformational change, since it may explain the higher heat stability observed for β-lac when ligands are present in the binding site.

Recently, we have looked at the mechanism of binding of a simple bitter molecule (phenylthiocarbamide, PTC) to one of the G-protein coupled receptors (GPCR) that have been identified as being involved in bitter taste perception (Euston and Hughes, unpublished data). Ultimately, our aim is to use this type of model to investigate the mechanisms that control bitterness in the hop acid components of alcoholic beverages. To test the methodology, we have looked at PTC binding to the T2R38 GPCR.

The perception of flavour compounds is a complex process. It is believed that various sweet and bitter flavour compounds are sensed through their ability to bind to one (or more) of a number of receptor molecules, GPCRs, which are found embedded in the membrane of cells in the taste buds on the tongue. To date, some 30 GPCRs (Behrens and Meyerhof, 2009) have been identified that are sensitive to bitter compounds, including the T2R38 receptor that is sensitive to PTC. Other receptors have been identified that respond to the hop bitter acids in beer (Intelmann et al., 2009), in particular the T2R1, T2R2 and T2R40 GPCRs. Since GPCRs are transmembrane proteins, it is difficult to isolate and determine their structure. Thus, the T2R38 structure has been inferred from the primary sequence by homology with the structure of rhodopsin (Floriano et al., 2006). Using the structure for the T2R38 GPCR, we have simulated the free energy of binding of the PTC molecule using an umbrella sampling methodology (Torrie and Valleau, 1977). In this, the PTC molecule is held at various positions within the GPCR binding pocket along a line connecting the most probable binding site with the outside of the binding pocket.

The PTC molecule is constrained within a parabolic potential defined by a spring constant of known energy. The PTC is then moved a small displacement (typically 0.1 nm) along the line and constrained in this new position. In this way a potential of mean force can be built up that describes the forces experienced by the PTC molecule at various positions in the binding pocket, and an estimate of the free energy of binding can be made.

Figure 14.3 is a snapshot of the T2R38 GPCR with a bound PTC in the most probable binding position, as determined by Floriano et al. (2006) using HierDock. Superimposed over the top of this is a plot of potential of mean force versus position in the binding pocket for the PTC molecule. The free energy determined using umbrella sampling is about –15 kcal/mole. This is of the same order of magnitude as the binding energy found by Dai et al. (2011), who used Autodock to study the binding of three bitter peptides Gly-Phe (GF), Gly-Leu (GL) and Gly-Gly (GG) to the human bitter receptor hTAS2R1. They found that the order of decreasing binding energy was –8.25 (GF), –6.69 (GL) and –5.63 (GG) kcal/mole, which was in agreement with the order of their bitterness. Floriano et al. (2006) have used HierDock to predict the binding free energy of PTC to human bitter receptor and found a value of –22.1 kcal/mol. These authors point out that predictions using docking programmes do not include entropic terms or enthalpic temperature corrections, which may explain the difference between their binding energy and the one we have predicted from MD simulation.

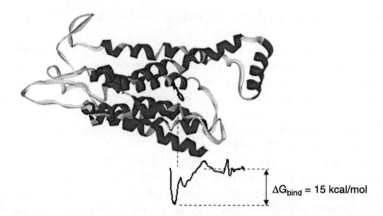

$\Delta G_{bind} = 15$ kcal/mol

Fig. 14.3 Snapshot conformations from the simulation of PTC binding to the human bitter GPCR receptor T2R38. View showing the position of PTC in the binding site of T2R38, with a potential of mean force, determined using umbrella sampling, superimposed onto the image. The helix bundle that defines the binding pocket is clearly visible.

14.4 Food biomolecular structure and function: carbohydrates and triglycerides

14.4.1 Carbohydrates

Monosaccharides

MD simulation has been used extensively to probe the structural stability of monosaccharides and their properties such as sweetness and glass transition. The first MD simulations of monosaccharides in water were carried out in the late 1980s (Grigera, 1988; Koehler *et al.*, 1988; Brady, 1989; van Eijck and Kroon, 1989). Some of the early simulations of sugars in solution are excellent illustrations of the usefulness of the method and how they can complement experimental and other theoretical techniques. In particular, the ability to explicitly include water in carbohydrate MD simulations has allowed investigations of the effects of the hydration shell around sugars on the stability of the different isomeric forms. Quantum mechanical calculations of the stability of, for example D-glucopyranose (Polavarapu and Ewig, 1992), predict that the α-anomer is more stable than the β-anomer, a situation not supported by the experimental evidence. MD simulations reveal that the hydration layer around the two anomers plays an important role in their stability (Brady, 1989; Ha *et al.*, 1991). Differences in the density of water in the hydration shell around the anomeric carbon lead to an extra stabilising effect for the β-anomer. This is not predicted in quantum chemical calculations, since only steric effects and not hydration effects are included. MD simulation can also be used to explain the apparently anomalous stability of anomers of other sugars such as D-xylose (Schmidt *et al.*, 1996).

The glassy state of monosaccharides is important in many foods, and in this the interaction between water and sugar molecules is also important. In a glassy state, water and monosaccharides are able to participate in a cooperative hydrogen-bonded network (Franks and Grigeria, 1985) that has rheological properties similar to polymers (Ferry, 1980). This imparts structure in some food materials. The water in the glassy state has a lower diffusion coefficient, and this helps to reduce water activity and helps preserve the food (Roos, 2010). Simulation has been used widely to study the glassy state. Caffarena and Grigera (1997, 1999) have simulated the glassy state of β-D-glucose by following the change in water diffusion coefficient, with the aim of determining the usefulness of MD in predicting glass transition temperature (T_g), particularly at high glucose concentrations where experimental measurements can be difficult. They were able to show a depression in T_g and in water melting temperature as the glucose concentration increased. Caffarena and Grigera (1997) considered the agreement between simulation and experiment to be very good for the T_g, particularly at high solute concentrations. For the melting of water, the agreement was qualitative, with the predicted temperatures being lower than experimental ones. The latter effect is thought to be an artefact of the way in which initial conditions for the simulation were defined, with the water molecules surrounding the glucose being in a more disordered state than they are in a real system.

Caffarena and Grigera's (1997) original model did not account for the anomeric effect in glucose. To remedy this, they introduced a model where the glucose molecules were distributed with the experimental ratio of the α- and β-anomers (50:50). This they applied to a high concentration glucose solution (85% w/w) to study the water dynamics around the glucose molecules in the rubbery state (40 K above the T_g) and the glassy state (40 K below the T_g). The dynamics of water molecules in the hydration layer around the glucose was observed to slow down dramatically in the glassy state compared to the rubbery state. In foods this can be linked to a reduced reactivity in reactions that require water. The mean residence time for water molecules in the hydration layer increased from 2.00 ps in the rubbery state to 5.75 ps in the glassy state, and their motion was more restricted in the glassy state. It was also observed that water close to the anomeric site had a longer residence time than at other hydration sites, which supports the findings of Brady et al. (1989) and Ha et al. (1991).

Roberts and Debenedetti (1999) also used MD simulation to look at the structure and dynamics of solutions of β-D-glucose, β-D-mannose and D-fructose over the concentration range 0 to 80% sugar. Several interesting features of the solutions were observed. At low sugar concentrations, the hydrogen bonding in the system was dominated by a percolating water–water H-bond network characteristic of bulk water. As the concentration of sugar was increased, this changed to a more stable H-bond network with tetrahedral order. Finally, at high sugar concentrations, a percolating sugar–sugar H-bond network formed. A percolating water phase was observed to form only below about 60% sugar concentration, whilst above this the water became trapped in 'cages' bounded by the sugar molecules and showed much reduced diffusion kinetics.

A further feature of the diffusion of water in the systems was the presence of so-called jump diffusion. In this, the water molecules show hindered diffusion for relatively long periods of time as they become trapped in restricted regions of the system by surrounding sugars molecules. Then they display a jump as they escape from this region and diffuse into another region of the system. Interestingly, these jump kinetics occur at sugar concentrations as low as 29%, much lower than the observed caging of water molecules that occurs above 60%. This suggests that jump kinetics have more to do with the sugar molecules in the hydrogen bond network of water rather than caging effects. Roberts and Debenedetti (1999) found that these observations are qualitatively the same for the three carbohydrates used in the simulations, although the magnitude and details of the effects were sugar dependent, which was attributed to the differing effect that each sugar has on the hydrogen bond network of water.

Mollineri and Goddard (2005) have used a CG MD model to investigate the origin of the water diffusion via jumps in a glucose glass. This utilised the M3B CG model for carbohydrate water mixtures, where the water was represented as a single bead and a glucose molecule as three beads in a triangular conformation (Mollineri and Goddard, 2004). The diffusion of water through a sugar glass had been observed before, and it had been hypothesised that this was due to diffusion of the molecules through free volume channels that form due to the packing of the sugar molecules (Hills *et al.*, 2001). Mollineri and Goddard (2005) ruled this out as a diffusion mechanism in their simulations by measuring the size of free volume pores (for a 12.2% water system) and showing that the diameter of these pores was smaller than the size of a water molecule. Then they were able to show that water diffusion was via two similar mechanisms, either via jumping into the position of another water molecule or into the position of a glucose bead. The jumping of water into the positions occupied by glucose beads was surprising, since no translational diffusion of glucose was observed at this concentration of glucose. This implied that water jumps were linked to other motions in the glucose molecule, and indeed they observed coupling between rotational motion of the glucose and water jump diffusion.

Polysaccharides
Polysaccharides are long-chain polymers of monosaccharides, and so it is no surprise that their interaction with water molecules has been the subject of a number of simulation studies. This is all the more important, as polysaccharide molecules are often used as thickeners in foods and this is linked intimately to the way in which they interact with water. Studies on disaccharides (Edge *et al.*, 1990) have shown that the presence of waters of hydration around the glycosidic bond lead to a damping of molecular motion, which in turn influences conformation adopted in solution. Similarly, simulations on the structure of oligosaccharides of α-1-4 linked and β-1-4 linked glucose have illustrated the role that differences in hydration layer structure around the glycosidic link play in the rigidity of the cellulose molecule (Umemura *et al.*, 2005).

Several similar studies have looked at the structure of various other polysaccharides. These have been summarised in a previous review (Euston *et al.*, 2010). More recent studies have tended to look at polysaccharides such as β-glucans, which are of interest because they have been linked to bioactive effects in human nutrition (Wood, 2001, 2003). β-glucan is the generic name given to the group of glucose polymers that contain β(1-4), β(1-3) or a mixture of the two glycosidic links. The structure–function relationship in bioactive mixed β-glucans is not well understood, at least in part because of a lack of knowledge on the detailed structure of glucans. Glucans, such as the β(1-4) cellulose, form flat structures that self associate into fibrils (Marchessault and Sundararajan, 1983), whilst β(1-3) linked glucans form triple helices (Deslandes *et al.*, 1980). The structure of the mixed linkage glucans is much less clear, and this has spurred the use of molecular modelling to build molecular models that can be used to investigate their bioactivity. What has been known for some time is that glucans contain repeating β(1-4) linked cellotriosyl or cellotetrosyl units separated by single β(1-3) bonds (Roubroeks et al., 2001). It is the cellotriosyl units that are believed to confer bioactive properties such as cholesterol binding (Lazaridou *et al.*, 2004).

Christensen *et al.* (2010) used NMR and MD simulation to explore the energy minima of β(1-4) linked methyl-cellobioside and β(1-3) linked methyl β-laminarabioside. They identified the average glycosidic dihedral angles and then used these to construct polymers for a cellulose (β(1-4) linked glucose repeats), cellotriose repeats linked β(1-3) to each other and cellotetrose repeats linked β(1-3) (Fig. 14.4). The cellulose chain adopted a straight chain conformation as expected (Fig. 14.4). Introducing β(1-3) links into the chain introduces kinks that

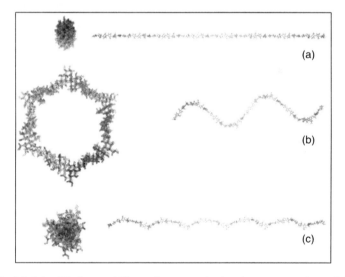

Fig. 14.4   Models of β-glucans (48-mers) generated using the average values of glycosidic dihedral angles observed at the global MD simulation minima: (a) the cellulose polymer; (b) the mixed-linkage cellotetraosyl polymer; and (c) the mixed-linkage cellotriosyl. Right panels: side views with the major principal axis of each polymer horizontally aligned. Left panels: top views from the reducing end of each polymer along the major principal axis. The top view of (b) deviates slightly from the view along the major principal axis c, since the present depiction allows for better appreciation of the 6-fold symmetry. Reprinted from Christensen *et al.* (2010), with permission from Elsevier.

cause the chain to twist and adopt a helical structure. The cellotriose-based polymer has more β(1-3) links and so adopts a tighter helical structure than the more open one for cellotetrasoyl repeats (which has almost 6-fold symmetry) (Fig. 14.4).

In a similar study, Li et al. (2012) have used the rotational isomeric state Metropolis MC simulation method to deduce essentially the same structure for a cellotriosyl repeat polymer. They found that the polymer adopted a three-fold helix with a pitch of 41.35 Å. They used the model to simulate polymers with realistic molecular weights corresponding to β-glucan fractions found in oat and wheat. Predictions of the intrinsic viscosity, radius of gyration and persistence length for these polymers was in close agreement with experimental results and suggests that such models may be useful in studying the structure–function relationships in β-glucans.

The simulations of Christensen et al. (2010) predicted a straight chain conformation for short chain cellulose-like polymers, and they point out that this allows for interaction between the chains to form fibres, which accounts for their insolubility. The interactions that govern the aggregation of cellulose chains into fibres have also been studied using MD simulation, by calculating the potential of mean force between cellulose chains (Bergenståhle et al., 2010). These authors simulated the potential of mean force between pairs of glucose, cellobiose, cellotriose and cellotetraose. Not surprisingly, they found that both hydrophobic and hydrogen bonding favoured the formation of ordered crystals over free cellulose chains in solution. What was more surprising is that by comparing the systems with increasing chain length, they found that the conformational entropy did not favour the free chains in solution. This suggests that cellulose does not adopt a random coil structure in solution but is extended and rigid.

14.4.2   Triglycerides

Triglycerides are the most difficult of the three major food biomolecules to simulate. The most interesting triglyceride state is the solid state. Simulations of triglyceride solidification from the melt have mostly been unsuccessful. Triglycerides display polymorphism in the solid state, and this leads to simulations becoming trapped in disordered or partially ordered metastable states as the system is cooled. Engelsen et al. (1994) tried to simulate the solidification of trilaurin from the melt, but were unable to reach the ordered state due to the prohibitively long simulation that would have been required. Other attempts at modelling triglycerides have had more success by starting the simulation in the ordered state, and using this model to investigate the effects of changing parameters such as acyl chain length, on the ordering in the system (Chandrasekhar and van Gunsteren, 2001, 2002).

Greater progress has been made using a CG representation of the triglyceride structure (Brasiello et al., 2010, 2011). These authors have used several CG representations of the structure of tridecanan, where each acyl chain has been represented by two, three or four beads. All models were able to reproduce physical properties (such as density) compared to real tridecanoin, but only the four-bead model exhibited a liquid–solid phase transition at a temperature close to the experimental value. This latter study suggests that any future progress in modelling the phase behaviour of triglycerides will come from CG simulations. This is confirmed by the success that CG representations for lipids have had in the modelling of cell membrane structure and processes (Orsi *et al.*, 2007). However, care must be taken to choose appropriate CG schemes and to parameterise the force field in an appropriate manner.

14.5   Adsorption of food biomolecules

Most if not all foods are multi-component systems, with the interface between different components important in determining the structure, texture and sensory properties. The interfacial region is a site for the accumulation of surface active food components, which aid the formation and stabilisation of the separate component domains. We will look at two general examples of the adsorption of biomolecules that are important to food systems, protein adsorption at fluid interfaces and bile salt adsorption at oil–water interfaces. The former is important in stabilisation of dispersed food colloids systems, with examples being the stabilisation of fat droplets in the form of an emulsion (e.g. milk, sauces, beverage emulsions) and air bubbles in the form of a foam (e.g. whipped cream, meringues, bread, cakes). Bile salt adsorption does not contribute to the structure of food; rather they are a part of the digestive system in humans that aids the breakdown and adsorption of triglycerides. Interest in bile salt surface chemistry has arisen because of the desire to control fat digestion, with the aim of controlling the energy density of foods.

14.5.1   Protein adsorption at fluid interfaces

In food colloids, the surface of dispersed oil droplets or air bubbles is stabilised by an adsorbed layer that contains proteins and other surface active ingredients. The adsorbed protein molecules provide a protective layer at the air–water or oil–water interface that prevents separation of the droplets or bubbles through various mechanisms (McClements, 2004). Proteins stabilise interfaces in foods via steric stabilisation, although if the ionic strength is not too high and the pH is not too close to the isoelectric point of the proteins, electrostatic repulsion can play a role. The properties of the adsorbed protein layer depend in part on the structure that the molecule adopts at the surface. This in turn is linked to the stability of the native conformation of the protein in solution, the nature of the interface and the environmental conditions. Two general types of food protein are commonly encountered, the globular proteins and the random coil (or intrinsically disordered proteins). Globular proteins have a compact, folded tertiary structure and will adsorb intact at an interface, and then unfold slowly across the surface. The extent of folding depends on the conformational stability of the molecule and the surface density of the adsorbed molecules (Graham and Phillips, 1979a,b). Disordered proteins lack a tertiary structure, have limited secondary structure and spread rapidly at interfaces.

Early attempts at modelling protein adsorption treated them as linear chain molecules, and were a progression from models for polymer chain adsorption (Dickinson and Euston, 1992a). These authors used a linear chain representation to investigate β-casein adsorption at a model oil–water interface in an MC simulation on a 3D lattice (Dickinson and Euston, 1992a). In this simulation, each amino acid residue was represented as a single chain segment, and was assigned to one of three types (hydrophobic, polar and charged) that determined how they interacted with the surface and with each other. The model was able to produce a protein surface density plot normal to the interface that was similar to those derived for the same protein from neutron reflectance at a hydrocarbon–water interface (Dickinson et al., 1993).

To model adsorption of globular proteins requires a different approach, so as to capture the globular nature and the unfolding of the structure at the interface. Dickinson and Euston (1990, 1992b) introduced two simple MC models, where the globular protein was treated as deformable. In one model the proteins were represented by cyclic 2D lattice chains, which were an extension of the author's earlier models for suspensions of deformable emulsion-like particles (Dickinson and Euston, 1989). In this model the proteins have no internal structure, but the globular nature is achieved through repulsive or attractive inter-chain interactions that allow it to adopt a bulbous or collapsed folded structure (Fig. 14.5). The figure shows snapshot conformations for a cyclic lattice chain model for adsorbing globular proteins.

The conformational stability of the globule is controlled by a weak attractive interaction (E_i) between segments on the cyclic chain, which causes the molecule to adopt a compact folded conformation in bulk solution. The globule is allowed to interact with a hard surface with varied surface–segment interaction energies (E_s). Adsorption only occurs when E_s exceeds a critical value, which depends on E_s, and for the conditions in Fig. 14.5 this is $-0.8\,kT$ per segment. As the strength of the adsorption energy increases, the globule adsorbs and starts to unfold by exchanging segment–segment interactions for segment–surface interactions. Different values of E_i lead to a differing adsorption behaviour. Stronger segment–segment interactions give a more stable globule folded conformation and it unfolds less at the surface for a given value of E_i. Conversely, weaker (or no) attraction between segments leads to a globule that adsorbs at lower E_s and adopts a flatter conformation. The model demonstrated that the degree of unfolding of the 'protein' at the interface depends on the balance between protein–surface and protein–protein interactions.

A second model, also based on a deformable particle representation, treated the proteins as an aggregate of subunits that rather than being bonded in a linear fashion were allowed to rearrange their conformation subject to the condition that they always remained linked together (Dickinson and Euston, 1992b). This model had the advantage that by changing the magnitude of the attractive forces between subunits, the flexibility of the protein could be changed. This model was then used

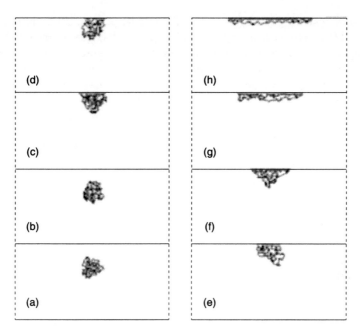

Fig. 14.5 Images of typical configurations of isolated 100-segment particles with $E_i = -0.05\,kT$ as a function of the adsorption energy E_s: (a) $0.0\,kT$; (b) $-0.8\,kT$; (c) $-1.2\,kT$; (d) $-1.6\,kT$; (e) $-2.0\,kT$; (f) $-2.5\,kT$; (g) $-3.0\,kT$; (h) $-4.0\,kT$. Reproduced by permission of the Royal Society of Chemistry.

to investigate the co-adsorption of flexible and globular molecules at 2D lattice surfaces (Dickinson and Euston, 1992b). Euston and Naser (2005) extended the deformable globule model to a 3D lattice and used this to simulate the surface equation of state for model proteins with varying degrees of flexibility. Single deformable globules were allowed to adsorb to a solid surface, and then a radial pressure field of known strength applied to the molecule to mimic the presence of other surrounding molecules. By varying the pressure and measuring the area occupied by the molecule, a surface equation of state was constructed. The resulting simulated surface pressure-area isotherms bear a strong resemblance to experimental isotherms for globular and disordered proteins, including such features as 'kinks' in the isotherm corresponding to rearrangements in the protein orientation at the surface, although the corresponding collapse pressure could not be simulated due to the method of application of pressure in the model.

Lattice chain models use a CG representation of the proteins lacking in molecular detail. Combining these with MC simulations gives information on the gross equilibrium unfolded structures of proteins, but says little about the kinetics and nothing about atomic level structural changes. To take full account of the secondary structure elements and tertiary fold of proteins requires a more detailed approach.

To investigate the adsorption of a small globular protein, barley lipid transfer protein (LTP) at fluid interfaces, Euston *et al.* (2008a,b) have used the GROMACS MD program, with an atomic level representation of the protein structure. Barley LTP is a small protein (9 kDa, 91 amino acids) that is important in the formation and stabilisation of the foam (head) in beer (Evans and Sheehan, 2002), and is therefore expected to adsorb to the air–water interface. Its structure is relatively complex for a small protein. It contains four alpha helical regions that are organised to form a four helix bundle that defines a binding pocket for lipid molecules (Heinemann *et al.*, 1996). Its high conformational stability in solution (its denaturation temperature is reported to be around 100 °C (Lindorff-Larsen and Winther, 2001)) is maintained by four disulphide bonds (Heinemann *et al.*, 1996).

LTP is extracted from barley during the mashing process of brewing, where milled barley is mixed with hot water and enzyme action releases fermentable sugars from starch. During the subsequent wort boiling process of brewing, the disulphide bonds in LTP may be reduced. The adsorption of barley LTP to a water–vacuum interface has been simulated using MD (Euston *et al.*, 2008a,b; Euston, 2010). In this model, the LTP was included in a simulation box of explicit water where the box was expanded to allow for the formation of two water–vacuum interfaces. Over the course of a 23 ns simulation, the LTP molecule adsorbs to the vacuum–water interface and stays bound to the surface (Fig. 14.6) (Euston *et al.*, 2008a,b) without penetrating into the vacuum phase. The mechanism of adsorption appears to be displacement of water molecules from the interfacial region.

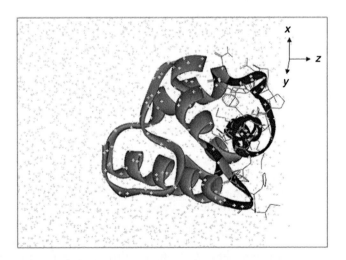

Fig. 14.6   Snapshot conformation of native LTP adsorbing at a water–vacuum interface after a 23 ns MD simulation. The global fold of the LTP is shown as a solid ribbon. Only adsorbed amino acid residue side chains are shown. Amino acid residues that are adsorbed in the interfacial region are highlighted in black. Reprinted with permission from Euston *et al.* (2008), © American Chemical Society.

Given that LTP is often found in a reduced form, Euston *et al.* (2008a,b) also simulated the adsorption of a fully reduced LTP molecule where the four-disulphide bonds were broken. The reduced molecule showed a lowered tendency to adsorb to the surface, probably due to an increase in the conformational entropy of the molecule, which opposed the initial adsorption of the molecule to the interface, with the molecule sitting further from the interface when it did adsorb. Analysis of the conformational changes in the adsorbed LTP shows that the secondary structure is conserved for both native and reduced LTP at the water–vacuum interface. There is evidence of a small degree of tertiary structure spreading at the interface from the root mean square displacement (RMSD) and radius of gyration (R_g) of the molecules, but nothing substantial. This is not surprising given the short (ns) simulation times and known high conformational stability of LTP. Simulations of the adsorption of native and reduced LTP at the decane–water interface (Euston *et al.*, 2008a,b) show that the LTP exhibits a high degree of penetration into the decane phase interface and shows a greater degree of conformational change at the surface (Fig. 14.7). It is evident that this conformational change is due to adsorption, since changes in the RMSD and R_g only start after the LTP molecule has adsorbed to the decane surface.

Simulations at the decane–water interface have been carried out for up to 500 ns (Euston, 2010), and show that the conformational change in the tertiary structure of the molecule continues over the course of this time scale, whilst the secondary structure is conserved throughout. The main change in the tertiary structure appears to be the increased flexibility in the N-terminal loop of the LTP protein (Fig. 14.7), which penetrates a considerable distance into the decane phase. The molecule is also observed to orientate at the interface, such that the four α-helices lie almost parallel to the decane phase (Fig. 14.7), a result that confirms the experimental ATR spectroscopy results of Subirade *et al.* (1995, 1996) for wheat and maize LTP adsorption to phospholipid monolayers. Euston (2010) has also used a CG model for LTP adsorption to a decane surface, to generate equilibrium unfolded LTP conformations over a longer time scale (a scaled time of 1200 ns). These conformations are formed in a much shorter (real) time simulation (12 hr compared to several months for an equivalent atomistic simulation).

The results of the CG model, simulated using the Martini CG force field, produce results that are qualitatively similar to the atomistic model in some respects, but differ in others. The CG model also reproduces conformations where the helical regions of the molecule align nearly parallel to the interface. However, the penetration of the LTP into the decane surface is not observed. The differences between the atomistic and CG simulation for LTP adsorbed to a decane–water interface highlight some of the difficulties encountered when simulating adsorbed proteins. A full atomistic simulation of the system would require prohibitively long computational times to generate equilibrium unfolded conformations. However, the CG simulations that can easily achieve these equilibrium conformations do not reproduce the same adsorption behaviour as the atomistic models. The differences lie in the CG mapping used in the protein model (a 4:1 atom:CG particle mapping) and the parameterisation of the forces fields involved.

(a)

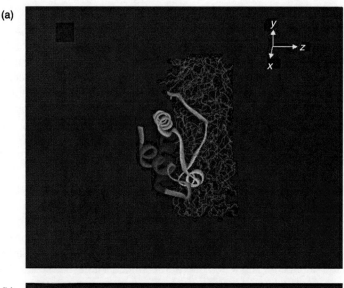

(b)

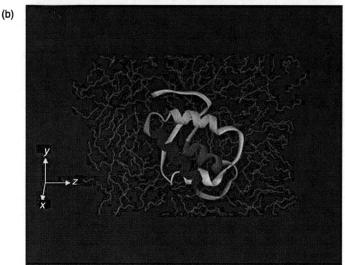

Fig. 14.7 Simulated adsorbed conformation for all-atom LTP adsorbed at the decane–water interface: (a) view looking parallel to the decane–water interface; and (b) view looking perpendicular to the decane–water interface. Water molecules have been omitted for clarity. Reprinted with permission from Euston, (2010). © American Chemical Society.

Whilst CG simulations will almost certainly prove useful in the study of protein adsorption, the results of Euston (2010) suggest that more work may need to be done to produce reliable CG models and/or force fields for protein adsorption studies.

MD simulation has also been applied to the study of another group of hydrophobic proteins that are of great interest to food scientists, namely the hydrophobins. These are small (7–10kDa) proteins that are produced by filamentous fungi (Cox and Hooley, 2009). They are natural amphiphiles and surfactants, and appear to have a number of functions in fungal growth and interaction with the environment (Linder *et al.*, 2005). The surface activity of hydrophobins has led to their investigation as foam stabilisers, and for the formation of air-filled emulsions as a mechanism for fat reduction (Cox and Hooley, 2009). They are classified into class I and class II hydrophobins, based on their solubility (Wessels, 1994). Class I hydrophobins have a poor solubility in water compared to class II. All hydrophobins share an octet of conserved cysteine residues, which form four disulphide bonds, although there are differences in the amino acid sequence between class I and class II molecules in other parts of the sequence.

The first MD simulation of hydrophobin adsorbed at an air–water interface was carried out by Zangi *et al.* (2002) on the type I hydrophobin SC3 from *Schizophyllum commune* before the tertiary structure of hydrophobins was determined. They used a fully extended conformation of the SC3 primary sequence. The four disulphide bridges in the molecule were formed by setting distance constraints between the cysteines during the simulation, and slowly reducing the constraint distance to an –S-S- distance of 0.21 nm, after which the constraints were replaced by an –S-S- bond. The folding of the SC3 molecule was followed at the water–hexane interface, as well as in water and hexane as a solvent. The SC3 molecule was found to fold rapidly at the water–hexane interface over a 100 ns time scale, and formation of β-sheet secondary structure occurred as a consequence of the folding process (Fig. 14.8).

Fan *et al.* (2006) carried out a further study of SC3 adsorption at interfaces using a more sophisticated model for SC3 based on homology, with the known tertiary structure of the type II hydrophobin HBFII and the assumption that the pairings of

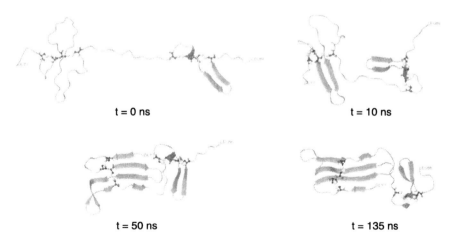

t = 0 ns   t = 10 ns

t = 50 ns   t = 135 ns

Fig. 14.8 Structures of SC3 from the simulation at the water–hexane interface after 0, 10, 50 and 135ns. Reprinted from *Biophysical Journal*, **83**, Zangi *et al.* (2002), © with permission from Elsevier.

cysteine residues in the disulphide bonds were conserved. Based on these assumptions, they proposed a model for the tertiary structure of the SC3 molecule in aqueous solution that contained mainly β-strand and random coil regions. This structure was used in an MD model to simulate SC3 adsorption at an air–water, oil–water and a solid–water interface. Fan *et al.* (2006) found that a loop comprised of amino acids 14-47 (a region between the 3rd and 4th cysteine residues) preferentially adsorbed to all interfaces. In all cases, this was also associated with an increase in the proportion of α-helix in the protein, which agrees with experimental observations of SC3 adsorption to interfaces (Wang *et al.*, 2004).

Our own simulations of hydrophobins have looked at the adsorption of the class II hydrophobin HFBI from *Trichoderma reesei* using the crystal structure determined by Hakanpää *et al.* (2006). The HFBI structure contained 4 β-sheets, which are arranged in a small β-barrel and an α-helix. The HFBI structure also exhibited the same conserved pattern of cysteine residues, as seen in another class II hydrophobin, HFBII, the structure of which had been determined previously (Hakanpää *et al.* 2006a,b). These authors identified a hydrophobic patch in the HFBI, which comprised 13 residues that were also found in the hydrophobic patch of HFBII. This hydrophobic patch was found to be responsible for the strong tendency of hydrophobins to form oligomers in solution. When the HFBI molecule is simulated adsorbing to an air–water interface over an 80 ns MD simulation, which is long enough to see rotational as well as translational diffusion, the HFBI molecule adsorbs with the hydrophobic patch orientated into the air phase (Fig. 14.9; Euston, unpublished data). A similar situation is observed if two HFBI molecules are simulated in a water box, where the two molecules form a dimer connected via the association between the hydrophobic patches on each molecule (Fig. 14.10; Euston, unpublished data).

Fig. 14.9 Snapshot conformation after 80 ns of an MD simulation of the hydrophobin HFBI at an air–water interface. Water molecules are represented as grey dots in the simulation cell. The hydrophobic patch on HFBI identified by Hakanpää *et al.* (2006a,b) is denoted by black residues in the conformation. Clearly the hydrophobic patch is adsorbed at the interface.

Fig. 14.10 Snapshot conformation after 80 ns of an MD simulation of two HFBI hydrophobins in a water box. After a few nanoseconds, the HFBI molecules associate via the hydrophobic patches on each molecule (black residues) and remain aggregated throughout the simulation.

14.5.2   Bile salt adsorption at fluid interfaces

The second example where simulation has been used to model adsorption of molecules relevant to food science is in the study of bile salt adsorption to oil–water interfaces (Euston *et al.*, 2011). Bile salts are biological surfactants that are synthesised in the liver from cholesterol and are secreted into the duodenum. They adsorb at the surface of emulsified fat, displace adsorbed proteins and peptides and facilitate the lipolysis of triglycerides by promoting the adsorption and enzymic action of lipases. In addition, they are able to form micelles in solution and it is believed that this aids in the transport and subsequent adsorption of fatty acids.

Compared to low molecular weight emulsifiers that are commonly added to foods (e.g. phospholipids, mono and diglycerides, Tweens, Spans, etc.), bile salts have an unusual and complex structure (Maldonado-Valderrama *et al.*, 2011). They have a sterol ring structure with substituted hydroxyl and methyl groups, and often a conjugated amino acid (glycine or taurine) attached. The sterol ring is almost flat, and the hydroxyl and methyl groups are arranged on opposite sides of the ring to give the molecule a planar amphilicity, that is, bile salts have a hydrophilic and hydrophobic face. Various studies of the adsorption of bile salts to fluid (usually phosholipid) interfaces suggest that they adsorb flat to the interface based on estimates of the molecular area they occupy (Ulmius *et al.*, 1982; Fahey *et al.*, 1995; Wenzel and Cammenga, 1998; Tiss *et al.*, 2001). This has led to the hypothesis that they adsorb with the sterol ring parallel to the interface, the ring methyl groups orientated towards the hydrophobic phase and the hydroxyl groups towards the water phase (Ulmius *et al.*, 1982; Fahey *et al.*, 1995; Wenzel and Cammenga, 1998; Tiss *et al.*, 2001). In contrast, studies at an octanol–water interface (Vadnere and Lindenbaum, 1982) suggest that they adsorb in a similar manner to normal amphiphilic surfactants, with the hydrophobic sterol ring and the charged carboxyl end group penetrating the decane and water phases, respectively.

Obviously, if bile salts play a role in the binding of lipases to oil–water interfaces during digestion, then the conformation they adopt at the interface may be important. To further understand bile salt adsorption at oil–water interfaces, Euston *et al.* (2011) have used MD simulation to probe the structure and free energy of adsorption of two bile salts, sodium cholate and sodium deoxycholate. These differ by one hydroxyl group on the sterol ring, with cholate being a trihydroxy and deoxycholate a dihydroxy bile salt. Neither bile salt is observed to adopt a flat conformation at a simulated water–decane interface. Instead they adopt conformations that are tilted at an average angle of close to 45 degrees with the sterol ring penetrating into the decane phase, and the charged carboxyl end group into the aqueous phase where it can form hydrogen bonds with water (Fig. 14.11).

The deoxycholate molecule has one less hydroxyl group on the sterol ring and is therefore more hydrophobic than cholate. The deoxycholate penetrates slightly further into the decane phase (Fig. 14.11b), and has a free energy of adsorption, determined using umbrella sampling, that is greater than for the cholate (Fig. 14.11c; Euston *et al.*, 2011).

Atomic force microscopy (AFM) studies of co-existing adsorbed films of β-lactoglobulin and bile salts have demonstrated that they displace proteins via an orogenic mechanism (Maldonado-Valderrama *et al.*, 2008), similar to that hypothesised for other food grade surfactants (Mackie *et al.*, 1999). In orogenic displacement, surfactants are believed to adsorb into gaps in an adsorbed protein layer and form surfactant domains that grow as further surfactant adsorbs. These surfactant domains exert a surface pressure on the adsorbed protein film as they grow, and eventually the protein layer buckles and lifts off the interface. The mechanism of orogenic displacement has been simulated by Pugnaloni *et al.* (2003) using a BD model. Snapshot conformations of simulated protein displacement and experimental AFM images of protein displacement by surfactant (Pugnaloni *et al.*, 2004) display a high degree of similarity.

The simulations of Euston *et al.* (2011) showed that the bile salts have a tendency to cluster together at the decane–water interface, possibly by forming reverse micelle like structures, which supports the idea that orogenic displacement may be due to the formation and growth of bile salt domains.

More recently, Euston *et al.* (unpublished work) have extended their simulations to look at the adsorption of sodium cholate to different oil–water interfaces, to try

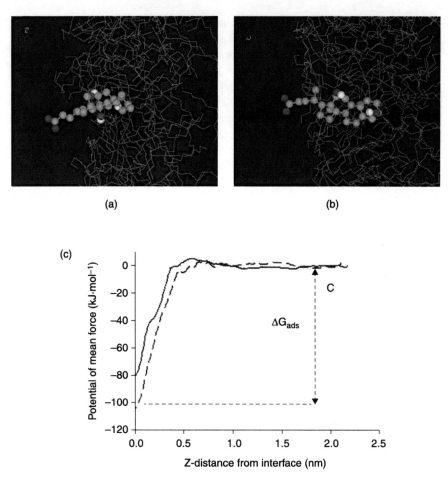

Fig. 14.11 Snapshot conformations for: (a) sodium cholate; and (b) sodium deoxycholate adsorbed at the decane–water interface. Water molecules are omitted for clarity. The conformations clearly show the penetration of the sterol ring into the decane phase, and the localization of the carboxyl end group in the water phase. In (c), the computed potential of mean force is plotted against distance from the interface for both bile salts. Euston *et al.* (2011), reproduced by permission of The Royal Society of Chemistry.

to understand the role of interface structure on bile salt adsorption. It appears that the orientation of the bile salt depends on the degree of order at the interface. For dipalmitoyl phosphatidyl choline (DPPC) bilayers, the cholate molecule will initially adsorb flat to the phospholipid–water interface and remain in that orientation for a few ns, regardless of the temperature. Then, if the simulation is at a temperature (323 K) above the gel phase transition, the bile salt sterol ring will eventually penetrate into the bilayer structure and sit normal to the interface. However, if the temperature of the simulation (300 K) is below the gel transition temperature, the bile salt sits parallel (flat) to the surface and does not enter the bilayer.

A similar situation is observed for an interface formed between the triglyceride tripalmitin and water. Here the equilibrated tripalmitin–water interface consists of aligned acyl chains. At 323 K, the cholate is able to enter the interfacial region and sit between the acyl chains. However, at 300 K, the cholate is excluded from the interfacial region and sits parallel to the tripalmitin layer. At the triolein–water interface, a similar ordering of the acyl chain is seen, although the double bond present in the acyl chains leads to a greater degree of disorder in the interfacial region than for tripalmitin or DPPC. Consequently, the cholate molecule is able to penetrate into the triolein layer at both 323 K and 300 K.

From these results it would seem that the conformation adopted by the bile salt sodium cholate depends on the temperature, which affects the degree of ordering of the lipid molecules in the interfacial layer. A highly ordered DPPC and triplamitin layer, at 300 K, does not have enough free space for the cholate to fit between the acyl chains. At the higher temperate of 323 K, the lipid interfacial layer is more fluid and gaps in the layer form through thermal fluctuations that are large enough to allow the cholate sterol ring to squeeze into. Obviously, the hydrophobic sterol ring of the cholate molecule would prefer to be wetted by the lipid layer, but there is an energy barrier to penetration of the layer. Only at a high enough temperature can this barrier be overcome.

For the more disordered triolein layer, the energy barrier to cholate penetration must be lower than for tripalmitin and DPPC and is not sufficient to prevent cholate penetration. For the decane–water interface used by Euston *et al.* (2011) in their earlier simulations, the degree of ordering of this interface is very low, and so cholate is free to enter into the oil phase. The fact that the degree of ordering in the lipid region of the oil–water interface is important in determining bile salt adsorbed conformation suggests that the barrier to penetration is entropic.

14.6   Simulation of food colloids

14.6.1   Protein gelation

Gels form when proteins aggregate and form a continuous network structure that traps aqueous phase in pores to form a soft solid. In food technology the phenomenon is exploited to impart a semi-solid texture in an initially liquid food. There have been many studies of the structure and formation of food-like gels, with the origin of many of these studies being traced back to the seminal work on the fractal structure of particle aggregates in the 1980s. Several researchers (Meakin, 1983; Kolb and Hermann, 1987) identified two idealised mechanisms of particle aggregation in dilute solution, which have been called diffusion limited aggregation (DLA) and reaction limited aggregation (RLA). This arose from the finding that the radius of gyration (R_g) of an aggregate scaled as a power law with the number of individual particles in the aggregate (N) i.e.

$$R_g \sim N^{\frac{1}{d_f}}$$

[14.8]

The exponent d_f is the fractal dimension. For DLA the particles stick together irreversibly as soon as they touch and cannot move positions. Under these conditions, d_f has been found to equal 1.8 (Meakin, 1983). However, for RLA, the particles do not necessarily stick irreversibly when they touch. This gives a d_f of about 2.1 (Kolb and Hermann, 1987).

However, in food gelation, the mechanism of aggregation proceeds via a different route and this leads to a divergence from the fractal behaviour observed for simple systems. These differences arise from a number of complicating factors that lead to structural differences between fractal aggregates and food gels. Food gels are usually formed from aggregated proteins or polysaccharides. For food applications, these are often formed at a much higher concentration than the dilute systems where fractal behaviour is observed. As a consequence, many studies of the fractal behaviour of gels will yield a fractal exponent greater than 2.1, which strictly is outside the fractal regime. It is therefore questionable whether the fractal concept is useful for the majority of food gels.

The mechanism of gelation in food gels is also somewhat different to that observed for the idealised fractal systems. It has been proposed that two types of gelation mechanism and gel type are important in food systems (Dickinson, 2000), namely reversible and irreversible gelation. If a heat-denatured protein solution is quench cooled, it can form a transient gel forming via weak attractive interactions that can break and reform, which allows the gel to undergo a time-dependent rearrangement to a more compact state. This type of gel occurs when the protein solution is cooled into a metastable spinodal region of its phase diagram. Sometimes the rearrangement of transient gels leads to spinodal decomposition (Binder, 1991) or the systems become trapped in a stable gelled state (de Hoog and Tromp, 2003). Dickinson et al. have studied the formation of transient gels using a BD simulation model, where the interactions between particles are attractive but non-bonded (Bijsterbosch et al., 1995; Dickinson, 2000). The initial gel structure that formed was observed to rearrange over time, but instead of an expected increase in fractal dimension, the d_f decreased with time. This suggested that although rearrangement led to a short-range compaction of the gel structure, the longer-range structure becomes more open and diffuse.

A second type of aggregation model has been introduced by Euston and Costello (2006) that reproduces features of the transient gel state. This is an MC 3D lattice model, where the 'proteins' are represented as deformable globules of subunits rather than as hard spheres. This allows for the denaturation step of protein gelation to be included in the model, by varying the interactions between subunits as a function of reduced temperature. At low temperature, below a critical unfolding temperature, the globules are compact and folded. Above a critical temperature, the globule becomes unfolded, open and chain-like.

Two forms of globule were used in the model, a homo-globule where all of the subunits are of the same kind and a hetero-globule where a proportion of the subunits are hydrophobic and are found at the centre of the globule surrounded by the rest of the subunits, which are polar. Further complexity was introduced into the model by varying the interaction between different types of subunit, where hydrophobic subunits had a strong attractive interaction with their own type and were allowed to interact with a weak attraction or no interaction with polar subunits. Polar subunits were allowed to interact attractively with their own type. Similarly, cross interactions between subunits on different globules were allowed between hydrophobic–hydrophobic, hydrophobic–polar or polar–polar subunits and these were set as attractive or zero. The hetero-globule is a simple representation of the structure of a globular protein.

By simulating systems of the globules at different volume fractions and reduced temperature, Costello and Euston (2006) were able to construct state diagrams that defined a number of states of the system separated by state boundaries, which depend on the allowed interactions between the subunits. The unfolding temperature was defined as the reduced temperature at which the globule changed from a folded to an open conformation. A gelation line was defined as the temperature concentration combination at which a percolating network was formed. Both homo- and hetero-globules exhibited a temperature-dependent gelation concentration between volume fractions of 4 to 6%. As the complexity of the globule structure and the number of allowed interactions increased, the state diagram became more complicated. Even for the simplest case of a homoglobules with no inter-globule interactions, the state diagram showed four states that were bounded by the denaturation and gelation curves (Figs 14.12 and 14.13).

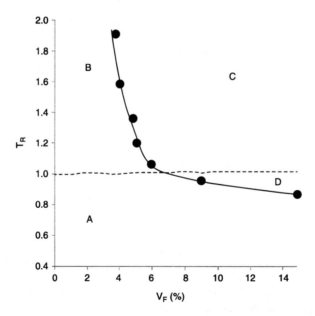

Fig. 14.12 State diagram for homoglobules with no interglobule interactions. The dashed line represents the folding/unfolding curve; ●, the gelation line for systems heated from the folded state. Different states are marked with capital letters A–D, and the lower case letters in Fig. 14.13 represents the snapshot conformations in the states. Reprinted with permission from Costello and Euston (2006), © American Chemical Society.

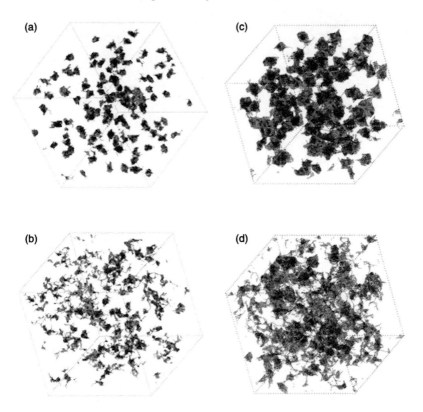

Fig. 14.13 Snapshot conformations for non-interacting homoglobule systems from the state diagram in Fig. 14.12. The following volume fraction (V_F) and reduced temperature (T_R) combinations define the conformations: (a) $V_F = 2\%$, $T_R = 0.82$; (b) $V_F = 2\%$, $T_R = 1.23$; (c) $V_F = 10\%$, $T_R = 0.82$; (d) $V_F = 10\%$, $T_R = 1.23$. The box side lengths are 73 lattice sites for $V_F = 2\%$ and 43 lattice sites for $V_F = 10\%$. Reprinted with permission from Costello and Euston (2006), © American Chemical Society.

For interacting globules and hetero-globules, the state diagram is more complex, but also more interesting. For interacting heteroglobules, it shows seven states (Figs 14.14 and 14.15), ranging from systems that resemble entanglement type gels to dense phase separated gels that are fibril-like in structure. Rich and complex phase behaviour is observed experimentally for heat denatured gels formed from globular proteins (van der Linden and Foegeding, 2009). For the milk whey protein β-lac, varying the protein concentration, pH and ionic strength can produce structures that are fibrillar in nature (strong interactions between the proteins at low pH away from the pI and low ionic strength) or particulate (weak interactions, pH close to the pI and high ionic strength), which differ in physic-chemical properties such as opacity and water-holding capacity (Langton and Hermansson, 1992). The pH induced transition from particulate to fibrillar gel is believed to be a microphase separation, and fibrillar gels to be a phase separated system (Ako et al., 2009).

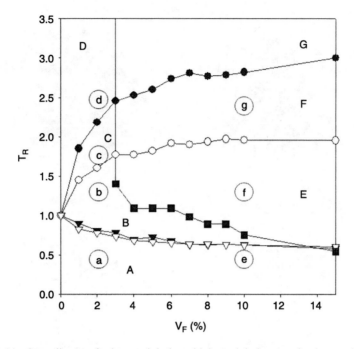

Fig. 14.14   State diagram for heteroglobules with interglobule attractive interactions. ■ = the gelation/crossover line; ● = the local unfolding line for type 1 subunits; ○ = the local unfolding line for type 2 subunits; ▼ = the global unfolding line for type 1 subunits; ▽ = the global unfolding line for type 2 subunits. Different states are marked with capital letters A–G, and the lower case letters a–g represents the region of the state diagram represented by snapshot conformations in Fig. 14.15. Reprinted with permission from Costello and Euston (2006), © American Chemical Society.

Irreversible gelation has also been simulated using models where permanent bonds are formed between the protein particles in the system, usually with a torsional component to the bonding force added so that the gel can rearrange but does not undergo spinodal decomposition. Whittle and Dickinson (1997) introduced such a model, where the probability of two protein particles forming a permanent bond (P_b), and the strength of non-bonded interactions are used as parameters to control gel structure. By varying these parameters, irreversible gels of varying density and degree of phase separation can be formed. Phase separation can be arrested by having gels that form quickly (high P_b) (Whittle and Dickinson, 1998). A particularly interesting application of this model was to study the effect of added particles on the rheological response of the gel. The results were in qualitative agreement with experimental studies from the same laboratory on the rheological properties of emulsion gels (Chen and Dickinson, 1999), in that when the particles were allowed to interact with the gel matrix, a strengthening of the gel was observed (simulated storage and loss moduli increased), whilst non-interacting particles weakened the gel structure (Whittle and Dickinson, 1997).

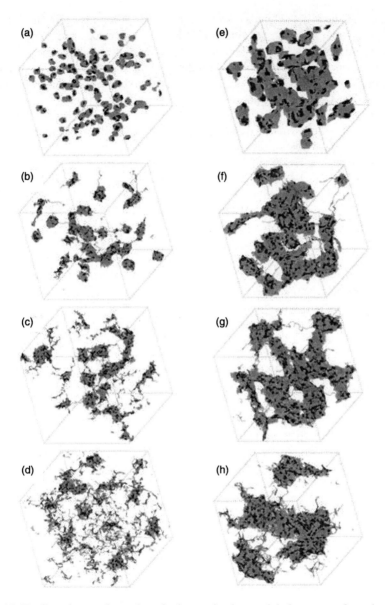

Fig. 14.15 Snapshot conformations for interacting heteroglobule systems from the state diagram in Fig. 14.14. The following volume fraction (V_F) and reduced temperature (T_R) combinations define the conformations: (a) $V_F=2\%$, $T_R=0.7$; (b) $V_F=2\%$, $T_R=1.23$; (c) $V_F=2\%$, $T_R=1.96$; (d) $V_F=10\%$, $T_R=0.7$; (e) $V_F=10\%$, $T_R=1.23$; (f) $V_F=10\%$, $T_R=1.96$. The box side lengths are 73 lattice sites for $V_F=2\%$ and 43 lattice sites for $V_F=10\%$. Reprinted with permission from Costello and Euston (2006), © American Chemical Society.

The tendency for globular proteins to form different aggregate and gel structures under differing conditions is of interest to the food technologist, since these aggregates may be exploitable as food ingredients (Iordache and Jelen, 2003). Recent research has focused on the controlled aggregation of proteins to form fibrillar aggregates or particulate aggregates (Ikeda and Morris, 2002; Gosal *et al.*, 2004). These are of interest to the food scientist, as they may offer ways to form novel food structures or improved protein functional properties. Both processes have been studied using simple molecular models (Zhang *et al.*, 2008; Adamcik *et al.*, 2010).

The ability of several food globular proteins to form fibrils has been known for some time. The mechanism of formation appears to be similar to that found with the amyloid proteins responsible for diseases such as Alzheimers, BSE, Creutzfeld-Jakob and Huntingdon's disease, although the details of the mechanism of formation are still uncertain. The most studied of the food proteins to form fibrils is β-lac (Gosal *et al.*, 2004). With β-lac (and other globular food proteins), fibrils form at low pH, low ionic strength and heating at 80 °C for several hours. The mechanism seems to require partial denaturation of the protein, acid hydrolysis to form peptides that then self-associate into filaments, which then aggregate further into fibrils. It appears that only certain peptides are found preferentially in the fibrils (Akkermans *et al.*, 2008). The fibrils themselves can have a range of thicknesses (Adamcik *et al.*, 2010).

It is now believed that individual filaments are twisted around each other to from a helical fibril (Adamcik *et al.*, 2010). These authors have carried out a very elegant CG MD simulation of filament aggregation into fibrils, where the proteins are represented as spheres interacting via Lennard–Jones potentials and intra-molecular potentials via a Finitely Extensible Non-linear Elastic (FENE) potential (Warner, 1972). They have used this to explain why AFM observations clearly show fibrils with differing helical repeat length, and to propose that fibrils are formed of differing numbers of filaments arranged in a twisted ribbon-like structure (Adamcik *et al.*, 2010; Fig. 14.16). AFM images show that there is a

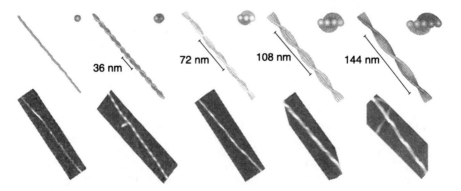

Fig. 14.16 AFM images and corresponding coarse-grain molecular dynamics reconstructions of left-handed helical fibril formation from the twisting of multi-stranded ribbons, with the number of filaments ranging between 1 and 5. Reprinted from Adamcik *et al.* (2010) with permission from Macmillan Publishers Ltd, © 2010.

correlation between the maximum height of fibrils and the repeat distance of the helix. The simulations suggest this arises from the differing number of filaments (Fig. 14.16), which lead to a change in the degree of twisting and thus the period of the fibril repeat.

At the other end of the structural spectrum of globular protein aggregates (and gels) are the particulate aggregates. Far from being linear and fine-stranded, these are dense, roughly spherical particles that form under different conditions to fibrils (pH 4–6, high ionic strength) close to the isoelectric point under conditions where electrostatic repulsion is weak (Ikeda and Morris, 2002). Aggregation in this regime has been used to produce micro-particulated (Iordache and Jelen, 2003) and partially denatured whey protein products (Holst et al., 1996; Campbell, 2007), which have found application as, for example, fat replacers. One of the problems associated with making aggregated protein products is how to control the size of aggregates formed. This is also a more general problem in the biotechnology industry, where proteins are susceptible to aggregation when they are processed downstream of the production, and suppression of this becomes an issue.

Recently, Jones and McClements (2010) have explored a method for making heat-denatured whey protein particles that are stabilised by a layer of anionic polysaccharide that adsorbs to the particle surface after they are formed. Although this has not been simulated specifically, Zhang et al. (2008) have used an MC lattice chain model that is relevant to protein-polysaccharide co-aggregates. In their model, they look at the effect of introducing a weakly hydrophobic polymer to a system of model proteins that in the absence of the second polymer would undergo aggregation. The polymer has to be of sufficient hydrophobicity and chain length that it can wrap around the hydrophobic regions of the folded and unfolded proteins and protect them against aggregation by segregating them into discrete protein rich regions. Although the interactions in whey-protein/pectin co-aggregates are electrostatic in nature, the principle is similar in that over-aggregation of whey protein is suppressed and controlled by the adsorption of pectin to the exposed hydrophobic surface of the aggregates. The simulations may also hint at other mechanisms of stabilisation of globular protein aggregates that involve the use of weakly adsorbing hydrophobic proteins or polysaccharides.

14.6.2   Food emulsions

Emulsions are dispersions of one liquid in a second liquid. In foods they are most often dispersions of triglyceride oil in an aqueous continuous phase, although fat continuous systems such as butter and margarine contain water droplets dispersed through the fat phase. Emulsions are stabilised by an adsorbed layer of surface active material, typically a protein or mixtures of protein and other surface-active ingredients, which sit at the oil–aqueous interface. Food emulsions are trapped in a kinetically stable state, since the lowest free energy state is one where the oil and aqueous phase are demixed. Thus, like gelled protein systems, they will undergo time-dependent changes in structure. The role of the food scientist is to control the instability, such that products of adequate shelf-life are ensured.

A feature of emulsions that they do not share with dispersions of solid particles is that oil-droplets are able to deform. This influences their structure by allowing them to pack more densely so that oil high volume fraction products such as mayonnaise can be formed and also influences their rheological properties, since fluid droplets couple with the aqueous disperse phase in a different manner to solid particle. Modelling of deformable particles is not that simple and as a consequence emulsion-like systems were not simulated until the 1980s. Early deformable particle models were invariably 2D lattice-based MC simulations, which incorporated a simple representation of the oil droplets as a deformable particle (Dickinson, 1984; Barker and Grimson, 1987; Dickinson and Euston, 1989, 1992; Pakula, 1991a,b). These were either lattice chain representations of the oil-droplet (Dickinson, 1984; Dickinson and Euston, 1989, 1992; Pakula, 1991a,b) or deformable globules (Barker and Grimson, 1987).

Dickinson (1984) introduced a simple cyclic lattice chain model for deformable particles, which was investigated in more detail by Dickinson and Euston (1989, 1992). Pakula (1991a,b) introduced a similar model that used a co-operative motion algorithm, which allowed simulations to very high area fractions. In all of these models, a clear link between particle deformability and system structure was observed. Deformability of the droplets was shown to lead to liquid-like behaviour, even at droplet concentrations well above the close-packed area fraction for rigid 2D disks of about 0.7 (Dickinson and Euston, 1989, 1992; Barker and Grimson, 1987; Pakula, 1991a,b). Pakula's model was capable of simulating highly deformable globules at an area fraction of 1.0, and even at this density, liquid-like behaviour was seen (Pakula, 1991a,b).

Since the early attempts at simulating emulsions, the MC approach has fallen out of favour and given way to models that use dynamic simulation techniques such as BD and DPD, and also the LBM. Urbina-Villalba and García-Sucre (2000) used a BD method that also incorporated creaming (gravitational) forces to simulate coalescence in emulsions. In their model, droplets were allowed to coalesce and form a larger single droplet under the action of attractive and repulsive forces. They also included the effect of adsorbed surfactant at the drop surface, by calculating the change in surfactant adsorption, surfactant density and thus the change in charge density at the droplet surface as they coalesced. Using this model they were able to investigate the effect of varying surfactant concentration and/or creaming on the final number and size of droplets. Good agreement was observed with analytical theories for droplet coalescence under some conditions. The authors point out that hydrodynamic interactions were ignored in this model, as was any effect of hydrodynamics on droplet shape and interactions. They considered that these would need to be included to improve simulation predictions compared to theory and experiment.

DPD is well-suited to modelling of multi-component systems such as emulsions, and it has been used in a number of studies to look at the effect of adsorbing molecules on the stability of oil or water droplets in emulsions. These have mainly been carried out on hydrocarbon oil emulsions with synthetic copolymers as the adsorbing molecules, but the methodology and the general results are relevant to food emulsions. Lin *et al.* (2012) used DPD to simulate the formation and time evolution of n-octadecane droplets in water in the presence of adsorbing styrene-maleic acid copolymers using a CG model. Octadecane molecules quickly separate and the copolymer adsorbs to the oil–water interface. As the simulation progresses the system coarsens, since the copolymer surface coverage is too low to stabilise the oil droplets and they coalesce. Eventually only a single droplet remains.

Alvarez *et al.* (2011) have simulated a more complex system, where the oil phase is a mixture of asphaltenes, resins, aromatic hydrocarbons and saturated hydrocarbons that reflects the composition of Mexican crude oil. They studied the effect of ethylene oxide-propylene oxide-ethanolamine triblock copolymer de-emulsifiers on the stabilisation of two water droplets suspended in the crude oil continuous phase. They observed a strong dependence of the stability on copolymer chain length, with coalescence occurring below a critical polymer chain length. They also observed an upper limit of chain length where the polymer became large enough to bridge between the two water droplets and then acted as a channel for water migration between the beads.

Rekvig *et al.* (2004) have used a similar DPD approach to study the effect of model low molecular weight surfactants on the coalescence stability of two approaching oil droplets dispersed in water. They looked at the approach of two water droplets through an oil phase and two oil droplets through a water phase. For the interaction of water droplets with an adsorbed layer of surfactant, repulsion between the hydrophobic tails was seen that had the effect of stabilising the interface. Conversely, for the approach of two oil droplets, the steric stabilising effect of the charged head group was minimal and coalescence of the two oil phases occurred.

Figure 14.17 is a snapshot confirmation from a simulation of the coalescence mechanism for two oil droplets in the presence of a five subunit surfactant (one

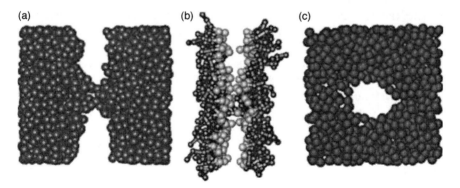

(a)   (b)   (c)

Fig. 14.17   Snapshots from film rupture: (a) the two oil phases projected onto the y–z plane; (b) the surfactant monolayers projected onto the y–z plane, headgroups are in light grey and tail groups in dark grey; (c) the water film projected onto the x–y plane. The surfactant had one head group and hydrophobic tail subunits. Reprinted with permission from Rekvig *et al.* (2004), © American Chemical Society.

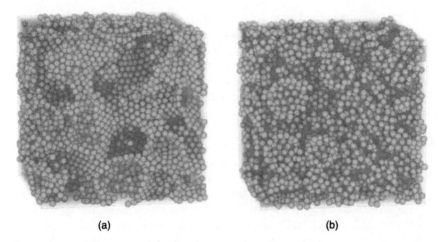

(a)   (b)

Fig. 14.18   Snapshot conformations after 250 000 lattice–Boltzmann steps for a simulation of a 2-phase mixture in the presence of adsorbing particles. For both conformations the volume fraction of the two fluid phases (dark and light grey regions) is constant. In conformation (a), the conditions (contact angle, adsorbing particle concentration) are such that adsorbing particles span the cell and trap or jam the fluid phases in a bicontinuous structure. In conformation (b), at a lower adsorbing particle concentration, the particles adsorb to the fluid phase and stabilise it in droplets as a Pickering stabilised emulsion. Reprinted from Figs 14.8 and 14.12, with permission from Jansen and Harting (2011), http://link.aps.org/doi/10.1103/PhysRevE.83.046707. © American Physical Society.

polar head group and four hydrophobic tail subunits) at a surface area coverage fraction of 0.6. As the surfaces of the two droplets approach, they reach a point where they are close enough for the water film to rupture. At this point, a hole is seen in the water phase in a projection onto the x–y plane (Fig. 14.17). The surfactant layer starts to bend and this forms a channel through which the oil particles can flow and start to mix between droplets. The results of this simulation were in qualitative with the predictions of theories for emulsion coalescence, such as the channel nucleation theory (Kabalnov and Wennerström, 1996).

Lattice–Boltzmann modelling has also been applied to emulsion systems that have a relevance to foods. Jansen and Harting (2011) have used LBM to investigate the transition from a bijel (bicontinuous interfacially jammed emulsion gel) (Stratford *et al.*, 2005) to a Pickering stabilised emulsion (Dickinson, 2006), by varying the particle concentration, contact angle and ratio of the oil and aqueous phases. Figure 14.18 illustrates the difference between the bijel and Pickering emulsion states and demonstrates the potential for LBM to be applied to food systems.

14.7   Conclusion

It appears that, in spite of continuing advances in methodology, hardware and software over the past few years, the uptake of molecular simulation methods in the study of food systems is still relatively slow. A large part of this can be attributed to the complex multi-component nature of foods. However, progress is being made and the next few years promise further advances. The increased use of non-atomistic methods such as CG MD simulation, DPD and in particular LBM, promises to open up simulation to larger, more complex (and computationally more demanding) systems.

In addition, there is a growing realisation that the so-called soft-matter approach to studying condensed phase biological systems is extremely useful in the study of food systems. Van der Sman (2009), in a recent review, listed 55 separate researchers who apply soft matter concepts to food systems, 9 of whom directly use simulation techniques of one form or another. The Royal Society of Chemistry has also organised a Faraday Discussion (http://www.rsc.org/ConferencesAndEvents/RSCConferences/FD158/index.asp) on the application of soft-matter physics to food systems, to encourage greater uptake of these ideas by the food science community. It is to be hoped that once soft-matter scientists realise the challenge that food systems offer, they will turn their attention to these and this will drive an increased use of simulation methods in food science.

14.8   Acknowledgements

The author would like to thank Wely Floriano, from Lakehead University, Canada for kindly providing a pdb file of the structure for T2R38. Prof. Raffaele Mezzenga, ETH Zurich is thanked for kindly providing a copy of Fig. 14.16.

14.9   References

ADAMCIK, J., JUNG, J. M., FLAKOWSKI, J., DE LOS RIOS, P., DIETLER, G. and ENGA, R. (2010), Understanding amyloid aggregation by statistical analysis of atomic force microscopy images, *Nature Nanotechnology*, **5**, 423–8.

AKKERMANS, C., VENEMA, P., VAN DER GOOT, A. J., GRUPPEN, H. *et al.* (2008), Peptides are building blocks of heat-induced fibrillar protein aggregates of β-lactoglobulin formed at pH 2, *Biomacromolecules*, **9**, 1474–9.

ALDER, B. J. and WAINWRIGHT, T. E. (1957), Phase transition for a hard sphere system, *Chem. Phys.*, **27**, 1208–10.

ALDER, B. J. and WAINWRIGHT, T. E. (1959), Studies in molecular dynamics. Part I: General method, *J. Chem. Phys.*, **31**, 459–66.

ALVAREZ, F., FLORES, E. A., CASTRO, L. V., HERNÁNDEZ, J. G., LÓPEZ, A. and VÁZQUEZ, F. (2011), Dissipative particle dynamics (DPD) study of crude oil-water emulsions in the presence of a functionalized co-polymer, *Energy Fuels*, **25**, 562–7.

AKO, K., NICOLAI, T., DURAND, D. and BROTONS, G. (2009), Micro-phase separation explains the abrupt structural change of denatured globular protein gels on varying the ionic strength or the pH, *Soft Matter*, **5**, 4033–41.

AMADEI, A., LINSSEN, A. B. M. and BERENDSEN H. J. C. (1993), Essential dynamics of proteins, *Proteins: Structure, Function, and Bioinformatics*, **17**, 412–25.

BARKER, G. C. and GRIMSON, M. J. (1987), A model suspension of deformable particles, *Mol. Phys.*, **62**, 269–81.

BEHRENS, M. and MEYERHOF, W. (2009), Mammalian bitter taste perception, *Results Probl. Cell Differ.*, **47**, 203–20.

BERGENSTRÅHLE, M., WOHLERT, J., HIMMEL, M. E. and BRADY, J. W. (2010), Simulation studies of the insolubility of cellulose, *Carbohydrate Research*, **345**, 2060–6.

BINDER, K. (1991), Spinodal decomposition, in, P. Haasen, *Phase Transformations in Materials*, Weinheim, VCH, 405–72.

BIJSTERBOSCH, B. H., BOS, M. T. A., DICKINSON, E., VAN OPHEUSDEN, J. H. J. and WALSTRA, P. (1995), Brownian dynamics simulation of particle gel formation: from argon to yoghurt, *Faraday Discussions of the Chemical Society*, **101**, 51–64.

BRADY, J. W. (1989), Molecular dynamics simulations of α-D-glucose in aqueous solution, *J. Am. Chem. Soc.*, **111**, 5155–65.

BRASIELLO, A., RUSSO, L., SIETTOS, C., MILANO, G. and CRESCITELLI, S. (2010), Multi-scale modelling and coarse-grained analysis of triglycerides dynamics, in, S. Pierucci and B. G. Ferraris, *Computer-Aided Chemical Engineering*, **28**, 625–30.

BRASIELLO, A., CRESCITELLI, S. and MILANO, G. (2011), Development of a coarse-grained model for simulations of tridecanoin liquid–solid phase transitions, *Phys. Chem. Chem. Phys.*, **13**, 16618–28.

CAFFARENA, E. R. and GRIGERA, J. R. (1997), Glass transition in aqueous solutions of glucose. Molecular dynamics simulation, *Carbohydrate Research*, **300**, 51–7.

CAFFARENA, E. R. and GRIGERA, J. R. (1999), Hydration of glucose in the rubbery and glassy states studied by molecular dynamics simulation, *Carbohydrate Research*, **315**, 63–9.

CAMPBELL, L. J. (2007), Fat replacement material and method of manufacture thereof, US Patent No. 7166316.

CARROTTA, R., BAUER, R., WANINGE, R. and RISCHEL, C. (2001), Conformational characterization of oligomeric intermediates and aggregates in β-lactoglobulin heat aggregation, *Protein Sci.*, **10**, 1312–18.

CHANDRASEKHAR, I. and VAN GUNSTEREN, W. F. (2001), Sensitivity of molecular dynamics simulations of lipids to the size of the ester carbon, *Current Science India*, **81**, 1325–7.

CHANDRASEKHAR, I. and VAN GUNSTEREN, W. F. (2002), A comparison of the potential energy parameters of aliphatic alkanes: molecular dynamics simulations of triacylglycerols in the alpha phase, *European Biophysics Journal*, **31**, 89–101.

CHEN, J. and DICKINSON, E. (1999), Effect of surface character of filler particles on rheology of heat-set whey protein emulsion gels, *Colloids and Surfaces B: Biointerfaces*, **12**, 373–81.

CHRISTENSEN, N. J., HANSEN, P. I., LARSEN, F. H., FOLKERMAN, T., MOTAWIA, M. S. and ENGELSEN, S. B. (2010), A combined nuclear magnetic resonance and molecular dynamics study of the two structural motifs for mixed-linkage β-glucans: methyl β-cellobioside and methyl β-laminarabioside, *Carbohydrate Research*, **345**, 474–86.

COSTELLO, G. and EUSTON, S. R. (2006), A Monte Carlo simulation of the denaturation, aggregation, phase-separation and gelation of model globular molecules, *Journal of Physical Chemistry B*, **110**, 10151–64.

COX, P. W. and HOOLEY, P. (2009), Hydrophobins: new prospects for biotechnology, *Fungal Biology Reviews*, **23**, 40–7.

DAI, W., YOU, Z., ZHOU, H., ZHANG, J. and HU, Y. (2011), Structure-function relationships of the human bitter taste receptor hTAS2R1: Insights from molecular modelling studies, *J. Recept. Signal Transduct. Res.*, **31**, 229–40.

DAY, R., BENNION, B. J., HAM, S. and DAGGETT, V. (2002), Increasing temperature accelerates protein unfolding without changing the pathway of unfolding, *J. Mol. Biol.*, **322**, 189–203.

DESLANDES, Y., MARCHESSAULT, R. H. and SARKO, A. (1980), Triple-helical structure of (1,3)-ß-D-glucan, *Macromolecules*, **13**, 1466–71.

DICKINSON, E. (1984), Statistical-model of a suspension of deformable particles, *Physical Review Letters*, **53**, 728–31.

DICKINSON, E. (2000), Structure and rheology of simulated gels formed from aggregated colloidal particles, *J. Colloid Interface Sci.*, **225**, 2–15.

DICKINSON, E. (2006), Interfacial particles in food emulsions and foams, in B. P. Binks and T. S. Horozov, *Colloidal Particles at Liquid Interfaces*, Cambridge, Cambridge University Press, 298–327.

DICKINSON, E. and EUSTON, S. R. (1989), Statistical study of a concentrated dispersion of deformable particles modelled as an assembly of cyclic lattice chains, *Molecular Physics*, **66**, 865–86.

DICKINSON, E. and EUSTON, S. R. (1990), Simulation of adsorption of deformable particles modelled as cyclic lattice chains, *Journal of the Chemical Society Faraday Transactions 2*, **86**, 85–9.

DICKINSON, E. and EUSTON, S.R. (1992a), Monte Carlo simulation of colloidal systems, *Advances in Colloid and Interface Science*, **42**, 89–148.

DICKINSON, E. and EUSTON, S.R. (1992b), A statistical model for the simulation of adsorption from a mixture of deformable particles, *Journal of Colloid and Interface Science*, **152**, 562–72.

DICKINSON, E. and KRISHNA, S. (2001), Aggregation in a concentrated model protein system: a mesoscopic simulation of β-casein self assembly, *Food Hydrocolloids*, **15**, 107–15.

DICKINSON, E., HORNE, D. S., PHIPPS, J. S. and RICHARDSON, R. M. (1993), A neutron reflectivity study of the adsorption of β-Casein at fluid interfaces, *Langmuir*, **9**, 242–8.

DYSON, H. J. and WRIGHT, P. E. (2005), Intrinsically unstructured proteins and their functions, *Nat. Rev. Mol. Cell Biol.*, **6**, 197–208.

EDGE, C. J., SINGH, U. C., BAZZO, R., TAYLOR, G.L.,DWEK, R. A. and RADEMACHER, T. W. (1990), 500-picosecond molecular simulation in water of Man α-1-2 Man a-glycosidic linkage present in Asn-linked oligomannan-type structures in glycoproteins, *Biochemistry*, **29**, 1971–4.

ENGELSEN, S. B, BRADY, J. W. and SHERBON, J. W. (1994), Simulations of the aqueous solvation of trilaurin, *Journal of Agricultural and Food Chemistry*, **42**, 2099–107.

ERMAK, D. L. and MCCAMMON, J. A. (1978), Brownian dynamics with hydrodynamic interactions, *J. Chem. Phys.*, **69**, 1352–60.

ESPAÑOL, P. and WARREN, P. B. (1995), Statistical-mechanics of dissipative particle dynamics, *Europhys. Lett.*, **30**, 191–6.

EUSTON, S. R. (2010), Molecular dynamics simulation of protein adsorption at fluid interfaces: a comparison of all-atom and coarse-grained models, *Biomacromolecules*, **11**, 2781–7.

EUSTON, S. R. (2012), Molecular dynamics simulation of the effect of heat on the conformation of bovine β-lactoglobulin A: a comparison of conventional and accelerated methods, *Food Hydrocolloids*, **30**, 519–30.

EUSTON, S. R. and HORNE, D. S. (2005) Simulating the self-association of caseins, *Food Hydrocolloids*, **19**, 379–86.

EUSTON, S. R and NASER, M. A. (2005), Modelling the equation of state of globular proteins adsorbed at a surface, *Langmuir*, **21**, 4227–35.

EUSTON, S. R. and NICOLOSI, M. (2007), Simulating the self-association of caseins: towards a model for the casein micelle, *Le Lait – Dairy Science and Technology*, **87**, 389–412.

EUSTON, S. R., UR-REHMAN, S. and COSTELLO, G. (2007), Denaturation and aggregation of ß-lactoglobulin – A preliminary molecular dynamics study, *Food Hydrocolloids*, **21**, 1081–91.

EUSTON, S. R., HUGHES, P., NASER MD. A. and WESTACOTT, R. (2008a), Comparing the conformation of barley LTP-1 adsorbed at a water-vacuum and a water-decane interface: a molecular dynamics study, *Biomacromolecules*, **9**, 1443–53.

EUSTON, S. R., HUGHES, P., NASER MD. A. and WESTACOTT, R. (2008b), Molecular dynamics simulation of the co-operative adsorption of barley lipid transfer protein and *cis* iso-cohumulone at the vacuum–water interface, *Biomacromolecules*, **9**, 3024–32.

EUSTON, S. R., BELLSTEDT, U., SCHILLBACH, K. and HUGHES, P. S. (2011), The adsorption and competitive adsorption of bile salts and whey protein at the oil–water interface, *Soft Matter*, **7**, 8942–51.

EVANS, D. E. and SHEEHAN, M. C. (2002), Do not be fobbed off, the substance of beer foam, a review, *J. Am. Soc. Brew. Chem.*, **60**, 47–57.

FAHEY, D. A., CAREY, M. C. and DONOVAN, J. M. (1995), Bile acid/phosphatidylcholine interactions in mixed monomolecular layers: Differences in condensation effects but not interfacial orientation between hydrophobic and hydrophilic bile acid species, *Biochemistry*, **34**, 10886–97.

FAN, H., WANG, X., ZHU, J., ROBILLARD, G. T. and MARK, A. E. (2006), Molecular dynamics simulations of the hydrophobin SC3 at a hydrophobic/hydrophilic interface, *Proteins*, **64**, 863–73.

FARRELL JR, H. M., QI, P. X., BROWN, E. M., COOKE, P. H., TUNICK, M. H. *et al.* (2002), Molten globule structures in milk proteins: implications for potential new structure–function relationships, *J. Dairy Sci.*, **85**, 459–71.

FERRY, J. D. (1980), *Viscoelastic Properties of Polymer Solutions*, 3rd edition, New York: John Wiley & Sons.

FLORIANO, W. B., HALL S., VAIDEHI, N., KIM, U, DRAYNA, D. and GODDARD, W. A. III (2006), Modeling the human PTC bitter-taste receptor interactions with bitter tastants, *J. Mol. Model.*, **12**, 931–41.

FRENKEL, D. and SMIT, B. (2002), *Understanding Molecular Simulation: From Algorithms to Applications*, 2nd edition, London, Academic Press.

GOSAL, W. S., CLARK, A. H. and ROSS-MURPHY, S. B. (2004), Fibrillar beta-lactoglobulin gels. Part I: Fibril formation and structure, *Biomacromolecules*, **5**, 2408–19.

GRAHAM, D. E. and PHILLIPS, M. C. (1979a), Proteins at liquid interfaces. Part I: Kinetics of adsorption and surface denaturation, *Journal of Colloid and Interface Science*, **70**, 403–14.

GRAHAM, D. E. and PHILLIPS, M. C. (1979b), Proteins at liquid interfaces Part III: Molecular structures of adsorbed films, *Journal of Colloid and Interface Science*, **70**, 427–39.

GRIGERA, J. R. (1988), Conformation of polyols in water: Molecular-dynamics simulations of mannitol and sorbitol, *J. Chem. Soc. Perkin Trans. 1*, **84**, 2603–8.

GROOT, R. D. and WARREN, P. B. (1997), Dissipative particle dynamics: bridging the gap between atomistic and mesoscopic simulation, *J. Chem. Phys.*, **107**, 4423–35.

GU, W. and BRADY, J. W. (1992), Molecular dynamics simulations of the whey-protein β-lactoglobulin, *Protein Engineering*, **5**, 17–27.

HA, S., GAO, J., TIDOR, B., BRADY, J. W. and KARPLUS, M. (1991), Solvent effect on the anomeric equilibrium in D-glucose: a free energy simulation analysis, *Journal of the American Chemical Society*, **113**, 1553–7.

HAKANPÄÄ, J., LINDER, M., POPOV, A., SCHMIDT, A. and ROUVINEN J. (2006a), Hydrophobin HFBII in detail: Ultrahigh-resolution structure at 0.75 Å, *Acta Crystallogr. D. Biol. Crystallogr.*, **62**, 356–67.

HAKANPÄÄ, J., SZILVAY, G. R., KALJUNEN, H., MAKSIMAINEN, M., LINDER, M. and ROUVINEN, J. (2006b), Two crystal structures of Trichoderma reesei hydrophobin HFBI – The structure of a protein amphiphile with and without detergent interaction, *Protein Science*, **15**, 2129–40.

HEINEMANN, B., ANDERSEN, K. V., NIELSEN, P. R., BECH, L. M. and POULSEN, F. M. (1996), Structure in solution of a four-helix lipid binding protein, *Protein Sci.*, **5**, 13–23.

HEINZ, V. and BUCKOW, R. (2010), Food preservation by high pressure, *Journal of Consumer Protection and Food Safety*, **5**, 73–81.

HILLS, B. P., WANG, Y. L. and TANG, H. R. (2001), Molecular dynamics in concentrated sugar solutions and glasses: an NMR field cycling study, *Mol. Phys.*, **99**, 1679–87.

HOCKNEY, R. W. (1970), The potential calculation and some applications, *Methods in Computational Physics*, **9**, 136–211.

HOFFMANN, M. A. M. and VAN MIL, P. J. J. M. (1997), Heat-induced aggregation of β-lactoglobulin: Role of the free thiol group and disulfide bonds, *J. Agric. Food Chem.*, **45**, 2942–8.

HOLST, H. H., CHRISTENSEN, A., ALBERTSEN, K., JENSEN, L. D., PEDERSEN, M. C. *et al.* (1996), Partially denatured whey protein product, US Patent No. 5494696.

HOLT, C. and SAWYER, L. (1993), Caseins as rheomorphic proteins, *J. Chem. Soc. Faraday Trans.*, **89**, 2683–90.

DE HOOG, E. H. A. and TROMP, R. H. (2003), On the phase separation kinetics of an aqueous biopolymer mixture in the presence of gelation: the effect of the quench depth and the effect of the molar mass, *Colloids and Surfaces A*, **213**, 221–34.

HOOGERBRUGGE, P. J and KOELMAN, J. M. V. A. (1992), Simulating microscopic hydrodynamic phenomena with dissipative particle dynamics, *Europhys. Lett.*, **19**, 155–60.

HORNE, D. S. (1998), Casein interactions: casting light on the black boxes, the structure in dairy products, *International Dairy Journal*, **8**, 171–7.

IAMETTI, S., DEGREGORI, B., VECCHIO, G. and BONOMI, F. (1996), Modifications occur at different structural levels during the heat denaturation of β-lactoglobulin, *Eur. J. Biochem.*, **237**, 106–12.

IKEDA, S. and MORRIS, V. J. (2002), Fine-stranded and particulate aggregates of heat-denatured whey proteins visualized by atomic force microscopy, *Biomacromolecules*, **3**, 382–9.

INTELMANN, D., BATRAM, C., KUHN, C., HASELEU, G. *et al.* (2009), Three TAS2R bitter taste receptors mediate the psychophysical responses to bitter compounds of hops (*Humulus lupulus* L.) and beer, *Chemosensory Perception*, **2**, 118–32.

IORDACHE, M. and JELEN, P. (2003), High pressure microfluidization treatment of heat denatured whey protein for improved functionality, *Innovative Food Sci. Emerging Technol.*, **4**, 367–76.

JANSEN, F. and HARTING, J. (2011), From bijels to Pickering emulsions: a lattice Boltzmann study, *Phys. Rev. E*, **83**, 046707.

JONES, O. G. and MCCLEMENTS, D. J. (2010), Biopolymer nanoparticles from heat-treated electrostatic protein–polysaccharide complexes: Factors affecting particle characteristics, *J. Food Sci.*, **75**, N36–43.

JUNG, J-M., SAVIN, G., POUZOT, M., SCHMITT, C. L. and MEZZENGA, R. (2008), Structure of heat-induced β-lactoglobulin aggregates and their complexes with sodium-dodecyl sulfate, *Biomacromolecules*, **9**, 2477–86.

KABALNOV, A. and WENNERSTRÖM, H. (1996), Macroemulsion stability: the oriented wedge theory revisited, *Langmuir*, **12**, 276–92.

KOEHLER, J. H. E., SAENGER, W. and VAN GUNSTEREN, W. F. (1988), Conformational differences between alpha-cyclodextrin in aqueous solution and in crystalline form. A molecular dynamics study, *J. Mol. Biol.*, **203**, 241–50.

KOELMAN, J. M. V. A. and HOOGERBRUGGE, P. J. (1993), Dynamic simulations of hard-sphere suspensions under steady shear, *Europhys. Lett.*, **21**, 363–8.

KOLB, M. and HERMANN, H. J. (1987), Surface fractals in irreversible aggregation, *Physical Review Letters*, **59**, 454–7.

KUMOSINSKI, T. F., BROWN, E. M. and FARRELL, H. M. (1991), Three-dimensional molecular modelling of bovine caseins: α_{s1}-casein, *Journal of Dairy Science*, **74**, 2889–95.

KUMOSINSKI, T. F., BROWN, E. M. and FARRELL, H. M. (1993a), Three-dimensional molecular modelling of bovine caseins: an energy minimized β-casein structure, *Journal of Dairy Science*, **76**, 931–45.

KUMOSINSKI, T. F., BROWN, E. M. and FARRELL, H. M. (1993b), Three-dimensional molecular modelling of bovine caseins: a refined, energy-minimized κ-casein structure, *Journal of Dairy Science*, **76**, 2507–20.

LADD, A. J. C. and VERBERG, R. (2001), Lattice–Boltzmann simulations of particle-fluid suspensions, *Journal of Statistical Physics*, **104**, 1191–251.

LAIO, A. and GERVASIO, F. L. (2008), Metadynamics: a method to simulate rare events and reconstruct the free energy in biophysics, chemistry and material science, *Rep. Prog. Phys.*, **71**, 126601.

LANGTON, M. and HERMANSSON, A. M. (1992), Fine-stranded and particulate gels of β-lactoglobulin and whey protein at varying pH, *Food Hydrocolloids*, **5**, 523–39.

LAZARIDOU, A., BILIADERIS, C.G., MICHA-SCRETTAS, M. and STEELE, B. R. (2004), A comparative study on structure-function relations of mixed-linkage (1-3), (1-4) linear-D-glucans, *Food Hydrocolloids*, **18**, 837–55.

LI, W., CUI, S. W., WANG, Q. and YADA, R. Y. (2012), Study of conformational properties of cereal β-glucans by computer modelling, *Food Hydrocolloids*, **26**, 377–82.

LIN, S. XU, M. and YANG. Z. (2012), Dissipative particle dynamics study on the mesostructures of n-octadecane/water emulsion with alternating styrene–maleic acid copolymers as emulsifier, *Soft Matter*, **8**, 375–84.

LINDER, M. B., SZILVAY, G. R., NAKARI-SETALA T. and PENTTILA, M. E. (2005), Hydrophobins: the protein amphiphiles of filamentous fungi, *FEMS Microbiol. Rev.*, **29**, 877–96.

LINDORFF-LARSEN, K. and WINTHER, J. R. (2001), Surprisingly high stability of barley lipid transfer protein, LTP1, towards denaturant, heat and proteases, *FEBS Letters*, **488**, 145–8.

LOPEZ, C. A., RZEPIELA, A., DE VRIES, A. H., DIJKHUIZEN, L., HUENENBERGER, P. H. and MARRINK, S. J. (2009), The Martini coarse grained force field: extension to carbohydrates, *J. Chem. Theory Comp.*, **5**, 3195–210.

MACKIE, A. R., GUNNING, A. P., WILDE, P. J. and MORRIS, V. J. (1999), The orogenic displacement of protein from the air/water interface by competitive adsorption, *J. Colloid Interface Sci.*, **210**, 157–66.

MALDONADO-VALDERRAMA, J., WOODWARD, N. C., GUNNING, A. P., RIDOUT, M. J., HUSBAND, F.A. *et al.* (2008), Interfacial characterization of beta-lactoglobulin networks: displacement by bile salts, *Langmuir*, **24**, 6759–67.

MALDONADO-VALDERRAMA, J., WILDE, P., MACIERZANKA, A. and MACKIE, A. (2011), The role of bile salts in digestion, *Adv. Colloid Interface Sci.*, **165**, 36–46.

MARCHESSAULT, R. H. and SUNDARARAJAN, P. R. (1983), 'Cellulose', in, G. O. Aspinal, *The Polysaccharides*, vol. 2, New York, Academic Press, 11–95.

MARRINK, S. J., RISSELADA, H. J., YEFIMOV, S., TIELEMAN, D. P. and DE VRIES, A. H. (2007), The Martini force field: Coarse-grained model for biomolecular simulations, *J. Phys. Chem. B*, **111**, 7812–24.

MCCAMMON, J. A., GELIN, B. R. and KARPLUS, M. (1977), Dynamics of folded proteins, *Nature*, **267**, 585–90.

MCCARTHY, A. N. and GRIGERIA, J. R. (2006), Effect of pressure on conformation of proteins. A molecular dynamics simulation of lysozyme, *Journal of Molecular Graphics and Modelling*, **24**, 254–61.

MCCLEMENTS, D. J. (2004), *Food Emulsions: Principles, Practices, and Techniques*, 2nd edition, Boca Raton, FL, CRC Press.

MEAKIN, P. (1983), Formation of fractal clusters and networks by irreversible diffusion-limited aggregation, *Physical Review Letters*, **51**, 1119–22.

METROPOLIS, N. and ULAM, S. (1949), The Monte Carlo method, *Journal of the American Statistical Association (American Statistical Association)*, **44**, 335–41.

METROPOLIS, N., ROSENBLUTH, A. ROSENBLUTH, M. TELLER, A. H. and TELLER, E. (1953), Equation of state calculations by fast computing machines, *Journal of Chemical Physics*, **21**, 1087–92.

MOLINERO, V. and GODDARD III, W. A. (2004), M3B: A coarse grain force field for molecular simulations of malto-oligosaccharides and their water mixtures, *J. Phys. Chem. B*, **108**, 1414–27.

MOLINERO, V. and GODDARD III, W. A. (2005), Microscopic mechanism of water diffusion in glucose glasses, *Phys. Rev. Lett.*, **95**, 045701.

MONTICELLI, L., KANDASAMY, S. K., PERIOLE, X., LARSON, R. G., TIELEMAN, D. P. and MARRINK, S. J. (2008), The Martini coarse-grained forcefield: Extension to proteins, *J. Chem. Theory Comp.*, **4**, 819–34.

NORTON, T. and SUN D-W. (2007), An overview of CFD applications in the food industry, in D-W. Sun, *Computational Fluid Dynamics in Food Processing*, Boca Raton, FL, CRC Press, 1–41.

ORSI, M., SANDERSON, W. and ESSEX, J. W. (2007), Coarse-grain modelling of lipid bilayers: a literature review, in M. G. Hicks and C. Kettner, *Molecular Interactions – Bringing Chemistry to Life*, Frankfurt, Beilstein-Institut, 185–205.

OSEEN, C. W. (1924), *Hydrodynamik*, Leipzig, Akademische Verlagsgesellschaft.

PAKULA, T. (1991a), A model for dense colloidal systems with deformable, incompressible particles, *Journal of Chemical Physics*, **94**, 2104–9.

PAKULA, T. (1991b), A model for dense systems with deformable particles or molecules, *Journal of Non-Crystalline Solids*, **131**, 289–92.

PEREZ, D., UBERUAGA, B. P., SHIM, Y., AMAR, J. G. and VOTER, A. F. (2009), Accelerated molecular dynamics methods: Introduction and recent developments, *Ann. Rep. Comput. Chem.*, **5**, 79–98.

POLAVARAPU, P. L. and EWIG, C. S. (1992), *Ab-initio* computed molecular structures and energies of the conformers of glucose, *J. Comput. Chem.*, **13**, 1255–61.

PUGNALONI, L. A., ETTELAIE, R. and DICKINSON, E. (2003), Growth and aggregation of surfactant islands during the displacement of an adsorbed protein monolayer: a Brownian dynamics simulation study, *Colloids and Surfaces B: Biointerfaces*, **31**, 149–57.

PUGNALONI, L. A., DICKINSON, E., ETTELAIE, R. MACKIE, A. R. and WILDE, P. J. (2004), Competitive adsorption of proteins and low-molecular-weight surfactants: computer simulation and microscopic imaging, *Advances in Colloid and Interface Science*, **107**, 27–49.

QI, X. L., BROWNLOW, S., HOLT, C. and SELLERS, P. (1995), Thermal denaturation of beta-lactoglobulin: effect of protein concentration at pH 6.75 and 8.05, *Biochim. Biophys. Acta*, **1248**, 43–9.

QI, X. L., HOLT, C., MCNULTY, D., CLARKE, D. T., BROWNLOW, S. and JONES, G. R. (1997), Effect of temperature on the secondary structure of beta-lactoglobulin at pH 6.7, as determined by CD and IR spectroscopy: a test of the molten globule hypothesis, *Biochem. J.*, **324**, 341–6.

REKVIG, L., HAFSKJOLD, B. and SMIT, B. (2004), Molecular simulations of surface forces and film rupture in oil/water/surfactant systems, *Langmuir*, **20**, 11583–93.

ROBERT, C. P. and CASELLA, G. (2004), *Monte Carlo Statistical Methods*, London, Springer.

ROBERTS, C. J. and DEBENEDETTI, P. G. (1999), Structure and dynamics in concentrated, amorphous carbohydrate-water systems by molecular dynamics simulation, *J. Phys. Chem. B*, **103**, 7308–18.

ROOS, Y. H. (2010), Glass transition temperature and its relevance in food processing, *Annual Review of Food Science and Technology*, **1**, 469–96.

ROTNE, J. and PRAGER, S. (1969), Variational treatment of hydrodynamic interaction in polymers, *J. Chem. Phys.*, **50**, 4831–7.

ROUBROEKS, J. P., ANDERSSON, R., MASTROMAURO, D. I., CHRISTENSEN B. E. and ÅMAN, P. (2001), Molecular weight, structure and shape of oat (1-3),(1-4)-β-D-glucan fractions obtained by enzymatic degradation with (1-4)-β-D-glucan 4-glucanohydrolase from *Trichoderma reesei*, *Carbohydr. Polym.*, **46**, 275–85.

SASAHARA, K., SAKURAI, M. and NITTA, K. (2001), Pressure effect on denaturant-induced unfolding of hen egg white lysozyme, *Protein Struct. Funct. Genet.*, **44**, 180–7.

SAWYER, L. and KONTOPIDIS, G. (2000), The core lipocalin, bovine beta-lactoglobulin, *Biochim. Biophys. Acta*, **1482**, 136–48.

SCHMIDT, R. K., KARPLUS, M. and BRADY, J. W. (1996), The anomeric equilibrium in D-xylose: free energy and the role of solvent structuring, *Journal of the American Chemical Society*, **118**, 541–6.

STRATFORD, K., ADHIKARI, R., PAGONABARRAGA, I., DESPLAT, J.-C. and CATES, M. E. (2005), Colloidal jamming at interfaces: a route to fluid-bicontinuous gels, *Science*, **309**, 2198.

SUBIRADE, M., MARION, D. and PÉZOLET, M. (1996), Interaction of two lipid binding proteins with membrane lipids: Comparative study using the monolayer technique and IR spectroscopy, *Thin Solid Films*, **284**, 326–9.

SUBIRADE, M., SALESSE, C., MARION, D. and PÉZOLET, M. (1995), Interaction of a nonspecific wheat lipid transfer protein with phospholipid monolayers imaged by fluorescence microscopy and studied by infrared-spectroscopy, *Biophys. J.*, **69**, 974–88.

SUGITA, Y. and OKOMOTO, Y. (1999), Replica exchange molecular dynamics method for protein folding, *Chem. Phys. Lett.*, **314**, 141–51.

SWOPE, W. C., ANDERSEN, H. C., BERENS, P. H. and WILSON, K. R. (1982), A computer simulation method for the calculation of equilibrium constants for the formation of physical clusters of molecules: Application to small water clusters, *J. Chem. Phys.*, **76**, 637–49.

TISS, A., RANSAC, S., LENGSFELD, H., HADVARY, P., CAGNA A. and VERGER, R. (2001), Surface behaviour of bile salts and tetrahydrolipstatin at the air–water and oil–water interfaces, *Chem. Phys. Lipids*, **111**, 73–85.

TORRIE, G. M. and VALLEAU, J. P. (1977), Nonphysical sampling distributions in Monte Carlo free-energy estimation: Umbrella sampling, *Journal of Computational Physics*, **23**, 187–99.

ULMIUS, J., LINDBLOM, G., WENNERSTROM, H., JOHANSSON, L. B.-A., FONTELL, K. *et al.* (1982), Molecular organization in the liquid-crystalline phases of lecithin-sodium cholate-water systems studied by nuclear magnetic resonance, *Biochemistry*, **21**, 1553–60.

UMEMURA, M., YUGUCHI, Y. and HIROTSU, T. (2005), Hydration at glycosidic linkages of malto- and cello-oligosaccharides in aqueous solution from molecular dynamics simulation: Effect of conformational flexibility, *Journal of Molecular Structure (Theochem)*, **730**, 1–8.

URBINA-VILLALBA, G. and GARCÍA-SUCRE, M. (2000), Brownian Dynamics simulation of emulsion stability, *Langmuir*, **16**, 7975–7985.

VADNERE, M. and LINDENBAUM, S. (1982), Distribution of bile salts between 1-octanol and aqueous buffer, *J. Pharm. Sci.*, **71**, 875–81.

VAN DER LINDEN, E. and FOEGEDING, E. A. (2009), gelation: principles, models and applications to proteins, in, S. Kasapis, I. T. Norton and J. B. Ubbink, *Modern Biopolymer Science: Bridging the Divide Between Fundamental Treatise and Industrial Application*, London, Elsevier, 29–92.

VAN DER SMAN, R. G. M. (2012), Soft matter approaches to food structuring, *Advances in Colloid and Interface Science*, **176–7**, 18–30.

VAN EIJCK, B. P. and KROON, J. (1989), Molecular-dynamics simulations of β-D-ribose and β-D-deoxyribose solutions, *J. Mol. Struct.*, **195**, 133–46.

VERLET, L. (1967), Computer 'experiments' on classical fluids. Part I: Thermodynamical properties of Lennard-Jones molecules, *Phys. Rev.*, **159**, 98–103.

WANG, L. and SUN, D-W. (2003), Recent developments in numerical modelling of heating and cooling processes in the food industry – a review, *Trends in Food Science and Technology*, **14**, 408–23.

WANG, X., GRAVELAND-BIKKER, J. F., DE KRUIF, C. G. and ROBILLARD, G. T. (2004), Oligomerization of hydrophobin SC3 in solution: from soluble state to self-assembly, *Protein Sci.*, **13**, 810–21.

WARNER, H. R. (1972), Kinetic theory and rheology of dilute suspensions of finitely extendible dumbbells, *Ind. Eng. Chem. Fundamentals*, **11**, 379–87.

WENZEL, P. B. and CAMMENGA, H. K. (1998), Equilibrium penetration of DMPC monolayers by sodium cholate, *J. Colloid Interface Sci.*, **207**, 70–7.

WESSELS, J. G. H. (1994), Developmental regulation of fungal cell-wall formation, *Annu. Rev, Phytopathol.*, **32**, 413–37.

WHITTLE, M. and DICKINSON, E. (1997), Brownian dynamics simulation of gelation in soft sphere systems with irreversible bond formation, *Molecular Physics*, **90**, 739–58.

WHITTLE, M. and DICKINSON, E. (1998), Large deformation rheological behaviour of a model particle gel, *Journal of the Chemical Society Faraday Transactions*, **94**, 2453–62.

WOOD, P. (2001), Cereal β-glucans: Structure, properties and health claims, in B. McCleary and L. Prosky, *Advanced Dietary Fibre Technology*, Oxford, Blackwell Science, 315–27.

WOOD, P. J. (2003), Relationships between solution properties of cereal β-glucans and physiological effects – a review, *Trends in Food Science and Technology*, **13**, 313–20.

ZANGI, R., DE VOCHT, M. L., ROBILLARD, G. T. and MARK, A. E. (2002), Molecular dynamics study of the folding of hydrophobin SC3 at a hydrophilic/hydrophobic interface, *Biophysical Journal*, **83**, 112–24.

ZHANG, L., LU, D. and LIU, Z. (2008), How native proteins aggregate in solution: a dynamic Monte Carlo simulation, *Biophysical Chemistry*, **133**, 71–80.

Appendix: Electron microscopy: principles and applications to food microstructures

K. Groves, Leatherhead Food Research, UK and M. L. Parker, Institute of Food Research, UK

DOI: 10.1533/9780857098894.2.386

Abstract: Electron microscopy encompasses a range of different techniques and used to be considered as mainly useful for the traditional sciences of biology, materials and physics. In the last 30 years, the use of electron microscopy to further the understanding of foods and the development of new food products has increased hugely and it is now accepted that the techniques can give genuine benefits. This chapter describes the traditional electron microscopy techniques of transmission and scanning electron microscopy and the typical sample preparation methods used for foods. Examples of where studies using electron microscopy have given new understanding of food structures and the technological advances are included, with some images illustrating both the techniques and the product structures. Background references are given for further reading and examples of the information that can be obtained are included.

Key words: electron microscopy, scanning, transmission, food structure, microstructure, shelf-life, nanotechnology.

A1.1   Introduction

Over recent times the eating habits of consumers in the developed world have changed from using largely home-cooked meals of meat and vegetables, which were purchased regularly in small amounts from local shops, to the consumption of ready meals manufactured and transported in bulk to supermarkets. This, together with an improvement in affluence, has also meant a larger choice of foods and a subsequent increase in the number of foods on offer to the consumer. These changes are the result of increasing population and changes in lifestyle and have resulted in a real need to understand how foods can be made safely and of high quality. The necessary understanding has come about by applying science to the manufacture of ingredients and food products. Food science has advanced from a basic knowledge of recipe combinations and safety to a level where methods used for medicine or traditional sciences are readily used to understand how foods are put together and how this affects their flavour, texture and stability on the shelf. This understanding has then been used to change the manufacturing processes and help in the development of new foods. Techniques such as electron microscopy and other structural analytical methods have been combined with techniques in other fields such as rheology, sensory and microbiology to achieve this.

Changes in ingredients, in processing and in the foods during their shelf-life, result in differences in properties, often unwanted. These differences can sometimes be seen by eye but there is a need to get a closer look at the structure in greater detail to understand what the changes mean. Light and electron microscopes can give this increase in structural detail or resolution. The resolution of an object, that is the ability to discriminate between two close points, is determined by Abbe's equation, which defines the resolution in terms of the wavelength of the illumination. Using light will typically give a resolution of about 0.2 μm, although in practice it is difficult to see much below 1 μm. The electron microscope is able to image much finer detail in the structure of a sample than the light microscope, as it uses a beam of electrons for imaging and these have a smaller wavelength than photons.

The possibility that electrons could be used to form an image came into being when Hertz (1857–94) suggested that cathode rays had a wavelike motion. In 1899, Weichert demonstrated that cathode rays could be brought to a small spot by a magnetic field produced by a long solenoid. In 1926, Hans Busch showed that theoretically an electron beam could be focused with a solenoid in a similar way to light that is focused by glass lenses. This important principle allowed the development of the first prototype electron microscope in transmission form, designed by Ernst Ruska and Max Knoll in 1931. Many years later, Ruska was awarded the Nobel Prize for this important invention. A few years later, Von Ardenne (1938a,b) constructed a scanning transmission electron microscope (STEM) by adding scan coils to a transmission electron microscope (TEM), and produced an image of a zinc oxide crystal at a resolution of about 50 to 100 nm. The development of the scanning electron microscope (SEM) came from this pioneering science and was described by Zworykin *et al.* (1942).

The resolution of the first SEM was in the order of 50 nm, and at that time the TEM was producing a resolution of about 5 nm, therefore many considered the imaging ability of the SEM to be inferior to the TEM. The main advantage of the SEM was not shown until 1953, when a team working under Oatley in Cambridge were developing the SEM. They produced the first image of a 3D solid object (McMullan, 1953). The applications were initially not fully appreciated but the research of Smith and Oatley (1955) clearly could see the wider applications. The timeline of the developments in SEM is described by Breton, McMullen and Smith and gathered together by Hawkes (2004).

Since then the technique has progressed in many ways. The resolution of the instruments has been improved, with increasingly high-accelerating voltages to 'pull' the electron beam faster and therefore shorten the wavelength. Improved lens designs and much improved vacuum pumps also increase the resolution and quality of the image, reducing aberration. The detectors have been developed, especially for SEM, and the filament design and filament material have also become specialised, with the development of the field emission microscope (Pawley 1997).

The electron microscope is considered to be a key instrument in the evaluation of structure at the molecular and micro level in all areas of science including food. It can give an idea of surface or internal structure overall (especially the SEM) and can also reveal the role of molecules such as proteins, fat crystals or polysaccharides, for example in the emulsion, gel or foam matrix of a food product. However, as it is an expensive instrument and needs space, preparation equipment and expertise to interpret the images, it is usually only available in universities, or the research laboratories of global companies or contract R&D laboratories. Smaller desktop SEMs are now on the market, but it is true to say that electron microscopy is still a specialist technique.

This chapter concentrates on the applications of both TEM and SEM in food science, together with the use of X-ray analysis in combination with electron microscopy imaging. However, there are other forms of electron microscopy that need to be considered, such as scanning transmission electron microscopy (STEM), low voltage SEM and other more analytical forms of electron microscopy. These are discussed towards the end of the chapter under 'Developments in electron microscopy'.

The drive towards examination of samples without the need to fix and dehydrate has led to the development of the Environmental SEM (ESEM) and similar techniques such as the variable vacuum SEM. This is a hugely important development in electron microscopy, with an increasing number of applications, including looking at dynamic changes in samples, and as such is given a chapter in its own right in this book.

A1.2   Techniques and sample preparation

When considering electron microscopy, the two main techniques in conventional use are transmission electron microscopy (TEM) and scanning electron microscopy (SEM). Scanning coils can be fitted to a TEM allowing the electron beam to be scanned across the thin preparation of the sample and then imaged through the scanning imaging photomultiplier setup. This technique is known as scanning transmission electron microscopy (STEM) and further information is given by Colliex and Mory (1994).

Although both techniques use electrons to image the sample structure, they generate very different types of information. The image produced by SEM in conventional secondary detection mode is more readily interpreted by the non-microscopist, since it appears similar to the image seen by the eye, only magnified. However, the image produced by TEM is more difficult to understand and interpret, especially those produced by the freeze fracture technique. They complement each other and provide alternative facets of information on the ultrastructure. The SEM gives a 3D view of the sample, placing the position of the ingredients or components clearly on view and allowing changes in sensory properties or processing to be understood in terms of differences in structure. However, the TEM mainly gives a view of a thin slice through the sample or a copy of a fractured surface. These images are 2D and need to be considered in terms of how they fit in the 3D product. High voltage electron microscopy (HVEM) on sections up to 1 μm thick can be photographed to produce stereo pairs, which can be viewed to give a 3D image.

A1.2.1   Transmission electron microscopy (TEM)

The TEM is similar in principle to the transmitted light microscope, but uses a beam of electrons to form the image rather than using light. The electron beam is generated in the electron gun, traditionally using a tungsten filament (but also more recently field emission sources) in combination with a high voltage differential to accelerate the electrons from the filament. These electrons are focused, using electrostatic and electromagnetic lenses, through the column and onto and through a very thin preparation of the sample. Contrast in the sample is obtained by using metal, either as stains that bind to the sample, or in coating parts of the sample to impede the transmission of the electrons. The image is usually formed and seen after the beam passes through the sample by focusing the transmitted beam onto a fluorescent screen, photographic film or, more usually these days, digitally. A very simple schematic is shown in Fig. A1.1.

The main developments in TEM design to improve resolution have been in the generation of high voltages to increase the speed of the electrons and consequently the resolution, and in correction of aberrations, mainly spherical and chromatic. Advances in sample preparation have helped, and the use of low temperature (cryo) stages and microtomes opened a new door to visualisation of cell structure in the biological sciences. This technique was not adopted for food research to the same extent that cryo stages for SEM were, but has been developed more especially for immuno-labelling.

The use of TEM in food research has never been as popular as SEM. One of the reasons is that the preparation of the sample is time-consuming, involving chemical fixation, dehydration and usually embedding. Detailed descriptions of the processes are given later in this appendix. In addition, the preparation usually takes several days and requires extra equipment such as an ultra microtome, which adds to the expense. However, the effort is worthwhile as the type of information gained on the ultrastructure is not obtainable with any technique except STEM. It can be difficult to gain an idea of the scale of the structural detail revealed by TEM and so it is usually preceded by light microscopy, and often compared with SEM. It is best to use light microscopy at first to show the gross structure and then the TEM to show the finer detail. Examples of the images obtained from the oil gland of mint are shown in Figs A1.2a–c. Figure A1.2a shows the light microscope view of a section through the mint leaf, Fig. A1.2b the view by TEM and Fig. A1.2c the view by SEM.

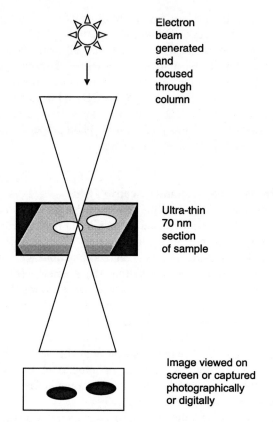

Electron
beam
generated
and
focused
through
column

Ultra-thin
70 nm
section
of sample

Image viewed on
screen or captured
photographically
or digitally

Fig. A1.1   Simple schematic for imaging by TEM. Reproduced with permission, ©
Leatherhead Food Research.

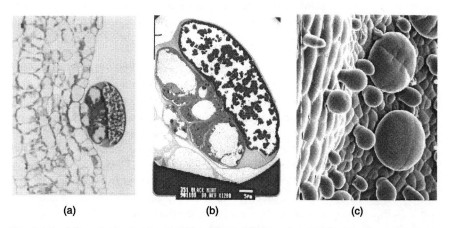

(a)   (b)   (c)

Fig. A1.2   Mint oil gland: (a) and (b) have been fixed in glutaraldehyde and osmium then
resin embedded and show resin sections by light microscopy and TEM, respectively;
(c) mint that has been prepared by CryoSEM. © IFR Norwich.

A1.2.2   Scanning electron microscopy (SEM)

Scanning electron microscopy (SEM) is by far the most frequently used electron microscopy technique for studying foods. It can give an image of the sample within hours, if the sample is dry and stable to the electron beam.

The imaging of the SEM could be compared with the use of a stereo light microscope, but using electrons. For SEM, the beam is generated in a similar way to the TEM from a filament or using field emission mode, focused using similar lenses but then scanned across the solid sample. The beam interacts with the sample, producing a number of different resulting outputs, and electrons scattered or emitted by the sample are collected by detectors and used to form the image (Fig. A1.3).

The typical topographical image seen under the SEM and easily recognisable by the eye is produced by the emission and imaging of secondary electrons, which are produced from the surface of the sample through inelastic interactions between the electron beam and the atoms of the sample. These secondary electrons are collected by a detector and accelerated towards a scintillator/photomultiplier and then displayed on a screen. The accelerating voltage of the electron beam has a strong effect on the imaging properties and sample integrity. Higher voltages give higher resolution, but can degrade the sample. Lower voltages are less damaging and produce secondary electrons from nearer the surface. Thus with low voltage imaging, delicate structures are retained. An example of this is the imaging of sorbitol crystals in the SEM. Sorbitol is used as a low calorie alternative to sugar in food products and Groves *et al.* (1996) showed that the properties of sorbitols manufactured by different suppliers were linked to their crystalline structure. These were easily damaged by conventional electron beam voltages of 15 or 20 kV, but could be seen using 5 kV or less (Fig. A1.4). A further advantage of low

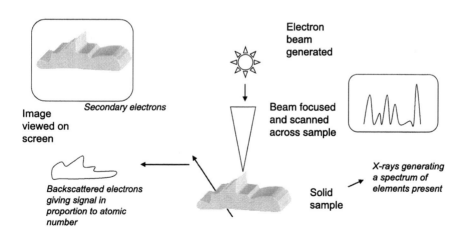

Fig. A1.3   Simple schematic for imaging by SEM. Reproduced with permission, © Leatherhead Food Research.

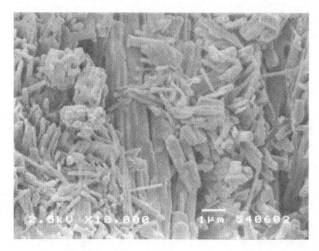

Fig. A1.4   Crystals of sorbitol by SEM. The alignment and density of crystals affects the properties of the sorbitol in products. Reproduced with permission, © Leatherhead Food Research.

voltage SEM is that structures on the surface can be missed by using conventional voltages but imaged at these lower voltages. Using conventional imaging, low contrast objects or detail only on the very surface of samples will be lost, unless low voltage SEM is used.

There are other emissions caused by the electron beam interaction with the sample that are used in SEM. One of these is the emission of backscattered electrons. These are electrons elastically scattered after collisions between the electron beam and the sample. Larger atoms (higher atomic number) in the sample have a higher probability of producing elastically backscattered electrons. Therefore the number of these electrons, and hence the strength of the signal to the detector, is proportional to the atomic number of the elements in the sample. Organic matrices typically found in foods (proteins, sugars, etc.) produce a low number of backscattered electrons and a dark signal, whereas higher atomic number elements, such as salt (sodium chloride), metals or additives such as calcium phosphate or titanium dioxide, will give a strong bright signal. An example of the difference in image appearance between the two forms of emission is shown in Figs A1.5a and b. Here the topographic image of the sample is shown using the secondary electron detector and the atomic number contrast using the backscattered electron detector. The salt can be seen as bright areas in the backscattered electron image, due to the higher atomic number of the sodium and chlorine.

When a sample is irradiated by the electron beam in the SEM, X-rays are produced. The energy of these X-rays is related to the elements in the sample and by using an X-ray detector, a spectrum of the energies of X-rays can be produced from the sample and the corresponding image (EDX-ray analysis). This information can be displayed as a spectrum of the whole image, as a spectrum from a point or region on the image, or as an elemental map. A map provides a

(a)

(b)

Fig. A1.5   SEM salt crystals showing the difference in image type with detector: (a) is the conventional secondary electron detector; and (b) shows the same area by backscattered electron imaging, which gives a signal related to the atomic number of the elements in the sample; the higher the atomic number the brighter the image. Reproduced with permission, © Leatherhead Food Research.

typical SEM image and at the same time with separate maps of the elements present. An element can be selected and overlaid on the SEM image with a colour to show the distribution of that element in the sample. This technique transforms the SEM from a purely imaging instrument to one which provides chemical analysis. An example of this is shown in Fig. A1.6, where sugar containing an anti-caking agent is imaged. The phosphorus in the anti-caking agent has been mapped and appears bright here but can be coloured to show the distribution among the sugar crystals. This is useful to highlight problems of distribution or aggregation in a mixed powder, which might cause problems on storage.

There is an alternative detector which responds to the wavelength of the X-rays and measures one wavelength at a time (WDX-ray analysis). This is more sensitive and has better resolution, but is more expensive. It is generally used in conjunction with the energy dispersive analyser. For food research, the energy dispersive analysis is by far the most commonly used technique.

(a)

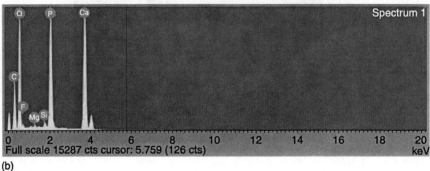

(b)

Fig. A1.6 (a) Sugar crystals containing a permitted anti-caking calcium phosphate salt viewed by SEM using a backscattered electron detector. The anti-caking agent appears bright and a spectrum from one of the particles (b) shows the calcium, phosphorus and oxygen peaks from that particle. Reproduced with permission, © Leatherhead Food Research.

A1.2.3 Sample preparation
The main ways to prepare the sample for TEM are:

- fixation, dehydration, drying and resin embedding;
- freeze-fracture replica technique;
- negative staining;
- ultra-thin cryo sectioning.

The main ways to prepare a sample for SEM are related and can be listed as:

- fixation, dehydration, drying;
- freezing in liquid nitrogen slush or similar, and examination while frozen using a cryo stage in the microscope;

- no preparation (only possible for dry, fairly low fat, samples, but always worth considering first).

Fixation, dehydration, resin embedding and thin sectioning for TEM
As electrons are relatively weak, the inside of the microscope and the environment around the sample are usually under a high vacuum. This allows the maximum resolution of the instrument to be achieved but has consequences for the preparation of the sample. In addition, the concentration of the electron beam is intense and can cause damage to the sample. Therefore preparation techniques need to preserve the structure unchanged as much as possible and to 'fix' or remove volatiles in the sample. In addition, for TEM the sample needs to be very thin to allow penetration of the electrons and high resolution of the structure.

Most, if not all, developments in instrumentation and sample preparation for electron microscopy have been developed for biological or medical research. These have then been applied and modified to suit foods. Natural foods such as cereals, fruits, vegetables and meat lend themselves to methods devised for cells and tissues of plants and animals. For processed foods, particularly products such as spreads, butter and cheeses which are high in fat, a modified sample preparation is required.

A typical regime for preserving the ultrastructure of the sample is first to sub-sample the structure using a sharp blade and then to 'fix' or chemically cross-link the proteins, fats and carbohydrates in suitable fixatives. The sample is then dehydrated to remove the water then embedded in a resin which is hardened by polymerisation. This process will then allow ultra-thin sections of 70 to 100 nm to be cut using a specialised ultra-microtome. The sections are usually stained with heavy metal stains that bind to the components and appear black on the screen of the electron microscope. All these steps can affect the quality and resolution of the structure of the sample.

The classical methods for fixation are described by many authors, one example being Hayat (1981). These involve cross-linking reagents such as formaldehyde or permanganate, which have been used for histology for many years. The development of the use of glutaraldehyde as an alternative giving improved rate of fixation or cross-linking of proteins was taken up after its introduction by Sabatini *et al.* (1963). Glutaraldehyde is a very effective fixative for electron microscopy but not suitable for paraffin wax embedding for light microscopy. An alternative that became very popular was a mixture of formaldehyde and glutaraldehyde described by Karnovsky (1965).

Nowadays, samples are typically fixed in glutaraldehyde followed by post-fixation in osmium tetroxide. The use of osmium was known as a fixative for cells, and Wrigglesworth (1957) described its action as one of polymerisation of unsaturated lipids. It is thought to act on the lipid fractions of membranes as well as on fats in food products. Long fixation in osmium tetroxide can be used to good effect on high fat foods. Lewis (1981), in a review of microscopy work at Leatherhead Food Research, described the examination of osmium-fixed cocoa butter by TEM. The harder crystals of cocoa butter were not fixed but the softer liquid fraction was. This resulted in the crystals being removed during subsequent

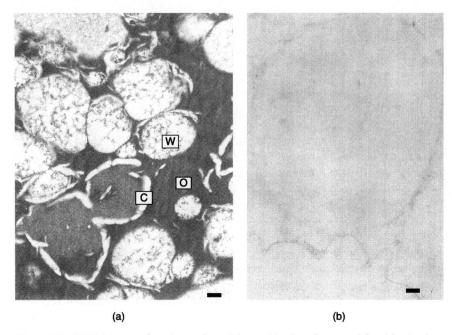

(a)   (b)

Fig. A1.7 TEM images of sections of emulsions: (a) a low fat spread fixed in osmium tetroxide where the fat crystals C are formed around the water phases W in the oil continuous phase O; and (b) an emulsion where the oil has been removed during preparation for microscopy leaving the interfacial protein visible. Magnification bar = 500 nm. Reproduced with permission, © Leatherhead Food Research.

processing, but their outlines remained visible in the thin sections. The method is very useful for spreads and has shown that the water droplets in a water-in-oil spread are stabilised by fat crystals at the oil-water interface (Fig. A1.7a).

Fixation in glutaraldehyde alone without osmium is also used, even with fatty foods. The fat is removed during the process, but as a result the delicate protein interface at the surface of the oil droplets becomes visible (Groves, 2006). An example of this type of imaging is shown in Fig. A1.7b.

Freeze fracture replica technique
This is not a commonly used method these days, but was used extensively in the early days of TEM in food research. The method involves rapid freezing of the sample and then under a vacuum evaporating a thin layer of metal, usually platinum or gold palladium, from an angle onto the surface of the sample (referred to as shadowing due to the effect produced on the surface of the sample and hence the replica), followed by a thin layer of carbon from directly overhead. Usually it is the internal structure of the sample which is of interest, in which case the frozen sample is fractured immediately after freezing, then shadowed with metal and coated with carbon. The sample is then removed from the vacuum and dissolved away leaving a replica of the surface or fracture surface behind. This is

collected onto a TEM grid and examined in the electron microscope. The images are often difficult to interpret and relate to the food products and properties, but this technique can give information that is difficult to obtain with other techniques.

The technique has been used to look at crystallisation of fat in bakery shortenings and the results produced evidence for a mechanism to explain the behaviour of shortening in products and how they stabilise the air bubbles. It has also been used to examine cocoa butter and produced images of the different polymorphic forms, which were related to separate X-ray diffraction studies. Following this, a study was made on chocolate, examining the surfaces and internal structures of chocolate made before and after tempering and also after heat-produced bloom. This was ground-breaking research at the time, showing that typical chocolate bloom was in fact due to the formation of form VI cocoa butter crystals and that cocoa butter in stable chocolate was usually in the form V. Examples of the images of chocolate form V and VI are given in Figs A.8a and b.

This research and the use of replica techniques led to the understanding and refinement of the chocolate tempering process. A review of this research is given by Lewis (1984a,b), where the role of this technique and electron microscopy in a number of products is explained. The freeze fracture replica technique has also been used to look at the formation and stabilisation of fat droplets in toffees and caramels. Dodson *et al.* (1984a,b) looked at the specific role of milk proteins and minerals in the properties and structure of toffees. This knowledge enabled the industry to understand the control of caramel viscosity and stability during depositing and storage.

The ability of the method to follow changes in gelation of food macromolecules was shown by studies on soy proteins by Hermansson and Buchheim (1981). The

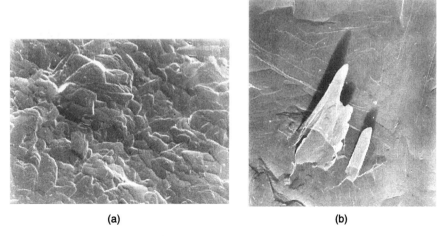

(a)   (b)

Fig. A1.8   Freeze fracture replicas of chocolate surfaces: (a) surface of well-tempered chocolate with form V cocoa butter crystals; and (b) surface of bloomed chocolate showing form VI crystals. Reproduced with permission, © Leatherhead Food Research.

technique revealed that differences in gel structure depended on the level of salt and they were able to compare the results with those from other electron microscopy methods. Soya is used in many foods including meat products, and as these contain salt, the effects of salt on the functionality of soy proteins is key to understanding its role in the food product.

The freeze-fracture replica technique is a difficult technique, both to carry out successfully and to interpret, so it is understandable that it is not used much currently. For many products, the development of the cryo stage for SEM made this technique redundant, but it is a pity as there are advantages to it that CryoSEM cannot give.

Negative staining

This technique has not been used a great deal for food research. The method involves drying down a dilute solution of the sample, usually macromolecules, onto a TEM grid coated with a thin film. A drop of a solution of a metal salt, such as phosphotungstic acid or uranyl acetate, is then dried down onto the grid over the sample. If the concentrations are correct, the metal salt will dry around the molecules or structures in the sample, outlining them so they can be imaged in the TEM. An example of negative staining of a pair of Campylobacter bacteria is shown in Fig. A1.9a, and can be compared with images of the same organism viewed by thin sectioning in the TEM and also by SEM (Figs A1.9b and c).

Typically, negative staining is used to study viruses or bacteria, and applications for food product research have mainly been in the study of these organisms; however, there are some studies on macromolecules from foods. A good review of the technique is given by Harris and Horne (1994), who discuss the relevance of the technique in the light of newer electron microscopy methods such as cryo electron microscopy. In the review, they show images of bacteria and protein molecules under different conditions and illustrate the uses of the method in following protein crystallisation and structural changes. Walker and Trinick (1986, 1988) investigated the structure of myosin, a muscle protein, using negative staining. They observed structural changes induced by ionic salts such as calcium that could be related to the action of the molecule in the muscle. Although this research was not extended into the food area as such, the behaviour of myosin in meat production is thought to affect the texture of beef and so studies such as this have relevance to the issues in food manufacture.

Cryo TEM

Like most aspects of science experimentation, sample preparation affects the results. The use of chemical fixation and drying inevitably leads to the production of artefacts and these are a concern for some. In actuality the production of artefacts can be helpful in understanding the differences between samples, and as long as the microscopist is aware of the effects of processing the sample on the resultant image and that the samples have been prepared in a similar way, then it is valid to use these well-established techniques. There are circumstances when it is necessary to visualise the structure without fixation, and the use of freezing, sectioning and examining in the TEM has been developed. It is a parallel technique to the use of CryoSEM described below, but with more complexity.

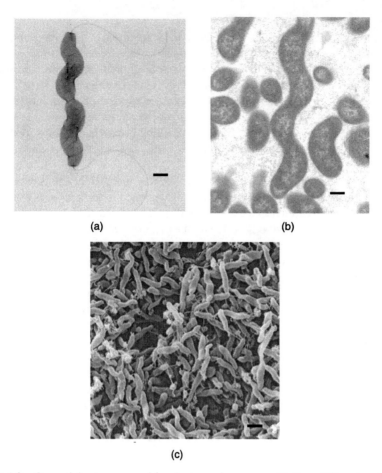

(a)   (b)

(c)

Fig. A1.9   Campylobacter prepared for electron microscopy with three different methods: (a) negative staining and TEM, bar = 500 μm; (b) aqueous fixation, dehydration and thin sectioning and TEM, bar = 200 μm; and (c) aqueous fixation, dehydration and critical point drying and SEM, bar = 1 μm. © IFR Norwich.

The method involves the very rapid freezing of the hydrated sample so that the water becomes glassy and no ice crystals are formed. It is necessary to ensure that the sample is small enough to freeze rapidly, and liquids such as liquid ethane cooled by liquid nitrogen can be used to achieve the fast freezing rate. An alternative is ultrahigh pressure freezing, and ultramicrotomy to produce frozen ultra-thin sections, which are examined frozen in the TEM.

There are advantages in that immuno-labelling and morphological studies are available without the use of chemical fixation and dehydration. There are also disadvantages in that the sample is more sensitive to the electron beam and there is less contrast produced.

Immuno-labelling

Changes in structure, which are related to differences in texture, stability or processing, are very important for the treatment or development of foods. As well as this aspect, there is a need to be able to identify ingredients or structures more precisely. Identification can be achieved by using recognisable structures or selective staining, but these are limited in their application. Immuno-labelling provides a useful tool for following specific objects or molecules as they change in the ultrastructure of the food. Immuno-labelling is now becoming widely used in light and confocal microscopy to achieve this, but is less commonly applied using electron microscopy.

However, it is a powerful technique for labelling specifically and is developing in the research field. Typically it involves labelling the sample with a primary antibody to the required molecule or object, such as a bacterium, and then using a gold-labelled second antibody to locate the primary antibody. The gold particle (usually <20 nm in size) being electron dense, can be seen in the electron microscope. The technique is either used directly on the sample (usually followed by negative staining or SEM) or is applied to thin sections of the sample and viewed in the TEM. It is widely used in biological or microbiological research, but to a much less extent in food applications. It has been applied to locate proteins and hydrocolloids in dairy products (Armbruster and Desai, 1995; Hillbrick *et al.*, 1999) and for labelling glycoproteins on the milk fat globule membrane (Heertje and Paques, 1995). Using TEM and antibody labelling, the presence of soya protein molecules was shown inside the muscle fibres in meat products (Groves, 2006). Also in meat it has been used to investigate the role of specific molecules in the structure and action of muscle (connectin and titin (Wang and Greaser, 1985; Suzuki *et al*, 2001)) and the localisation of cathepsin (Kubo *et al.*, 2002). An example of gold labelling is given later in the case studies.

Preparation for SEM

Typically the samples are fixed in a cross-linking agent such as glutaraldehyde in buffer, as for TEM described above. They can be post-fixed in osmium tetroxide to preserve fat or protein/membrane structures and then dehydrated as for TEM. For good image quality, the samples are conventionally fractured and dried using critical point drying before mounting on a sample holder and coated with a conductive coating such as platinum or gold-palladium. They can then be examined at high vacuum in the SEM.

Many food ingredients or products, especially those that are fairly dry, such as food powders, biscuits, snacks, etc., can be prepared for SEM without fixation in an aqueous buffer. If they are high in fat, this can be removed with a fat solvent and the sample imaged without the need for further drying. Examples of typical foods prepared in this way and imaged by SEM are shown in Figs A1.10a to d.

Alternatively, for higher fat samples, fixation in osmium tetroxide vapour retains and stabilises the fat without dissolving the rest of the structure. This is particularly useful in combination with backscattered electron imaging, revealing the distribution of the osmium and hence the fat. An example of this is shown in

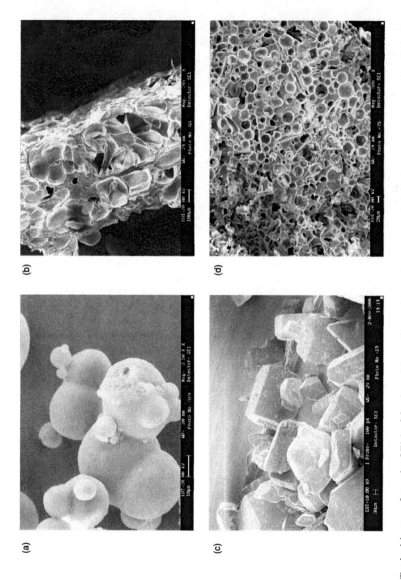

Fig. A1.10 Typical images from the SEM of foods and ingredients: (a) milk powder and (b) a fracture through a crisp. In both images, the fat has been removed using a solvent. Some air is present inside the milk powder particles as well as holes where the fat was present. The highly aerated structure of the crisp caused by the frying process determines the texture of the product. (c) Granulated sugar and (d) a fracture through freeze dried coffee. Both of these have been coated and imaged without any sample preparation. Reproduced with permission, © Leatherhead Food Research.

(a)   (b)

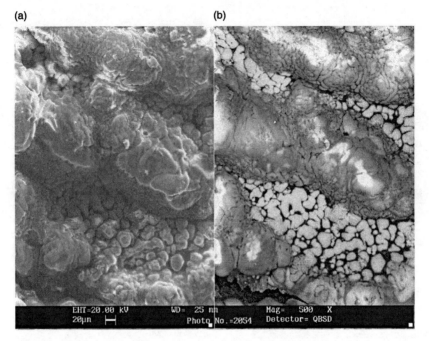

EHT=20.00 kV   WD= 25 mm   Mag= 500 X
20µm ⊢—⊣   Photo No.=2054   Detector= QBSD

Fig. A1.11 SEM images of osmium vapour-fixed raw puff pastry: (a) the secondary electron image; and (b) the same area viewed with the backscattered electron detector. The bright areas in this image are due to the osmium present mainly in the fat. Reproduced with permission, © Leatherhead Food Research.

Figs A1.11a and b, where pastry has been fixed in osmium vapour and imaged by secondary and backscattered electron imaging. The secondary electron image shows the nature of the structure of the pastry and the layering of the fat can be seen in the backscattered electron image. By understanding the role of fat in the product, it becomes possible to design foods with lower fat but retain the desired texture and mouth feel.

For samples where this procedure would be difficult or unwanted, such as ice cream or high fat products, CryoSEM is used. Here the sample is not fixed but frozen very rapidly in liquid nitrogen slush under vacuum and then transferred to a specially cooled stage attached to the microscope. In this, the sample can be held at liquid nitrogen temperatures or just above, and fractured and coated before being inserted into the SEM for imaging. Usually before coating, the temperature is raised a little to allow the ice formed in the sample to sublime and this makes it easier to see the ice matrix and where the water had been in the sample. An example of this is the visualisation of water in low fat mayonnaise by CryoSEM in Fig. A1.12. The starch granules take the place of some of the oil and thicken the water phase to give some stability.

CryoSEM is useful for other fatty foods such as chocolate and frozen fatty foods such as ice cream. Figure A.13 shows the nature of the form VI bloom crystals on the surface of a badly bloomed chocolate sample. A comparison of this image, with

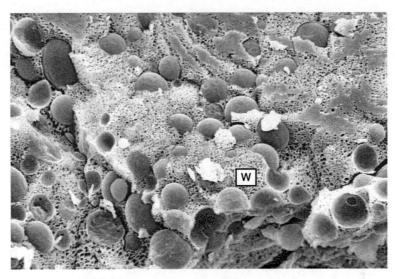

Fig. A1.12 A CryoSEM image of low fat mayonnaise in which starch replaces some of the oil in the oil-in-water emulsion W, which appears as a lacy matrix after sublimation. Magnification bar = 50 μm. Reproduced with permission, © IFR Norwich.

Fig. A1.13 CryoSEM of bloomed chocolate showing clusters of form VI cocoa butter crystals on the surface. Magnification bar = 100 μm. Reproduced with permission, © Leatherhead Food Research.

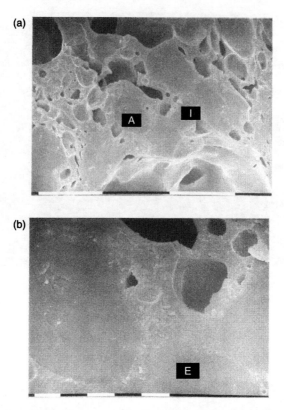

Fig. A1.14   CryoSEM of ice cream. The distribution of air (A) and ice crystals (I) can be seen in (a) and the higher magnification of (b) shows the emulsion (E) between the ice crystals. Reproduced with permission, © Leatherhead Food Research.

that of the replica of chocolate bloom in Fig. A1.8b, shows the characteristic images produced by the two different techniques. Figures A1.14a and b show images of ice cream, where the sublimed water reveals the size and distribution of ice crystals. These, together with the air and the fat emulsion, determine the sensory mouth feel of the product. During storage, the ice crystals grow larger and the air bubbles coalesce, eventually forming an almost continuous network of air among the matrix of aggregated ice crystals. The ice crystal aggregates behave in the mouth like very large ice crystals, giving the typical gritty mouth feel of old ice cream.

A1.3   Applications of electron microscopy (EM) to the understanding of food product structure

A1.3.1   Fruits and vegetables
The main application of microscopy to fruits and vegetables has been in the identification of the different cell types, and more importantly for food product

development, in the relationship of structure to texture. A useful general review is given by Jewell (1979), who discusses the structure of the different cell types and then gives examples of the effects of different processing methods on the structures of a range of foods. Light microscopy, scanning and TEM have been used to demonstrate cell changes that occur, particularly on cooking. These involve changes to the cell wall, softening and expansion, as well as cell-cell separation. Using examples of the changes in potato after treatment with enzymes, Holgate (1984) revealed that removal of pectic material produced some loss of cell wall integrity but mainly loss of cell-cell binding. Cellulose was less important in the binding of cells together, but more involved with the strength of the cell walls overall. Softening during cooking produced changes in the structure as described above and examples are shown with potato by Holgate and Lewis (1985). They also used SEM to show that as the potato softened there came a point where the tissues were soft enough so that on fracturing they separated between the cells rather than through the middle of each cell. The fracture plane is determined by the line of least resistance, and can be likened to consumer's teeth biting through the food. In raw or crisp potatoes or other fruits and vegetables, the fracture plane passes through the centre of the cells because the walls and cell-cell binding are strong. As the food softens or if it is mealy, then the fracture plane passes between the cells in the middle lamella region.

An example of this is shown in the scanning electron micrograph in Fig. A1.15, illustrating the result of fracturing a raw bean. As the raw sample is firm and the cell walls rigid with strong binding between the cells, the fracture plane has passed through the walls and through the centre of the cells revealing their contents. These are oval starch grains embedded in a protein matrix within the cells. In a cooked bean, the fracture passes between the cells as the binding between the

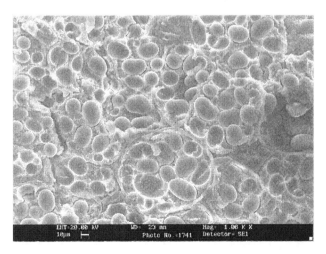

Fig. A1.15 SEM of raw haricot bean showing cells with starch granules embedded in a protein matrix. Reproduced with permission, © Leatherhead Food Research.

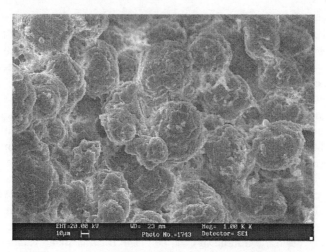

Fig. A1.16   SEM of cooked haricot bean showing rounded separated cells. Reproduced with permission, © Leatherhead Food Research.

walls is weakened. Figure A1.16 shows the outside surface of rounded and separated cells in a cooked bean, where the fracture has gone through the middle lamella. This type of research has enabled the understanding of the processes used in manufacturing fruit and vegetable products, allowing the development of new cooking, freezing and storage conditions.

Early work on pickling of cauliflower (Saxton *et al.*, 1969) and onions (Jewell, 1972) showed degradation of the organelles and structure inside the cells, followed by eventual cell wall breakdown. Microscopy has also been used to follow the effects of pickling on cucumber (Smith *et al.*, 1979; Walter *et al.*, 1985) and SEM and CryoSEM for carrots (Massa *et al.*, 1998; Llorca *et al.*, 2001). The role of pH and ionic content initially considered as part of the pickling process has lead to the investigation of the effects of treatment with other ions to alter texture. Holgate and Lewis (1986) examined varying ionic conditions and salts on the texture and structure of cucumber and carrots, observing that pre-soaking in calcium lengthened the required cooking time and pre-soaking in tripolyphosphate (a calcium sequester) shortened the cooking time. The importance and influence of calcium in the textural changes that occur during processing led to the investigation of its use to prevent or delay softening caused by freezing, blanching and canning (Burke *et al.*, 1993).

An understanding of more unusual processes, such as popping of maize to make popcorn, has benefitted from the use of the electron microscope (Parker *et al.*, 1999). During the popping process, the small amount of water vapour in each starch granule becomes superheated steam and melts the starch. The pressure inside the maize kernel is at first contained by the outer coat, but when the pressure becomes too much, the outer coat suddenly bursts. The steam inflates each starch granule forming a thin-walled foam, which expands rapidly and flows out of the

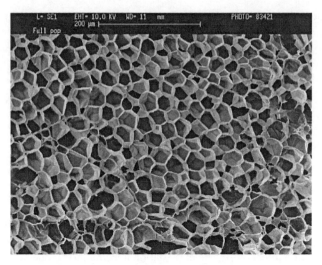

Fig. A1.17   SEM of popcorn showing the foam-like structure made up of inflated starch granules. © IFR Norwich.

kernel (Fig. A1.17). Because the cells walls are shattered, and the starch and protein cooked, other popped cereals make a simple and nutritious snack requiring little fuel to prepare. They can readily be made into transitional foods for babies and easily digestible foods for those who need them.

The changes associated with fruits or vegetables are not always due to processing, but can be affected by storage conditions or ripening, and there are many studies involving microscopy to follow these changes. Examples include tomato ripening (Gross and Wallner, 1979; Sozzi *et al.*, 2001), apple ripening and cultivar (Kovacs *et al.*, 1999) and storage conditions of blueberries (Allan-Wojtas *et al.*, 2001).

A1.3.2   Meat

The structure of meat and the changes that occur post-mortem affect its texture after cooking, and these changes as seen by microscopy, are well reviewed by Voyle (1979). The fine structure of muscle is similar for all animals, and consists simplistically of long multinucleated cells in bundles known as myofibres, surrounded by membranes of connective tissue (collagen and elastin). Within each myofibre are a number of fibrous proteins assembled in a complex array called a myofibril. For skeletal muscle (the bulk of the muscle used for food), these myofibrils consist of thick and thin filaments of protein that slide together when the muscle contracts. There are other proteins involved in the structure, as well as enzymes and ions essential for the contraction process. The chemistry of contraction and relaxation is complex, but can be affected by stress or conditions before death as well as after, resulting in a change of texture or properties of the food (Dutson, 1983).

In addition, the way the meat is treated in the production of products, the level of salts, chopping process and other factors, will affect the texture and water holding capacity of the meat. This is discussed in more detail by Lewis (1979), who reviews the use of microscopy techniques to understand the changes that occur in meat products after processing. This review covers the effects of salt and phosphate, two commonly used ingredients that alter the texture of meat on cooking by increasing dispersion of proteins from the myofibrillar structure. The dispersed proteins act to form a gel, holding water on cooking, and the remaining myofibrillar structure is tenderised by the removal of the proteins. The reasons for the effects of polyphosphate in particular were further investigated by Offer and Trinick (1983) and, in a later paper, were shown using TEM to be mainly due to the high pH produced by this chemical in the brine solution (Lewis *et al.*, 1986).

Lewis (1979) also discussed the reasons why using different fatty tissues result in changes in free fat in the product and subsequent fat loss on cooking. Using light and electron microscopy, the explanation appeared to be due to the thickness of the fat cell wall and also the solid fat content. Thicker cell walls and lower solid fat result in less free fat in the product and lower fat loss on cooking. It was suggested that the thicker cell walls are more resistant to the blades in the chopping process and the lower fat content made the cells softer and inclined to bend under the blades. These observations illustrated the complex nature of animal products, and that effects of structure on texture can be due to changes at both the molecular and the macro level. An example of this is shown in a report by Jeffery and Lewis (1983), where the texture of beef burgers, as assessed by a taste panel, was found to be affected by the temperature of mincing. Meat that was minced at –5 °C gave a less firm, less chewy and more tender product than that minced at 0 °C or +3 °C. This was shown by light microscopy and SEM to be due to the increased disruption of the myofibrillar structure when minced frozen. An example of the effects of freezing on muscle structure of cod is shown in Figs A1.18a and b, where the result of fast freezing at –80 °C produced smaller ice crystals (and hence a better texture when thawed) than a slower freeze at –20 °C. A similar example of the effects of macro changes in properties of meat was shown by Allan-Wojtas and Poste (1992), who reported that the stickiness of roast pork slices from pigs in an experimental breeding and feeding study was due to dispersed collagen, a factor absent in the control samples.

These examples illustrate that the answer to a problem is not always a complex one, and that the use of a microscope can reveal relevant changes at all levels. A combination of light microscopy with electron microscopy can show the relationships between fine structural differences and the changes seen in the processed products.

Other publications include a study of the gelation properties of a range of different meat species. Montejano *et al.* (1984) used SEM to evaluate texture-structure relationships of the gelation properties of the products. Koolmees *et al.* (1993) used light microscopy to follow changes in the structure and distribution of components in meat batters during heating.

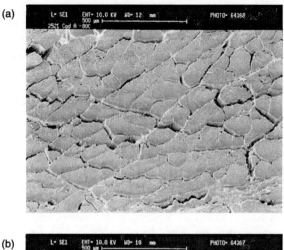

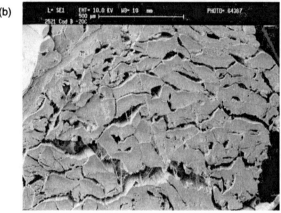

Fig. A1.18   SEM of cod muscle frozen at –80 °C (a) and –20 °C (b), showing the increased damage to the muscle by the large ice crystals formed at the slower freezing rate at –20 °C.
© IFR Norwich.

The effects of pressure as a tenderiser of meat have been studied and the changes in the ultrastructure followed using microscopy (Macfarlane *et al.*, 1986; Berry *et al.*, 1986; Macfarlane and McKenzie, 1986).

A1.3.3   Dairy products

Most of the early food-related electron microscopy work was performed on dairy products, mainly yoghurt and cheese. Useful reviews are given by Brooker (1979), Kalab (1979a,b,c, 1981, 1993), Holcomb (1991) and Schmidt and Buchheim (1992).

Milk constituents and dairy products exhibit a wide range of physical states including liquids, gels, foams, emulsions, plastic and thermoplastic materials and solid powders. They range from products with high water content (liquid milk), to dry products (milk powders) and from low fat (yoghurt) to high fat (butter).

Structuring elements may comprise fat globules, water droplets, air spaces, colloidal protein aggregates and crystals. The textural behaviour of these materials is greatly dependent on their microstructural elements and their physicochemical interactions. Dairy processing, for example evaporation, shearing or heating, has profound effects on the microstructures, and hence textures, of dairy foods. Understanding the relationship between microstructure and a dairy product's functional and sensory attributes can aid the design of novel products.

Many dairy products have been studied using SEM; these include milk gels and various cheese types including cream cheese, Cheddar, Mozzarella and Gouda (Kalab, 1981, 1993; Schmidt and Buchheim, 1992).

Cryo-SEM, where frozen hydrated bulk samples can be directly imaged under the electron beam, has been successfully used to study dairy spreads, mayonnaise and ice cream (Sargent, 1988), although freezing-induced artefacts can occur (Schmidt, 1982; Kalab, 1984). Freeze fracturing allows visualisation of internal structures, such as air and fat distribution in whipped cream (Schmidt and Van Hooydonk, 1980; Brooker, 1985).

Creamer et al. (1978) used the negative staining technique and TEM to study casein micelles, their subunit structure and their association with whey proteins.

Thin sectioning of resin embedded samples and viewing by TEM is a popular technique for dairy products, as it reveals a great deal of information on the state of the protein and interfacial structures. It has been applied to milk (Henstra and Schmidt, 1970), yoghurt (Kalab et al., 1993) and various cheese types including Cheddar, Mozzarella, Gouda, cream cheese and processed cheese (Green et al., 1980; Rayan et al., 1980; Kalab, 1977; 1993). Image analysis has been used on thin sections of yoghurt (Skriver et al., 1997), whey protein gels (Langton and Hermansson, 1996) and Mozzarella cheese (Cooke et al., 1995), in order to provide quantitative data, which might lead to the establishment of a pattern.

A1.3.4 Bakery

The use of electron microscopy to study bakery products and starch is extensive, especially using SEM. A good early review is given by Angold (1979), where he covers the methodology as well as examples of applications. Examples of some of the many uses of electron microscopy in bakery are a study on the role of proving in bread dough, which has been explored by Gan et al. (1990) using SEM to show the development of gas cells. The effects of different starches and processing of starches has been followed by Yan and Zhengbiao (2010) and the relationship between structure, cooking and digestibility is examined by Duodu et al. (2002). These are a few examples to show the diversity of applications of this technique to starch-based products.

A1.3.5 Confectionery products

Microscopy has been used extensively in the study of confectionery products, with a large number of publications on chocolate, particularly cocoa butter crystallisation, ice cream and the role of air and fat in the structure and texture, and sugar-based jellies or foamed products.

A review of microscopy in the study of chocolate at Leatherhead Food International is given by Lewis (1981). In this he covers work from 1922, in which light microscopy was used to study the crystals formed by cocoa butter on cooling, relating these to the formation of bloom, through to 1970 when TEM revealed the different crystalline forms of the polymorphs of cocoa butter and established that bloom in chocolate was due to the formation of form VI as discussed earlier.

The process of conching chocolate has also been followed by electron microscopy. Hoskin and Dimick (1980) used SEM to look at changes in the structure of the ingredients of chocolate during the process, and the effects of lecithin on these. A description of the use of microscopy to follow ingredients in chocolate is also given by Groves and Subramaniam (2008). Katsuragi and Sato (2001) used SEM to look at the effects of emulsifiers as antibloom agents in chocolate. They found that the emulsifiers delayed the onset of bloom. SEM has also been used to distinguish between fat and sugar bloom on chocolate (Bindrich and Franke, 2002). Subramaniam and Groves (2002) used macro photography, stereo light microscopy and Cryo-SEM to show that changes in gloss of chocolate coatings was related to the hardness of the coating, fat type and crystalline appearance of the surface.

Different aspects of research into chocolate using electron microscopy show the variety of uses the technique gives to confectionery research. Examples include the work of Duffett and Firth (2001), who used electron microscopy to demonstrate changes in crystal formation of spray crystallised chocolate powder, and Jackson *et al.* (1995), who combined electron microscopy with NMR to follow changes when water was added to chocolate. Electron microscopy was also used to examine fungal growth in chocolate (Kinderlerer, 1997).

Elemental X-ray analysis (EDX) has been used to characterise the cocoa solids and milk protein in chocolate (Brooker, 1990), although this is an under-used application of the technique in food research. Finally microscopy, both light and electron, has been used to show the relationship between the structure of the milk particles and the viscosity of chocolate. The combination of light microscopy and SEM usefully showed that the shape, size and internal structure of the particles, as well as the way they broke during manufacture, all affected the resulting viscosity of the chocolate (Dodson *et al.*, 1984c; Pepper and Holgate, 1985).

A1.4   Case studies

A1.4.1   Shelf-life of foods

The shelf-life of foods is an area that is little understood and there is a pressing need for both further research into the changes that occur during storage of food products and also tests that can predict these changes. If the shelf-life can be accurately stated, then one of the benefits would be much less wastage of food.

The shelf-life of foods is not solely dependent on the microbiological safety of the product, although this is a critical area that is addressed first in assessing the use-by dates. It is also very much a quality issue and this is a much more difficult property to assess and predict. The main reason for the difficulty lies in the variable nature of food products. They can be relatively simple, such as fruit or vegetables, bread, cake and cereals. The main indicator for these would be moisture changes, and that is also an indicator for more complex products such as pizza, ready meals, biscuits, desserts, chocolates, etc. However, the moisture changes in these are much more difficult to predict as the different components will have their own moisture values and the changes will be driven by how they are situated in the product. In addition, fat and the changes associated with shelf-life for fat, such as development of rancidity or other flavour differences, can have severe detrimental effects on the quality of the product. Finally the combination of all the ingredients in a product is never a 'dry mix', but involves a reaction between the ingredients in a number of different ways. How changes in the individual ingredients and in the way they interact together in complex products is not really well known.

Until they are better understood, it will not be possible to predict how a product will behave with time. This makes formulation changes very difficult for a food company, when they need to assess how recipe changes will affect shelf-life. The only way to really know the true shelf-life of a product is to store it and assess in real time. For many products with a shelf-life of many months or more, this is not practically feasible and so accelerated storage tests have been introduced as a means of more accurately estimating the true shelf-life. These accelerated tests are best run by first assessing the real-time changes in a product over the shelf-life. These changes can then be used as indicators of the type of decline in the product during the accelerated test. The conditions for the accelerated tests will depend on the product type, but could involve higher temperatures, cycling of temperature or humidity, or light. Assessments would include texture, sensory, microscopy and chemical analysis. As an example of how electron microscopy can add value and understanding in this area, the use of CryoSEM to look at changes in chocolate with time and assess the main quality properties of appearance and gloss has been used by Subramaniam and Groves (1999, 2002, 2003). In these studies the development and time of appearance of fat bloom crystals in chocolates made with different antibloom fats gave information both on the relative shelf-life properties of the antibloom fats added and on the type of bloom crystals produced. Usually fat bloom on chocolate is due to cocoa butter form VI crystals forming. However, the migration of filling fats in chocolates or even sometimes the addition of antibloom fats to the chocolate can also lead to unusual bloom structures on storage. An example of this is shown in Fig. A1.19. Here, typical form VI crystals are present, but also there are long thin needle-like crystals that protrude from the surface and are often mistaken for mould.

As a second example in this area of shelf-life, a project carried out at Leatherhead Food Research looked at the changes in real-time shelf-life of biscuits and analytical predictors of changes (publication in preparation). The use of

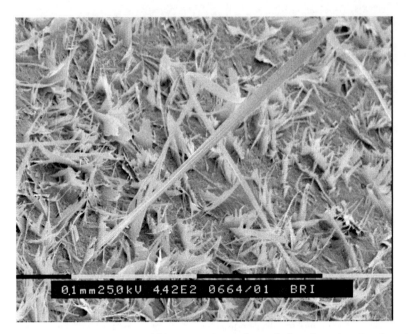

Fig. A1.19 CryoSEM image of bloom on the surface of chocolate showing form VI cocoa butter bloom as the typical thick finger-like crystals, but also long thin crystals that protrude several millimetres above the surface and are sometimes mistaken for mould. Magnification bar = 10 μm. Reproduced with permission, © Leatherhead Food Research.

osmium vapour to follow the fat showed remarkable changes in the fat distribution with time. Figure A1.20 shows the freshly prepared biscuits viewed by conventional secondary electron detector imaging, as well as the backscattered imaging showing the location of the osmium and hence the fat. Figure A1.21 shows the same types of electron microscopy imaging on the biscuits after storage for three months. It can be seen that the freshly made biscuits show an even distribution of fat and in fact a different crumb structure overall compared to the aged biscuits. With time, the crumb structure of the biscuit became coarser and the fat more coalesced. This is probably due to moisture movement causing changes in the starch and subsequent fat redistribution. It is surprising how much the structure does change with time and an example of how mobile solid food products are.

A1.4.2   Healthier products
There is a strong drive to produce healthier foods that are low in fat, sugar or salt. This is in response both to government pressure to improve the health of the nation and to the consumer who wants a healthier lifestyle and also to enjoy the foods that are 'unhealthy'. It is difficult to produce lower-fat chocolate, but sugar-free chocolate or confectionery products are available and have been the result of

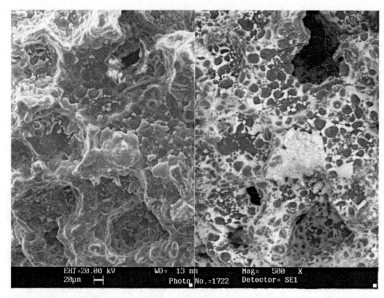

Fig. A1.20   SEM images of a freshly made biscuit fixed in osmium vapour. The image on the left shows the secondary electron image of the crumb structure and the image on the right the backscattered electron image where the fat appears white. Reproduced with permission, © Leatherhead Food Research.

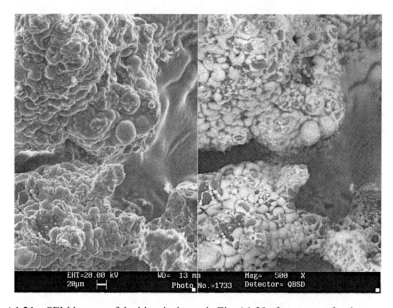

Fig. A1.21   SEM images of the biscuit shown in Fig. A1.20 after storage for three months, fixed in osmium vapour. The image on the left shows the secondary electron image of the crumb structure and the image on the right the backscattered electron image where the fat appears white. Comparison with Fig. A1.20 shows the fat is more coalesced in the aged sample and the crumb structure less aerated. Reproduced with permission, © Leatherhead Food Research.

research and product development. In a collaborative study to lower sugar (Narain *et al.*, 2007), the research looked at coating an inert safe material such as a permitted calcium salt with sugar. The thinking behind this was that the sugar is present in the product for taste reasons and if the surface was sweet then this would be the first taste sensation received. The bulk of the particle would be the calcium salt and therefore both less calorific and also beneficial to bones and health generally. Figure A1.22 shows these particles after production. The use of X-ray mapping allowed the efficiency of the coating and distribution of the calcium to be revealed.

There are also pressures to reduce the level of salt in the consumers' diet for health reasons. However, removal of salt has not been easy, since the salt has several functions: a food safety function in preventing microbial growth, a taste function, and also a functional effect in food products such as bakery, meat products and snacks. To understand how to reduce the salt in those cases, where functionality is dependent on salt, it is necessary to understand the role of the salt in the product. This will vary from product to product and as such considerable research is needed in these different products.

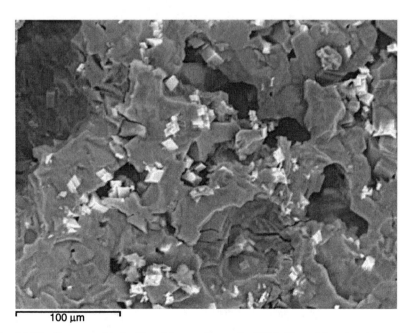

100 µm

Fig. A1.22   Sugar co-dried with calcium carbonate by SEM. The calcium is bright against the organic sugar background. Much of the calcium is coated with a thin layer of sugar and so is not bright. This sample was fairly good at reducing the sugar level but maintaining sweetness, even though the aim of coating the calcium was not completely fulfilled.
Reproduced with permission, © Leatherhead Food Research.

Where salt is added dry, it has been shown (Angus *et al.*, 2005) that reducing the crystal size increases the speed that the salt is tasted and also the perceived intensity. This is partly due to a faster dissolution in the mouth for the smaller salt crystals but also probably linked to the 'number' of salt hits on the palate. In the electron microscope, the size of typical table salt compared to 'microsalt' shows the considerable difference in structure and actual size (Figs A1.23a and b). On the surface of a product such as a potato chip (Fig. A1.24), using a smaller size of salt crystal would allow more particles of salt over the surface, giving a faster taste of salt to the consumer. The end result would in fact deliver a salt reduction if it was possible to control the amount shaken onto the product to allow for the smaller size crystals. In snacks produced commercially, this control of dosage is possible. The presence of fat on the surface of the product will affect the sensory properties of the salt and it is probable that the smaller salt crystals would be enveloped in fat and thus the benefits of the faster dissolution negated. It is not an easy problem to solve.

(a)

(b)

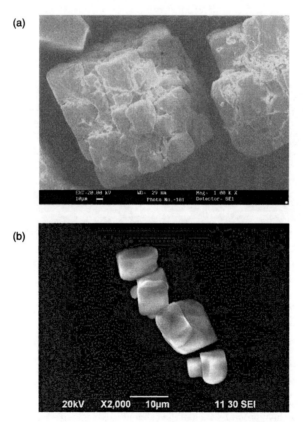

Fig. A1.23 SEM images of salt (sodium chloride) made to different crystal sizes. The magnification bar is 10 µm: (a) shows typical table salt and (b) shows micro salt. Reproduced with permission, © Leatherhead Food Research.

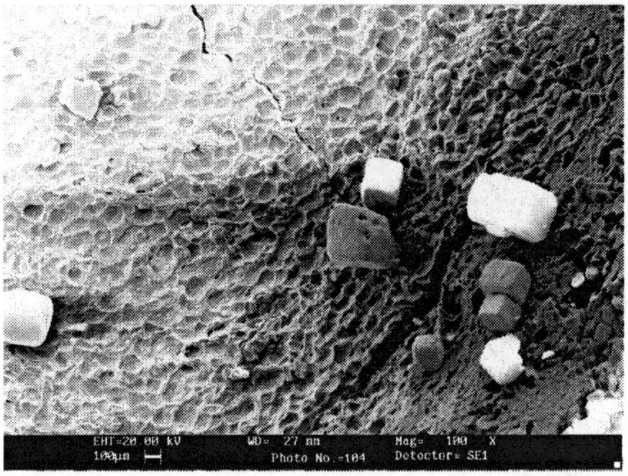

Fig. A1.24 Surface of a fried potato chip with table salt crystals on the surface. This was taken from an actual meal where the consumer had added the salt in the usual way. Reproduced with permission, © Leatherhead Food Research.

A1.4.3 Food safety

There are three main areas of contamination in food products, namely microbial, chemical and physical. Electron microscopy has been used in the microbiological area of food safety in many ways, covering areas of rapid identification and research into the colonisation and growth of microorganism in foods and on food preparation equipment. One area of concern is in the contamination and growth on salad ingredients. SEM studies have shown that although bacteria will attach to the surface of lettuce leaves, they do not remain in any numbers and find it difficult to colonise the surfaces of intact leaves. However, in shredded leaf products they can utilise the nutrients released from the cut surfaces of cells and multiply rapidly, accelerating spoilage and becoming cause for concern (Figs A1.25a and b).

Rapid detection of food poisoning organisms or location of small numbers of bacteria in particularly difficult products such as chocolate is an area of special interest, and the combination of electron microscopy with immuno-labelling is able to play a role in these areas of research. Specific antibodies can be used to target organisms and these can be visualised in the electron microscope by using gold labelled secondary antibodies (Fig. A1.26).

Electron microscopy is also used in the investigation of physical contaminants, which are a food safety issue. These contaminants can be anything from particles such as twigs, insects or stones that might be brought in with the ingredients to glass, plastic or metals fragments from machinery in the production process. It is important that the identity of the particle is determined so that the likely origin and cause of the contamination can be found. This will enable the manufacturer of the food product to know whether there is an issue in the production or whether the contamination occurred after the product left the factory. If the foreign body

(a)

(b)

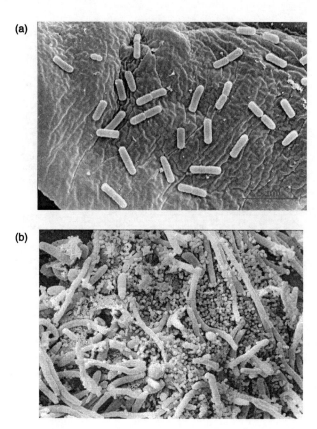

Fig. A1.25   A few bacteria on the surface of intact lettuce; in (b), the cut open cells have become populated with many yeasts and bacteria. © IFR Norwich.

was present in the production process, then steps have to be put in place to ensure it does not happen again. To identify the object, light microscopy is the first step. After this, combining SEM with X-ray analysis will give the elements present and often give the origin of the foreign material. An example is the analysis of glass fragments that are sometimes found in foods. The surface appearance and shape provides information on the possible origin and the spectrum obtained from the SEM and X-ray analysis, particularly the levels of magnesium, potassium and calcium, can be used to match the glass from a database of known glass objects. Figure A1.27 shows typical spectra from a domestic Pyrex container and a glass bottle as examples of differences.

A1.5   Developments in EM techniques and future prospects

There are always developments and improvements in the imaging and resolution of microscopes. The advent of Environmental SEM instruments with detectors to

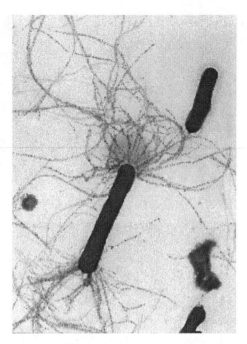

Fig. A1.26 A negatively stained bacterium in the TEM with gold labelled antibody to the flagellae. The bacteria are approximately 2 to 3 μm long. © IFR Norwich

allow imaging of wet samples has re-energised the arena with truly different benefits. Other areas of interest that are thought to become widespread are in the use of X-rays to image samples and in the development of X-ray tomography techniques.

Electron microscopes have used the properties of X-rays for many years, to obtain chemical compositional information from samples. The use of X-rays to image structure was developed some years ago, but was generally superseded by electron microscopy. More recently, new advances in X-ray sources, optics and preparation techniques have lead to the increased use of X-ray microscopes to image samples. X-ray microscopy has some advantages over conventional electron microscopy in that it can image thick hydrated samples with good resolution (30 nm or better). The depth of field depends on the energy of the X-rays, soft X-rays giving a range in microns, but higher energy X-rays give a depth of millimetres. In addition to imaging, X-ray microscopy can be used to map elements present in the sample, and provide spectro-microscopy information of a similar standard to conventional electron energy loss spectroscopy (EELS) instruments.

A disadvantage of high-resolution X-ray microscopes has been cost and accessibility. However, commercial instruments are becoming more available, even though still expensive.

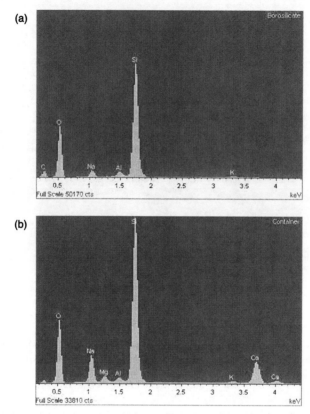

Fig. A1.27 Spectra of glass from SEM and X-ray analysis: (a) borosilicate or heat resisting glass with no calcium and (b) container glass. Reproduced with permission, © Leatherhead Food Research.

The ability to provide 3D structural information using tomography on wet intact specimens is an extremely useful property that is finding applications in the food area. A useful review of soft X-ray microscopy is given by Jacobsen (1999), with details on the potential of 3D X-ray microscopy by Reimann *et al.* (1994). The use of X-ray tomography for imaging confectionery product structures is demonstrated by Decker and Ziegler (2002).

The combination of spectroscopy and electron microscopy is expected to develop further. Typical examples of these developments are given by Kanemaru *et al.* (2009), with the use of a hybrid fluorescence SEM to look at biological samples, but the technique could be applied to foods. James and Smith (2009) used electron microscopy as well as separately X-ray photoelectron spectroscopy in the study of chocolate bloom to show that tempered chocolate bloomed with cocoa butter crystals, whereas poorly tempered chocolate bloomed to give a mix of sugar and fat bloom.

Finally, integrative microscopy is growing, with systems designed to allow the visualisation of samples in the light microscope and direct comparative images obtained from the electron microscope. No applications for foods have been published as yet, but an example of the technique can be seen in the work of Faas *et al.* (2013) looking at fluorescently labelled biological samples by light and electron microscopy.

A1.6   New challenges and nanotechnology

There are many new challenges for the food industry in terms of sustainability, understanding the relationship between diet and health, and modern living demands on food safety and quality, to name a few. Emerging technologies, especially in the arena of research outside food, will play an important part in helping the industry to respond to these challenges. These include GM, synthetic biology and nanotechnology. Nanotechnology is in fact a multidisciplinary area of many technologies and the potential of nanotechnology in food applications has been in the spotlight many times recently, both in terms of the exciting benefits that it could bring, as well as concerns over possible safety issues.

Nanotechnology is concerned with the manipulation of structures at the atomic or molecular level and as such has a direct relationship to electron microscopy as well as other techniques, which can image small structures, such as scanning probe microscopies (SPM). There is considerable debate as to the definition of nanotechnology and to date there is no agreed legislative definition for food ingredients or products, but there is a recommended definition within the EU guidelines.

The 2011 Commission Recommendation on the definition of nanomaterials defines 'nanomaterial' as:

> . . . a natural, incidental or manufactured material containing particles, in an unbound state or as an aggregate or as an agglomerate and where, for 50% or more of the particles in the number size distribution, one or more external dimensions is in the size range 1 nm to 100 nm. In specific cases and where warranted by concerns for the environment, health, safety or competitiveness, the number size distribution threshold of 50% may be replaced by a threshold between 1 and 50%. [. . .]

This size range of 1 to 100 nm includes a large number of natural nanostructures in foods, such as the cellulose fibrils in plant cell walls, the protein filaments in the muscle fibres in meat, and the crystalline structures in starch grains. During manufacturing or cooking of foods, nanostructures can also be produced. Examples of these include the production of interfacial structures in emulsions such as mayonnaise, and in foams such as meringues and marshmallows; the formation of gels, for example in egg white on cooking, and the formation of very small nanoparticles in the production of powdered ingredients such as sugar. Morris (2010, 2011), in reviews of the emerging roles of engineered nanomaterials in the food industry, listed three classes of nanomaterials; first, naturally occurring nanomaterials such as those described above. Second, Morris considered there are engineered particulate nanomaterials that are either metabolised completely or excreted, and finally persistent engineered nanomaterials whose components are not broken down but are retained and accumulate in the body.

Engineered nanomaterials made with accepted food ingredients, which are normally metabolised, are not of much concern. However, there are reservations over the safety of nanomaterials, which are not readily metabolised and might accumulate in the body. These persistent engineered particulate nanomaterials include nano silver particles, nano silica and nano titanium. Little is known about the level of accumulation of these materials and the consequences of absorption. The considerations of these risks are discussed by Cushen *et al.* (2012), a list of persistent particles in products is given, and the regulatory perspective overviewed.

The images shown in this chapter all include nanostructures, both natural and engineered or formed through processing. The manipulation of these structures will control the texture and quality of foods. It is by understanding how to control the formation of these nanostructures and their assembly into the larger-scale food products that the industry will be allowed to make efficient use of ingredients and energy processes and provide healthy nutritious foods. Understanding and controlling the formation of structures of ingredients in foods has been the subject of food research for many years and the manipulation of this first category of nanomaterials still has a great deal of value to bring to the food industry. Studies on designing the second type of nanomaterials have begun to expand, such as the production of protein structures as delivery vehicles for nutraceuticals (Livney, 2010) and the formation of structures by self-assembly (Ipsen and Otte, 2007). Self-assembly of food ingredients and the control of these structures at the atomic or molecular level is an area that ultimately will allow the ability to design food structures and use them for specific purposes such as lowering fat or sugar. An example is the manufacture of water-in-oil-water emulsions to simulate the mouthfeel of or an alternative approach to reducing fat, which is described by Chaudry and Groves (2010). Here, part of the water phase in mayonnaise-type emulsions is emulsified inside the oil droplets, producing water-in-oil in water emulsions, allowing the total fat to be reduced without the need for additives to thicken the water. An interesting and challenging way to lower fat is also described by Morris (2011), where the fat in the food is structured so that the rate of fat digestion in the gut is slowed.

A1.7   Conclusion

The electron microscope was developed to be able to visualise nano- and micro-structures and as such has been at the forefront of nanotechnology for many years. Its use to further the understanding of food product development and support the industry has been of undoubted benefit. The advent of techniques such as AFM opened up the possibilities for design of nanostructures, and with the information from the electron microscope allowed a way of exploring the understanding of changes in structures of foods that lead to attributes the consumer desires. It is a technique that is under-used considering the benefits that can come from it. The ability to interpret the structures is essential and the combination with other techniques outside microscopy especially beneficial.

A1.8   References

ALLAN-WOJTAS, P. and POSTE, L. M. (1992), Microstructural manifestations of two unusual phenomena detected in experimental roast pork: a scanning and transmission electron microscope study, *Meat Science*, **31**, 103–20.

ALLAN-WOJTAS, P. M., FORNEY, C. F., CARBYN, S. E. and NICHOLAS, K. U. K. G. (2001), Microstructural indicators of quality-related characteristics of blueberries-an integrated approach, *Lebensmittel Wissenschaft und Technologie*, **34**, 23–32.

ANGOLD, R. (1979), Cereals and bakery products, in J. G. Vaughan, *Food Microscopy*, London, Academic Press, 273–331.

ANGUS, F., PHELPS, T., CLEGG, S., NARAIN, C., DENRIDDER, C. and KILCAST, D. (2005) *Salt in Processed Foods*, Leatherhead Food International Collaborative Research Report, Surrey, UK, Leatherhead Food International.

ARMBRUSTER, B. L. and DESAI, N. (1995), Identification of milk proteins in dairy products prepared at low temperature, *Scanning*, **17**(Suppl. V), V100–1.

BERRY, B. W., SMITH, J. J., SECRIST, J. L. and ELGASIM, E. A. (1986), Effects of pre-rigor pressurisation, method of restructuring and salt level on characteristics of restructured beef steaks, *J. Food Science*, **51**, 781–5.

BINDRICH, U. and FRANKE, K. (2002), Bloom on chocolate: distinguishing fat bloom and sugar bloom, *Zucker und Susswaren Wirtschaft*, **55**, 27–8.

BROOKER, B. E. (1979), Milk and its products, in, J. G. Vaughan, *Food Microscopy*, London, Academic Press, 273–331.

BROOKER, B. E. (1985), Observations on the air-serum interface of milk foams, *Food Microstructure*, **4**, 289–96.

BROOKER, B. E. (1990), Identification and characterisation of cocoa solids and milk proteins in chocolate using X-Ray microanalysis, *Food Structure*, **9**, 9–21.

BURKE, O. C., CLUTTON, A. P., ARNABY-SMITH, F. M., SUBRAMANIAM, P. J., GROVES, K. N. M. and JONES, S. A. (1993), *Manipulation of Fruit and Vegetable Cell Wall Structure and its Implications for Food Processing*, Food Research Association Research report 709, Surrey, UK, Leatherhead Food International.

CHAUDRY, Q. and GROVES, K. (2010), Nanotechnology. Applications for food ingredients, additives and supplements, in Q. Chaudry, L. Castle and R. Watkins (eds), *Nanotechnologies in Food, RSC Nanoscience & Nanotechnology*, **14**, 69–85.

COLLIEX, C. and MORY, C. (1994), Scanning transmission electron microscopy of biological structures, *Biology of the Cell*, **80**, 175–80.

COOKE, P. H., TUNICK, M. H., MALIN, E. L., SMITH, P. W. and HOLSINGER, V. H. (1995), Electron-density patterns in low-fat mozzarella cheeses during refrigerated storage, in E. L. Malin and M. H. Tunick, *Chemistry of Structure–Function Relationships in Cheese*, New York, Plenum Press, 311–20.

CREAMER, L. K., BERRY, G. P. and MATHESON, A. R. (1978), The effect of pH on protein aggregation in skim milk, *New Zealand Dairy Science and Technology*, **13**, 9–15.

CUSHEN, M., KERRY, J., MORRIS, M., CRUZ-ROMERO, M. and CUMMINS, E. (2012), Nanotechnologies in the food industry – Recent developments, risks and regulation, *Trends in Food Science & Technology*, **24**, 30–46.

DECKER, N. R. and ZIEGLER, G. R. (2002), The structure of aerated confectionery, *Manufacturing Confectioner*, **82**, 101–8.

DODSON, A. G., BEACHAM, J., WRIGHT, S. J. C. and LEWIS, D. F. (1984a), *Role of Milk Proteins in Toffee Manufacture. Part I: Milk Powders, Condensed Milk and Wheys*, Leatherhead Food RA Research Report No. 491, Surrey, UK, Leatherhead Food International.

DODSON, A. G., BEACHAM, J., WRIGHT, S. J. C. and LEWIS, D. F. (1984b), *Role of Milk Proteins in Toffee Manufacture. Part II: Effect of Mineral Content and Casein to Whey Ratios*, Leatherhead Food RA Research Report No. 492, Surrey, UK, Leatherhead Food International.

DODSON, A. G., LEWIS, D. F., HOLGATE, J. H. and RICHARDS, S. P. (1984c), *Role of Milk Proteins in Chocolate-flavoured Coatings*, Leatherhead Food RA Research Report No. 495, Surrey, UK, Leatherhead Food International.

DUFFETT, B. and FIRTH, M. (2001), At last! A stable real chocolate powder, *Focus on Innovation*, **3**, 36–7.

DUODU, K. G., NUNES, A., DELGADILLO, I., PARKER, M. L., MILLS, E. N. C. *et al.* (2002), Effect of grain structure and cooking on sorghum and baize *in vitro* protein digestibility, *Journal of Cereal Science*, **35**, 161–74.

DUTSON, T. R. (1983), Relationship of pH and temperature to disruption of specific muscle proteins and activity of lysosomal proteases, *J. Food Biochemistry*, **7**, 223–45.

FAAS, F. G. A., BARCENA, M., AGRONSKAIA, A. V., GERRITSEN, H. C., MOSCICKA, K. V. *et al.* (2013), Localization of fluorescently labeled structures in frozen-hydrated samples using integrated light electron microscopy, *J. Structural Biology*, **181**, 283–90

GAN, Z., ANGOLD, R. E., WILLIAMS, M. R., ELLIS, P. R., VAUGHAN, J. G. and GALLIARD, T. (1990), The microstructure and gas retention of bread dough, *Journal of Cereal Science*, **12**, 15–24

GREEN, M. L., TURVEY, L. A. and HOBBS, D. G. (1980), Development of structure and texture in cheddar cheese, *Journal of Dairy Research*, **48**, 343–55.

GROSS, K. and WALLNER, S. J. (1979), Degradation of cell wall polysaccharides during tomato fruit ripening, *Plant Physiology*, **63**, 117–21.

GROVES, K. H. M. (2006), Microscopy: a tool to study ingredient interactions in foods, in A. G. Gaonlar and A. McPherson (eds), *Ingredient Interactions: Effects on Food Quality*, New York, CRC Press, 21–48.

GROVES, K. H. M. and SUBRAMANIAM, P. J. (2008), The influence of ingredients on the microstructure of chocolate, in, *Focus on Chocolate, Supplement to AgroFOOD Industry Hi-tech*, May/June. **19(3)**, 8–10.

GROVES, K. H. M., JONES, H. F., ROBERTS, C. and JONES, S. A. (1996), *Physical Properties of Sorbitol Powders and their Relationship to Performance in Confectionery Products*, Leatherhead Food International Research Report 736, Surrey, UK, Leatherhead Food International.

HARRIS, J. R. and HORNE, R. W. (1994), Negative staining: a brief assessment of current technical benefits, limitations and future possibilities, *Micron*, **25(1)**, 5–13.

HAWKES, P. (2004), *Advances in Imaging and Electron Physicss: Sir Charles Oatley and the Scanning Electron Microscope*, Elsevier Academic Press, 133.

HAYAT, M. A. (1981), *Principles and Techniques of Electron Microscopy. Biological Applications*, 2nd edition, vol. 1, Baltimore, University Park Press.

HEERTJE, I. and PAQUES, M. (1995), Advances in electron microscopy, in *New Physico-chemical Techniques for the Characterization of Complex Food Systems*, Glasgow, Blackie Academic and Professional, 1–52.

HENSTRA, S. and SCHMIDT, D. G. (1970), Ultrathin sections from milk using the microcapsule method, *Naturewissenschaften*, **57**, 247–8.

HERMANSSON, A-M. and BUCHHEIM, W. (1981), Characterization of protein gels by scanning and transmission electron microscopy, *Journal of Colloid and Interface Science*, **81(2)**, 519–30.

HILLBRICK, G. C., MCMAHON, D. J. and MCMANUS, W. R. (1999), Microstructure of indirectly and directly heated ultra-high-temperature (UHT) processed milk examined using transmission electron microscopy and immunogold labelling, *Lebensmittel Wissenschaft und Technologie*, **32**, 486–94.

HOLCOMB, D. N. (1991), Structure and rheology of dairy products: a compilation of references with subject and author indexes, *Food Structure*, **10**, 45–108.

HOLGATE, J. H. (1984), *Electronic Microscopy of Plant Tissues. Part I: Studies on the Effects of Enzymes on Plant Cell Walls*, Leatherhead Food Research Association, Research Report 476, Surrey, UK, Leatherhead Food International.

HOLGATE, J. H. and LEWIS, D. F. (1985), *Electron Microscopy of Plant Tissues. Part II: Structure and Texture Changes Occurring on Heating of Potato Tissue*, Leatherhead Food Research Association, Research Report 528, Surrey, UK, Leatherhead Food International.

HOLGATE, J. H. and LEWIS, D. F. (1986), *Electron Microscopy of Plant Tissues. Part III: Effect of Altering Ionic Conditions on Cell Wall Structure*, Leatherhead Food Research Association Research Report 542, Surrey, UK, Leatherhead Food International.

HOSKIN, J. M. and DIMICK, P. S. (1980), Observations of chocolate during conching by scanning electron microscopy and viscometry, *J Food Sci.*, **45**, 1541–5.

IPSEN, R. and OTTE, J. (2007), Self assembly of partially hydrolysed α lactalbumin, *Biotechnol. Adv.*, **25**, 602–5.

JACKSON, C., WEILER, R., SMART, M. and CAMPBELL, B. (1995), Water-macromolecular interactions in chocolate, in P. S. Belton., I. Delgadillo, A. M. Gi and G. A. Webb, *Magnetic Resonance in Food Science, Proceedings of the 2nd International Conference on Application of Magnetic Resonance in Food Science*, Portugal, September 1994, Cambridge, UK, Royal Society of Chemistry, 243–56.

JACOBSEN, C. (1999), Soft X-ray microscopy, *Trends in Cell Biology*, **9**, 44–7.

JAMES, B. J. and SMITH, B. G. (2009), Surface structure and composition of fresh and bloomed chocolate analysed using X-ray photoelectron spectroscopy, cryo-scanning electron microscopy and environmental scanning electron microscopy, *LWT – Food Science and Technology*, **42**(5), 929–37.

JEFFERY, A. B. and LEWIS, D. F. (1983), *Studies on beefburgers. Part I: Effect of Mincing Plate Size and Temperature of the Meat in the Production of Beefburgers*, Leatherhead Food R.A. Research Report No. 439, Surrey, UK, Leatherhead Food International.

JEWELL, G. G. (1972), Structure and textural changes in brown onions during pickling, *J. Food Technology*, **7**, 387.

JEWELL, G. G. (1979), Fruits and vegetables, in J. G. Vaughan (ed.), *Food*, London, Academic Press, 1–34.

KALAB, M. (1977), Milk gel structure. Part VI: Cheese texture and microstructure, *Milchwissenschaft*, **32**, 449–58.

KALAB, M. (1979a), Scanning electron microscopy of dairy products: an overview, *Scanning Electron Microscopy*, **III**, 261–72.

KALAB, M. (1979b), Microstructure of dairy foods. Part I: Milk products based on protein, *Journal of Dairy Science*, **62**, 1352–64.

KALAB, M. (1979c), Microstructure of dairy foods. Part II: Milk products based on fat, *Journal of Dairy Science*, **68**, 3234–48.

KALAB, M. (1981), Scanning electron microscopy of dairy products: a review of techniques, *Scanning Electron Microscopy*, **III**, 453–72.

KALAB, M. (1984), Artefacts in conventional scanning electron microscopy of some milk products, *Food Microstructure*, **3**, 95–111.

KALAB, M. (1993), Practical aspects of electron microscopy in dairy research, *Food Structure*, **12**, 95–114.

KANEMARU, T., HIRATA, K., TAKASU, S., ISOBE, S., MIZUKI, K. *et al.* (2009), A fluorescence scanning electron microscope, *Ultramicroscopy*, **109**, 344–9.

KARNOVSKY, M. J. (1965), A formaldehyde-glutaraldehyde fixative of high osmolality for use in electron microscopy, *Journal of Cell Biology*, **27**, 137A–8A.

KATSURAGI, T. and SATO, K. (2001), Effects of emulsifiers on fat bloom stability of cocoa butter, *Journal of Oleo Science*, **50**, 243–8.

KINDERLERER, J. L. (1997), Chrysosporium species, potential spoilage organisms of chocolate, *J. Applied Microbiology*, **83**, 771–8.

KOOLMEES, P. A., WIJNGAARDS, G., TERSTEEG, M. H. G., VAN LOGTESTIJN, J. G. *et al.* (1993), Changes in the microstructure of a comminuted meat system during heating, *Food Structure*, **12**, 427–41.

KOVACS, E., SASS, P. and AL-ARIKI, K. (1999), *Cell Wall Analysis of Different Apple Cultivars*, W. Plocharski, International Society for Horticultural Science, 219–24.

KUBO, T., GERELT, B., HAN, G. D., SUGIYAMA, T., NISHIUMI, T. and SUZUKI, A. (2002), Changes in immunoelectron microscopic localization of cathepsin D in muscle induced by conditioning or high pressure treatment, *Meat Science*, **61**, 415–18.

LANGTON, M. and HERMANSSON, A-H. (1996), Image analysis of particulate whey protein gels, *Food Hydrocolloids*, **10**, 179–91.

LEWIS, D. F. (1979), Meat products, in J. G. Vaughan (ed.), *Food Microscopy*, London, Academic Press, 233–72.

LEWIS, D. F. (1981), The use of microscopy to explain the behaviour of foodstuffs, Leatherhead Food Research Association, *Scanning Electron Microscopy*, **III**, 391–404.

LEWIS, D. F., GROVES, K. H. M. and HOLGATE, J. H. (1986), Action of polyuphosphates in meat products, *Food Microstructure*, **5**, 53–62.

LIVNEY, Y. D. (2010), Milk proteins as vehicles for bioactives, *Curr. Opin. Coll. Interface Sci.*, **15**, 73–83.

LLORCA, E., PUIG, A., HERNANDO, I., SALVADOR, A., FISZMAN, S. M. and LLUCH, M. A. (2001), Effect of fermentation time on texture and microstructure of pickled carrots, *J. Science of Food and Agriculture*, **81**, 1553–60.

MACFARLANE, J. J. and MCKENZIE, I. J. (1986), Pressure accelerated changes in the proteins of muscle and their influence on Warner-Bratzler shear values, *J. Food Science*, **51**, 516–25.

MACFARLANE, J. J., MCKENZIE, I. J. and TURNER, R. H. (1986), Pressure-heat treatment of meat: changes in myofibrillar proteins and ultrastructure, *Meat Science*, **17**, 161–76.

MASSA, A., PUIG, A. and LLUCH, M. A. (1998), SEM and cryo-SEM observed microstructure of pickled carrot (*Daucus carota*, L.). Salt penetration during the process, *Polish Journal of Nutrition and Food Sciences*, **7–48**, 31–6.

MCMULLAN, D. (1953), An improved scanning electron microscope for opaque specimens, *Proc. Inst. Electr. Engrs*, **100**(II), 245–9, reprinted in: *Selected Papers in Electron Optics*, vol. MS94, P. W. Hawkes (ed.), SPIE Milestones (1994), 186–200.

MONTEJANO, J. G., HAMANN, D. D. and LANIER, T. C. (1984), Thermally induced gelation of selected comminuted muscle systems: rheological changes during processing, final strengths and microstructure, *J. Food Science*, **49**, 1496–505.

MORRIS, V. J. (2010), Natural food nanostructures, in, Q. Chaudry, L. Castle and R. Watkins (eds), *Nanotechnologies in Food, RSC Nanoscience & Nanotechnology*, **14**, 50–68.

MORRIS, V. J. (2011), Emerging roles of engineered nanomaterials in the food industry, *Trends in Biotechnology*, **29**(10), 509–16.

NARAIN, C., ANGUS, F., KILCAST, D., PHELPS, T. and CLEGG, S. (2007), *Sugar Minimisation in Processed Foods, Leatherhead Food International, Collaborative Study for the Sugar Minimisation*, Research Group, Surrey, UK, Leatherhead Food International.

OFFER, G. and TRINICK, J. (1983), On the mechanism of water holding in meat: the swelling and shrinking of myofibrils, *Meat Science*, **8**, 245–81.

PARKER, M. L., GRANT, A., RIGBY, N. M., BELTON, P. S. and TAYLOR, J. R. N. (1999), Effects of popping on the endosperm cell walls of sorghum and maize, *J. Cereal Science*, **30**, 209–16.

PAWLEY, J. (1997), The development of field-emission scanning electron microscopy for imaging biological surfaces, *Scanning*, **19**, 324–36.

PEPPER, T. and HOLGATE, J. H. (1985), *Role of Milk Protein in Chocolate*, Leatherhead Food RA Research Report No 524, Surrey, UK, Leatherhead Food International.

RAYAN, A. A., KALAB, M. and ERNSTROM, C. A. (1980), Microstructure and rheology of process cheese, *Scanning Electron Microscopy*, **III**, 635–43.

REIMANN, D. A., HAMES, S. and FLYNN M. J. (1994), An instrument for three dimensional X-ray microscopy, 103rd Annual Meeting, Toledo, Ohio, 22–24 April, *The Ohio Journal of Science*, **94**(2), 36.

SABATINI, D. D., BENSCH, K. and BARRNETT, R. J. (1963), Cytochemistry and electron microscopy: the preservation of cellular ultrastructure and enzymatic activity by aldehyde fixation, *Journal of Cell Biology*, **17**: 19–58.

SARGENT, J. A. (1988), The application of cold stage scanning electron microscopy to food research, *Food Microstructure*, **7**, 123–35.

SAXTON, C. A., JEWELL, G. G. and DAKIN, J. C. (1969), *Structural and Textural Changes in Cauliflower during Pickling*, Leatherhead Food Research Association Research Report 142, Surrey, UK, Leatherhead Food International.

SCHMIDT, D. G. and VAN HOOYDONK, A. C. M. (1980), A scanning electron microscopical investigation of the whipping of cream, *Scanning Electron Microscopy*, **III**, 653–8.

SCHMIDT, D. G. (1982), Electron microscopy of milk and milk products: problems and possibilities, *Food Microstructure*, **1**(2), 151–21.

SCHMIDT, D. G. and BUCHHEIM, W. (1992), The application of electron microscopy in dairy research, *Journal of Microscopy*, **167**, 105–21.

SKRIVER, A., HANSEN, M. B. and QVIST, B. (1997), Image analysis applied to electron micrographs of stirred yoghurt, *Journal of Dairy Research*, **64**, 135–43.

SMITH, K. C. A. and OATLEY, C. W. (1955), The scanning electron microscope and its fields of application, *Br. J. Appl. Phys. I.*, **6**, 391–9.

SMITH, K. R., FLEMING, H. P., VAN DYKE, C. G. and LOWER, R. L. (1979), Scanning electron microscopy of the surface of pickling cucumber fruit, *Journal of the American Society for the Horticultural Science*, **104**, 528–33.

SOZZI, G. O., FRASCHINA, A. A. and CASTRO, M. A. (2001), Ripening-associated microstructural changes in antisense ACC synthase tomato fruit, *Food Science and Technology International*, **7**, 59–71.

SUBRAMANIAM, P. J. and GROVES, K. H. M. (1999), *A Study of Fat Bloom and Anti-Bloom Agents*, Leatherhead Food International Ltd, Research Report RR759, Surrey, UK, Leatherhead Food International.

SUBRAMANIAM, P. J. and GROVES, K. H. M. (2002), *A Study of Gloss Characteristics of Chocolate Coatings*, Leatherhead Food International Research Report No. 783, Surrey, UK, Leatherhead Food International.

SUBRAMANIAM, P. J. and GROVES, K. H. M. (2003), *A Study of Anti-Bloom Fats for Delaying Migration-induced Bloom*, Leatherhead Food International Ltd, Research Report RR830, Surrey, UK, Leatherhead Food International.

SUZUKI, A., HOMMA, Y., KIM, K., IKEUCHI, Y., SUGIYAMA, T. and SAITO, M. (2001), Pressure induced changes in the connectin/titin localization in the myofibrils revealed by immunoelectron microscopy, *Meat Science*, **5**, 193–7.

VON ARDENNE, M. (1938a), Das Elektronen-Rastermikroskop, *Theoretische Grundlagen, Z. Phys.*, **109**, 553–72.

VON ARDENNE, M. (1938b), Das Elektronen-Rastermikroskop, Praktische Ausführung, *Z. Tech, Phys.*, **19**, 407–16

VOYLE, C. A. (1979), Meat, in *Food Microscopy*, J. G. Vaughan (ed.), London, Academic Press, 193–232.

WALKER, M. and TRINICK, J. (1986), Electron microscope study of the effect of temperature on the length of the tail of the myosin molecule, *J. Mol. Biol.*, **192**(3), 661–7.

WALKER, M. and TRINICK, J. (1988), Visualization of domains in native and nucleotide-trapped myosin heads by negative staining, *J. Muscle Res. Cell Motil.*, **4**, 359–66.

WALTER, W. M., TRIGIANO. R. N. and FLEMING, P. (1985), Comparison of the microstructure of firm and stem-end softened cucumber pickles preserved by brine fermentation, *Food Microstructure*, **4**, 165–72.

WANG, S-M. and GREASER, M. L. (1985), Immunocytochemical studies using a monoclonal antibody to bovine cardiac titin on intact and extracted myofibrils, *J. Muscle Res. Cell Motil.*, **6**, 293–313.

WIGGLESWORTH, V. B. (1957), The use of osmium in the fixation and staining of tissues, *Proceedings of the Royal Society of London. Series B, Biological Sciences*, **147**(927), 185–99.

YAN, H. and ZHENGBIAO, G. U. (2010), Morphology of modified starches prepared by different methods, *Food Research International*, **43**(3), 767–72.

ZWORYKIN, V. A., HILLIER, J. and SNYDER, R. L. (1942), A scanning electron microscope, *ASTM Bull.*, **117**, 15–23.

Index